# Modeling Groundwater Flow and Pollution

Theory and Applications of Transport in Porous Media

*A Series of Books Edited by Jacob Bear,*
*Technion — Israel Institute of Technology, Haifa, Israel*

Jacob Bear

*Albert and Anne Mansfield Chair in Water Resources,
Technion — Israel Institute of Technology, Haifa, Israel*

and

Arnold Verruijt

*Delft University of Technology, Delft, The Netherlands*

# Modeling Groundwater Flow and Pollution

*With Computer Programs for Sample Cases*

## D. Reidel Publishing Company

A MEMBER OF THE KLUWER ACADEMIC PUBLISHERS GROUP

Dordrecht / Boston / Lancaster / Tokyo

Library of Congress Cataloging in Publication Data

Bear, Jacob.
    Modeling groundwater flow and pollution,

    (Theory and applications of transport in porous media)
    Bibliography: p.
    Includes index.
    1. Groundwater flow — Mathematical models.   2. Water, Underground — Pollution — Mathematical models.   3. Groundwater flow — Data processing.   4. Water, Underground — Pollution — Data processing.   I. Verruijt, A.   II. Title.   III. Series.
TC176.B38    1987           551.49          87—16389
ISBN 1—55608—014—X
ISBN 1—55608—015—8 (pbk.)

---

Published by D. Reidel Publishing Company,
P.O. Box 17, 3300 AA Dordrecht, Holland.

Sold and distributed in the U.S.A. and Canada
by Kluwer Academic Publishers,
101 Philip Drive, Norwell, MA 02061, U.S.A.

In all other countries, sold and distributed
by Kluwer Academic Publishers Group,
P.O. Box 322, 3300 AH Dordrecht, Holland.

All Rights Reserved

© 1987 by D. Reidel Publishing Company, Dordrecht, Holland
No part of the material protected by this copyright notice may be reproduced or utilized in any form or by any means, electronic or mechanical including photocopying, recording or by any information storage and retrieval system, without written permission from the copyright owner

Printed in The Netherlands

*Dedicated to
Professor Gerard de Josselin de Jong,
Scientist, Teacher, Artist and Friend*

# Table of Contents

| | |
|---|---:|
| PREFACE | xi |
| | |
| **Chapter 1. INTRODUCTION** | 1 |
| 1.1. Groundwater and Aquifers | 1 |
| 1.2. Management of Groundwater | 7 |
| 1.3. Groundwater Modeling | 11 |
| 1.4. Continuum Approach to Porous Media | 17 |
| 1.5. Horizontal Two-Dimensional Modeling of Aquifers | 21 |
| 1.6. Objectives and Scope | 23 |
| | |
| **Chapter 2. GROUNDWATER MOTION** | 27 |
| 2.1. Darcy's Law and its Extensions | 27 |
| 2.2. Aquifer Transmissivity | 43 |
| 2.3. Dupuit Assumption | 45 |
| | |
| **Chapter 3. MODELING THREE-DIMENSIONAL FLOW** | 53 |
| 3.1. Effective Stress in Porous Media | 53 |
| 3.2. Mass Storage | 56 |
| 3.3. Fundamental Mass Balance Equation | 60 |
| 3.4. Initial and Boundary Conditions | 65 |
| 3.5. Complete Statement of Mathematical Flow Model | 76 |
| 3.6. Modeling Soil Displacement | 78 |
| | |
| **Chapter 4. MODELING TWO-DIMENSIONAL FLOW IN AQUIFERS** | 85 |
| 4.1. Aquifer Storativity | 85 |
| 4.2. Fundamental Continuity Equations | 88 |
| 4.3. Initial and Boundary Conditions | 102 |
| 4.4. Complete Statement of Aquifer Flow Model | 104 |
| 4.5. Regional Model for Land Subsidence | 105 |
| 4.6. Streamlines and Stream Function | 114 |
| | |
| **Chapter 5. MODELING FLOW IN THE UNSATURATED ZONE** | 123 |
| 5.1. Capillarity and Retention Curves | 123 |
| 5.2. Motion Equations | 138 |
| 5.3. Balance Equations | 145 |
| 5.4. Initial and Boundary Conditions | 149 |
| 5.5. Complete Statement of Unsaturated Flow Model | 152 |

## Chapter 6. MODELING GROUNDWATER POLLUTION — 153
- 6.1. Hydrodynamic Dispersion — 155
- 6.2. Advective, Dispersive, and Diffusive Fluxes — 159
- 6.3. Balance Equation for a Pollutant — 167
- 6.4. Initial and Boundary Conditions — 179
- 6.5. Complete Statement of Pollution Model — 184
- 6.6. Pollution Transport by Advection Only — 186
- 6.7. Macrodispersion — 190

## Chapter 7. MODELING SEAWATER INTRUSION — 196
- 7.1. The Interface in a Coastal Aquifer — 196
- 7.2. Modeling Seawater Intrusion in a Vertical Plane — 198
- 7.3. Modeling Regional Seawater Intrusion — 208

## Chapter 8. INTRODUCTION TO NUMERICAL METHODS — 216
- 8.1. Analytical versus Numerical Solutions — 216
- 8.2. Survey of Numerical Methods — 217
- 8.3. Computer Programming — 223

## Chapter 9. THE FINITE DIFFERENCE METHOD — 225
- 9.1. Steady Flow — 225
- 9.2. Unsteady Flow — 233
- 9.3. Accuracy and Stability — 239
- 9.4. Generalizations — 242

## Chapter 10. THE FINITE ELEMENT METHOD — 247
- 10.1. Steady Flow — 247
- 10.2. Steady Flow in a Confined Aquifer — 257
- 10.3. Steady Flow with Infiltration and Leakage — 268
- 10.4. Steady Flow through a Dam — 272
- 10.5. Unsteady Flow in an Aquifer — 276
- 10.6. Generalizations — 281

## Chapter 11. TRANSPORT BY ADVECTION — 285
- 11.1. Basic Equations — 285
- 11.2. Semi-Analytic Solution — 286
- 11.3. System of Wells in an Infinite Field — 288
- 11.4. System of Wells in an Infinite Strip — 296
- 11.5. Numerical Solution in Terms of the Piezometric Head — 299
- 11.6. Numerical Solution in Terms of the Stream Function — 300
- 11.7. Tracing Particles Along a Stream Line — 311

TABLE OF CONTENTS ix

Chapter 12. TRANSPORT BY ADVECTION AND DISPERSION — 316
  12.1. Dispersion in One-Dimensional Flow — 316
  12.2. Numerical Dispersion — 323
  12.3. A Finite Element Model for Two-Dimensional Problems — 326
  12.4. Random Walk Model — 336

Chapter 13. NUMERICAL MODELING OF SEAWATER INTRUSION — 344
  13.1. Model for Flow in a Vertical Plane — 344
  13.2. Basic Equations for a Regional Model of Seawater Intrusion — 351
  13.3. Finite Element Model for Regional Interface Problems — 355

Appendix. SOLUTION OF LINEAR EQUATIONS — 364

REFERENCES — 381

PROBLEMS — 386

INDEX OF SUBJECTS — 409

# Preface

Groundwater constitutes an important component of many water resource systems, supplying water for domestic use, for industry, and for agriculture. Management of a groundwater system, an aquifer, or a system of aquifers, means making such decisions as to the total quantity of water to be withdrawn annually, the location of wells for pumping and for artificial recharge and their rates, and control conditions at aquifer boundaries. Not less important are decisions related to groundwater quality. In fact, the quantity and quality problems cannot be separated. In many parts of the world, with the increased withdrawal of groundwater, often beyond permissible limits, the quality of groundwater has been continuously deteriorating, causing much concern to both suppliers and users. In recent years, in addition to general groundwater quality aspects, public attention has been focused on groundwater contamination by hazardous industrial wastes, by leachate from landfills, by oil spills, and by agricultural activities such as the use of fertilizers, pesticides, and herbicides, and by radioactive waste in repositories located in deep geological formations, to mention some of the most acute contamination sources.

In all these cases, management means making decisions to achieve goals without violating specified constraints. In order to enable the planner, or the decision maker, to compare alternative modes of action and to ensure that the constraints are not violated, a tool is needed that will provide information about the response of the system (the aquifer) to various alternatives. Good management requires the ability to forecast the aquifer's response to planned operations, such as pumping and recharging. The response may take the form of changes in water levels, changes in water quality, or land subsidence. Any planning of mitigation, clean-up operations, or control measures, once contamination has been detected in the saturated or unsaturated zones, requires the prediction of the path and the fate of the contaminants in response to the planned activities. Any monitoring or observation network must be based on the anticipated behavior of the system.

The necessary information about the response of the system is provided by a model that describes the behavior of the considered system in response to excitation. The model may take the form of a well posed mathematical problem, or that of a set of algebraic equations, referred to as a numerical model, solved with the aid of a computer. For most practical problems, because of the heterogeneity of the considered domain, the irregular shape of its boundaries, and the nonanalytic form of the various source functions, only a numerical model can provide the required forecasts.

However, given a certain problem, one cannot go directly to the computer, pick up a program, push some buttons, and expect a solution to the problem. A most important part of the modeling procedure is a thorough understanding of the system, and the processes that take place in it. It is important to identify those parts of the system's behavior that are relevant to the considered problem, while other parts may be neglected. On the basis of this understanding, summarized as a conceptual model of the given problem, a numerical model is constructed and a program is prepared for its solution by means of a computer.

Unfortunately, at many universities much of this material is not taught as part of the regular curriculum in relevant disciplines, such as civil engineering, agricultural engineering, and geology. This is so in spite of the growing needs for professional activities associated with water resources (quantity and quality) management, and the prevention and control of pollution. To supplement the education of those who are already active in this area, or who wish to join it, we have developed and taught many regular and short (usually one-week) courses, in many parts of the world, on mathematical and numerical modeling of groundwater flow and pollution. Typically, the objectives of such courses are to enable the student

— to understand the meaning of models, learn the modeling process, and the role models play in decision-making procedures,
— to understand the mechanisms that govern the movement and accumulation of water and pollutants in aquifers,
— to construct conceptual and mathematical models for flow and pollution problems,
— to understand various numerical methods, and to employ the major methods, of finite differences and finite elements, in order to construct numerical models for problems of practical interest, and to prepare programs for their solution by using computers. In recent years, special attention has been devoted to microcomputers, which are readily available to most students and practitioners.
— to gain 'hands on' computer experience in solving typical problems of practical interest.

This book is designed to tend to the needs of such courses. Its objectives are identical to those listed above. In addition, the book is written in a manner that should enable readers to achieve the same goals by studying the book on their own, and solving the problems included in it. Finally, the book, with the complete programs (written in BASIC) included in it, should serve the needs of many engineers and scientists who require such programs for their professional activities. To tend to the need of professionals, the programs are what one might call 'semiprofessional'. They facilitate the presentation of the educational aspects and emphasize the main issues of the modeling methodology, but at the same time they are sufficiently advanced and general for professional use. Although the programs have been tested extensively, and we believe them to be correct, we cannot accept

PREFACE                                                                                                          xiii

any responsibility for eventual errors, and we will not accept any liability for damages or losses that may be caused by the use of the programs.

The hardware required to run the programs is an IBM Personal Computer, or a similar computer, operating under MS-DOS, with a graphics monitor. Programs are given in Microsoft BASIC, and they all admit compilation by Microsoft's compilers. The programs can be copied by retyping them from the text. Alternatively, a diskette containing all programs can be obtained from the International Ground Water Modeling Center (IGWMC), which has branch offices in Indianapolis (IGWMC, Holcomb Research Institute, Butler University, Indianapolis, Indiana 46208, U.S.A.) and Delft (IGWMC, TNO-DGV Institute of Applied Geoscience, P.O. Box 285, 2600 AG Delft, The Netherlands). The IGWMC also distributes a variety of well tested programs for micro-, mini-, and mainframe computers. It is the policy of the IGWMC to sell programs at cost price.

With the above objectives in mind, the book is made up of two parts. The first part, Chapters 1 to 7, presents the conceptual and mathematical models of groundwater flow and pollution. Special attention is given to the assumptions underlying each model, to the possible simplifications, to the physical interpretation of the terms and coefficients that appear in the various equations, and to the boundary conditions that should reflect the conditions actually encountered, or to be anticipated, in reality.

Then, following a general introductory chapter on numerical methods, five chapters (9 to 13) introduce the details of the finite difference method and the finite element method, together with actual programs aimed at explaining and exemplifying the methods and, at the same time, training the student in solving problems of practical interest. Programs are included for the analysis of steady and unsteady flow in aquifers, flow through a dam, pollution transport, and seawater intrusion. It is suggested that the reader, when studying the book, runs all the programs, in order to obtain a complete understanding of their operation and facilities. In most of the programs, input data must be entered interactively, with the user responding to questions printed on the screen by the program. The prompting messages are usually self-explanatory, so that the meaning of the input variables is immediately clear. In some cases, with large amounts of input data, input is provided through a dataset. In those cases sample datasets are included. In general, the reader should be able to produce the same output as given in the book, either in the form of a listing of output data on the screen, or in the form of a figure or graph, produced on the screen of the computer. Many of the figures in the book were actually produced by the programs.

We hope that we have found the right balance between the theoretical background summarized in the form of mathematical models, and the numerical programs that are the actual tools for solving flow and pollution problems.

Finally, many thanks are due to many of our students, both at the Delft University of Technology and at the Technion — Israel Institute of Technology, for their contribution in developing, checking and validating the various models and to the McGraw-Hill Book Company, New York, for permission to adapt and use many of the figures, presented in Chapters 1 to 7, which originally appeared in Jacob Bear, *Hydraulics of Groundwater* (1972).

*Haifa*                                                                                           JACOB BEAR
*Delft*                                                                                         ARNOLD VERRUIJT

A $5\frac{1}{4}$-inch diskette containing the source code and compiled version of the programs described and listed in this book is available from

*IGWMC — Indianapolis*:

    Holcomb Research Institute
    Butler University
    4600 Sunset Avenue
    Indianapolis
    Indiana 46208, U.S.A.
    Phone: 317/283-9458

*IGWMC — Delft*:

    TNO-DGV Institute of Applied Geoscience
    P.O. Box 285
    2600 AG Delft
    The Netherlands
    Phone: 31-15-569330

The programs are written in Microsoft BASIC with Color Graphic Adapter, and operating systems PC DOS 2.0 or later. Diskette programs can be copied and modified by user. The cost of the diskette in US$10 and includes postage and handling.

CHAPTER ONE

# Introduction

The objective of this chapter is to set the stage for the modeling procedure and methodology presented in subsequent chapters: what is it that we wish to model and why and, especially, how do we model?

We start by presenting some basic definitions related to groundwater and to aquifers, focusing our attention on the flow of water in the saturated zone, noting that due to the heterogeneity of porous material at the pore, or grain size scale, it is practically impossible to *predict* and *observe* the flow at that scale. Instead, the continuum approach is proposed as a device for describing and observing flow in porous media, in the absence of information on the microscopic configuration of the solid matrix. As a further simplification, it is proposed to treat the flow in aquifers as approximately two-dimensional in the horizontal plane.

Planning and management of groundwater resources, and this includes pollution control and abatement, require a tool for predicting the response of the aquifer system to the planned activities. We shall discuss the content of this response and suggest the *model* as the tool for achieving this goal.

## 1.1. Groundwater and Aquifers

### 1.1.1. DEFINITIONS

*Subsurface water* is a term used to denote all the water found beneath the surface of the ground. Hydrologists use the term *groundwater* to denote water in the zone of saturation (Subsection 1.1.2). In drainage of agricultural lands, or agronomy, the term *groundwater* is sometimes used also to denote the water in the partially saturated layers above the water table. Practically all groundwater constitutes part of the *hydrologic cycle* (Figure 1.1; see any textbook on hydrology). Very small amounts, however, may enter the cycle from other sources (e.g., magmatic water).

An *aquifer* is a geological formation which (i) contains water and (ii) permits significant amounts of water to move through it under ordinary field conditions. Todd (1959) traces the term aquifer to its Latin origin: *aqui* comes from *aqua*, meaning water, and *-fer* from *ferre*, to bear.

In contradistinction, an *aquiclude* is a formation which may contain water (sometimes in appreciable quantities), but is incapable of transmitting significant quantities under ordinary field conditions. A clay layer is an example of an

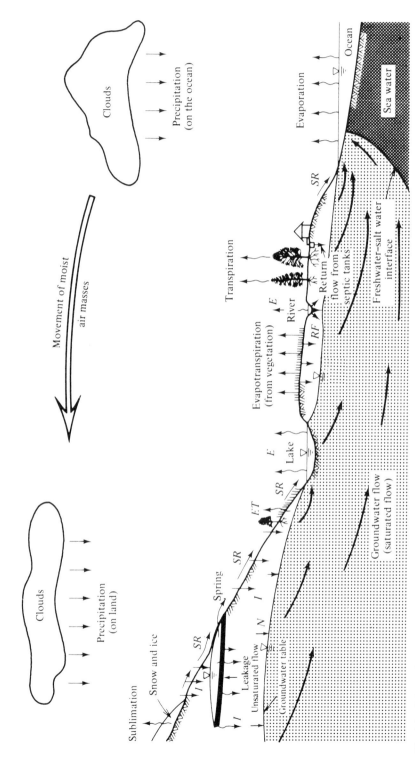

Fig. 1.1. Schematic diagram of hydrologic cycle (ET = Evapotranspiration, E = Evaporation, I = Infiltration, SR = Surface runoff, RF = Return flow from irrigation, N = Natural replenishment).

# INTRODUCTION

aquiclude. For all practical purposes, an aquiclude can be considered as an *impervious formation*.

An *aquitard* is a layer that is much less previous than the aquifer that underlies or overlies it, and often also much thinner. It thus behaves as a thin semi-pervious membrane through which leakage between aquifers separated by it is possible. An aquitard is often referred to as a *semi-pervious*, or a *leaky formation*.

An *aquifuge* is an impervious formation which neither contains nor transmits water.

That portion of the rock formation which is occupied by solid matter is called the *solid matrix*. The remaining part is called the *void space* (or *pore space*). The void space is occupied by one (water) or two (water and air) fluid phases. Only connected interstices can act as elementary conduits for fluid flow within the formation. Figure 1.2 (after Meinzer, 1942) shows several types of rock interstices. Interstices may range in size from huge limestone caverns to minute subcapillary openings in which water is held primarily by adhesive forces. The interstices of a rock formation can be grouped in two classes: original interstices (mainly in sedimentary and igneous rocks) created by geologic processes at the time the rock was formed, and secondary interstices, mainly in the form of fissures, joints and solution passages developed after the rock was formed.

## 1.1.2. MOISTURE ZONES

Subsurface formations containing water may be divided vertically into several horizontal zones according to the relative proportion of the pore space which is

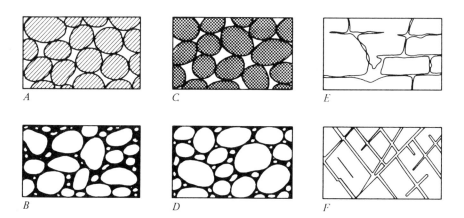

Fig. 1.2. Diagram showing several types of rock interstices. A Well-sorted sedimentary deposit having high porosity; B. Poorly sorted sedimentary deposit having low porosity; C. Well-sorted sedimentary deposit consisting of pebbles that are themselves porous, so that the deposit as a whole has a very high porosity; D. Well-sorted sedimentary deposit whose porosity has been diminished by the deposition of mineral matter in the interstices; E. Rock rendered porous by solution; F. Rock rendered porous by fracturing (*after Meinzer, 1942*).

occupied by water. Essentially, we have a *zone of saturation* in which the entire void space is filled with water, and an overlying *zone of aeration* (or unsaturated zone) in which the pores contain both gases (mainly air and water vapor) and water.

Figure 1.3 shows a schematic distribution of subsurface water in a homogeneous soil. Water (e.g., from precipitation and/or irrigation) infiltrates through the ground surface, moves downwards, and accumulates, filling all the interstices of the rock information in the zone above the impervious bedrock. The saturated zone in Figure 1.3 is bounded from above by a *phreatic surface*. We shall see below that under different circumstances, the upper boundary may be an impervious one. The term *groundwater* defined in Subsection 1.1.1 is used by groundwater hydrologists to denote the water in the zone of saturation. Wells, springs, and effluent streams act as outlets of water from the zone of saturation. The phreatic surface is an imaginary surface at all points of which the pressure is atmospheric (conveniently taken as $p = 0$). Actually, saturation (or almost so) extends a certain distance, called the *capillary fringe*, above the phreatic surface, depending on the type of soil (Subsection 5.1.4). The surface bounding the capillary fringe from above is usually called the *water table*.

The *zone of aeration* extends from the phreatic surface to the ground surface. It usually consists of three subzones: the *soil water zone* (or belt of soil water), the intermediate zone (or *vadose water zone*), and the capillary zone (or *capillary fringe*).

The capillary fringe extends from the phreatic surface up to the limit of capillary rise of water. Its thickness depends on the soil properties and on the homogeneity of the soil, mainly on the pore size distribution. The capillary rise ranges from practically nothing in coarse material, to as much as 2 to 3 m and more in fine materials (e.g., clay). Within the capillary, fringe, moisture decreases gradually with height above the phreatic surface. Just above the phreatic surface, the pores are practically saturated. Moving higher, only the smaller pores contain water. Still

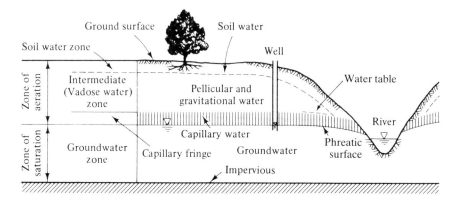

Fig. 1.3. Subsurface moisture zones.

INTRODUCTION 5

higher, only the smallest pores are filled with water. Hence, the upper limit of the capillary fringe has an irregular shape. For practical purposes, some average smooth surface — referred to as the water table — is taken as the upper limit of the capillary fringe, such that below it the soil is *assumed practically saturated* (say 75%). When the saturated zone below the phreatic surface is much thicker than the capillary fringe, the flow in the latter is often neglected as far as groundwater flow is concerned. Under such conditions, we shall use the terms phreatic surface and water table interchangeably. In drainage problems, the flow in the unsaturated zone may be of primary importance.

The *soil water zone* is adjacent to the ground surface and extends downward through the root zone. Vegetation depends on water in this zone. The moisture distribution in this zone is affected by conditions at the ground surface (seasonal and diurnal fluctuations of precipitation, irrigation, air temperature, and humidity), and by the presence of a shallow water table. A deep water table does not affect the moisture distribution in this zone. Water in this zone moves downward during infiltration (e.g., from precipitation, flooding of the ground surface, or irrigation), and upward by evaporation and plant transpiration. Temporarily, during a short period of excessive infiltration, the soil in this zone may be almost completely saturated.

After an extended period of gravity drainage without additional supply of water at the soil surface, the amount of moisture remaining in the soil is called *field capacity*. Below field capacity, the soil contains *capillary water* in the form of continuous films around the soil particles and meniscii between them, held by surface tension. Below the moisture content called the *hygroscopic coefficient* (= maximum moisture which an initially dry soil will absorb when brought into contact with an atmosphere of 50% relative humidity at 20°C), the water in the soil is called *hygroscopic water*. It also forms very thin films of moisture on the surface of soil particles, but the adhesive forces are very strong, so that this water is unavailable to plants.

1.1.3. CLASSIFICATION OF AQUIFERS

Aquifers may be classed according to their pressure system.

A *confined aquifer* (Figure[1] 1.4), also known as a *pressure aquifer*, is one bounded from above and below by impervious formations. In a well just penetrating such an aquifer, the water level will rise above the base of the upper confining formation; it may or may not reach the ground surface. This water level indicates the *piezometric head* (Subsection 2.1.1) at the center of the well's screen (when the latter is short relative to the aquifer's thickness).

The water levels in a number of observation wells tapping a certain aquifer define an imaginary surface called the *piezometric surface*. When the flow in an aquifer is essentially horizontal, such that equipotential surfaces are practically vertical, the depth of a piezometer's opening is immaterial; otherwise, a different

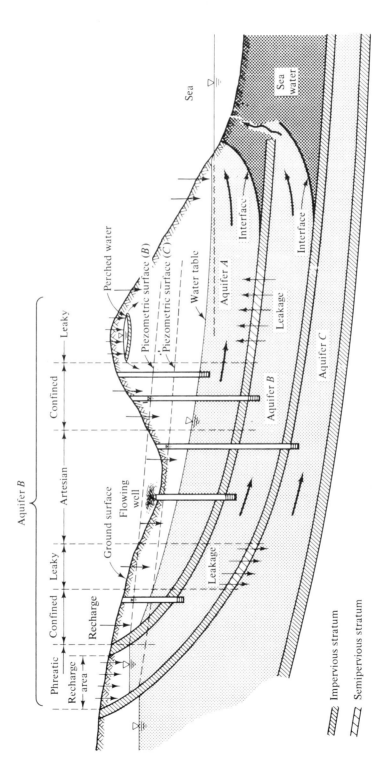

Fig. 1.4. Types of aquifers.

piezometric surface is obtained for piezometers which have openings at different elevations. Fortunately, except in the neighborhood of outlets such as partially penetrating wells or springs, the flow in aquifers is essentially horizontal.

An *artesian aquifer* is a confined aquifer (or a portion of it) where the elevation of the piezometric surface (say, corresponding to the base of the upper confining layer) is above the ground surface. A well in such an aquifer will flow freely without pumping (*artesian well, flowing well*). The name is derived from the county *Artois* (*Artesia*) in Northern France, where such wells were constructed by the monks of Lillers in the 12th century.

A *phreatic aquifer* (also called *unconfined aquifer, water table aquifer*) is one in which a water table (= *phreatic surface*) serves as its upper boundary. A phreatic aquifer is directly recharged from the ground surface above it, except where impervious layers, sometimes of limited areal extent, exist between the phreatic surface and the ground surface. Water enters a confined aquifer in a recharge area which is a phreatic aquifer formed where the confining strata terminate at, or close to the ground surface (Figure 1.4).

Aquifers, whether confined or unconfined, that can lose or gain water through aquitards bounding them from either above and/or below, are called *leaky aquifers*. Although these semipervious formations have a relatively high resistance to the flow of water through them, significant quantities of water may leak through them into or out of an aquifer over the large horizontal areas of contact involved. The amount and direction of leakage is governed in each case by the difference in the piezometric head which exists across the semipervious formation.

A phreatic aquifer (or part of it) which rests on a semipervious layer is a *leaky phreatic aquifer*. A confined aquifer (or part of it) which has at least one semipervious confining stratum is called a *leaky confined aquifer*.

Figure 1.4 shows several aquifers and observation wells. The upper phreatic aquifer (*A*) is underlain by two confined ones (*B* and *C*). In the recharge area, aquifer *B* becomes phreatic. Portions of aquifers *A, B,* and *C* are leaky, with the direction and rate of leakage determined by the elevations of the piezometric surface of each of these aquifers. The boundaries between the various confined and unconfined portions may vary with time as a result of changes in water table and piezometric surface elevations. A special case of a phreatic aquifer is the *perched aquifer* which is formed on an impervious (or semipervious) layer of limited areal extent located between the water table of a phreatic aquifer and the ground surface. Sometimes these aquifers exist only during a relatively short part of each year, as they drain to the underlying phreatic aquifer.

Sometimes we refer to the groundwater itself as confined, phreatic, or leaky, rather than to the geological formation.

## 1.2. Management of Groundwater

An aquifer, and especially when it constitutes an element in a water resource system, plays a number of roles.

*Source of water.* This is the more obvious function. The aquifer is replenished annually from precipitation over the region overlying it or, in a confined aquifer, over its *intake,* or *recharge, region.* In such cases, groundwater is considered as a renewable resource. Obviously, depending on the distribution of storms, land topography, cover, permeability of soil, etc., only a certain part of the precipitation infiltrates through the ground surface and replenishes the underlying phreatic aquifer. Aquifers can also be replenished from streamflows (with permeable beds) and from floods. In many arid regions, aquifers in the low lands are replenished during a very short period (once in several years) from flash floods originating in the mountains.

Aquifers usually contain water stored in them from times in the distant past, sometimes under different climatic conditions. Such groundwater should be regarded as a *nonrenewable resource.* The production of groundwater from such aquifers should be considered with great care.

The *yield* of an aquifer is a term which requires some attention. Obviously, *in the long run,* unless we wish to mine all or part of the volume of water in storage, the volume of water withdrawn from an aquifer cannot exceed the aquifer's replenishment. However, this is only one limiting factor to the annual rate of withdrawal. To conserve an aquifer (especially from the quality point of view), if we really wish to do so as part of a stated policy, we must limit the outflow of groundwater from it. This means maintaining certain minimal water levels, at least in the vicinity of outlets. Since inflows and outflows can be controlled by controlling water levels on which they depend (and artificial recharge is an example of a controlled input), *the rate of annual withdrawal is a decision variable*; it may vary within certain limits, according to our decision as to the rate of groundwater outflow we allow from the aquifer and to the inflow into the aquifer from different sources.

Special attention should be given to the quality of water in the aquifer and in the extracted water. The quality of the water in an aquifer is closely related to the magnitude of components of the water balance, e.g., natural replenishment and outflow. For example, increasing withdrawal, which means reduced outflow, reduces flushing and enhances the accumulation of contaminants. Another example is the movement of inferior quality water into areas of low water levels produced by large withdrawals. Thus, the quality and quantity problem are closely inter-related to each other.

Another important factor is the annual climatic fluctuations of natural replenishment. Because of the storage capacity of an aquifer, annual withdrawal is not directly related to the natural replenishment in the same year.

In addition to hydrological constraints, such as those imposed by (quality and quantity) conservation requirements, the rate of withdrawal may be constrained by economic, legal, and political (e.g., priority rights) considerations, directly, or through its effects on such constrained parameters as water levels, water quality, land subsidence, base flow, etc.

Altogether, the yield is a decision variable to be determined as part of the aquifer's management scheme. It is not necessarily a constant figure. It may vary from year to year, depending on the state of the aquifer (in terms of quantity and quality of water), on hydrological and nonhydrological constraints imposed on its operation and on the management objectives. We often use the term *optimal yield*, or *operational yield*.

*Storage reservoir*. Every water resource system requires storage, especially when replenishment is intermittent and is subject to random fluctuations. A large volume of storage is available in the void space of a phreatic aquifer. Just to give a rough idea, we can store $15 \times 10^6$ m$^3$ of water in a portion of an aquifer of $10 \times 10$ km, with a storativity of 15% (which is a rather low value) by raising the water table by 1 m. Using the technique of artificial recharge, large quantities of water can thus be stored in a phreatic aquifer. Obviously, the recharge should be accompanied by an appropriate scheme for withdrawing the stored water. Phreatic aquifers may be used as seasonal storage reservoirs, and even for shorter periods.

*A conduit*. Using the technique of artificial recharge, water can be introduced into an aquifer at one point and be withdrawn by pumping at another point (or several other points). The injected water will flow through the aquifer from the high water levels of the region of recharge to the region of pumping, where water levels are lower.

*A filter*. Using the technique of artificial recharge, an aquifer may serve as a filter and purifier for water of inferior quality injected into it. This may take the form of (i) the removal of fine suspended load from surface water recharged into aquifers through infiltration ponds, (ii) the removal of chemicals by chemical reactions, adsorption and ion-exchange phenomena on the surface of the solid matrix, especially when clay is present in the formation, and (iii) the mixing of injected water with indigenous water in the aquifer, due to the geometry of the flow pattern and to the mechanism of hydrodynamic dispersion.

*Control of base flow*. This can be achieved in springs and streams by controlling water levels in the aquifer supplying water to them.

*A water mine*. At any instant, a certain, rather large volume of water is stored in an aquifer. This volume, or part of it, can be *mined* as a *one-time reserve*. In some cases, this may lead to the establishment of a phreatic surface that is lower than the initial one. In other cases, withdrawal of this volume, or part of it, may lead to a complete destruction of the aquifer as a source of water, due to the invasion of water of inferior quality (e.g., seawater).

In general, the yield of an aquifer is a long term average of part of its replenishment (renewable resource). The yield is also constrained by requiring that the water quality be maintained below permissible levels of quality standards. However, under certain, usually economic conditions, albeit very rarely, we may plan to *mine* an aquifer (like any other nonrenewable resource), partly or completely.

We have thus summarized the roles that groundwater can play in the management of a regional water resource system. We have also suggested that the aquifer is a system which can perform different functions and that can be managed in order to achieve desired goals.

The main task of the groundwater hydrologist, water resources engineer, or planner who deals with a groundwater system, or with a water resource system of which groundwater is a component, is the *management of the groundwater system*. Simply stated, and using the terminology of systems analysis only loosely, management of a system means making various *decisions* (that is, assigning numerical values to decision variables) aimed at modifying the *state* of a considered system. Our reason for modifying the state of a considered system, that is, to bring it from its existing state to another one, is to achieve certain *goals* and *objectives*.

Typically, the same goal or goals can be achieved by making different sets of decisions (= *policies*). Management, therefore, includes the selection of the *best policy* which will lead to the achievement of a specified goal, or a number of goals simultaneously. This requires some measure of the relative effectiveness with which the different alternative policies meet, or approach, the specified goals. The scalar function of the decision variables which measures the efficiency of the different alternative policies is called an *objective function*. Not all policies are *feasible*; some violate specified social, economic, or technical *constraints* and should not be considered.

We refer to this decision-making activity as *solving the management problem*.

More specifically, management of a specified aquifer usually means determining the numerical values of certain *decision variables*, in order to maximize or minimize a certain *objective function*, or functions, subject to certain specified *constraints*.

Examples of *state variables*:
— Water levels.
— Concentration of specified species.
— Land subsidence.
— Sea water intrusion.

Examples of *decision variables*:
— Areal and temporal distributions of pumpage.
— Areal and temporal distributions of artificial recharge.
— Water levels in steams and lakes in contact with an aquifer.
— Quality of water to be used for artificial recharge.
— Quality of pumped water.
— Capacity of new installations for pumping and/or artificial recharge, their location, and time schedule of their construction.
— Location of wells for clean-up operations to remove contaminants.

Examples of *objective functions*:
— Total net benefits (or present worth of total net benefits, if timing of costs and benefits is taken into account), from operating the system during a specified period of time, and we wish to maximize the value of this function.

# INTRODUCTION

- Cost of clean-up operations to remove contaminants from an aquifer, and we wish to minimize this cost.
- Cost of a unit volume of water supplied to the consumer, and we wish to minimize the value of this function.
- Total consumption of energy, and we wish to minimize the value of this function.
- The sum of absolute values of the differences between certain desired water levels and actual ones (or sum of squares of differences), and we wish to minimize this sum, etc.

Examples of *hydrological constraints*:
- Water levels everywhere, or at specified locations, should not rise above specified maximum elevations.
- Water levels everywhere, or at specified locations, should not drop below specified minimum elevations.
- The discharge of a spring should not drop below a specified minimum.
- Base flow in a stream fed by groundwater emerging from an aquifer should not drop below a specified minimum.
- The concentrations of certain species in solution in the water pumped at specified locations should not exceed specified threshold values.
- Land subsidence should not exceed specified values.
- Total pumpage should at least satisfy the demand for water in a given region.
- Pumping (and/or artificial recharge) rates should not exceed installed capacity of pumping (and/or artificial recharge).
- The residence time for recharge water in an aquifer, before being pumped, should exceed a certain minimum period.
- The length of an intruding sea water wedge should not exceed a specified value.

In view of this statement of a groundwater management problem, it is obvious that forecasting of the response of an aquifer system is an intrinsic part of the procedure for determining any optimal management policy. We must know the future values of relevant state variables that will occur in an aquifer as a result of the implementation of a proposed set of decisions (i) in order to examine whether they violate specified constraints and (ii) in order to examine the value of the resulting level of achieving the objective function.

This book is dedicated to the tool that enables us to make the required forecasts.

## 1.3. Groundwater Modeling

As explained in Section 1.2, in the management of groundwater resources, and this includes pollution control and abatement, a tool is needed in order to predict the outcome of implementing management decisions. Depending on the nature of the management problem, its decision variables, its objective functions, and its constraints, the 'outcome' may take the form of future spatial distributions of water levels, water quality, land subsidence, etc. This tool is the *model*.

A model may be defined as a *simplified version of the real (here groundwater) system that approximately simulates the excitation-response relations of the latter*. The real system is very complicated and there is no need to elaborate on the need to simplify it for the purpose of planning and making management decisions. The simplification is introduced in the form of a set of assumptions that express *our* understanding of the nature of the system and its behavior. These assumptions will relate, among other factors, to the geometry of the investigated domain, to the way the effect of various heterogeneities will be smoothed out, to the nature of the porous medium (e.g., its homogeneity, isotropy, deformability), to the nature of the fluid (or fluids) involved and to the kind of flow regime that takes place. Because the model is a simplified version of the real system, *there exists no unique model* for a given groundwater system. Different sets of simplifying assumptions will result in different models, each approximating the investigated groundwater system in a different way.

In what follows, we shall limit the discussion to mathematical and numerical models, excluding physical, or laboratory ones. The latter were in use until the early 70's as practical tools for solving groundwater problems. With the introduction of computers and their application in the solution of numerical models, physical models and analogs have become redundant as tools for predicting future groundwater regimes.

The selection of the appropriate model to be used in any particular case depends on

(a) the objective, or objectives, of the investigations, and
(b) the available resources. These include such items as time, budget, skilled manpower, computers and codes.

The objectives dictate which features of the investigated problem should be represented in the model, and to what degree of accuracy. In some cases, we may be satisfied with averaged water levels taken over large areas, while in others we need water levels at specified points. In some cases, we may overlook land subsidence due to pumping, while in others the knowledge of land subsidence is an essential part of the studied problem. Natural replenishment may be introduced as monthly, seasonal, or annual averages. Pumping may be assumed to be uniformly distributed over large areas, or it may be represented as point sinks. Obviously, a more detailed model is more costly and requires more skilled manpower, more sophisticated codes and larger computers. It is, therefore, important to select the appropriate degree of simplification in each case.

Most models express nothing but a *balance of a considered extensive quantity*, e.g., mass of water, mass of a solute and heat.

The first step in the procedure of modeling is the construction of a *conceptual model* of the problem and the relevant aquifer domain. The conceptual model consists of a set of assumptions that reduce the real problem and the real domain to simplified versions that are acceptable in view of the objectives of the modeling and of the associated management problem. The assumptions should relate to such items as:

- the geometry of the boundaries of the investigated aquifer domain,
- the kind of material comprising the aquifer (with reference to its homogeneity, isotropy, etc.),
- the mode of flow in the aquifer (three-dimensional, or two-dimensional horizontal),
- the flow regime (laminar, or nonlaminar),
- the properties of the water (with reference to its homogeneity, compressibility),
- effect of dissolved solids and/or temperature on density and viscosity,
- the presence of assumed sharp fluid-fluid boundaries, such as a phreatic surface or a freshwater — saltwater interface,
- the relevant state variables and the area, or volume, over which the averages of such variables are taken,
- sources and sinks, of water and of relevant pollutants, within the domain and on its boundaries (with reference to their approximation as point sinks and sources, or distributed ones), and
- the conditions on the boundaries of the considered domain, that express the way the latter interacts with its surrounding.

Usually, the conceptual model is expressed in words as a set of assumptions. Actually, this set of assumptions constitutes the 'label' of the model being developed. In principle, we should not use a ready-made model for a given problem, unless we have examined the former's 'label' and decided that indeed our problem can be described by the same conceptual model.

In the second step, we express the conceptual model in the form of a *mathematical model*. The latter consists of (i) a definition of the geometry of the considered domain and its boundaries, (ii) an equation (or equations) that express the balance of the considered extensive quantity (or quantities), (iii) flux equations, that relate the flux(es) of the considered extensive quantity(ies) to the relevant state variables of the problem, (iv) constitutive equations that define the behavior of the particular materials — fluids and solids — involved, (v) initial conditions that describe the known state of the considered system at some initial time, and (vi) boundary conditions that describe the interaction of the considered domain with its environment, across the boundaries of the former.

In the continuum approach (Section 1.4) the balance equation takes the form of a partial differential equation written in terms of macroscopic state variables, each of which is an average taken over the representative elementary volume of the domain considered. In other cases, balances of extensive quantities are stated for various forms and sizes of aquifer cells. In such models, the stated variables are averages over the considered model cells. Boundary conditions are also expressed in mathematical forms. The most general boundary condition for any extensive quantity takes the form of equality of the flux of that quantity, normal to the boundary, on both sides of the latter. Certain simplified forms of this general condition are commonly derived and employed. (Sections 3.4, 4.3, 5.4, 6.4 and Subsection 7.3.2.)

A special case is that of the momentum balance. In the continuum approach, subject to certain simplifying assumptions (included in the conceptual model) as to the solid-fluid interaction, negligible internal friction in the fluid, and negligible inertial effects, *the averaged momentum balance equation reduces to the linear motion equation, known as Darcy's law, used as a flux equation for fluid flow in porous media.* With certain modification, it is also applicable to multiphase flows such as air-water flow in the unsaturated zone.

In the passage from the real system to the conceptual model and then to the mathematical one, various *coefficients of transport and storage* of the considered extensive quantities are introduced. The permeability of a porous medium (Section 2.1) aquifer transmissivity (Section 2.2), aquifer storativity (Section 4.1) and porous medium dispersivity (Section 6.2), may serve as examples of such coefficients. Permeability and dispersivity are examples of coefficients that express the macroscopic effects of the microscopic configuration of the solid-fluid interfaces of a porous medium. They are introduced in the passage from the microscopic level of description to the macroscopic, continuum, one. All these coefficients are *coefficients of the models*, and therefore, in spite of the similarity in their names in diferent models, their interpretation and actual values may differ from one model to the next.

Let us illustrate this point by an example. To obtain the drawdown in a pumping well and in its vicinity, we employ a conceptual model that assumes radially converging flow to an infinitesimally small well in a homogeneous, isotropic aquifer of constant thickness and of infinite areal extent. The same model is used to obtain the aquifer's storativity and transmissivity by conducting a pumping test and solving the model's equation for these coefficients. Note that, following common practice, we refer to these as aquifer's coefficients and not as coefficients of the aquifer's model. However, it is important to realize that the coefficients thus derived actually correspond to that particular model. One should refrain from employing these coefficients in a model that describes the flow in the same domain as one in a finite heterogeneous aquifer, with variable thickness and with nonradial flow in the vicinity of the well's location. Obviously, when we do use coefficients derived by employing one model in another model for a given domain, the magnitude of the error will depend on the differences between the two models. In principle, in order to employ a particular model, the values of the coefficients appearing in it should be determined, using some parameter identification technique for that particular model.

Obviously, no model can be employed in any particular case of interest in a specified domain, unless we know the numerical values of all the coefficients appearing in it. Estimates of natural replenishment and *a-priori* unknown location and type of boundaries, may be included in the list of model coefficients and parameters that have to be identified. We refer to the activity of identifying these model coefficients as the *identification problem*.

In principle, the only way to obtain the values of these coefficients for a considered model is to start by investigating the real aquifer system in order to find a period in the past for which information is available on (i) initial conditions of the system, (ii) excitations of the system, say in the form of pumping and artificial recharge (quality and quantity), natural replenishment, introduction of pollutants, or changes in boundary conditions, and (iii) observations of the response of the system, say in the form of temporal and spatial distributions of state variables, e.g., water levels, solute concentrations and land subsidence. If such period (or periods) can be found, we (i) impose the known initial conditions on the model, (ii) excite the model by the known excitations of the real system and (iii) derive the response of the model to these excitations. Obviously, in order to derive the model's response, we have to assume for it some trial values of the sought coefficients. We then compare the response observed in the real system with that predicted by the model. The sought values of the coefficients are those that will make the two sets of values of state variables identical. However, because the model is only an approximation of the real system, we should never expect these two sets of values to be identical. Instead, we search for the 'best fit' between them, according to some criterion. Various techniques exist for determining the 'best' or 'optimal' values of these coefficients, i.e., values that will make the predicted values and the measured ones sufficiently close to each other. Obviously, the values of the coefficients eventually accepted as 'best' for the model, depend on the criteria selected for 'goodness of fit' between the observed and predicted values of the relevant state variables. These, in turn, depend on the objective of the modeling. Some techniques use the basic trial-and-error method described above, while others employ more sophisticated optimization methods. In some methods, *a-priori* estimates of values to be expected for the coefficients as well as information about lower and upper bounds are introduced. In addition to the question of selecting the appropriate criteria, there still remains the question of the conditions under which the identification problem, also called the *inverse problem*, will result in a unique solution.

Once a mathematical model has been constructed in terms of relevant state variables, it has to be solved for cases of practical interest, for example, for planned pumping, or artificial recharge, or for anticipated spreading of a pollutant from a potential source of pollution in the considered aquifer domain. The preferable method of solution is the analytical one, because once such a solution is derived, it can be employed for a variety of planned, or anticipated situations. However, in most cases of practical interest, this method is not feasible because of the irregular shape of the domain's boundaries, the heterogeneity of the domain, expressed in the form of spatial distributions of its transport and storage coefficients, and the irregular temporal and spatial distributions of the various excitations, or sink-source, functions. Instead, numerical methods are employed for solving the mathematical model.

The main features of the various numerical methods are:

(a) The solution is sought for the numerical values of state variables only at specified points in the space and time domains defined for the problem (rather than their continuous variations in these domains).
(b) The partial differential equations that represent balances of the considered extensive quantities are replaced by a set of algebraic equations written in terms of the sought, discrete values of the state variables at the discrete points in space and time mentioned in (a).
(c) The solution is obtained for a specified set of numerical values of the various model coefficients (rather than as general relationships in terms of these coefficients).
(d) Because of the very large number of equations that have to be solved simultaneously, a *computer code* has to be prepared in order to obtain a solution, using a digital computer.

Sometimes, the term *numerical model* is used, rather than speaking of a 'numerical method of solution' (of the mathematical model). This is justified on the grounds that a number of assumptions are introduced, in addition to those underlying the mathematical model. This makes the numerical model, a model in its own right. It represents a different approximate version of the real system. It is sometimes possible to pass directly from the conceptual model to the numerical one, without first establishing a mathematical model. The numerical model has its own set of coefficients that have to be identified before the model can be used for any particular problem.

It is of interest to note that even those who consider the numerical model as one in its own right, very often validate it by comparing its predictions with those obtained analytically from a mathematical one (for relatively simple cases for which such solutions can be derived). One of the main reasons for such a validation is the wish to eliminate errors resulting from the numerical approximations alone.

Another important feature of modeling, closely associated with the problem of parameter identification is that of *uncertainty*. We are uncertain about whether the selected conceptual model (i.e., our set of assumptions) indeed represents what happens in the real aquifer system, albeit to the accepted degree of approximation. Furthermore, even when employing some identification technique, we are uncertain about the values of the coefficients to be used in the model. Possible errors in observed data used for parameter identification also contribute to uncertainty in model parameters. As a consequence, we should also expect uncertainty in the values of the state variables predicted by the model. These considerations pave the way to the development of *stochastic models*. In the latter, the information on coefficients appears in the form of probability distributions of values, rather than as deterministic ones. These probability distributions are derived by appropriate methods of solving the inverse problem, where the input data also appears in probabilistic forms. Probabilistic values of model coefficients, will yield probabilistic predicted values of state variables.

A large number of researchers are currently engaged in developing methods that incorporate the element of uncertainty in both the forecasting and the inverse problems. Hopefully, more such methods will be made available to the practising modeler in the future. In this book, beyond the comments made here, uncertainty and probabilistic modeling are not considered.

## 1.4. Continuum Approach to Porous Media

An aquifer is a domain occupied by a *porous medium*. Soil, sand, fissured rocks, sandstone and Karstic limestone are examples of porous media. However, ceramics, foam rubber, bread and animal and human tissue, are also regarded as porous media. Common to all these examples is the presence of both a persistent *solid matrix* and a *void space* within the porous medium domain. The void space is occupied by one or more fluid phases (e.g., water and air). Another common feature is that the solid matrix is distributed throughout the porous medium domain. This implies that samples of a sufficiently large volume, taken at different places within the domain, will always contain a solid phase. Let us refer to this volume as an *arbitrary elementary volume* (abbreviated to AEV).

In an aquifer, water flows through the complex network of pores and channels comprising the void space. This flow is bounded by the (microscopic) solid-water interface. In principle, the flow of a fluid in a porous medium may be treated at the *microscopic level*, at which we focus our attention at what happens at a point within the fluid, regarded as a *continuum* (i.e., overlooking its molecular structure). For example, for a single fluid that occupies the entire voids space, we could make use of the Navier—Stokes equations and solve them within the fluid domain, subject to boundary conditions on the solid-fluid interface that bounds this domain. However, this approach is usually impractical due to our inability to describe the complex configuration of this boundary. Moreover, even if we could solve for values of state variables, e.g., pressure, at the microscopic level, we could not verify these solutions by measurements at this level.

To circumvent these difficulties, another level of description is needed. This is the *macroscopic level*, at which quantities can be measured and boundary-value problems can be solved. To obtain the description of the flow at this level, we adopt the *continuum approach*. This is the same approach that is also used in order to pass from the molecular level of description to the microscopic one, at which each phase is regarded as a continuum. According to this approach, the real porous medium, in which each (solid, or fluid) phase occupies only a portion of the AEV mentioned above, is replaced by a *fictitious model* (Section 1.3) in which each phase is regarded as a continuum that fills up the entire AEV. We thus obtain within every AEV a set of overlapping and, possibly, interacting, continua. For each of these continua, average values, referred to as *macroscopic values*, can be taken over the AEV and assigned to its centroid, regardless of whether the latter falls within the solid or within one of the fluids that occupy the void space. By traversing the entire porous medium domain with a moving AEV, we obtain *fields* of macroscopic variables, which are differentiable functions of the space coordinates.

Thus, the macroscopic, continuum model of a porous medium obtained by employing the procedure described above eliminates the need for specifying the microscopic configuration of the individual phases and enables the solution of problems of flow through porous media by available methods of mathematical analysis. The configuration of the void-solid boundary and of interphase boundaries within the averaging volume, as wel as the effect of conditions that prevail on them, appear in the macroscopic description of the flow (= macroscopic model) in the form of *coefficients*. The numerical values of these coefficients have to be determined experimentally for any given porous medium.

We have still to select the appropriate size for the averaging volume. In principle, an AEV of any size may be used for that purpose. Once averaged values over an AEV, say of fluid density, or pressure, have been introduced in the macroscopic model, we shall construct instruments that measure these averaged values, i.e., that take averages of the corresponding microscopic values over the selected averaging volume. Within the range of erorr introduced by the conceptual model of the flow process, the predicted values and the measured ones must be the same. Obviously, for each selected averaging volume, the averaged values of the state variables will be different. Nevertheless, the question of which value is more 'correct' is meaningless. All values are correct; the size of the averaging volume to be employed depends on the purpose for which the macroscopic model is constructed.

The main drawback of the use of an AEV is that every averaged value must be accompanied by a label that specifies (like a yardstick) the volume over which this average was taken. To circumvent this difficulty, we need a universal procedure that (i) is applicable to all porous media and (ii) will ensure that the averaged values will remain, more or less, constant, at least for a certain range of averaging volumes, that corresponds to the range of variations in instrument sizes. This universal averaging volume is referred to as the *representative elementary volume* (abbreviated REV).

The size of the REV is selected such that the averaged values of all geometrical characteristics (= coefficients) of the microstructure of the void space, or the void-solid interface, at any point in a porous medium domain, be a single valued function (or almost so, within an acceptable error range) of the location of that point only, *independent of the size of the REV*.

INTRODUCTION 19

Bachmat and Bear (1986), in discussing the selection of the size of the REV that will satisfy the above requirement, start by discussing the distribution of solid and void space within an REV as a random phenomenon of the space coordinates. This distribution is *random* in the sense that given a point within a REV, it is not known prior to an observation, whether it falls in the void space, or in the solid. Furthermore, since it is impractical to observe all points within an REV, probability distributions are considered instead. Accordingly, the spatial configuration of the void space in the vicinity of a point in a given porous medium is described by the expected value, the variance and other moments of the void-solid distribution function. These moments, which characterize the geometry of the void space, are now *nonrandom* functions of the space coordinates. Bachmat and Bear (1986) show that under certain conditions, a good estimate of these geometrical characteristics can be obtained by spatial averaging over the volume of the REV. They state that, in fact, the volume of the REV should be selected such that the volumetric averages can be considered as satisfactory estimates of all the relevant statistical parameters of the void space configuration, i.e., estimates that are free of the effect of the size of the sample.

Denoting the characteristic dimension of the REV by $l$ (say, the diameter of a sphere) and the length characterizing the microscopic structure of the void space, by $d$ (say, the *hydraulic radius* which is equal to the reciprocal of the specific surface area of the void space), they show that a necessary condition for obtaining nonrandom estimates of the geometrical characteristics of the void space at any point within a porous medium domain is

$$l \gg d. \tag{1.4.1}$$

Another condition that sets an upper limit to the size of the REV is $l < l_{max}$, where $l_{max}$ is the distance beyond which the spatial distribution of the relevant macroscopic coefficients that characterize the configuration of the void space (e.g., porosity, permeability) deviates from the linear one by more than some acceptable value.

Finally, the selection of the size of the REV is also constrained by the requirement that

$$l \ll L \tag{1.4.2}$$

where $L$ is a characteristic length of the porous medium domain, over which significant changes in averaged (= macroscopic) quantities of interest occur.

To illustrate the determination of the size of an REV for a given porous medium domain, $(D)$, consider, as an example of a geometrical characteristic of the void space configuration, the ratio $U_v(\mathbf{x}_0)/U(\mathbf{x}_0)$, where $U(\mathbf{x}_0)$ is a volume of a sphere centered at an arbitrary point $\mathbf{x}$ within $(D)$ and $U_v(\mathbf{x}_0)$ is the volume of void space within $U(\mathbf{x}_0)$. Figure 1.5 shows the variations of the ratio $U_v(\mathbf{x}_0)/U(\mathbf{x}_0)$ as $U$ increases. For very small values of $U$, this ratio is one or zero, depending on whether $\mathbf{x}_0$ happens to fall in the void space or in the solid matrix. As $U$ increases, we note large fluctuations in this ratio due to the random distribution of void and

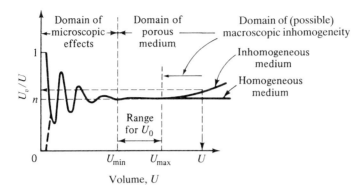

Fig. 1.5. Definition of porosity and Representative Elementary Volume.

solid within $U$. However, as $U$ is further increased, these fluctuations gradually decay until above some volume $U = U_{min}$ they reduce to some small value. If $U$ is further increased beyond some $U = U_{max}$, we may observe a trend in the considered ratio, due to a systematic variation in the latter, resulting from macroscopic heterogeneity of the porous medium. The size, $U_0$, of the REV that will make the considered ratio independent of the selected volume, albeit possibly dependent on $\mathbf{x}$, should be in the range $U_{min} < U_0 < U_{max}$. For such volume, the ratio $U_{0v}/U_0$ represents the medium *porosity, n*, at $\mathbf{x}_0$.

Once $U_0$ has been determined, it is used to derive the macroscopic (continuum) description of the flow by averaging the microscopic one over it. Obviously, the selected size of $U_0$ must be uniform over the entire porous medium domain. The macroscopic model obtained in this way describes the flow in terms of macroscopic or averaged quantities defined by

$$\overline{G}^\alpha_\alpha(\mathbf{x}, t) = \frac{1}{U_{0\alpha}} \int_{(U_{0\alpha}(\mathbf{x}))} G_\alpha(\mathbf{x}', t; \mathbf{x}) \, dU_\alpha(\mathbf{x}') \qquad (1.4.3)$$

where $G_\alpha$ is the state variable of the $\alpha$-phase, (such that its volumetric average is physically meaningful), $U_{0\alpha}$ is the volume of the $\alpha$-phase within $U_0$ and $\mathbf{x}'$ is a point in the REV centered at $\mathbf{x}$. From the discussion presented above, we are assured that the macroscopic geometrical characteristics (= coefficients) that appear in the macroscopic model represent properties of porous medium at $\mathbf{x}$. The average $\overline{G}^\alpha_\alpha$ of $G_\alpha$, as defined by (1.4.3) is called an *intrinsic phase average*.

Another type of average, called a *phase average*, defined by

$$\overline{G}^\alpha_\alpha(\mathbf{x}, t) = \frac{1}{U_0} \int_{(U_{0\alpha}(\mathbf{x}))} G_\alpha(\mathbf{x}', t; \mathbf{x}) \, dU_\alpha(\mathbf{x}') \qquad (1.4.4)$$

INTRODUCTION                                                                 21

is also often used. The two types of averages are related to each other by

$$\overline{G}_\alpha = \theta_\alpha \overline{G}_\alpha^\alpha \qquad (1.4.5)$$

where $\theta_\alpha (\equiv U_{0\alpha}/U_0)$ is the volumetric fraction of the $\alpha$-phase.

If a volume $U_0$ satisfying (1.4.1) and (1.4.2) cannot be found for a given porous medium domain, the latter cannot be treated as a continuum.

Instruments to measure the averaged, macroscopic, quantities must also be of a size in the range $(U_{min}, U_{max})$, in order to yield observations compatible with values calculated by the model.

In an analogous way, a *representative elementary area* (REA) should also be selected for a porous medium domain, to be used for averaging quantities for which only areal averages are meaningful.

Throughout this book, it is assumed that the porous medium can be considered as a continuum in the sense explained above.

## 1.5. Horizontal Two-Dimensional Modeling of Aquifers

In general, flow through a porous medium domain is three-dimensional. For example, the specific discharge vector **q** has the components $q_x$, $q_y$, and $q_z$ which may be all different from zero. Also, the piezometric head, $\phi$, usually varies in space, i.e., $\phi = \phi(x, y, z, t)$. However, since the geometry of most aquifers is such that they are *thin relative to their horizontal dimensions* (i.e., tens, or hundreds of meters, as compared to thousands of meters), a simpler approach can be introduced. According to this approach, we assume that the *flow in the aquifer is everywhere essentially horizontal* (*aquifer-type* flow), or that it may be approximated as such, neglecting vertical flow components. This is strictly true (not just an assumption) for flow in a horizontal, homogeneous, isotropic, confined aquifer, of constant thickness and with fully penetrating wells. Nevertheless, the approximation is still a good one when the thickness of the aquifer varies, but in such a way that the variations are much smaller than the average thickness (Figure 1.6a). Actually, we do not totally neglect vertical flow components; as the balance equations do take into account the effect of vertical flow, e.g., vertical accretion. What we do neglect is $\partial\phi/\partial z$.

Whenever justified on the basis of the geometry (i.e., thickness versus horizontal length) and the flow pattern, the assumption of mainly horizontal flow, which is introduced by assuming vertical equipotentials, $\phi = \phi(x, y, t)$, greatly simplifies the mathematical analysis of the flow in the aquifer. The error introduced by this assumption is small in most cases of practical interest (see the discussion in Section 2.2).

The assumption of essentially horizontal flow fails in regions where the flow has a large vertical flow component as, for example, in the vicinity of partially penetrating wells (Figure 1.6b), or outlets in the form of springs, rivers, etc. However, even in these cases, at some distance from the source or the sink, the

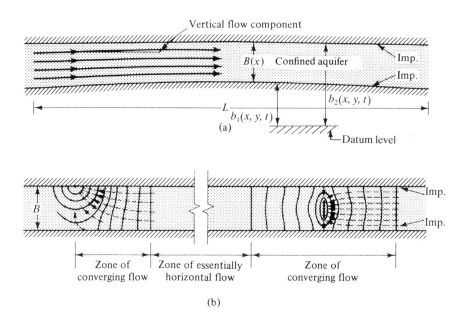
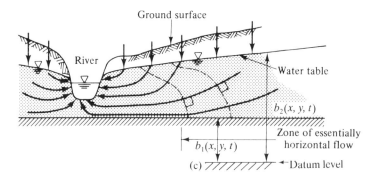
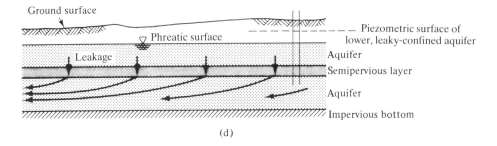

Fig. 1.6. Examples of the hydraulic approach to flow in aquifers. (a) Flow in a confined aquifer with variable thickness: $B(x) \ll L$. (b) Flow in a confined aquifer with partially penetrating wells. (c) Flow in a phreatic aquifer with accretion. (d) Flow in a leaky confined aquifer (Bear, 1979).

assumption of essentially horizontal flow is valid again. As a simple rule, one may assume aquifer-type flow at distances larger than 1.5 to 2 times the thickness of the aquifer in that vicinity. At smaller distances, equipotentials are no more vertical, the flow is three-dimensional and should be treated as such.

The assumption of essentially horizontal flow is applicable also to leaky aquifers (Figure 1.6d). When the hydraulic conductivity of the aquifer is much larger than that of the semipermeable layer, and the thickness of the first is much larger than that of the latter, it follows from the law of refraction of streamlines (e.g., Bear, 1972, p. 26) that the flow in the aquifer is essentially horizontal, while it is essentially vertical in the semipermeable layer. These assumptions, which in cases of practical interest introduce very small errors, greatly simplify the analysis of flow in leaky aquifers.

The approximation that the flow is essentially horizontal in phreatic aquifers is the basis for the Dupuit assumption presented in Section 2.3 (Figure 1.6c).

The procedure of representing the flow (and in, fact, other transport phenomena) as a model of essentially two-dimensional flow in the horizontal plane, is often referred to as the *hydraulic approach*, or the *aquifer approach*. In hydrology of groundwater, this approach is widely used in modeling flow in aquifers. In Section 2.2 and in Chapter 4, we shall show how a two-dimensional model can be obtained from a three-dimensional one, by averaging the latter (i.e., all the equations comprising it) over the aquifer's thickness, assuming vertical equipotentials and taking into account the conditions (of piezometric head and of flux) that exist at the upper and lower boundaries of the considered aquifer. As a consequence of this procedure, the state variables in an essentially horizontal flow model are a function of the horizontal coordinates (say $x$, $y$) and of time only. They are also obtained by averaging their three-dimensional counterpart over the aquifer's thickness. For example, the averaged piezometric head, in an aquifer is given by (Figures 1.6a and c)

$$\tilde{\phi} = \tilde{\phi}(x, y, t) = \int_{b_1(x, y, t)}^{b_2(x, y, t)} \phi(x, y, z, t)\, dz \tag{1.5.1}$$

where $b_2 = b_2(x, y, t)$ and $b_1 = b_1(x, y, t)$ denote the elevations of the upper and lower boundaries of the considered (confined, phreatic, or leaky) aquifer (see Figure 2.4).

## 1.6. Objectives and Scope

Two classes of problems have been introduced in Section 1.2: the *management problems* and the *forecasting problems*. In Section 1.3 we have also introduced the *identification problem*. We have emphasized that no management problem can be solved without solving first, or simultaneously, the forecasting problem, where forecasting refers to both quantity and quality of water. Obviously, we cannot forecast the future behavior of a groundwater system, unless we know its structure and coefficients of transport and storage. This information is obtained by solving the identification problem.

In Section 1.3 we have introduced models and suggested that they are the tool for making the required predictions. This book is devoted to mathematical and numerical modeling of aquifer systems and to the solution of the forecasting problem, primarily by numerical methods.

Section 1.4 introduced the continuum approach as a tool for circumventing difficulties inherent in any attempt to model what happens at points within the pore space of a porous medium. All the models presented and discussed in this book are continuum ones.

Section 1.5 introduced the approximation of *essentially horizontal flow* that greatly simplifies the modeling of flow and pollution transport in aquifers. In practice, this approach to modeling is employed in all regional studies, while three-dimensional modeling, i.e., models where the vertical flow component plays an important role, are employed in cases of a more localized nature. The horizontal two-dimensional model is obtained from the three-dimensional one by integration of the latter over the aquifer's thickness.

Six basic mathematical models are developed and discussed:

   (i) Three-dimensional saturated flow through porous media.
   (ii) Two-dimensional (in the horizontal plane) saturated flow in confined, phreatic and leaky aquifers.
   (iii) Land subsidence.
   (iv) Flow in the unsaturated zone.
   (v) Pollution in three-dimensional domains and in aquifers.
   (vi) Sea water intrusion in coastal aquifers.

The discussion of each model includes the following elements:

   (i) Understanding the phenomena involved.
   (ii) Flux equations and the coefficients appearing in them.
   (iii) Balance equations for the considered extensive quantities.
   (iv) Initial and boundary conditions.
   (v) Complete model.

Chapters 2 to 7 discuss mathematical models and Chapters 8 to 13 discuss numerical ones. Accordingly, Chapter 2 presents the basic equations of groundwater motion, first the motion equation for three-dimensional flow and then the integrated equation for flow in confined and phreatic aquifers.

Chapter 3 starts with the definitions of effective stress and specific storativity. The latter is used in developing the three-dimensional flow model based on the mass balanced equation. By inserting the appropriate flux equation into this balance, a flow equation is obtained in terms of a single dependant variable-pressure or piezometric head. The partial differential equation that represents the

INTRODUCTION 25

mass balance, together with appropriate initial and boundary conditions constitutes the complete model for flow through porous media.

A mathematical model describing land subsidence is included in this chapter (Section 3.6) as a modification and extension of the flow model.

Chapter 4 presents models of flow in confined, phreatic and leaky aquifers, based on the *essentially horizontal flow approximation*. Boundary and initial conditions are also based on this approximation.

Upon reaching this point, the reader should be able to construct complete models of groundwater flow, both in three-dimensional domains and in aquifers.

Chapter 5 considers flow in the unsaturated zone. This zone is important as replenishment passes through it on its downward movement to the aquifer and because pollutants also move through it with the water.

Changes in groundwater quality, as affected by the various transport and accumulation processes are discussed in Chapter 6. The main feature here is the introduction of hydrodynamic dispersion. The general equation of hydrodynamic dispersion (for both saturated and unsaturated flow) is developed. Following a discussion of the appropriate boundary and initial conditions, the complete, mathematical model of the problem of movement and accumulation of a pollutant is presented. The integrated equation describing pollutants' transport in aquifers is developed and discussed.

Chapter 7 deals with the problem of sea water intrusion into coastal aquifers. Again, the discussion leads to partial differential equations describing saltwater and freshwater balances, based on the concept of essentially horizontal flow in an aquifer. In this case, the horizontal flow is assumed to take place within each of the two zones — the freshwater zone and the saltwater one — with an assumed sharp interface separating them. The solution of the model will yield the shape and position of the interface.

This completes the presentation of the methodology of developing mathematical models, with application to the six model types listed above.

In view of the irregular shape of the aquifer boundaries and the possible heterogeneity of the aquifer material, analytical methods solutions of the models described above, although, in principle, superior to any other method of solution, can seldom be employed. Numerical techniques are the practical tools for solving these models in cases of practical interest. These techniques are described in Chapters 8 to 13. The objectives of these chapters are to present the methodology of constructing numerical models by employing the finite difference and the finite element methods, and to demonstrate their solution for many of the model types mentioned above. The complete computer codes (in BASIC) and examples of their application are also included in these chapter.

Particular care has been taken to ensure easy operation of the programs. This is accomplished by a unified presentation of the codes, and by writing all input procedures as interactive processes, in which the user is prompted by the program to enter all data in response to selfexplanatory questions.

All programs are presented in Microsoft BASIC, which seems to be the most widely available and portable computer language. Machine-dependent statements have been avoided, in order to ensure flexibility. All programs should run, with no or very little adaptation, on almost every computer. When using the BASIC interpreter, some of the programs may turn out to be rather slow, especially for large systems. Therefore, it is suggested that they be compiled. This will speed up performance by a factor 10 or 20. All programs in this book admit compilation by the Microsoft's BASCOM and QUICKBASIC compilers.

In Chapter 8 the general principles of numerical techniques are introduced, and a review is given of the main numerical methods. Although the examples in later chapters are mainly restricted to the finite difference and finite element methods, other numerical methods, e.g., the analytical element method and the boundary element method, are also briefly described.

Chapter 9 is devoted to a presentation of the finite difference method, applicable to problems of steady and unsteady groundwater flow. Because of the simplicity of this method, the computer programs are relatively simple. Both implicit and explicit methods are presented.

Chapter 10 describes the finite element method, which is usually more powerful and more flexible than the finite difference one. The theoretical foundation of the method is fully covered, and a variety of particular problems are considered. The presentation of each case includes the complete computer program. The problems considered are those of steady and unsteady flow in heterogeneous aquifers with distributed infiltration and local wells, and the problem of flow through a dam, with a free surface.

In Chapters 11 and 12, numerical models describing the transport of pollutants by moving groundwater are presented. In Chapter 11, the transport by advection, which is usually the main mechanism of transport, is considered. The numerical models include semi-analytical models, based upon an analytical solution of the groundwater flow problem, and fully numerical models, based on a finite element solution of the flow problem.

Chapter 12 is devoted to transport by dispersion. In addition to some simple solutions for one-dimensional transport, numerical models are given for general two-dimensional problems. Particular attention is paid to the errors that may be generated by the solution process itself (numerical dispersion). A random walk method, in which numerical dispersion is avoided, is also presented.

In Chapter 13, the problem of seawater intrusion is considered. Two models are presented, one for flow in a vertical plane, and one for regional flow problems. The model for flow in a vertical plane uses a flexible mesh of finite elements, which follows the motion of the interface between fresh and salt water. The model for regional flow applies to extended aquifers, in which the flow is mainly horizontal. A special feature of this model is that it can easily account for an interface intersecting the impermeable top or bottom boundaries of the aquifer.

CHAPTER TWO

# Groundwater Motion

As part of the *hydrologic cycle*, groundwater is always in motion from regions of natural and artificial replenishment to those of natural and artificial discharge. Bodies of stagnant, usually saline, water trapped in various porous geological formations do exist, but are of little interest to the groundwater hydrologist. When the salinity level of such water is acceptable, water can be mined from such nonreplenishable formations, until the resource is depleted.

Two major types of forecasting problems are encountered in the management of groundwater resources (Section 1.2). The first problem, often called the 'quantity, or the groundwater balance, problem' is one in which the objective is to predict changes in water levels in response to changes in groundwater withdrawal and artificial recharge. In the second problem, often referred to as the 'quality, or pollution problem', the objective is to predict future changes in groundwater quality. In both cases, the knowledge of groundwater motion is required.

This chapter deals with the basic laws that govern the motion of groundwater in aquifers, and with the porous matrix and aquifer properties appearing in these laws. The *continuum approach*, introduced in Section 1.4 is employed and all variables and parameters have already their average meaning in a porous medium regarded as a continuum.

## 2.1. Darcy's Law and its Extensions

### 2.1.1. EMPIRICAL, ONE-DIMENSIONAL FORM

In 1856, Henry Darcy (Darcy, 1856) investigated the flow of water in vertical homogeneous sand filters in connection with the fountains of the city of Dijon (France). From his experiments, Darcy concluded that the rate of flow (i.e., volume of water per unit time), $Q$, is (i) proportional to the cross-sectional area $A$, (ii) proportional to the difference in water level elevations in the inflow and exit reservoirs of the filter $(h_1 - h_2)$, and (iii) inversely proportional to the filter's length, $L$. When combined, these conclusions give the famous *Darcy formula (or law)*

$$Q = KA(h_1 - h_2)/L \qquad (2.1.1)$$

where $K$ is a coefficient of proportionality to be discussed in Subsection 2.1.2 below. The lengths $h_1$ and $h_2$ are measured with respect to some common datum level.

Figure 2.1 shows how Darcy's law (2.1.1) is extended to flow through an inclined homogeneous porous medium column. With the nomenclature of this figure, Darcy's law takes the form

$$Q = KA(\phi_1 - \phi_2)/L \qquad (2.1.2)$$

In (2.1.2), $\phi$ is the *piezometric head* (dimension length) defined by

$$\phi = z + p/\gamma \qquad (2.1.3)$$

where $z$ is the elevation of the point, $p$ is the pressure, and $\gamma$ is the volumetric weight of the water. The piezometric head expresses the sum of the potential energy and pressure energy, per unit weight of water.

The energy loss $\Delta\phi = \phi_1 - \phi_2$ is due to friction in the flow through the narrow tortuous paths of the porous medium. In Darcy's law, the kinetic energy of the water has been neglected as, in general, changes in the piezometric head along the flow path are much larger than changes in the kinetic energy. Inertial effects have also been neglected.

With the above definition of piezometric head, the quotient $(\phi_1 - \phi_2)/L$ is the *hydraulic gradient* (dimensionless). Denoting this gradient by $J$ and defining the specific discharge, $q$, as the volume of water flowing per unit time through a unit cross-sectional area normal to the direction of flow, we obtain

$$q = KJ \qquad (2.1.4)$$

where $q = Q/A$, as another form of Darcy's law.

Let us consider a point along the column's axis and a segment of the column of

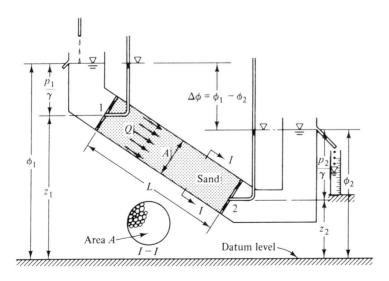

Fig. 2.1. Flow through an inclined sand column.

GROUNDWATER MOTION

length $s$ along the column's axis on both sides of the point. For this case

$$q_s = K \frac{\phi|_{s-(\Delta s/2)} - \phi|_{s+(\Delta s/2)}}{\Delta s} \qquad (2.1.5)$$

where the subscript in $q_s$ indicates that the flow is in the $s$-direction. In the limit, as $\Delta s \to 0$, converging on the point, we obtain

$$\lim_{\Delta s \to 0} \frac{\phi|_{s-(\Delta s/2)} - \phi|_{s+(\Delta s/2)}}{\Delta s} = -\frac{\partial \phi}{\partial s} \qquad (2.1.6)$$

and (2.1.5) reduces to

$$q_s = -K \frac{\partial \phi}{\partial s} \equiv KJ_s; \quad J_s = -\frac{\partial \phi}{\partial s}. \qquad (2.1.7)$$

This expression gives the component of the specific discharge in the direction $s$ at any point in a porous medium domain, given $K$ and the spatial distribution of the piezometric head, $\phi$.

It should be emphasized that Darcy's law (2.1.7), expressed in terms of the piezometric head, $\phi$, is valid only for a fluid of constant density, i.e., $\rho = $ const.

It is important to note that (2.1.2) states that the flow takes place from a higher piezometric head to a lower one and not from a higher to a lower pressure. For example, in the case shown in Figure 2.1 $p_1/\gamma < p_2/\gamma$, that is, the flow is in the direction of *increasing* pressure: however, it is in the direction of *decreasing* head. It is only in the special case of horizontal flow, where $z_1 = z_2$, that we may write

$$Q = KA(p_1 - p_2)/\gamma L. \qquad (2.1.8)$$

In (2.1.7), $q_s$ is considered positive in the positive direction of the $s$-axis.

Actually, flow takes place only through part of the cross-sectional area, $A$, of the column of porous medium shown in Figure 2.1, the remaining part being occupied by the solid matrix of the porous medium. Since it can be shown that *the average areal porosity is equal to the volumetric porosity, $n$,* the portion of the area $A$ available to flow is $nA$. Accordingly, the *average velocity, $V$,* of the flow through the column is given by

$$V = Q/nA = q/n. \qquad (2.1.9)$$

In (2.1.9), $V$ is the *intrinsic phase average velocity*, $\overline{V}^v$, defined, following (1.4.3), by

$$\overline{V}^v = \frac{1}{U_{0v}} \int_{(U_{0v})} V \, dU \qquad (2.1.10)$$

defined following the definition of an intrinsic phase average by (1.4.3). In (2.1.10), $U_{0v}$ is the volume of voids within an REV and $V$ is the microscopic velocity at points inside the void space. Similarly, the specific discharge, $q$, has the meaning of a *phase average velocity*, $\overline{V}$, as defined by (1.4.4). Needless to remind the reader that $\phi \equiv \overline{\phi}^v, p = \overline{p}^v$, etc.

Sometimes, even in the flow of a single homogeneous fluid, part of the fluid in the pore space is immobile (or practically so). This may occur when the flow takes place in a fine textured medium where adhesion (i.e., the attraction to the solid surface of the porous matrix of the fluid molecules adjacent to it) is important, or when the solid matrix includes a large portion of dead-end pores. In this case, one may define an *effective porosity with respect to the flow through the medium*, $n_{\text{eff}}$ ($< n$), such that

$$V = q/n_{\text{eff}}. \tag{2.1.11}$$

One should clearly distinguish between the specific discharge, to be used, for example, for determining the volume of fluid passing through a given cross-sectional area, and the average velocity, or simply 'velocity', to be used, for example, for front or particle movements. Both concepts should not be confused with the (actual, or microscopic) local velocity of the fluid at (microscopic) points inside the void space. Considering dimensions, one should note that $q$, $V$, and $K$ have the same dimensions

$$[q] = L/T; \qquad [V] = L/T; \qquad [K] = L/T. \tag{2.1.12}$$

2.1.2. HYDRAULIC CONDUCTIVITY

The coefficient of proportionality, $K$, appearing in Darcy's law (2.1.2) is called *hydraulic conductivity* of the porous medium. In an isotropic medium, it may be defined, using (2.1.4), as the specific discharge per unit hydraulic gradient. It is a scalar (dims. L/T) that expresses the ease with which a fluid is transported through the tortuous void space. It is therefore a coefficient that depends on both matrix and fluid properties. The relevant properties are the density, $\rho$, and the viscosity, $\mu$ (or in the combined form of the kinematic viscosity, $\nu$). The relevant solid matrix properties are mainly grain- (or pore-) size distribution, shape of grains (or pores), tortuosity, specific surface, and porosity. The hydraulic conductivity $K$ may be expressed as

$$K = k\rho g/\mu = kg/\nu \tag{2.1.13}$$

where $g$ is the acceleration of gravity and where $k$ (dims. L$^2$) − called the *permeability* of the porous medium − depends solely on the properties of the solid matrix.

Various formulas relating $k$ to the various properties of the solid matrix are presented in the literature. Some of these formulas are purely empirical, as, for example

$$k = cd^2 \tag{2.1.14}$$

where $c$ is a dimensionless coefficient, and $d$ is an effective grain diameter, say,

$d_{10}$. Krumbein and Monk (1943) suggest $c = 6.17 \times 10^{-4}$, so that for $\nu = 10^{-6}$ m²/s (corresponding to water at 20°C), one obtains $cg/\nu \simeq 62$/ms. As an average, the value of 100/ms is often used.

Another example is the Fair and Hatch (1933) formula developed from dimensional considerations and verified experimentally

$$k = \frac{1}{\beta} \left[ \frac{(1-n)^2}{n^3} \left( \frac{\alpha}{100} \sum_{(m)} \frac{P_m}{d_m} \right)^2 \right]^{-1} \qquad (2.1.15)$$

where $\beta$ is a packing factor, found experimentally to be about 5, $\alpha$ is a sand shape factor, varying from 6.0 for spherical grains to 7.7 for angular ones, $P_m$ is the weight percentage of sand held between adjacent sieves, and $d_m$ is the geometric mean diameter of the adjacent sieves.

Purely theoretical formulas for permeability are obtained from theoretical derivations of Darcy's law. Usually, such formulas include numerical coefficients which have to be determined empirically. An example is the Kozeny–Carman equation

$$k = C_0 \frac{n^3}{(1-n)^2 M_s^2} \qquad (2.1.16)$$

where $M_s$ is the specific surface area of the solid matrix (defined per unit volume of solid) and $C_0$ is a coefficient for which Carman (1937) suggested the value of $\frac{1}{5}$.

Under certain conditions, the permeability, $k$, may vary with time. This may be caused by external loads which change the structure and texture of the solid matrix by subsidence and consolidation, by the solution of the solid matrix (which over prolonged times may produce large channels and cavities), and by the swelling of clay, if present within the void space. When a soil contains argillaceous material, drying of the soil may shrink the clay, especially bentonite, causing the permeability to air of the dried soil to be higher than for water. Fresh water in a soil sample may cause the clay to swell as compared with salt water, thereby reducing the permeability. Biological activity in the medium may produe a growth which tends to clog the matrix, thus reducing $k$ with time. Clogging may also be caused by fines carried by the water (e.g., in artificial recharge).

Various units are used in the practice for the hydraulic conductivity $K$ (dims. L/T). Hydrologists prefer the unit m/day (meters per day). Soil scientists often use cm/sec. In the SI system, the unit m/s is used. In the U.S.A., as in many countries using the English system of units, two other units are commonly employed by hydrologists. One is a *laboratory*, or *standard, hydraulic conductivity*, defined as the total discharge $(Q)$ of water at 60°F, expressed in gallons per day, through a porous medium cross-sectional area $(A)$ expressed in ft² under a hydraulic gradient, $(\phi_1 - \phi_2)/L$, of 1 ft/ft. With this definition, the units of $K$ are gal/day ft². In a similar way, a *field, or aquifer, hydraulic conductivity* is defined as the discharge of water at field temperature, through a cross-sectional area of an aquifer one foot thick and one mile wide under a hydraulic gradient of 1 ft/mile. The unit is the same as for the laboratory $K$. Following are some conversions among these units.

$$1 \text{ US gal/day ft}^2 = 4.72 \times 10^{-5} \text{ cm/sec} = 4.08 \times 10^{-2} \text{ m/d}.$$

Permeability, $k$ (dims. L), is measured in the metric system in cm$^2$, or in m$^2$. In the English system, the unit is ft$^2$. For water at 20°C, we have the conversion (with $\nu = 10^{-6}$ m$^2$/s)

$$1 \text{ cm/sec is equivalent to } 1.02 \times 10^{-5} \text{ cm}^2 = 10^{-9} \text{ m}^2.$$

Reservoir engineers use the unit darcy defined by

$$1 \text{ darcy} = \frac{1 \text{ cm}^3/\text{sec}/\text{cm}^2 \times 1 \text{ centipoise}}{1 \text{ atmosphere/cm}}$$

with

$$1 \text{ darcy} = 9.8697 \times 10^{-9} \text{ cm}^2 = 1.062 \times 10^{-11} \text{ ft}^2$$

equivalent to

$$9.613 \times 10^{-4} \text{ cm/sec (for water at 20°C)}$$

or to

$$1.4156 \times 10^{-2} \text{ US gal/min ft}^2 \text{ (for water at 20°C)}.$$

Table 2.1 gives a summary of some values of hydraulic conductivity and permeability (Irmay, in Bear et al., 1968).

Table 2.1. Typical values of hydraulic conductivity and permeability[a]

| $-\log_{10} \cdot K$ (cm/sec) | −2 | −1 | 0 | 1 | 2 | 3 | 4 | 5 | 6 | 7 | 8 | 9 | 10 | 11 |
|---|---|---|---|---|---|---|---|---|---|---|---|---|---|---|
| Permeability | | Pervious | | | | Semipervious | | | | | Impervious | | | |
| Aquifer | | Good | | | | Poor | | | | | None | | | |
| Soils | | Clean gravel | | Clean sand or sand and gravel | | | Very fine sand, silt, loess, loam, solonetz | | | | | | | |
| Rocks | | | | | | Peat | | Stratified clay | | | Unweathered clay | | | |
| Rocks | | | | | | Oil rocks | | | Sandstone | | Good limestone, dolomite | | Breccia, granite | |
| $-\log_{10} \cdot k$ (cm$^2$) | | | 3 | 4 | 5 | 6 | 7 | 8 | 9 | 10 | 11 | 12 | 13 | 14 | 15 | 16 |
| $\log_{10} k$ (md) | | | 8 | 7 | 6 | 5 | 4 | 3 | 2 | 1 | 0 | −1 | −2 | −3 | −4 | −5 |

[a] From Bear et al. (1968).

## 2.1.3. RANGE OF VALIDITY

Column experiments indicate that the specific discharge, $q$, increases, the relationship between $q$ and the hydraulic gradient, $J$, gradually deviates from the linear relationship expressed by Darcy's law. Figure 2.2a shows this deviation. Therefore, it seems reasonable to define a range of validity for Darcy's linear law.

In flow through conduits, the *Reynolds number*, Re, which is a dimensionless number expressing the ratio of inertial to viscous forces acting on the fluid, is used as a criterion to distinguish between laminar flow occurring at low velocities and turbulent flow occurring at higher velocities (see any text of fluid mechanics). The critical value of Re between laminar and turbulent flow in pipes is around 2000. By analogy, a Reynolds number is defined also for flow through porous media

$$\text{Re} = qd/\nu \qquad (2.1.17)$$

where $d$ is some representative microscopic length characterizing the solid matrix, and $\nu$ is the kinematic viscosity of the fluid. Although by analogy to the Reynolds number for pipes, $d$ should be a length representing the cross-section of an elementary channel of the porous medium, it is customary (probably because of the relative ease of determining it) to employ for $d$ some representative length of the grains (in an unconsolidated porous medium). Often the mean grain diameter is taken as the length dimension $d$ in (2.1.17). Sometimes $d_{10}$, that is, the diameter such that 10% by weight of the grains are smaller than that diameter, is mentioned in the literature as a representative grain diameter. Collins (1961) suggested $d = (k/n)^{1/2}$, where $k$ is the permeability and $n$ is the porosity, as the representative length, $d$. On the basis of theoretical analysis (Bachmat and Bear, 1987), this seems to be a better choice.

In spite of the various definitions for $d$ in (2.1.17), practically all evidence indicates that *Darcy's law is valid as long as the Reynolds number does not exceed some value between 1 and 10.* Most groundwater flows occur in this range, except

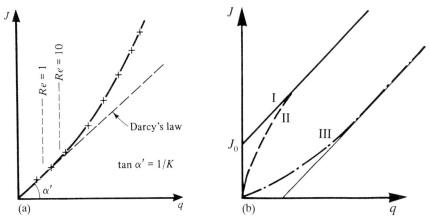

Fig. 2.2. Schematic curves relating $J$ to $q$.

in the very close vicinity of large pumping or recharging wells, or large (point) springs. Large Reynolds numbers may also be observed in very porous aquifers such as cavernous limestone, and in the flow through breakwaters constructed of gravel, or even larger stones.

In fine grained soil, e.g., clay, there are indications that there exists also a lower limit to the validity of Darcy's law. Figure 2.2b shows some typical schematic $J - q$ relationships at very low gradients.

Curve I in Figure 2.2b indicates the existence of some minimum, or *threshold gradient* $J = J_0$, below which there is practically no flow. Curve II in Figure 2.2b shows a faster growth of $q$ with $J$ than is indicated by Darcy's law. In curve III, the rate of growth of $q$ is smaller. For curve I, we have

$$q = 0, \qquad \text{for } J \leqslant J_0,$$
$$q = K(J - J_0), \quad \text{for } J > J_0. \tag{2.1.18}$$

Among explanations presented in the literature for the non-Darcy behavior in fine grained materials, (usually not in aquifers) we may mention (i) pores may be very small, such that water molecules in them are strongly influenced by the *double layer* effects of the clay particles. Because water molecules are polar, water near the electrically charged clay particles has a more crystalline structure which causes the viscosity to be higher than ordinary water. Under such conditions, a minimum hydraulic gradient is required to cause water movement, (ii) the effect of *streaming potential*. As water moves near the clay surface, it carries along some cations in the diffuse layer. The cations are electrically attracted to the clay particles, a fact that produces resistance to the movement of the cations. This, in turn, produces a drag on the moving water. The potential difference due to this migration of cations is called the streaming potential; it acts in the direction opposite to that of the flow, (iii) non-Newtonian behavior of the fluid in capillary spaces, and (iv) electroosmotic counterflow.

Some of these phenomena are important in connection with the thin water films that remain on soil particles in drained pores in unsaturated flow.

### 2.1.4. THREE-DIMENSIONAL MOTION EQUATION

The experimentally derived equation of motion in the form of Darcy's law (2.1.4) or (2.1.7), is limited to one-dimensional flow of a homogeneous incompressible fluid. When the flow is three-dimensional, the generalization of these equations is

$$\mathbf{q} = K\mathbf{J} = -K \text{ grad } \phi; \qquad \mathbf{V} = \mathbf{q}/n \tag{2.1.19}$$

where $\mathbf{V}$ is the velocity vector with components $V_x$, $V_y$, and $V_z$, $\mathbf{q}$ is the specific discharge vector, with components $q_x$, $q_y$, $q_z$ in the directions of the Cartesian, $xyz$, coordinates, respectively, and $\mathbf{J} = -\text{grad } \phi \equiv -\nabla \phi$ is the hydraulic gradient, with components $J_x = -\partial \phi/\partial x$, $J_y = -\partial \phi/\partial y$, $J_z = -\partial \phi/\partial z$, in the $xyz$ directions,

GROUNDWATER MOTION 35

respectively. When the flow takes place through a homogeneous isotropic medium, the coefficient $K$ is a constant scalar, and (2.1.19) may be written as three equations

$$q_x = KJ_x = -K\, \partial\phi/\partial x = nV_x; \qquad q_y = KJ_y = -K\, \partial\phi/\partial y = nV_y;$$
$$q_z = KJ_z = -K\, \partial\phi/\partial z = nV_z. \tag{2.1.20}$$

The flow in any direction indicated by the unit vector $\mathbf{1s}$ is given by

$$q_s = \mathbf{q} \cdot \mathbf{1s} = -K\, \partial\phi/\partial s = nV_s. \tag{2.1.21}$$

Equations (2.1.19) through (2.1.20) remain valid also for three-dimensional flow through *inhomogeneous media*, where $K = K(x, y, z)$, as long as the medium is also *isotropic*. Flow in anisotropic media is discussed in Subsection 2.1.6.

2.1.5. COMPRESSIBLE FLUID

For a homogeneous compressible fluid (i.e., no dissolved components) under isothermal conditions, $\rho = \rho(p)$, the piezometric head, $\phi^*$, was defined by Hubbert (1940) as

$$\phi^* = z + \int_{p_0}^{p} \frac{dp}{g\rho(p)}. \tag{2.1.22}$$

Often, $\phi^*$ is referred to as *Hubbert's potential*. For such a fluid, the motion equation (we may still refer to it as Darcy's law) is

$$\mathbf{q} = K\mathbf{J} = -K\,\mathrm{grad}\,\phi^*; \qquad \mathbf{J} = -\nabla\phi^*. \tag{2.1.23}$$

2.1.6. ANISOTROPIC POROUS MEDIA

A porous medium domain is said to be *homogeneous* with respect to its permeability, if the latter is the same at all its points. Otherwise, the domain is said to be *heterogeneous*. If, however, the permeability at a considered point is independent of direction, the porous medium is said to be *isotropic* at that point. Otherwise, the porous medium is said to be *anisotropic*. Similar definitions apply to other properties of porous media or of aquifers.

In many cases, aquifers are anisotropic. This may happen, for example, when the sediments comprising the aquifer are such (i.e., flat shaped mica particles) that when deposited, the resulting porous medium has a higher permeability in one direction (usually the horizontal, unless later tilting of the formation occurs) than in other directions. Both sedimentation and the stress produced by the material, cause flat particles to be oriented with their longest dimension parallel to the plane on which they settle. Later, the flow itself, in its predominant direction, produces more developed channels parallel to the bedding plane than in other directions, thus making the material anisotropic. In carbonate rocks, the flowing water dissolves the rock, producing solution channels that develop, often from very small fissures, in the direction of the predominant flow. In some soils, structural fissures develop more readily in one direction than in others, and the soil exhibits anisotropy.

An inhomogeneous material composed of alternating layers of different textures is equivalent in its overall behavior to an homogeneous anisotropic porous medium (Bear, 1972, p. 155). However this equivalence is valid only when the thickness of the individual layers is much smaller than lengths of interest within the porous medium domain. For example, it is meaningless to determine the equivalent anisotropic permeability of such a domain from a core whose size is smaller than the thickness of a single layer.

Consider the horizontal flow parallel to the layers in the stratified confined aquifer shown in Figure 2.3a. Equipotentials, $\phi$ = const. are vertical. The total discharge $Q$ is the sum of the discharge rates, $Q_i$, of the individual layers.

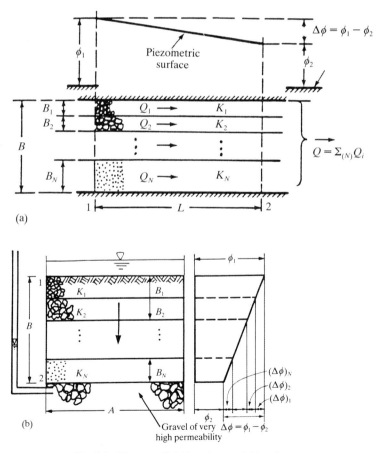

Fig. 2.3. Flow parallel (a) and normal (b) to layers.

GROUNDWATER MOTION

Employing the nomenclature of Figure 2.3a, we obtain

$$Q = \sum_{i=1}^{N} Q_i; \quad B = \sum_{i=1}^{N} B_i; \quad Q_i = K_i B_i \frac{\Delta \phi}{L},$$

$$Q = \frac{\Delta \phi}{L} \sum_{i=1}^{N} K_i B_i = K_{eq}^P B \frac{\Delta \phi}{L},$$

$$K_{eq}^P = \frac{1}{B} \sum_{i=1}^{N} K_i B_i \qquad (2.1.24)$$

where $K_{eq}^P$ is the equivalent hydraulic conductivity parallel to the layers in the aquifer.

For flow normal to the layers (Figure 2.3b), the equivalent hydraulic conductivity, $K_{eq}^N$, is obtained from

$$Q = \frac{(\Delta \phi)_1}{B_1} K_1 A = \frac{(\Delta \phi)_2}{B_2} K_2 A = \cdots = \frac{(\Delta \phi)_N}{B_N} K_N A = K_{eq}^N A \frac{\Delta \phi}{B},$$

$$\Delta \phi = \phi_1 - \phi_2 = \sum_{i=1}^{N} (\Delta \phi)_i = \frac{Q}{A} \sum_{i=1}^{N} \frac{B_i}{K_i} = \frac{QB}{K_{eq}^N A},$$

$$\frac{B}{K_{eq}^N} = \sum_{i=1}^{N} \frac{B_i}{K_i} \qquad (2.1.25)$$

It is of interest to note that if one of the $K_i$'s is zero in (2.1.25), $K_{eq}^N = 0$, i.e., the whole system is rendered impervious whereas in (2.1.4) such a layer simply does not contribute to the flow. However, it also follows from (2.1.25) that the flow is governed by the resistance (= $B/KA$) of the layers, i.e., by the layer of highest resistance, and not by their $K_i$'s and $B_i$'s separately.

From (2.1.24) and (2.1.25) it follows that $K_{eq}^P > K_{eq}^N$, i.e., the hydraulic conductivity is greater in the direction of stratification.

Bear (1972), p. 155, shows that similar results are obtained when the imposed hydraulic gradient makes any arbitrary angle with the layers. He uses this analysis to show that when the layers are alternating, i.e., $(K_1, B_1)$, $(K_2, B_2)$, $(K_1, B_1)$, $(K_2, B_2)$ ..., with each of them much thinner than the entire porous medium domain, the latter behaves as an equivalent homogeneous anisotropic porous medium.

When written for a homogeneous anisotropic porous medium, Darcy's law, which in the form of (2.1.20), is written for a homogeneous isotropic medium, takes the form

$$q_x = K_{xx}J_x + K_{xy}J_y + K_{xz}J_z,$$
$$q_y = K_{yx}J_x + K_{yy}J_y + K_{yz}J_z,$$
$$q_z = K_{zx}J_x + K_{zy}J_y + K_{zz}J_z. \quad (2.1.26)$$

In (2.1.26), $q_x$, $q_y$, and $q_z$ are the components in the $x$, $y$, and $z$ directions, respectively, of the specific discharge vector **q**; $J_x$, $J_y$, $J_z$ are the components of the hydraulic gradient vector **J**; $K_{xx}$, $K_{xy}$, ..., $K_{zz}$ are nine constant coefficients. In an inhomogeneous medium, each of these coefficients may vary in space.

The nine coefficients appearing in (2.1.26) are components of the *second rank tensor of hydraulic conductivity* of an anisotropic medium, **K**. Detailed discussions on the nature of second rank tensors and on methods for treating them, is beyond the scope of this book. The reader is referred to texts on tensor analysis (e.g., Morse and Feshbach, 1953; Spiegel, 1959; Aris, 1962). Symbolically, we write **K** in the matrix form

$$[\mathbf{K}] = \begin{bmatrix} K_{xx} & K_{xy} & K_{xz} \\ K_{yx} & K_{yy} & K_{yz} \\ K_{zx} & K_{zy} & K_{zz} \end{bmatrix}; \quad [\mathbf{K}] = \begin{bmatrix} K_{xx} & K_{xy} \\ K_{yx} & K_{yy} \end{bmatrix} \quad (2.1.27)$$

in three-and two-dimensional spaces, respectively.

The hydraulic conductivity tensor is symmetric, that is $K_{xy} = K_{yx}$; $K_{xz} = K_{zx}$ and $K_{yz} = K_{zy}$. This means that actually only six distinct components in three-dimensional flow, and three such components in two-dimensional flow, are needed for fully defining the hydraulic conductivity.

Equations (2.1.26) may be writen in several compact forms, for example

$$q_i = K_{ij}J_j; \quad \mathbf{q} = \mathbf{K} \cdot \mathbf{J},$$
$$q_i = -K_{ij}\frac{\partial \phi}{\partial x_j}; \quad \mathbf{q} = -\mathbf{K} \cdot \nabla \phi \quad (2.1.28)$$

where subscripts $i$ and $j$ stand for $x_i$, $x_j$ respectively, with $x_1 \equiv x$, $x_2 \equiv y$ and $x_3 \equiv z$. In (2.1.28), we have employed Einstein's summation convention (or the *double index convention*) according to which in any product of factors, a suffix (here subscript) repeated twice and only twice is held to be summed over the entire range of values (1,2,3 and 1,2 for three- and two-dimensional spaces, respectively). The component $K_{x_i x_j}$ ($\equiv K_{ij}$) may be interpreted as the contribution to the specific discharge $q_i$, in the $x_i$ direction, produced by a unit component of the hydraulic gradient, $J_{x_j}$, in the $x_j$ direction. The total specific discharge is the sum of partial specific discharges caused by $J_{x_1}$, $J_{x_2}$ and $J_{x_3}$.

While the hydraulic conductivity of a porous medium (expressed symbolically by **K**) is independent of the coordinate system which we happen to use, the magnitude of each component, $K_{ij}$, depends on the chosen coordinate system.

GROUNDWATER MOTION

Texts on tensor analysis give the rules of transformation of these components from one coordinate system to another. It is also shown in these texts that it is always possible to find *three mutually orthogonal directions* in space such that when these directions are chosen as the coordinate system for expressing the components $K_{ij}$, we find that $K_{ij} = 0$ for all $i \neq j$ and $K_{ij} \neq 0$ for $i = j$. These directions in space are called the *principal directions of the anisotropic porous medium* (actually of the permeability of the medium).

When the principal directions are used as the coordinate system, (2.1.27) becomes

$$[\mathbf{K}] = \begin{bmatrix} K_{xx} & 0 & 0 \\ 0 & K_{yy} & 0 \\ 0 & 0 & k_{zz} \end{bmatrix}; \quad [\mathbf{K}] = \begin{bmatrix} K_{xx} & 0 \\ 0 & K_{yy} \end{bmatrix} \quad (2.1.29)$$

and (2.1.26) reduces to

$$q_x = K_x J_x; \quad q_y = K_y J_y; \quad q_z = K_z J_z \quad (2.1.30)$$

where $K_{xx} \equiv K_x, K_{yy} \equiv K_y, K_{zz} \equiv K_z$.

It is of interest to note that from (2.1.30) it follows that *in an anisotropic porous medium, the specific discharge*, **q**, *and the hydraulic gradient*, **J**, *are not colinear* (except when **J** is along one of the principal directions). The angle, $\theta$, between them is given by

$$\cos \theta = \mathbf{q} \cdot \mathbf{J}/qJ, \quad q = |\mathbf{q}|, J = |\mathbf{J}| \quad (2.1.31)$$

Since **K** is a second-rank tensor, the transformation of its components $K_{x_i x_j}$ in an $x_i$ ($i = 1, 2, 3$) coordinate system, into components $K'_{x'_i x'_j}$ in another, $x'_i$ ($i = 1, 2, 3$), coordinate system, obtained from the latter by rotation, is given by

$$K'_{x'_i x'_j} = K_{ij} \cos(\mathbf{1x}'_i, \mathbf{1x}_i)\cos(\mathbf{1x}'_j, \mathbf{1x}_j) \quad (2.1.32)$$

where $\mathbf{1x}_i, \mathbf{1x}'_i, \mathbf{1x}'_j, \mathbf{1x}'_j$ indicate unit vectors along the respective axes. In fact, to establish that the nine entities $K_{ij}$ are indeed components of a second-rank tensor, we must show that they transform according to (2.1.32).

### 2.1.7. THE GENERAL MOTION EQUATION

In Subsection 2.1.1, following the work of Henry Darcy (1856), Darcy's law was presented as an empirical law. Over the years, and especially in the past 30 years, a number of researchers have attempted to derive Darcy's law or, more generally, the motion equation for a fluid phase in a porous medium, by theoretical consideration. A review of these works is beyond the scope of this book (see, for example, Bear (1972), Hassanizadeh and Gray (1979a, b), Bear and Bachmat (1986, 1987), Bachmat and Bear (1986)). Although a number of different approaches have been employed in these researches, most of them recognize that the motion equation for a fluid phase inside the void space of a porous medium, must be obtained by considering the *momentum balance equation* (often referred to as the *motion equation*) of that phase, regarded as a continuum. Accordingly, they derive the motion equation for a fluid present in the void space of a porous medium *by taking an average of the motion equation (= momentum balance equation) of that phase*, where the average is performed over the volume of the phase within a Representative Elementary Volume (Section 1.4) of the porous medium. By introducing a number of simplifying assumptions, and especially by assuming that (i) the inertial effects, and (ii) the internal friction inside the fluid, are negligible, in comparison with the drag produced at the fluid-solid interface, the following (macroscopic) motion equation has been obtained for the case of a single fluid (subscript $w$) that occupies the entire void space (i.e., saturated flow)

$$\mathbf{V}_w - \mathbf{V}_s = -\frac{\mathbf{k}}{n\mu_w}(\nabla p_w + \rho_w g \nabla z) \qquad (2.1.33)$$

where $\mathbf{V}_w$ and $\mathbf{V}_s$ denote the (intrinsic phase) average velocities of the fluid and of the solid (subscript $s$), respectively, $\rho_w$, $\mu_w$, and $p_w$ denote the fluid's density, viscosity, and pressure, respectively, $z$ denotes elevation and $\mathbf{k}$ denotes the possibly anisotropic permeability. For an anisotropic porous medium, the permeability is a second rank symmetric tensor which depends only on the porosity and on the (microscopic) configuration of the solid-fluid interface.

In (2.1.33), $-(\nabla p_w + \rho_w g \nabla z)$ represents the driving force, per unit volume of fluid, due to pressure gradient and to gravity. This force is balanced by the drag or resistance to the flow, at the solid-fluid interface, expressed by $n(\mathbf{V}_w - \mathbf{V}_s)\mu \mathbf{k}^{-1}$. Written in this form, Darcy's law states that the drag is proportional to the fluid's velocity, relative to the soil skeleton, proportional to the fluid's viscosity and inversely proportional to the property $k/n$ of the porous medium.

Thus (2.1.33), and actually any of the various forms of Darcy's law, or the motion equation, state that in the absence of the inertial effects, the resistance to the flow is linearly proportional to the specific discharge relative to the (possibly moving) solids. In indicial notation and Cartesian coordinates, (2.1.33) takes the form

$$V_{wi} - V_{si} = -\frac{k_{ij}}{n\mu_w}\left(\frac{\partial p_w}{\partial x_j} + \rho_w g \frac{\partial z}{\partial x_j}\right), \quad i,j = 1, 2, 3 \qquad (2.1.34)$$

where *Einstein's summation convention is employed.*

Equation (2.1.33) is the general motion equation (we may still refer to it as Darcy's law) for saturated flow of a single fluid in an anisotropic and inhomogeneous porous medium, i.e., where $\mathbf{k} = \mathbf{k}(x, y, z)$. In this equation, the fluid's density may depend on pressure, concentration of dissolved matter, and temperature. For $\rho_w = \rho_w(p_w)$, the right-hand side of (2.1.33) reduces to $-(1/n)\mathbf{K} \cdot \nabla \phi^*$ where $\mathbf{K} = \mathbf{k}\rho_w/\mu_w$. For $\rho_w = $ const, the right-hand side reduces to $-(1/n)\mathbf{K} \cdot \nabla \phi$.

An important feature of the theoretically derived motion equation (2.1.33) is

GROUNDWATER MOTION 41

the presence of the (intrinsic phase average) solid velocity, $\mathbf{V}_s$. From (2.1.33) it follows that Darcy's law gives the velocity of the fluid with respect to the (possibly moving) solid. In terms of specific discharge, $\mathbf{q} = n\mathbf{V}_w$ relative to a fixed coordinate system, and $\mathbf{q}_r = \mathbf{q} - n\mathbf{V}_s$ = specific discharge relative to the solid, we obtain

$$\mathbf{q}_r = -\frac{\mathbf{k}}{\mu}(\nabla p + \rho g \nabla z) \qquad (2.1.35)$$

where we have omitted the superscript $w$ and $\rho$, $\mu$, and $p$ represent (intrinsic phase) average values. When $\mathbf{V}_s = 0$, e.g., in a nondeformable porous medium, $\mathbf{q}_r = \mathbf{q}$. It may be noted that this theoretically derived form of Darcy's law is in agreement with Hubbert's (1940) formulation for flow of a compressible fluid (see Subsection 2.1.5).

In general, when dealing with groundwater flow problems, we may assume $\mathbf{V}_s \simeq 0$, although (as we shall see in Section 3.2) we do take into account $\mathbf{V}_s \neq 0$ in the considerations leading to the concept of specific storativity. When dealing with land subsidence due to pumping (Section 4.5), the effect of $\mathbf{V}_s \neq 0$ may have to be taken into account.

In conclusion, we shall use (2.1.33) whenever $\rho \neq$ const. Often we shall approximate $\mathbf{q}_r$ by $\mathbf{q}$. For $x$, $y$, $z$ that are principal directions, we shall often write

$$\mathbf{q} = -\frac{\mathbf{k}}{\mu}(\nabla p + \rho g \nabla z) \begin{cases} q_x = -\dfrac{k_{xx}}{\mu}\dfrac{\partial p}{\partial x} \\[6pt] q_y = -\dfrac{k_{yy}}{\mu}\dfrac{\partial p}{\partial y} \\[6pt] q_z = -\dfrac{k_{zz}}{\mu}\left(\dfrac{\partial p}{\partial z} + \rho g\right). \end{cases} \qquad (2.1.36)$$

### 2.1.8. FLOW AT LARGE Re

In the range of validity of Darcy's linear law, i.e., Re < 1–10, the viscous forces that resist the flow are predominant. As the flow velocity increases, a region of gradual transition is observed, from laminar flow, where viscous forces are predominant, to still laminar flow, but with inertial forces governing the flow. Often, the value of Re = 100 is mentioned as the upper limit of this transition region in which Darcy's law is not valid. Some authors explain the deviation from the linear law by the separation of the flow from the solid walls of the solid matrix caused at large Re by the inertial forces. This occurs at a gradually increasing number of (microscopic) points at the fluid-solid interface, where the flow diverges, or is cured. At still higher values of Re (say, Re 150–300), the flow becomes really turbulent.

Bachmat and Bear (1986, 1987), by averaging the (microscopic) momentum balance equation for a Newtonian incompressible fluid (i.e., Navier–Stokes equation), derive the (approximate) average momentum balance equation (= motion equation)

$$n\rho \left( \frac{\partial V_j}{\partial t} + V_i \frac{\partial V_j}{\partial x_i} \right) - \mu \left( \frac{\partial^2 q_{ri}}{\partial x_i \partial x_j} + \frac{\partial^2 q_{rj}}{\partial x_i \partial x_i} \right) +$$

$$+ n \left( \frac{\partial p}{\partial x_i} + \rho g \frac{\partial z}{\partial x_i} \right) T_{ij}^* + n\mu k_{jm}^{-1} T_{mi}^* q_{ri} = 0 \qquad (2.1.37)$$

where $T_{ij}^*$ is a coefficient that expresses the static moment of the oriented $S_{vv}$ areas with respect to planes passing through a point, per unit volume of the void space; the point serves as a centroid of an REV and $S_{vv}$ is the surface area of intersection of the void space with the (say, spherical) surface of the REV.

Each term in (2.1.37) represents a force. The first term represents the inertial force due to acceleration, both local and convective. The second term represents the viscous resistance due to internal friction inside the fluid phase. The third term represents the driving force due to pressure gradient and gravity. The last term represents the drag resistance caused by the fluid-solid interaction. When the first two terms are neglected (with respect to the remaining ones), we obtain Darcy's linear law, say (2.1.33). When only the second term is neglected, we obtain a motion equation that includes the effect of the inertial forces. This equation should be used at high Reynolds numbers.

There is no universally accepted simplified nonlinear motion equation (that is, relationship between **J** and **q**) which is valid for Re > 1–10. Many such relationships appear in the literature. Most of them have the general form (Forchheimer, 1901)

$$J = Wq + bq^2, \quad \text{or} \quad J = Wq + bq^m, \quad 1.6 \leqslant m \leqslant 2 \qquad (2.1.38)$$

where $J \equiv |\mathbf{J}|$ and $W$ and $b$ are constants. For example, Kozeny and Carman (see Scheidegger, 1960) suggested

$$J = 180\alpha \frac{(1-n)^2 v}{gn^3 d^2} q + \frac{3\beta(1-n)}{4gn^3 d} q^2 \qquad (2.1.39)$$

where $\alpha$ and $\beta$ are shape factors, $v$ is the fluid's kinematic viscosity and $d$ is the grain diameter. Ergun (1952) suggested a similar equation, but with 150 replacing 180 and 1.75 replacing $3\beta/4$. Ward (1964) proposed

$$J = \frac{v}{gk} q + \frac{0.55}{g\sqrt{k}} q^2, \quad k = d^2/360. \qquad (2.1.40)$$

Most groundwater flows occur at Re well within the laminar flow range, where the linear Darcy law is applicable. Flow at large Re may sometimes occur in

Karstic formations or in aquifers in the vicinity of outlets, i.e., very close to wells, springs, etc.

The effect of medium anisotropy on flow at large Re is more complicated. Barak and Bear (1981), who investigated this problem, suggested the following equation of motion

$$J_i = (\nu/g) w_{ij} q_j + \beta'_{ijkl} q_j q_k q_l / gq + \beta''_{ijk} q_j q_k / g \tag{2.1.41}$$

as a good approximation for a Newtonian fluid. In this equation, $w_{ij}$, $\beta''_{ijk}$ and $\beta'_{ijkl}$ are tensors of the second, third, and fourth orders, respectively, which represent matrix properties only. At low Re, the last two terms on the right-hand side of (2.1.41) vanish. The last term describes the effect of matrix nonsymmetry.

## 2.2. Aquifer Transmissivity

Let a confined aquifer's thickness, $B$, vary such that $B(x, y) = b_2(x, y) - b_1(x, y)$, where $b_1(x, y)$ and $b_2(x, y)$ are the elevations of its fixed bottom and ceiling (Figure 2.4). Employing (2.1.20), the total discharge per unit width through the entire thickness of the aquifer, can then be expressed by

$$Q'_x = \int_{b_1(x,y)}^{b_2(x,y)} q_x \, dz = - \int_{b_1(x,y)}^{b_2(x,y)} K(\partial \phi / \partial x) \, dz,$$

$$Q'_y = \int_{b_1(x,y)}^{b_2(x,y)} q_y \, dz = - \int_{b_1(x,y)}^{b_2(x,y)} K(\partial \phi / \partial y) \, dz. \tag{2.2.1}$$

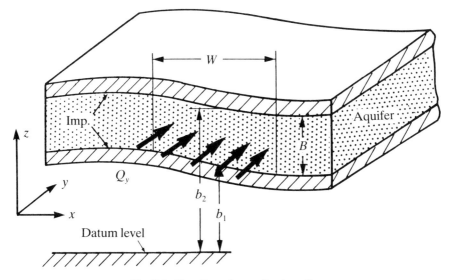

Fig. 2.4. Flow through a confined aquifer.

In the general case, $\mathbf{Q}' = \mathbf{Q}'(x, y)$, $\phi = \phi(x, y, z)$ and $K = K(x, y, z)$.

Whenever we have to integrate a derivative, or to differentiate an integral, we make use of *Leibnitz' rule* (see any text on advanced calculus)

$$\frac{\partial}{\partial x} \int_{b(x)}^{a(x)} f(x, t) \, dt = \int_{b(x)}^{a(x)} \frac{\partial f(x, t)}{\partial x} \, dt + f(x, a) \frac{\partial a}{\partial x} - f(x, b) \frac{\partial b}{\partial x}. \quad (2.2.2)$$

Applying this rule to (2.2.1), we obtain for the special case of $K = K(x, y)$

$$Q'_x = -KB \frac{\partial \tilde{\phi}}{\partial x} - K \left[ \tilde{\phi} \frac{\partial B}{\partial x} - \phi(x, y, b_2) \frac{\partial b_2}{\partial x} + \phi(x, y, b_1) \frac{\partial b_1}{\partial x} \right],$$

$$Q'_y = -KB \frac{\partial \tilde{\phi}}{\partial y} - K \left[ \tilde{\phi} \frac{\partial B}{\partial y} - \phi(x, y, b_2) \frac{\partial b_2}{\partial y} + \phi(x, y, b_1) \frac{\partial b_1}{\partial y} \right]. \quad (2.2.3)$$

Or, in the vector form

$$\mathbf{Q}' = -KB \nabla' \tilde{\phi} - K [\tilde{\phi} \nabla' B - \phi(x, y, b_2) \nabla' b_2 + \phi(x, y, b_1) \nabla' b_1] \quad (2.2.4)$$

where

$$\tilde{\phi}(x, y) = \frac{1}{B} \int_{b_1(x, y)}^{b_2(x, y)} \phi(x, y, z) \, dz \quad (2.2.5)$$

is the average piezometric head along a vertical line at point $(x, y)$. The prime symbol in $\mathbf{Q}'$ and in $\nabla'(\ )$ indicates that the vector and vector operation are in the $xy$-plane, only. Thus

$$\nabla'(\ ) = \frac{\partial (\ )}{\partial x} \mathbf{1x} + \frac{\partial (\ )}{\partial y} \mathbf{1y} \quad (2.2.6)$$

where $\mathbf{1x}$ and $\mathbf{1y}$ are unit vectors in the $x$ and $y$ directions, respectively.

If we now assume essentially horizontal flow, that is, vertical equipotentials, $\tilde{\phi} \simeq \phi(x, y, b_2) \simeq \phi(x, y, b_1)$, Equation (2.2.4) may be approximated by

$$\mathbf{Q}' = -T(x, y) \nabla' \tilde{\phi}, \qquad T(x, y) = K(x, y) B(x, y) \quad (2.2.7)$$

because $\nabla' B = \nabla' b_2 - \nabla' b_1$.

The product $KB$, denoted by $T$, which appears whenever the flow through the entire thickness of the aquifer is being considered, is called *transmissivity*. It is an aquifer characteristic which is defined by the rate of flow per unit width through the entire thickness of an aquifer per unit hydraulic gradient. The concept is valid only in two-dimensional, or aquifer-type flow. In three-dimensional flow through porous media, the concept of transmissivity is meaningless. It should be noted that the thickness of the aquifer, $B$, is not necessarily constant.

The error resulting from employing (2.2.7), based on the assumption of essentially horizontal flow, is given by the second term on the right-hand side of (2.2.4).

Another case of interest is $K = K(x, y, z)$, but we *immediately introduce the assumption of essentially horizontal flow*, that is $\phi = \phi(x, y)$ in (2.2.1). Then (2.2.4) leads to

$$\mathbf{Q}' = -T(x, y)\nabla'\phi, \qquad T(x, y) = \int_{b_1}^{b_2} K(x, y, z)\, dz \equiv \check{K}B. \qquad (2.2.8)$$

It can easily be shown that the assumption of vertical equipotentials, i.e., $\phi(x, y, b_1) \simeq \phi(x, y, b_2) \simeq \tilde{\phi}(x, y)$ leads to $\nabla'\tilde{\phi} = \overline{\nabla'\phi}$, i.e., the gradient of the average head is equal to the average of the head gradient. Equations (2.2.7) and (2.2.8) are identical if $\phi$ is understood to mean $\tilde{\phi}$.

For a layered aquifer with horizontal stratification, $K = K(z)$, Equation (2.2.8) becomes

$$\mathbf{Q}' = -T\nabla'\phi, \qquad T = \int_0^B K(z)\, dz \qquad (2.2.9)$$

where we have assumed essentially horizontal flow in the aquifer. When the aquifer is composed of $N$ distinct homogeneous layers, each with thickness $B_i$ and hydraulic conductivity $K_i$, Equation (2.2.9) reduces to

$$\mathbf{Q}' = -T\nabla'\phi, \qquad T = \sum_{i=1}^{N} B_i K_i. \qquad (2.2.10)$$

Following the discussion above, (2.2.10) is valid as an approximation also when $B_i = B_i(x, y)$ and $K_i = K_i(x, y)$.

We note that by assuming horizontal flow, we also assume a hydrostatic pressure distribution in the aquifer, i.e., $\partial p/\partial z = -\rho g$.

As indicated in Section 1.5, the assumption of essentially horizontal flow may be extended also to leaky aquifers (Figure 1.6d). Accordingly, the concept of transmissivity may also be extended to such aquifers, with $T$ defined by (2.2.7).

## 2.3. Dupuit Assumption

A phreatic aquifer has been defined in Section 1.2 as one in which a water table (or a phreatic surface) serves as its upper boundary. We have also introduced the fact that actually, above a phreatic surface, which is an imaginary surface, at all points of which the pressure is atmospheric, moisture does occupy at least part of the pore space (Figure 1.3). The capillary fringe was introduced as an approximation of the actual distribution of moisture in the soil above a phreatic surface.

Figure 2.5 shows how the actual moisture distribution is approximated by a step

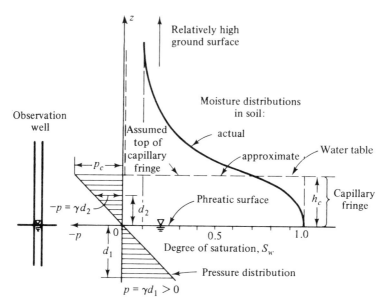

Fig. 2.5. Approximations of phreatic surface and capillary fringe.

distribution, assuming that no moisture is present in the soil above a certain level. This step defines the height, $h_c$, of the capillary fringe. Obviously, this approximation is justified only when the thickness of the capillary fringe thus defined is much smaller than the distance from the phreatic surface to the ground surface. In the capillary fringe (as in the entire aerated zone above the phreatic surface), pressures are negative; therefore, they cannot be monitored by observation wells which serve as piezometers. A special device, called *tensiometer*, is needed in order to measure the negative pressures in the aerated zone (Figure 5.5b and further details in Section 5.1). Water levels in observation wells that terminate below the phreatic surface give elevations of points on the phreatic surface. Using a sufficient number of such points, we can draw contours of this surface.

Thus, the *capillary fringe approximation* means that we assume a saturated zone up to an elevation $h_c$ above the phreatic surface, and no moisture at all above it. In this case, the upper surface of the capillary fringe may be taken as the *groundwater table*, as the soil is assumed saturated below it. However, when $h_c$ is much smaller than the thickness of an aquifer below the phreatic surface, and this is indeed the situation encountered in most aquifers, the hydrologist often neglects the capillary fringe. He then assumes that the (phreatic) aquifer is bounded from above by a phreatic surface. This is also the assumption underlying the presentation in this book.

An estimate of $h_c$ can be obtained, for example, from (Mavis and Tsui, 1939)

$$h_c = \frac{2.2}{d_H} \left( \frac{1-n}{n} \right)^{3/2} \qquad (2.3.1)$$

where $h_c$ is in inches, and $d_H$ is the mean grain diameter, also in inches and $n$ is porosity. Another expression is (Polubarinova-Kochina, 1952, 1962)

$$h_c = \frac{0.45}{d_{10}} \frac{1-n}{n} \qquad (2.3.2)$$

where both $h_c$ and the effective particle diameter are in centimeters. Silin-Bekchurin (1958) suggested a capillary rise of 2—5 cm in coarse sand, 12—35 cm in sand, 35—70 cm in fine sand, 70—150 cm in silt, and 2—4 m and more in clay. Equations (2.3.1) and (2.3.2) can be compared with the relationship $h = 2\sigma/r$ which expresses the rise of water in a capillary tube of radius $r$; $\sigma$ is the surface tension of the water.

Both $\phi$ and $\mathbf{q}$ vary from point to point within a phreatic aquifer. In order to obtain the specific discharge $\mathbf{q} = \mathbf{q}(x, y, z, t)$ at every point, we have to know the piezometric head $\phi = \phi(x, y, z, t)$ by solving the flow model in a three-dimensional space. An additional difficulty stems from the fact that the location of the phreatic surface, which serves as a boundary to the three-dimensional flow domain in the aquifer, is *a-priori unknown*. In fact its location is part of the sought solution. Once we solve for $\phi = \phi(x, y, z, t)$ within the flow domain, we use the fact that on the phreatic surface, the pressure is zero to obtain $\phi(x, y, z, t) = z$ on the phreatic surface. Hence, the equation that describes the phreatic surface is

$$F(x, y, z, t) \equiv \phi(x, y, z, t) - z = 0. \qquad (2.3.3)$$

From the above considerations it follows that this procedure is not a practical one for solving common problems of flow in phreatic aquifers.

In view of this inherent difficulty, Dupuit (1863) observed that in most groundwater flows, the slope of the phreatic surface is very small. Slopes of 1/1000 and 10/1000 are commonly encountered. In steady flow without accretion in the vertical two-dimensional $xz$-plane (Figure 2.6a), the phreatic surface is a streamline. At every point, $P$, along this streamline, the specific discharge is in a

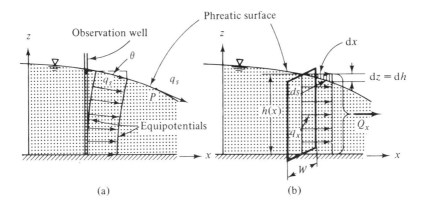

Fig. 2.6. The Dupuit assumption (Bear, 1979).

direction tangent to the streamline and is given by Darcy's law

$$q_s = -K \, d\phi/ds = -K \, dz/ds = -K \, \sin\theta \tag{2.3.4}$$

since along the phreatic surface $p = 0$ and $\phi = z$. As $\theta$ is very small, Dupuit suggested that $\sin\theta$ be replaced by the slope $\tan\theta = dh/dx$. The assumption of small $\theta$ is *equivalent* to assuming that equipotential surfaces are vertical (that is, $\phi = \phi(x)$ rather than $\phi = \phi(x, z)$), and the flow is essentially horizontal. Thus, the Dupuit assumption leads to the specific discharge expressed by

$$q_x = -K \, dh/dx, \quad h = h(x). \tag{2.3.5}$$

In general, $h = h(x, y)$ and we have

$$q_x = -K \, \partial h/\partial x, \quad q_y = -K \, \partial h/\partial y; \quad \mathbf{q} = -K\nabla' h. \tag{2.3.6}$$

Since $\mathbf{q}$ is thus independent of elevation, the corresponding total discharge through a vertical surface of width $W$ (normal to the direction of flow; Figure 2.6b) is

$$Q_x = -KWh \, \partial h/\partial x, \quad Q_y = -KWh \, \partial h/\partial y, \quad h = h(x, y) \tag{2.3.7}$$

or, in the compact vector form

$$\mathbf{Q} = -KWh\nabla' h. \tag{2.3.8}$$

Per unit width, we obtain

$$\mathbf{Q}' \equiv \mathbf{Q}/W = -Kh\nabla' h. \tag{2.3.9}$$

In (2.3.7) through (2.3.9), the aquifer's bottom is horizontal. It should be emphasized that the Dupuit assumption may be considered as a good approximation in regions where $\theta$ is indeed small and/or the flow is essentially horizontal. We note that the assumption of horizontal flow is equivalent to the assumption of *hydrostatic pressure distribution* $\partial p/\partial z = -\rho g$.

The important advantage gained by employing the Dupuit assumption is that the state variable $\phi = \phi(x, y, z)$ has been replaced by $h = h(x, y)$, that is, $z$ does no longer appear as an independent variable. Also, since at a point on the free surface, $p = 0$ and $\phi = h$, we assume that the vertical line through the point is also an equipotential line on which $\phi = h = $ const. In general, $h$ varies also with time so that $h = h(x, y, t)$. In this way, the complexity of the problem has been greatly reduced. It is two-dimensional, rather than three-dimensional and the unknown location of the phreatic surface is no longer an extra complication.

Another way of obtaining (2.3.9) is by integrating the point specific discharge along the vertical form the bottom of the aquifer, $\eta = \eta(x, y)$, which need not be horizontal, to the phreatic surface at elevation $h = h(x, y, t)$. For flow in the $+x$ direction, assuming $K = \text{const}$ or $K = K(x, y)$, we obtain

$$Q'_x = \int_{\eta(x,y)}^{h(x,y,t)} q_x \, dz = -K \int_{\eta(x,y)}^{h(x,y,t)} (\partial \phi / \partial x) \, dz$$

$$= -K \left\{ \frac{\partial}{\partial x} \int_\eta^h \phi \, dz - \phi \bigg|_h \frac{\partial h}{\partial x} + \phi \bigg|_\eta \frac{\partial \eta}{\partial x} \right\}$$

$$= -K \left\{ \frac{\partial}{\partial x} [(h - \eta)\tilde{\phi}] - \phi \bigg|_h \frac{\partial h}{\partial x} + \phi \bigg|_\eta \frac{\partial \eta}{\partial x} \right\} \quad (2.3.10)$$

where the average head is defined by

$$\tilde{\phi} = \frac{1}{h - \eta} \int_\eta^h \phi \, dz$$

and $\phi|_h \equiv h$ on the phreatic surface.

Equation (2.3.10) involves no approximation. If we now assume vertical equipotentials, i.e.,

$$\phi|_h (= h) \simeq \phi|_\eta \simeq \tilde{\phi}, \quad (2.3.11)$$

equation (2.3.11) reduces to

$$Q'_x = -K(h - \eta) \frac{\partial h}{\partial x}, \quad \text{or} \quad \mathbf{Q}' = -K(h - \eta)\nabla' h \quad (2.3.12)$$

which is the same as (2.3.9), written for a nonhorizontal bottom.

By comparing (2.3.12) with (2.3.10) for $\eta = 0$, we see that we have replaced $h\tilde{\phi} - h^2/2$ by $h^2/2$ in the flow equation based on the Dupuit assumption. The error reduces to zero as $\tilde{\phi} \to h$. Bear (1972), p. 363, gives an estimate of the error involved in replacing $\phi'' = h\tilde{\phi} - h^2/2$ by $h^2/2$

$$0 < \frac{h^2/2 - \phi''}{h^2/2} < \frac{i^2}{1 + i^2}, \quad i \equiv dh/dx \quad (2.3.13)$$

so that the error is small as long as $i^2 \ll 1$, where $i$ is the slope of the phreatic surface. When the medium is anisotropic, with $K_x \neq K_z$ ($x, z$ principal directions), $i$ in (2.3.13) should be replaced by $(K_x/K_z)i^2$.

The Dupuit assumption presented above is probably the most powerful tool for treating unconfined flows. In fact, it is the only simple tool available to most engineers and hydrologists for solving such problems.

As a simple example of the application of (2.3.12), consider the case of steady unconfined flow through a homogeneous formation between two reservoirs with vertical faces (Figure 2.7). Following the Dupuit assumption, the total discharge in the $x$ direction per unit width, through a vertical cross-section of height $h(x)$ is given by (2.3.9), i.e.,

$$Q'_x \equiv Q' = -Kh(x)\,dh/dx = \text{const}, \quad Q'\,dx = -Kh(x)\,dh. \quad (2.3.14)$$

By integrating this expression between the boundary at $x = 0$, where $h = h_0$, and any distance $x$, where $h = h(x)$, we obtain

$$Q' \int_{x^*=0}^{x} dx^* = -K \int_{h^*=h_0}^{h(x)} h^*(x^*)\,dh^*, \quad Q'x = K \frac{h_0^2 - h^2(x)}{2}. \quad (2.3.15)$$

Equation (2.3.15) describes a water table, $h = h(x)$, which has the shape of a parabola passing through $x = 0$, $h = h_0$. If we know $h(x)$ at some distance $x$, we can use (2.3.15) to derive $Q'$ (obviously, if $K$ is known). The boundary condition at the other end, $x = L$, however, is somewhat more complicated.

Whenever a phreatic surface approaches the downstream external boundary of a flow domain, it always terminates on it at a point that is located above the water table of the body of open water present outside the flow domain. Points $A$ in Figures 2.8a, b, and c are such points. The segment $AB$ of the boundary above the water table and below the phreatic surface is called the *seepage face*. Along the seepage face, water emerges from the porous medium into the external space, trickling down along the seepage face. In Figures 2.8c and b, the phreatic surface at $A$ is tangent to the external boundary; in Figure 2.8a, it is tangent to the vertical line at $A$.

Back to our problem, because of the presence of a seepage face which terminates at a point that is also a point on the (*unknown*) phreatic surface, $h_s$ in Figure 2.7 is *a-priori* unknown. Instead, whenever the Dupuit assumption is employed, we approximate the situation by overlooking the presence of the seepage face and assume that the water table at $x = L$ passes through $h = h_L$.

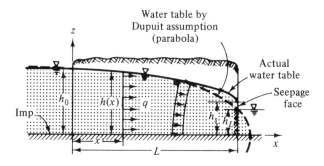

Fig. 2.7. Steady unconfined flow between two reservoirs (Bear, 1979).

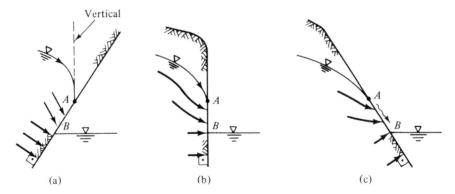

Fig. 2.8. The seepage face, AB (Bear, 1979).

Using this as the downstream boundary condition, we obtain from (2.3.15)

$$Q' = K \frac{h_0^2 - h_L^2}{2L} \qquad (2.3.16)$$

known as the *Dupuit—Forchheimer discharge formula*.

The parabolic water table is shown in broken line in Figure 2.7. Discrepancies exist at $x = 0$, where the water table should be tangent to the horizontal line, whereas the parabola has a slope of $dh/dx|_{x=0} = Q'/Kh_0$ and at $x = L$, where the seepage face is neglected. Otherwise, the discrepancy between the curves derived by the exact theory of the phreatic surface boundary and, by the Dupuit approximation, is negligible. A simple rule is that at distances from the downstream end larger than 1.5—2 times the average height of the flow domain, the solution based on the Dupuit assumption is sufficiently accurate for all practical purposes.

Moreover, it can be shown (Bear, 1972; p. 367) that (2.3.16) is accurate as far as the rate of discharge, $Q'$, is concerned, although (2.3.15) does not give the accurate water table elevators $h = h(x)$.

The Dupuit assumption should not be applied in regions where the vertical flow component is not negligible. Such flow conditions occur as a seepage face is approached (Figure 2.9c) or at a crest (*water divide*) in a phreatic aquifer with accretion (Figure 2.9b). Another example is the region close to the impervious vertical boundary of Figure 2.9a. It is obvious that the assumption of vertical equipotentials fails at, and in the vicinity of such a boundary. Only at a distance $x > \sim 2h_0$ have we equipotentials that may be approximated as vertical lines, or surfaces. It is important to note here that in cases with accretion, a horizontal (or almost so) water table is not sufficient to justify the application of the Dupuit assumption. One must verify that vertical flow components may indeed be neglected, before applying the Dupuit assumption. Another case to which the Dupuit assumption should be applied with care is that of unsteady flow in a decaying phreatic surface mound. Although no accretion takes place, yet at, and in

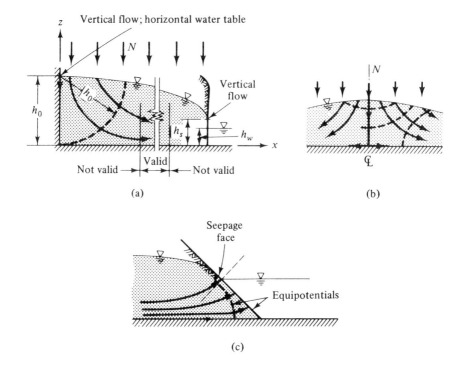

Fig. 2.9. Regions where Dupuit assumption is not valid (Bear, 1979).

the vicinity of, a crest the flow is vertically downward. At a distance of say 1.5—2 times the thickness of the flow, the approximation of vertical equipotentials is again valid.

In spite of what was said above, in regional studies, the Dupuit assumption, because of its simplicity and the relatively small error involved, is usually applied also to those (relatively small) parts of an investigated region where it is not strictly applicable. One should, however, be careful in making use of results (say, water levels) derived for these parts of an investigated region.

CHAPTER THREE

# Modeling Three-Dimensional Flow

The basic laws governing the flow of water were presented in the previous chapter. However, one cannot solve flow problems by using only these laws. Equation (2.1.19) is a single equation in two dependent variables: $\mathbf{q}(x, y, z, t)$ and $\phi(x, y, z, t)$. It can also be regarded as three equations in four unknowns $\phi$, $q_x$, $q_y$, $q_z$. This means that one additional equation is required in order to obtain a complete description of the flow within any given domain. Similarly, we have $Q'(x, y, t)$ and $\tilde{\phi}(x, y, t)$ in the single equation (2.2.7) and $Q'(x, y, t)$ and $h(x, y, t)$ in the single equation (2.3.9). The additional basic law that we have to invoke is that of *mass conservation*, or *mass balance*.

Our objective in what follows is to develop the mass conservation equations for three-dimensional domains and for different types of aquifers. The distribution of $\phi = \phi(\mathbf{x}, t)$ in a specified flow domain is obtained by solving these equations, subject to appropriate boundary and initial conditions.

We shall first consider the basic mass balance equation and boundary conditions for three-dimensional flows. Then, the equations will be developed for flow in confined, leaky, and phreatic aquifers. We shall derive these (integrated, or averaged) aquifer equations by integrating the point mass balance, or continuity equation over the vertical thickness of the aquifer. In this way, the conditions on the confined, leaky, or phreatic, upper and lower boundaries of the aquifers will be incorporated into the resulting integrated equations in a natural and unified way.

With the material presented in this section, one should be able to formulate the mathematical model of any groundwater flow problem. It is needless to remind the reader that the mathematical model is based on the conceptual model of the investigated problem.

Solving the *forecasting problem* (Chapter 1) means solving a model in order to obtain the future distribution of water levels, or of piezometric heads, produced in a specified aquifer (i.e., with known geometry and properties) by any anticipated natural replenishment and by any planned schedule of future pumping and artificial recharge, as envisaged in a proposed management scheme.

## 3.1. Effective Stress in Porous Media

The concept of *effective stress*, or *intergranular stress*, was first introduced in soil mechanics by Terzaghi (1925). Essentially, this concept assumes that in a granular porous medium, the pressure in the water, that almost completely surrounds each solid grain, produces in the latter a stress of equal magnitude, without contributing to the deformation of the skeleton, which is determined mainly by the contact forces. These are transmitted from grain to grain at the contact points. Thus, the *strain producing stress*, or *intergranular stress*, is obtained by subtracting the pore water pressure from the stress in the solid material. Here, both the pressure and the stress are average values.

Basic to the notion of effective stress is the observation that the deformation of granular materials, as a result of stress changes, is much larger than can be explained by compression of the material itself. This suggests that deformation is mainly produced by a rearrangement of the assembly, with localized slipping and rolling. Laboratory investigations also support the thesis that during soil deformation, the grains slip and roll. This means that the deformation process is governed by what happens in the localized contact points, where a concentrated normal and shear force are transmitted from grain to grain. These contact forces are not affected by a change in the pore pressure. Therefore, a change in pore pressure, with an equal change in total stress, produces no deformation and, hence, should produce no change in the effective stress.

To illustrate the concept of effective stress in a simple way, limiting the discussion for the moment to vertical forces only, consider the vertical cross-section through a confined aquifer and the horizontal unit area, $AB$, shown in Figure 3.1. At every instant, the overburden load above $AB$, due to soil, water and

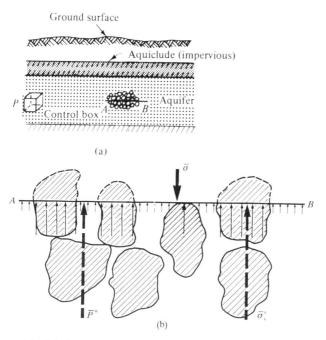

Fig. 3.1. Nomenclature for the definition of effective stress.

whatever load is added on the natural ground surface, produces a (macroscopic) stress, $\bar{\sigma}$ (= total force per unit area of porous medium), acting on the upper side of $AB$. This force must be in equilibrium with two forces per unit area acting on $AB$ from below: a force $n\bar{p}^w$, resulting from the (average) pressure, $\bar{p}^w$, in the water acting on the water portion of $AB$, and a force $(1-n)\bar{\sigma}_s^s$, resulting from the (average) stress, $\bar{\sigma}_s^s$, in the solid skeleton, acting on the solid portion of $AB$. Note that both $\bar{p}^w$ and $\bar{\sigma}_s^s$ are intrinsic phase averages, while $\bar{\sigma}$ is a volume average. (These averages are defined in Section 1.4.) It can be shown (e.g., Bachmat and Bear, 1983) that these intrinsic phase averages are equal to intrinsic areal averages, i.e., $\bar{p}^w$ and $\bar{\sigma}_s^s$ also express force per unit area of water and per unit area of solid, respectively.

With stresses $\bar{\sigma}$ and $\bar{\sigma}_s^s$ taken as positive for compression (as is customary for the fluid pressure, $\bar{p}^w$, but not always for the stress in the solid), the above statement of equilibrium can be expressed in the form

$$\bar{\sigma} = (1-n)\bar{\sigma}_s^s + n\bar{p}^w. \tag{3.1.1}$$

As noted above, $\bar{\sigma}_s^s$ is not the strain producing intergranular stress, $\overline{\sigma_s^*}$, which, by Terzaghi's concept stated above is expressed by $(1-n)(\bar{\sigma}_s^s + \bar{p}^w)$. In order to express (3.1.1) in terms of $\overline{\sigma_s^*}$, we add and substract $(1-n)\bar{p}^w$ on the right-hand side of (3.1.1), obtaining

$$\bar{\sigma} = (1-n)\bar{\sigma}_s^s + n\bar{p}^w - (1-n)\bar{p}^w + (1-n)\bar{p}^w = \overline{\sigma_s^*} + \bar{p}^w \tag{3.1.2}$$

where

$$\overline{\sigma_s^*} = (1-n)\overline{\sigma_s^{*s}} = (1-n)(\bar{\sigma}_s^s - \bar{p}^w) \tag{3.1.3}$$

is the intergranular stress, or effective stress (expressed in terms of force per unit area of porous medium). This is the stress that produces porous medium deformation.

Although (3.1.2) and (3.1.3) are based on the simplification of *vertical stress only*, they can easily be extended to three-dimensional expressions. For the total stress at a point within a saturated (three-dimensional) porous medium domain, we obtain from (1.4.4)

$$\bar{\sigma} = \frac{1}{U_0}\int_{(U_0)} \sigma \, dU = \frac{1}{U_0}\sum_{\alpha=w,s}\int_{(U_{0\alpha})}\sigma_\alpha \, dU_\alpha = \bar{\sigma}_w + \bar{\sigma}_s \tag{3.1.4}$$

where $\bar{\sigma}, \bar{\sigma}_w$ and $\bar{\sigma}_s$ are second rank symmetric tensors.

The (phase average) stress in the water, $\bar{\sigma}_w$ is related to the *shear stress*, $\tau$, and to the pressure in the water, $\bar{p}^w$, by

$$\bar{\sigma}_w^w = \bar{\tau}^w + \bar{p}^w \mathbf{I}$$

From (3.1.2) written for a three-dimensional stress field, it follows that with $\overline{\boldsymbol{\tau}}^w \approx 0$

$$d\overline{\boldsymbol{\sigma}} = d\overline{\boldsymbol{\sigma}}_s^* + d\overline{p}^w \mathbf{I} \tag{3.1.5}$$

where $\mathbf{I}$ is the unit tensor. This relation, between total stress, effective stress and pore water pressure, substantiates the statement made above that a change in pore water pressure, combined with an equal change in total stress, produces no change in the effective stress. We usually assume that the shear stress in the fluid is negligible, so that (3.1.4) reduces to

$$\overline{\boldsymbol{\sigma}} = \overline{\boldsymbol{\sigma}}_s + \overline{p}\mathbf{I}. \tag{3.1.6}$$

Making use of (1.4.4), we now rewrite (3.1.6) in the form

$$\overline{\boldsymbol{\sigma}} = (1-n)(\overline{\boldsymbol{\sigma}}_s^s - \overline{p}^w \mathbf{I}) + \overline{p}^w \mathbf{I} = \overline{\boldsymbol{\sigma}}_s^* + \overline{p}^w \mathbf{I} \tag{3.1.7}$$

where $\overline{\boldsymbol{\sigma}}_s^*$ denotes the effective stress tensor and $\overline{p}^w$ is the pressure in the water. We note that (3.1.7) has been derived without the assumption that grains are almost completely surrounded by water. We also note that there is no need to limit the discussion only to granular materials.

Verruijt (1984) makes a distinction between *effective stress* and *intergranular stress*. He modifies (3.1.7) to the form

$$\overline{\boldsymbol{\sigma}} = \overline{\boldsymbol{\sigma}}_s^* + (1-\gamma')\overline{p}^w \mathbf{I} \tag{3.1.8}$$

where $\gamma'$ is a coefficient, such that the deformation of the solid skeleton is fully determined by the effective stress. He shows that $\gamma' = \beta_s/\alpha'$ where $\beta_s$ is the compressibility of solid, related to the intergranular stress, and $\alpha'$ is a bulk porous medium compressibility. Verruijt (1982) shows that $\gamma'$ is very small in most natural soils, so that the effective stress is identical to the intergranular stress.

Henceforth, we shall omit the overbar symbols that indicate intrinsic phase and phase averages, i.e., $\overline{\boldsymbol{\sigma}} \equiv \boldsymbol{\sigma}$, $\overline{\boldsymbol{\sigma}}_s^* \equiv \boldsymbol{\sigma}_s^*$ and $\overline{p}^w \equiv p$.

## 3.2. Mass Storage

In a saturated porous medium domain, the mass of water present a in unit volume of porous medium, is expressed by the product $n\rho$. As flow takes place, the pressure, $p$, in the water varies with time. Even if the overburden load, producing the total stress, $\boldsymbol{\sigma}$, remains unchanged, by (3.1.5), the effective stress, $\boldsymbol{\sigma}_s^*$, also varies with time. In the general case, the water is assumed to be compressible and the solid matrix is deformable and, hence, the changes in both $p$ and $\boldsymbol{\sigma}_s^*$ will cause $n\rho$ to vary with time.

Accordingly, the rate of change of the mass of the fluid per unit volume of porous medium is given by

$$\lim_{\Delta t \to 0} \frac{(n\rho)|_{t+\Delta t} - (n\rho)|_t}{\Delta t} = \frac{\partial(n\rho)}{\partial t}$$

# MODELING THREE-DIMENSIONAL FLOW

with

$$\frac{\partial(n\rho)}{\partial t} = n\frac{\partial\rho}{\partial t} + \rho\frac{\partial n}{\partial t}. \tag{3.2.1}$$

The general equation of state for a fluid phase is $\rho = \rho(p, c, T)$, which states that the fluid's density depends on the pressure, $p$, the concentration of various components, $c$, say dissolved solids, and absolute temperature, $T$. Under isothermal conditions, the general relationship is reduced to $\rho = \rho(p, c)$. If the fluid is also homogeneous or single component, the general equation of state is further reduced to $\rho = \rho(p)$.

Thus, under isothermal conditions

$$d\rho = \left.\frac{\partial\rho}{\partial p}\right|_{c,\,T\,=\,\text{const}} dp + \left.\frac{\partial\rho}{\partial c}\right|_{p,\,T\,=\,\text{const}} dc = \rho(\beta_p\, dp + \beta_c\, dc) \tag{3.2.2}$$

where

$$\beta_p = \frac{1}{\rho}\left.\frac{\partial\rho}{\partial p}\right|_{c,\,T}$$

is the *coefficient of compressibility* of the fluid, at constant concentration and temperature, and

$$\beta_c = \frac{1}{\rho}\left.\frac{\partial\rho}{\partial c}\right|_{p,\,T}$$

is a coefficient that introduces the effect of concentration.

In certain ranges of $p$, $c$ and $T$, the coefficients $\beta_p$ and $\beta_c$ are constants, and (3.2.2) takes the form

$$\begin{aligned}\rho &= \rho_0 \exp\{\beta_p(p - p_0) + \beta_c(c - c_0)\} \\ &= \rho_0\{1 + \beta_p(p - p_0) + \beta_c(c - c_0) + \cdots\}\end{aligned} \tag{3.2.3}$$

where $\rho = \rho_0$ for $p = p_0$ and $c = c_0$.

Some fluids obey the empirical relationship

$$\rho = \rho_0\{1 + \beta_p(p - p_0) + \beta_c(c - c_0)\} \tag{3.2.4}$$

which, for all practical purposes is equal to (3.2.3).

For an *incompressible fluid* $\partial\rho/\partial p = 0$, or $\beta_p = 0$. Henceforth, we shall use the symbol $\beta$ to denote $\beta_p$. For fluids containing small amounts of gas (e.g., air), the compressibility is significantly increased by the possible compression of the air bubbles (e.g., Verruijt, 1969).

Making use of (3.2.3), with $c = c_0$, and under isothermal conditions, the first term on the right-hand side of (3.2.1) becomes

$$n\frac{\partial\rho}{\partial t} = n\frac{\partial\rho}{\partial p}\frac{\partial p}{\partial t} + n\frac{\partial\rho}{\partial c}\frac{\partial c}{\partial t} = n\rho\beta\frac{\partial p}{\partial t} + n\rho\beta_c\frac{\partial c}{\partial t}. \tag{3.2.5}$$

In order to relate the second term on the right-hand side of (3.2.1) to the rate of change in the pressure, let us follow Jacob (1940) and assume that there are no horizontal displacements in the soil. All deformations are only in the vertical direction and all forces and resulting stresses act also only in the vertical direction.

With the above assumption (removed in Section 3.6), and omitting the symbols for averages, (3.1.6) reduces to

$$\sigma = \sigma_s^* + p \tag{3.2.6}$$

where now $\sigma$ and $\sigma_s^*$ are in the vertical direction.

If we assume that the total stress remains unchanged, i.e., $d\sigma = 0$, then

$$d\sigma_s^* = -dp, \tag{3.2.7}$$

i.e., any increase in pressure in the water is accompanied by an equal increase in the effective stress.

Now, let $U_b$ denote the bulk volume of a soil sample at some point within a porous medium domain. Then

$$U_b = U_w + U_s, \qquad U_s = (1-n)U_b, \qquad U_w = nU_b \tag{3.2.8}$$

where $U_s$ and $U_w$ denote the volumes of solids and of water, respectively, within $U_b$. We shall assume that as the volume $U_b$ deforms as a result of changes in the effective stress, $\sigma_s^*$, *the volume $U_s$ within it remains unchanged*. This is consistent with the assumption in Section 3.2 that the grains are incompressible.

Hence, in view of (3.2.7) and (3.2.8)

$$\frac{\partial U_s}{\partial \sigma_s^*} = 0, \qquad \frac{\partial}{\partial \sigma_s^*}(1-n)U_b = 0,$$

$$\frac{1}{U_b}\frac{\partial U_b}{\partial \sigma_s^*} = \frac{1}{1-n}\frac{\partial n}{\partial \sigma_s^*} = -\frac{1}{1-n}\frac{\partial n}{\partial p}. \tag{3.2.9}$$

At this point *we assume that we deal with relatively small volume changes, so that the soil is assumed to behave as an elastic material* with a constant coefficient of soil compressibility, $\alpha$, defined by

$$\alpha = -\frac{1}{U_b}\frac{\partial U_b}{\partial \sigma_s^*}. \tag{3.2.10}$$

By combining (3.2.9) with (3.2.10), we obtain

$$\alpha = \frac{1}{1-n}\frac{\partial n}{\partial p} \tag{3.2.11}$$

where $n = n(\sigma_s^*)$ or $n = n(p)$ only. By employing (3.2.11), we now obtain

$$\rho\frac{\partial n}{\partial t} = \rho\frac{\partial n}{\partial p}\frac{\partial p}{\partial t} = \rho(1-n)\alpha\frac{\partial p}{\partial t}. \tag{3.2.12}$$

MODELING THREE-DIMENSIONAL FLOW 59

By inserting (3.2.5) and (3.2.12) into (3.2.1), we now obtain for a homogeneous fluid ($c$ = const.)

$$\frac{\partial(n\rho)}{\partial t} = \rho\{n\beta + (1-n)\alpha\}\frac{\partial p}{\partial t}. \qquad (3.2.13)$$

Consider now the vicinity of a point in an aquifer, where water pressure is reduced by pumping. This results in an increase in the intergranular compressive stress transmitted by the solid skeleton of the aquifer. This, in turn, causes the aquifer to be compacted, reducing its porosity. At the same time, as a result of pressure reduction, the water will expand. Together, the two effects — the slight expansion of water and the small reduction in porosity — cause a certain amount of water to be released from storage in the aquifer. Altogether, a reduction in water pressure is accompanied by a release of water from storage in the aquifer. Conversely, in response to adding water to a unit volume of the aquifer, the pressure in it will rise, accompanied by a reduction in the intergranular compressive stress, which, in turn, increases the porosity. If we assume both water and solid matrix to be perfectly elastic, within the range of the considered changes, the two processes are reversible. In reality, however, changes in a granular matrix are irreversible. Such irreversible deformations are outside the scope of this book.

Based on the above considerations, we can now define a coefficient of (*fluid*) *mass storage* $S_{0p}^*$, in an elastic porous medium of an aquifer as the mass of water released from storage (or added to it) in a unit volume of aquifer per unit decline (or rise) in pressure

$$S_{0p}^* = \Delta m_w / U_b \Delta p. \qquad (3.2.14)$$

Since the left-hand side of (3.2.13) expresses added mass of water per unit volume of porous medium per unit time, by combining (3.2.13) with (3.2.14), we obtain

$$S_{0p}^* = \rho\{n\beta + (1-n)\alpha\} \qquad (3.2.15)$$

as a relationship between $n$, $\beta$, $\alpha$ and $S_{0p}^*$.

Since $\partial \phi^*/\partial t = (1/\rho g)\partial p/\partial t$, we could also define a coefficient of mass storage, $S_{0\phi}^*$, as the mass of water released or added to a unit volume of porous medium per unit change in the piezometric head, $\phi^*$ (defined by (2.1.22))

$$S_{0\phi}^* = \Delta m_w / U_b \Delta \phi^* = \rho g S_{0p}^*. \qquad (3.2.16)$$

In Section 3.6, we shall return to the question of changes in the quantity of water stored in a porous medium, this time without the assumption of vertical stress only and with $d\sigma \neq 0$.

It is important to note that in developing (3.2.13), which eventually leads to the definitions (3.2.15) and (3.2.16), we have considered the *change in fluid mass within a control box*, allowing water and solids to move freely through the walls of this box. We shall return to this question at the end of the next section.

## 3.3. Fundamental Mass Balance Equation

The core of the mathematical model that describes the transport of any extensive quantity, e.g., mass or energy, in a porous medium domain is the balance equation of that quantity. In the continuum approach, the balance equation takes the form of a partial differential equation, each term of which describes a change in the amount of the considered extensive quantity per unit volume of porous medium per unit time. Here, we shall focus our attention on the mass of water that completely fills the void space (i.e., *saturated flow*).

### 3.3.1. THE BASIC MASS BALANCE EQUATION

Consider a *control volume* (or *control box*) having the shape of a rectangular parallel-piped box of dimensions $\delta x$, $\delta y$, $\delta z$ centered at some point $P(x, y, z)$ inside the flow domain in an aquifer (Figures 3.1 and 3.2). A control box may have any arbitrary shape, but once its shape and position in space have been fixed, they remain unchanged during the flow, although the amount and identity of the material in it may change with time. In the present analysis, water and solids enter and leave the box through its surfaces, and our objective is to write a balance, or a statement of conservation, for the mass of water entering, leaving, and being stored in the box. In hydrodynamics, this is called the *Eulerian approach*.

Let the vector $\mathbf{J}^* = \rho \mathbf{q}$ denote the mass flux (i.e., mass per unit area per unit time) of water of density $\rho$ at point $P(x, y, z)$. Referring to Figure 3.2, the excess of inflow over outflow of mass during a short time interval $\delta t$, through the surfaces which are perpendicular to the $x$ direction, may be expressed by the difference

$$\delta t \{ J_x^* |_{x - \delta x/2, y, z} - J_x^* |_{x + \delta x/2, y, z} \} \, \delta y \, \delta z.$$

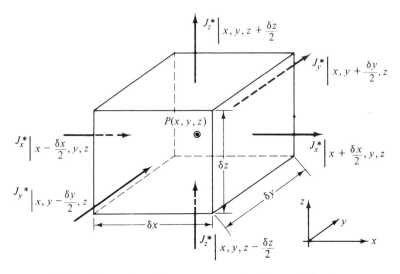

Fig. 3.2. Nomenclature for mass conservation for a control volume.

## MODELING THREE-DIMENSIONAL FLOW

Similar expressions may be written for the $y$ and $z$ directions. By adding the three expressions, for all three directions, we obtain an expression for the total excess of mass inflow over outflow during $\delta t$

$$\delta t \left\{ \frac{J_x^*|_{x-\delta x/2, y, z} - J_x^*|_{x+\delta x/2, y, z}}{\delta x} + \right.$$

$$+ \frac{J_y^*|_{x, y-\delta y/2, z} - J_y^*|_{x, y+\delta y/2, z}}{\delta y} +$$

$$\left. + \frac{J_z^*|_{x, y, z-\delta z/2} - J_z^*|_{x, y, z+\delta z/2}}{\delta z} \right\} \delta x \, \delta y \, \delta z$$

where $\delta x \, \delta y \, \delta z = \delta U$ is the volume of the box. By dividing the above expression by $\delta U$ and $\delta t$, we obtain the *excess of mass inflow over outflow per unit volume of porous medium and per unit time*. Then, by letting the box converge on the point $P$, that is, by lettering $\delta x, \delta y, \delta z \to 0$, the excess of inflow over outflow per unit volume of medium (around $P$) and per unit time becomes

$$-(\partial J_x^*/\partial x + \partial J_y^*/\partial y + \partial J_z^*/\partial z) \quad \text{or} \quad -\text{div } \mathbf{J}^* (\equiv -\nabla \cdot \mathbf{J}^*).$$

We may draw a general conclusion from the above development, namely, that *the excess of efflux over influx of any extensive quantity, per unit volume and per unit time, is always expressed by the divergence of the flux vector of that quantity.*

Since $\partial n\rho/\partial t$ expresses the increase in water mass per unit volume of porous medium per unit time (see Section 3.2), we may now write the complete balance in the form

$$-\nabla \cdot \rho \mathbf{q} = \frac{\partial n\rho}{\partial t} \tag{3.3.1}$$

where $\partial n\rho/\partial t$ may be expressed by (3.2.1), or by (3.2.13), combined with (3.2.15), and we recall that $\mathbf{q}$ denotes the specific discharge with respect to a fixed coordinate system, while Darcy's law, say (2.1.35), expresses the specific discharge with respect to the (possibly moving) solids.

For a homogeneous, incompressible fluid, and a nondeformable porous medium (or for steady flow of a homogeneous fluid), $\nabla \rho = 0$, $\partial n\rho/\partial t = 0$, and (3.3.1) reduces to

$$\nabla \cdot \mathbf{q} = 0. \tag{3.3.2}$$

In (3.3.1), the total mass flux is made up of the advective flux $\mathbf{J}^* (= \rho \mathbf{q})$ only. Had we developed the mass balance equation by starting from the microscopic mass balance equation, written for a point within the void space, and then averaged it to obtain its macroscopic counterpart, we would have obtained a total mass flux made up of three parts: an advective part, a dispersive one, and a diffusive one (see Section 6.2). Thus, (3.3.2) should be regarded as an *approximate* mass balance equation, in which the sum of the mass dispersive and diffusive fluxes is assumed to be much smaller than the advective one, $\rho\mathbf{q}$.

If distributed sinks of strength $P_w = P_w(x, y, z, t)$ (= volume of water withdrawn per unit volume of porous medium per unit time) are present in the flow domain, we modify (3.3.1) to the form

$$-\nabla \cdot \rho\mathbf{q} - \rho P_w(x, y, z, t) = \frac{\partial n\rho}{\partial t}. \qquad (3.3.3)$$

When, instead of distributed sinks, we have point sinks of strength $P'_w(x_i, y_i, z_i, t)$ located at isolated points $(x_i, y_i, z_i)$, $i = 1, 2\ldots$, we modify (3.3.1) to the form

$$-\nabla \cdot \rho\mathbf{q} - \sum_{(i)} \rho P'_w(x_i, y_i, z_i, t)\, \delta(x - x_i, y - y_i, z - z_i) = \frac{\partial n\rho}{\partial t} \qquad (3.3.4)$$

where $P'_w$ is the discharge of a sink (dims. $L^3 T^{-1}$) located at point $(x_i, y_i, z_i)$ and $\delta$ is the *Dirac delta function*.

All the balance equations in this section are developed from the 'Eulerian point of view'. In an *Eulerian Formulation*, we observe what happens at a *fixed* point (and in its vicinity). The balance over a control box is, typically, the way to derive a balance equation (for any extensive quantity) from the Eulerian point of view.

### 3.3.2. ADDITIONAL FORMS OF THE MASS BALANCE EQUATION

For the sake of simplicity, we shall delete the terms expressing sinks or sources present in the flow domain. They can be added whenever necessary. We shall also continue to assume that $\rho$ is unaffected by changes in solute concentration. This effect can also be added when necessary, making use of (3.2.5).

Let us rewrite (3.3.1) in terms of the relative specific discharge, $\mathbf{q}_r$ and the solid's velocity, $\mathbf{V}_s$, in the form

$$\nabla \cdot \rho(\mathbf{q}_r + n\mathbf{V}_s) + \rho\{n\beta + (1-n)\alpha\}\frac{\partial p}{\partial t} = 0 \qquad (3.3.5)$$

where we have made use of (3.2.13), or

$$\nabla \cdot \rho\mathbf{q}_r + \rho n \nabla \cdot \mathbf{V}_s + n\mathbf{V}_s \cdot \nabla\rho + \rho\mathbf{V}_s \cdot \nabla n$$
$$+ \rho\{n\beta + (1-n)\alpha\}\frac{\partial p}{\partial t} = 0 \qquad (3.3.6)$$

with $\nabla \cdot \mathbf{V}_s$ defined by (3.6.9) below and $\alpha$ defined by (3.2.11), this equation becomes

$$\nabla \cdot \rho\mathbf{q}_r + \rho(\alpha + n\beta)\frac{d^s p}{dt} = 0; \quad \frac{d^s(\ )}{dt} = \frac{\partial(\ )}{\partial t} + \mathbf{V}_s \cdot \nabla(\ ). \qquad (3.3.7)$$

Or, with $|\partial p/\partial t| \gg |\mathbf{V}_s \cdot \nabla p|$

$$\nabla \cdot \rho \mathbf{q}_r + \rho(\alpha + n\beta) \frac{\partial p}{\partial t} = 0. \tag{3.3.8}$$

For $|n\, \partial \rho/\partial t| \gg |\mathbf{q}_r \cdot \nabla \rho|$, we obtain

$$\nabla \cdot \mathbf{q}_r + (\alpha + n\beta) \frac{\partial p}{\partial t} = 0. \tag{3.3.9}$$

Another form of (3.3.9) is

$$\nabla \cdot \mathbf{q}_r + \rho g(\alpha + n\beta) \frac{\partial \phi^*}{\partial t} = 0. \tag{3.3.10}$$

At this point, we introduce the definition of *specific storativity*, $S_0$, defined by

$$S_0 = \rho g(\alpha + n\beta). \tag{3.3.11}$$

We recall that this definition is based on the assumption of no change in total stress, i.e., $d\sigma = 0$, that underlies the development of (3.2.13). In Section 3.6, we shall return to the definition of specific storativity for the general case of $d\sigma \neq 0$ and without the constraint of vertical stresses only.

With specific storativity as defined by (3.3.11), we may now rewrite (3.3.8) in the form

$$\nabla \cdot \rho \mathbf{q}_r + \rho S_0 \frac{\partial \phi^*}{\partial t} = 0 \tag{3.3.12}$$

and (3.3.10) in the form

$$\nabla \cdot \mathbf{q}_r + S_0 \frac{\partial \phi^*}{\partial t} = 0. \tag{3.3.13}$$

If a source, $I = I(\mathbf{x}, t)$, is present (= added volume per unit volume of porous medium, per unit time), then (3.3.13) becomes

$$-\nabla \cdot \mathbf{q}_r + I = S_0 \frac{\partial \phi^*}{\partial t}. \tag{3.3.14}$$

In soil mechanics, an *undrained test* is an experiment carried out so fast that there is practically no motion of the water relative to the solids, i.e., $\mathbf{q}_r = 0$. Then, it is obvious from (3.3.14), that the added water goes only into (elastic) storage, raising the piezometric head. We may, therefore, define the *specific storativity*, $S_0$, as the *volume of water added to storage, per unit volume of porous medium, per unit rise in piezometric head* under the conditions equivalent to those determined by an undrained test.

With these assumptions, and the definition (3.3.11) for $S_0$, equation (3.3.8) becomes

$$\nabla \cdot \left\{ \rho \frac{\mathbf{k}}{\mu} \cdot (\nabla p + \rho g \nabla z) \right\} = \frac{1}{g} S_0 \frac{\partial p}{\partial t}. \tag{3.3.15}$$

Or, if, $\rho = \rho(p)$ only

$$\nabla \cdot (\rho \mathbf{K} \cdot \nabla \phi^*) = \rho S_0 \frac{\partial \phi^*}{\partial t}. \tag{3.3.16}$$

Note that (3.3.15) is a single equation in terms of $\rho$ and $p$. However, together with information on the equation of state, $\rho = \rho(p)$, it can be solved for the single-state variable $p = p(\mathbf{x}, t)$.

Let us further simplify (3.3.16) by assuming

(a) that, the hydraulic conductivity, $\mathbf{K}$, is practically independent of pressure changes, although $\rho = \rho(p)$.
(b) that $S_0$, as defined by (3.3.11) and $\mathbf{K}$, are practically unaffected by variations in the porosity $n$, although we have assumed a deformable porous medium.
(c) that $|\mathbf{q}_r \cdot \nabla \rho| \ll |n \, \partial \rho / \partial t|$ so that $\nabla \cdot \rho \mathbf{q}_r \simeq \rho \nabla \cdot \mathbf{q}_r$, i.e., we assume that the spatial variations in $\rho$ are much smaller than the local, temporal ones.

Under these assumptions, (3.3.16) reduces to

$$\nabla \cdot \{\mathbf{K} \cdot \nabla \phi^*\} = S_0 \frac{\partial \phi^*}{\partial t} \tag{3.3.17}$$

which is a *single* equation in the single variable $\phi^* = \phi^*(x, y, z, t)$. Often $\phi^*$ is replaced by $\phi$ in (3.3.17).

For a homogeneous isotropic porous medium, (3.3.17) reduces to

$$K \nabla^2 \phi^* = S_0 \frac{\partial \phi^*}{\partial t}. \tag{3.3.18}$$

In Cartesian coordinates (3.3.18) takes the form

$$K \left( \frac{\partial^2 \phi^*}{\partial x^2} + \frac{\partial^2 \phi^*}{\partial y^2} + \frac{\partial^2 \phi^*}{\partial z^2} \right) = S_0 \frac{\partial \phi^*}{\partial t}. \tag{3.3.19}$$

For an isotropic, but inhomogeneous medium, (3.3.17) becomes, in Cartesian coordinates,

$$\frac{\partial}{\partial x} \left( K \frac{\partial \phi^*}{\partial x} \right) + \frac{\partial}{\partial y} \left( K \frac{\partial \phi^*}{\partial y} \right) + \frac{\partial}{\partial z} \left( K \frac{\partial \phi^*}{\partial z} \right) = S_0 \frac{\partial \phi^*}{\partial t}. \tag{3.3.20}$$

For a nonhomogeneous, anisotropic medium, where the principal axes are in the $x$, $y$, and $z$ directions, (3.3.17) becomes

$$\frac{\partial}{\partial x} \left( K_x \frac{\partial \phi^*}{\partial x} \right) + \frac{\partial}{\partial y} \left( K_y \frac{\partial \phi^*}{\partial y} \right) + \frac{\partial}{\partial z} \left( K_z \frac{\partial \phi^*}{\partial z} \right) = S_0 \frac{\partial \phi^*}{\partial t}. \tag{3.3.21}$$

Obviously, in (3.3.20) and (3.3.21), $K = K(x, y, z)$ must be continuous and have a continuous first derivative everywhere in the considered flow domain (see discussion in Section 3.4).

Finally, if the flow is steady and/or when both water and solid matrix are assumed to be incompressible, the right-hand side of (3.3.17) through (3.3.21) vanishes. For example, (3.3.18) reduces to the well-known *Laplace equation*

$$\nabla^2 \phi^* \equiv \frac{\partial^2 \phi^*}{\partial x^2} + \frac{\partial^2 \phi^*}{\partial y^2} + \frac{\partial^2 \phi^*}{\partial z^2} = 0. \qquad (3.3.22)$$

For a homogeneous, isotropic porous medium, the Laplace equation (3.3.22) can be also obtained from (3.3.2), by assuming $\mathbf{q} \simeq \mathbf{q}_r$.

Under the same conditions of $\alpha = \beta = 0$, Equations (3.3.9) reduce to

$$\nabla \cdot \mathbf{q}_r = 0 \qquad (3.3.23)$$

This, for example is the case where $\rho = \rho(c)$.

If we now introduce Darcy's law (2.1.35) into (3.3.23), we obtain

$$\nabla \cdot \frac{k}{\mu} (\nabla p + \rho g \nabla z) = 0. \qquad (3.3.24)$$

or

$$\nabla \cdot \left( \frac{k}{\mu} \nabla p \right) = -\frac{\partial}{\partial z} \frac{k\gamma}{\mu} \qquad (3.3.25)$$

where the right-hand side plays the role of distributed source (de Josselin de Jong, 1969).

In Subsection 7.2.4, and especially in Subsection 13.1, this equation will be employed as part of a model describing sea water intrusion in a coastal aquifer.

When using any balance equation, one should always bear in mind the various assumptions made along their development. For problems of practical interest in groundwater flow (and this is the point of view of this book), (3.3.17) through (3.3.22) should be sufficient. However, whenever the situation calls for it, it is always possible to remove any of the assumptions made here, and arrive at different, usually more complicated, equations.

## 3.4. Initial and Boundary Conditions

Each of the basic equations presented in the previous section is a second-order partial differential equation which describes a *class of phenomena*; the equations themselves are merely expressions of mass balance (including the special case of constant density), and contain no information related to any specific case of flow, not even the shape of the domain within which this flow occurs. Therefore, each equation has an infinite number of possible solutions, each of which corresponds to a particular case of flow through a porous medium domain.

To obtain from this multitude of possible solutions one particular solution corresponding to a certain specific problem of interest, it is necessary to provide supplementary information that is not contained in the equation. The supplementary information that, together with the partial differential equation, defines the model of a specific problem, should include a specification of *initial conditions* and of *boundary conditions*. The former describes the distribution of the values of the considered state variable at some initial time, usually taken as $t = 0$, at all points within the considered domain, $D$. For example

$$\phi = \phi(x, y, z, 0) = f(x, y, z) \quad \text{in } D \tag{3.4.1}$$

where $f(x, y, z)$ is a known function.

*Boundary conditions* express the way the considered domain interacts with its environment. In other words, they express the conditions, e.g., *known* water fluxes, or *known* values of state variables, such as piezometric head, that (what happens in) the external domain imposes on the considered one.

Different boundary conditions result in different solutions. Hence, the importance of stating the correct boundary conditions. It should be clear that although, as part of a mathematical model, boundary conditions are expressed in mathematical terms, their content obviously expresses a physical reality, as visualized by the modeler.

As we shall see below, all boundary conditions take the form of equalities between either the values of state variables, or of fluxes, on both 'sides' of the points on a considered boundary. In such equalities, *the information related to the external side must be known*. It is obtained from actual measurements or by assuming, on the basis of past experience and subject to *a-posteriori* verification, the future situation that will prevail on the external side of the boundary. A lack of information related to the external side of a considered boundary, requires a simultaneous solution for the considered domain and for the one that is external to it.

### 3.4.1. THE BOUNDARY

We recall that our entire disucssion is at the macroscopic level, viz., one that is obtained from the microscopic one by averaging over an REV (Section 1.4).

The boundary may take the form of an arbitrary mathematical surface that separates an investigated domain from its environment, within a large porous medium domain. Often, however, it coincides with a surface of discontinuity, i.e., a surface across which the porosity of the porous medium and/or other coefficients, such as permeability, undergo an abrupt change. The boundary between a porous medium domain and the external space devoid of solid matrix and a boundary

# MODELING THREE-DIMENSIONAL FLOW

between a porous medium domain and an external domain having no void space (i.e., an impervious body), may serve as examples of such boundary.

Figure 3.3 shows four regions of different media: a region with no void space, two regions composed of solid matrices of different porosities and a region with no solid matrix. When rigorously employing the methodology of averaging over an REV in order to determine the porosity, we note that the latter varies *gradually*, as we move along the $x$-axis. No abrupt change occurs. In principle, it is possible to regard the entire domain, composed of the four regions, as a *single heterogeneous medium*, with no internal boundaries. Conditions have then to be specified at infinity. However, this approach requires information on the variation of porosity across the transition zones. Usually, this information is not available and we do not wish to introduce it in one approximate form or another. Instead, we replace the actual variation in porosity by an *idealized boundary that takes the form of a surface across which an abrupt change in porosity takes place*. We locate this boundary somewhere, say in the middle, of the transition zone. The continuum approach is then applied to the domains on either side of the boundary. The values of porosity (or other coefficients) on both sides of the boundary are obtained by extrapolating the spatial trend of the porosity (or the other coefficients) as the boundary is approached from within each subdomain.

In this way we hypothesize the existence of regular continuum domains up to the boundary surface on both its sides. However, within the distance corresponding to half a REV from a boundary, a strip exists within which, strictly speaking, the macroscopic description is not valid. Values of state variables calculated by a

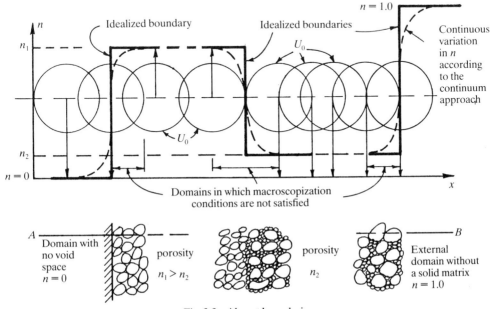

Fig. 3.3. Abrupt boundaries.

continuum model for points within this strip should not be compared with ones measured in the real system (by instruments that take an average over a REV).

In Sections 6.4 and 7.1 we shall apply similar considerations to boundaries between miscible and immiscible fluids, respectively.

The boundary surface itself, whether stationary or moving, can be described by an equation that has the general form

$$F(x, y, z, t) = 0. \tag{3.4.2}$$

With the vector $\mathbf{u}(x, y, z, t)$ denoting the velocity of points belonging to this surface, it follows that its material derivative vanishes, i.e.,

$$\frac{dF}{dt} \equiv \frac{\partial F}{\partial t} + \mathbf{u} \cdot \nabla F = 0. \tag{3.4.3}$$

This statement stems from the observation that $F$ is a quantity that is conserved as the surface moves. From (3.4.3) it follows that

$$\mathbf{u} \cdot \nabla F = -\frac{\partial F}{\partial t} \tag{3.4.4}$$

and that the outward unit vector normal to the surface $F = 0$ is given by

$$\boldsymbol{\nu} = \frac{\nabla F}{|\nabla F|}. \tag{3.4.5}$$

### 3.4.2. GENERAL BOUNDARY CONDITION

*All* boundary conditions are derived from a single general one which states that, unless sources or sinks exist on the boundary, the component normal to the boundary of *the total flux of any extensive quantity, relative to the possibly moving boundary, remains unchanged as the boundary is crossed.* Examples of extensive quantities are mass of a phase, mass of a phase component, and momentum and energy of a phase. This condition is often referred to as the *no-jump condition*. Since the porous medium is a multiphase material body, the total flux of a considered extensive quantity is not necessarily conserved within a single phase as it is transported across the boundary. Interactions between phases on the boundary may exist, recalling that we are considering a hypothetical boundary across which an abrupt change in volumetric (and, hence, also in areal) porosity is assumed to take place. The term 'total flux' used above indicates the sum of advective, dispersive, and diffusive fluxes (Section 6.2). The mathematical form of the general boundary condition is given by (Bear and Shapiro, 1986)

$$S^{\alpha\alpha}[\theta_\alpha \psi_\alpha (\mathbf{V}_\alpha - \mathbf{u}) + \theta_\alpha \mathbf{J}_\alpha^*]_{1,2} \cdot \boldsymbol{\nu} -$$

$$- \sum_{(\beta \neq \alpha)} S^{\beta\alpha}[\theta_\beta \psi_\beta (\mathbf{V}_\beta - \mathbf{u}) + \theta_\beta \mathbf{J}_\beta^*]_{1,2} \cdot \boldsymbol{\nu} = 0 \tag{3.4.6}$$

where   $\alpha$ = a considered phase,
       $\beta$ = all other phases,
$S^{\alpha\alpha}$, $S^{\beta\alpha}$ = the $\alpha - \alpha$ and $\beta - \alpha$ fractions, respectively, of the boundary surface area,
       $\psi$ = density (per unit volume),
       $\theta$ = volumetric fraction of a phase,
       $\mathbf{u}$ = velocity of points on the boundary,
       $\mathbf{\nu}$ = the unit vector normal to the surface pointing away from the phase,
       $[(\ )]_{1,2} = (\ )|_1 - (\ )|_2$, i.e., jump in ( ) in crossing from side 1 to side 2 of the boundary,
       $\mathbf{J}^*$ = sum of diffusive and dispersive fluxes (Section 6.2).

By introducing various simplifying assumptions, the general no-jump condition is reduced to the boundary conditions that are commonly employed in mathematical models of transport phenomena in porous media.

We have introduced this brief discussion on the general boundary condition as a background to the development of the boundary condition for the case in hand, in which (i) we have only two phases — solid and fluid — that occupy the entire space, and (ii) the extensive quantity is the mass of the fluid, with the special case of a fluid of constant density. In each case, we shall attempt to indicate the assumptions that underlie the passage from the general condition to a particular one. In Chapter 6 we shall apply the general condition to another extensive quantity — the mass of a component of a phase.

Of special interest is the assumption that the nonequilibrium processes under consideration (and we mean mass, momentum, solute, and energy transport), proceed through a series of states of a local thermodynamic equilibrium at all points of a considered domain, including points on the boundary. We shall refer to this assumption as the one of *thermodynamic equilibrium*. A direct consequence of this assumption is that

(a) $[p]_{1,2} = 0$,    (b) $[c]_{1,2} = 0$,
(c) $[\rho_f]_{1,2} = 0$,    (d) $[T_f]_{1,2} = 0$   on $B$    (3.4.7)

where $B$ denotes the boundary, $p$ is pressure, $c$ is concentration, $\rho$ is mass density, $T$ is temperature, and $f$ is a subscript that denotes the fluid phase.

For the particular case of the saturated flow of a fluid's mass, ($\theta_\alpha = n$), assuming the validity of (3.4.7c), and neglecting the sum of dispersive and diffusive mass fluxes as much smaller than the advective ($= n\mathbf{V}_f$) one, (3.4.6) reduces for the fluid ($\alpha = f$) to

$$[n(\mathbf{V}_f - \mathbf{u})]_{1,2} \cdot \mathbf{\nu} = 0, \qquad (3.4.8)$$

where $n\mathbf{V}_f = \mathbf{q}$ is the specific discharge, and for a (deformable) solid ($\alpha = s$) to

$$[(1-n)(\mathbf{V}_s - \mathbf{u})]_{1,2} \cdot \mathbf{\nu} = 0. \qquad (3.4.9)$$

In order to obtain this condition from (3.4.6), we noted that the solid-fluid portion of the boundary, $S^{sf}$, is a *material surface* for both the solid and the fluid.

When the boundary is a material surface for the solid, i.e., no solid crosses it from one side to the other (and we shall always make this assumption), we have

$$(\mathbf{V}_s - \mathbf{u})|_1 \cdot \boldsymbol{\nu} = (\mathbf{V}_s - \mathbf{u})|_2 \cdot \boldsymbol{\nu} = 0. \tag{3.4.10}$$

In such case, by combining (3.4.10) with (3.4.8), we obtain the condition

$$[\mathbf{q}_r]_{1,2} \cdot \boldsymbol{\nu} = 0. \tag{3.4.11}$$

Since $n\mathbf{V}_s \ll \mathbf{q}$, we often make the approximation $\mathbf{q} \simeq \mathbf{q}_r$.

### 3.4.3. BOUNDARY OF PRESCRIBED PRESSURE

When the fluid pressure is specified as a known function of space and time, say $f_1(\mathbf{x}, t)$, at all points of the external side of a boundary segment, independent of what happens in the domain itself, we employ the first no-jump condition in (3.4.7) to write the boundary condition

$$p(\mathbf{x}, t) = f_1(\mathbf{x}, t) \quad \text{on } B \tag{3.4.12}$$

where $p(\mathbf{x}, t) \equiv p(\mathbf{x}, t)|_1$ and $f_1(\mathbf{x}, t) \equiv p(\mathbf{x}, t)|_2$.

When the piezometric head, $\phi = z + p/\gamma$, is used as the dependent variable, the boundary condition becomes

$$\phi(\mathbf{x}, t) = f_2(\mathbf{x}, t) \quad \text{on } B \tag{3.4.13}$$

where $f_2(\mathbf{x}, t)$ is a known function.

In the theory of partial differential equations, the condition which specifies the values of a dependent variable at all points of a boundary segment is referred to as a *boundary condition of the first kind*, or *a Dirichlet condition*.

A boundary of this kind occurs whenever the porous medium flow domain is in contact with a body of open water. Segments *AB* and *EF* in Figure 3.4a are examples of boundaries of prescribed piezometric head. On *AB*, the head is $\phi = H_1(t)$. On *FE*, the condition is $\phi = H_2(t)$. Obviously $H_1$ and $H_2$ may also vary along the rivers. A special case of this kind of boundary is the *equipotential boundary*

$$\phi = \text{constant} \quad \text{on } B. \tag{3.4.14}$$

### 3.4.4. BOUNDARY OF PRESCRIBED FLUX

Let the known total (volume) flux of the fluid normal to the boundary be denoted by $f_3(\mathbf{x}, t)$. Making use of (3.4.11), with 1 and 2 denoting the internal and external sides of the boundary, we obtain the boundary condition

$$\mathbf{q}_r \cdot \boldsymbol{\nu} = f_3(\mathbf{x}, t) \quad \text{on } B \tag{3.4.15}$$

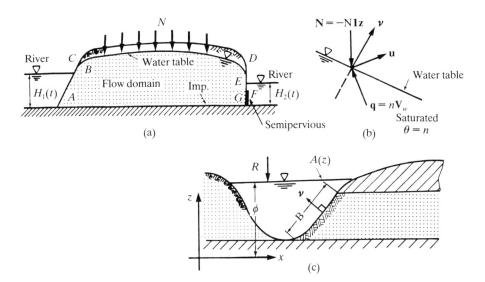

Fig. 3.4. Flow domain between two rivers.

where, as we recall, $\mathbf{q}_r$ is the specific discharge of the fluid relative to the possibly moving solid. Employing (2.1.35) and (3.4.5), we rewrite (3.4.15) in the form

$$-\frac{k}{\mu}(\nabla p + \rho_f g \nabla z) \cdot \nabla F = |\nabla F| f_3(\mathbf{x}, t) \quad \text{on } B. \tag{3.4.16}$$

Or, in terms of $\phi$ (when $\rho_f = $ const)

$$-\mathbf{K} \cdot \nabla \phi \cdot \nabla F = |\nabla F| f_3(\mathbf{x}, t) \quad \text{on } B. \tag{3.4.17}$$

For an impervious boundary, (e.g., $AG$ in Figure 3.3a), $f_3(\mathbf{x}, t) = 0$, and the boundary condition (3.4.17) reduces to

$$\mathbf{K} \cdot \nabla \phi \cdot \nabla F = 0 \quad \text{on } B. \tag{3.4.18}$$

If the porous medium is anisotropic, with $x$, $y$, $z$ its principal directions, (3.4.18) becomes

$$K_x \frac{\partial \phi}{\partial x}\frac{\partial F}{\partial x} + K_y \frac{\partial \phi}{\partial y}\frac{\partial \phi}{\partial y} + K_z \frac{\partial \phi}{\partial z}\frac{\partial F}{\partial z} = 0 \quad \text{on } B. \tag{3.4.19}$$

For an isotropic material, (3.4.18) becomes

$$\frac{\partial \phi}{\partial x}\frac{\partial F}{\partial x} + \frac{\partial \phi}{\partial y}\frac{\partial F}{\partial y} + \frac{\partial \phi}{\partial z}\frac{\partial F}{\partial z} = 0 \quad \text{on } B. \tag{3.4.20}$$

Thus, here we have conditions that specify the pressure gradient, or the hydraulic gradient, or a linear combination of their components, as a function of location, and possibly time, on the boundary.

In the theory of partial differential equations, this type of boundary condition is referred to as *boundary condition of the second kind* or a *Neumann condition*.

### 3.4.5. SEMIPERVIOUS BOUNDARY

This type of boundary occurs when the porous medium domain is in contact with a body of water continuum (or another porous medium domain) through a relatively thin semipervious layer separating the two domains. (*FG* in Figure 3.4a). Let $\phi$ denote the piezometric lead in the considered domain and $\phi_0$ (= $H_2(t)$ in the example of Figure 3.4a) denote that in the external one. If we *assume no change in water storage in the semipermeable thin layer*, then the flux normal to the boundary may be expressed by $f_3(\mathbf{x}, t) = (\phi - \phi_0)/c$, where $c = B'/K'$ = the ratio of thickness to hydraulic conductivity of the semipervious membrane, and (3.4.17), recalling all the assumptions underlying it, becomes

$$\mathbf{K} \cdot \nabla \phi \cdot \nabla F = |\nabla F|(\phi_0 - \phi)/c \quad \text{on } B. \tag{3.4.21}$$

For $x, y, z$ that are principal directions of an anisotropic porous medium

$$\left( K_x \frac{\partial \phi}{\partial x} \frac{\partial F}{\partial x} + K_y \frac{\partial \phi}{\partial y} \frac{\partial F}{\partial y} + K_z \frac{\partial \phi}{\partial z} \frac{\partial F}{\partial z} \right) +$$

$$+ \frac{\phi |\nabla F|}{c} = \frac{\phi_0 |\nabla F|}{c} \quad \text{on } B. \tag{3.4.22}$$

This type of condition, which contains information on the relationship between the state variable and its derivatives, is called a *mixed boundary condition, boundary condition of the third kind*, or a *Cauchy condition*.

### 3.4.6. PHREATIC SURFACE WITH ACCRETION

The phreatic (or free) surface has already been discussed in Section 1.1, where it was defined as the surface on which $p = 0$. Here, we shall neglect the capillary fringe above this surface.

Again, the condition of no-jump in the total mass flux is employed. For the sake of simplicity, we shall assume that the accretion (and the same is true for evaporation which may be considered as negative accretion) has the same density as the aquifer's water. We shall also continue to assume that the dispersive and diffusive fluxes of the fluid's mass may be neglected in the saturated zone below the phreatic surface. Under these conditions, the no-jump condition takes the form

$$\{n(\mathbf{V}_w - \mathbf{u})\}|_{\text{sat}} \cdot \boldsymbol{\nu} = \mathbf{N}' \cdot \boldsymbol{\nu} \quad \text{on } \mathbf{B} \tag{3.4.23}$$

MODELING THREE-DIMENSIONAL FLOW

where $\{\ \}|_{sat}$ indicates that $\{\ \}$ is evaluated at the phreatic surface, $B$, as it is approached from within the saturated domain and $\mathbf{N}'$ denotes the (volume) rate of water added to the saturated zone from the partially saturated zone above it, per unit area of the phreatic surface. This flux is *independent* of the displacement of the phreatic surface (at a velocity $\mathbf{u}$), and of the quantity of water present in the void space above it.

Equation (3.4.23) can be used to determine the velocity, $\mathbf{u}$, and the displacement $\mathbf{u} \cdot \mathbf{\nu}$ of the phreatic surface in a numerical model.

Assuming that the water moisture is at the *irreducible moisture content* $\theta_{w0}$ in the unsaturated zone above the phreatic surface (Section 5.1), the strength of the water source at the phreatic surface can then be expressed by

$$\mathbf{N}' = \{\theta_{w0}(\mathbf{V}_w - \mathbf{u})\}|_{unsat} \equiv \mathbf{N} - \theta_{w0}\mathbf{u} \tag{3.4.24}$$

where $\mathbf{N} = \theta_{w0}\mathbf{V}_w|_{unsat}$. For downward vertical infiltration at a rate of $N$, we have $\mathbf{N} = -N\,\nabla z$. By inserting (3.4.23) into (3.4.24), we obtain (Figure 3.4b)

$$(\mathbf{q}_w - n\mathbf{u}) \cdot \mathbf{\nu} = (\mathbf{N} - \theta_{w0}\mathbf{u}) \cdot \mathbf{\nu} \quad \text{on } B.$$

or

$$(\mathbf{q}_w - \mathbf{N}) \cdot \nabla F + (n - \theta_{w0}) \frac{\partial F}{\partial t} = 0 \quad \text{on } B \tag{3.4.25}$$

where $\mathbf{q}_w = n\mathbf{V}_w|_{sat}$ and we have employed (3.4.3) and (3.4.4) to express $\mathbf{u} \cdot \mathbf{\nu}$. In (3.4.25), the difference $n_e = n - \theta_{w0}$ is the *effective porosity*, or *specific yield* often denoted as $S_y$. In order to express this boundary condition in terms of $\phi$, we make use of (2.1.28), assuming $n\mathbf{V}_s \ll \mathbf{q}_w$, $\mathbf{q}_w \simeq \mathbf{q}_{rw}$.

The location and shape of the free surface are *a-priori* unknown. In fact, their determination constitutes part of the required solution. Following the usual procedure, however, we must specify for the boundary being considered (i) its geometry, and (ii) the condition to be satisfied at all points along it.

Since the pressure at all points of the free surface, $B$, is taken as $p = 0$, we have from $\phi(x, y, z, t) = z + p(x, y, z, t)/\gamma$

$$\phi(x, y, z, t) = z \quad \text{or} \quad \phi(x, y, z, t) - z = 0 \quad \text{on } B. \tag{3.4.26}$$

Thus, the shape of the interface is described by the equation

$$F(x, y, z, t) \equiv \phi(x, y, z, t) - z = 0. \tag{3.4.27}$$

We may now combine (3.4.25), (2.1.28), and (3.4.27) to yield the condition

$$(\mathbf{K} \cdot \nabla\phi + \mathbf{N}) \cdot \nabla(\phi - z) = n_e \frac{\partial \phi}{\partial t} \quad \text{on } B \tag{3.4.28}$$

or with $x$, $y$, $z$ principal directions

$$n_e \frac{\partial \phi}{\partial t} = K_x \left( \frac{\partial \phi}{\partial x} \right)^2 + K_y \left( \frac{\partial \phi}{\partial y} \right)^2 +$$

$$+ K_z \left\{ \left( \frac{\partial \phi}{\partial z} \right)^2 - \left( \frac{\partial \phi}{\partial z} \right) \right\} - N \left( \frac{\partial \phi}{\partial z} - 1 \right) \quad \text{on } B. \quad (3.4.29)$$

We recall that in this expression, $\phi(x, y, z, t)$ is unknown until the problem is solved. On the other hand, the problem cannot be solved unless the shape of the phreatic surface and the boundary condition on it are known. This vicious circle, prohibits an analytical solution of a problem with a phreatic surface boundary, except for very special cases. The numerical solution of such problem is discussed in Section 10.4.

It may be noted that we actually have two boundary conditions along the phreatic surface: (3.4.26) and (3.4.29). The extra condition compensates for the lack of *a-priori* information on the location of the boundary.

### 3.4.7. SEEPAGE SURFACE

The *seepage surface* (or *seepage face*) was presented in Section 2.3. Segments $BC$ and $DE$ in Figure 3.4a are examples of seepage surfaces.

The phreatic surface is tangent to the boundary of the porous medium at points $C$ and $D$ (Figure 3.4a). Along a seepage surface, water emerges from the flow domain, trickling downward to the adjacent body of water.

Being exposed to the atmosphere, the pressure along a seepage face is atmospheric ($p = 0$) and, hence, the boundary condition along such a surface is

$$\phi(x, y, z, t) = z. \quad (3.4.30)$$

The geometry of the seepage face is known (as it coincides with the boundary of the porous medium), except for its upper limit (points $C$ and $D$ in Figure 3.4a) which is also lying on the (*a priori*) unknown phreatic surface. The location of this point is, therefore, part of the required solution.

### 3.4.8. BOUNDARY WITH A FINITE VOLUME RESERVOIR

Let the total discharge from an aquifer enter a reservoir, such that as water enters the reservoir its water level rises (Figure 3.4c). Similarly, as water flows from the reservoir, its water level drops. The reservoir may also be fed by an external supply at a rate $R$. Thus, the reservoir's water level governs the piezometric head on the aquifer's boundary. Equality of total fluxes dictates the condition

$$A(z) \frac{\partial \phi}{\partial t} = \int_{(B)} \mathbf{q} \cdot \mathbf{v} \, dB = -\int_{(B)} \mathbf{K} \cdot \nabla \phi \cdot \frac{\nabla F}{|\nabla F|} \, dB \qquad (3.4.31)$$

where $B$ denotes the contact area between the aquifer and the reservoir. In this condition, we see a combination of spatial and temporal derivatives of $\phi$.

### 3.4.9. BOUNDARY BETWEEN TWO POROUS MEDIA

Along such a boundary, we have discontinuities in both the porosity and the permeability.

Equations (3.3.20) and (3.3.21) describe flow in inhomogeneous domains. However, in order for these equations to have analytical solutions, within the framework of a mathematical model, the distributions $K = K(x, y, z)$, (or $\mathbf{K} = \mathbf{K}(x, y, z)$, in an anisotropic domain) and its first derivatives must be continuous. If a discontinuity in $K$ or in $\nabla K$ exists along certain surfaces (or lines in a two-dimensional domain), the only way to solve the problem analytically is to divide the flow domain by these surfaces (or lines) into subdomains, such that no discontinuities occur within each subdomain. In order to have a well-posed problem for each subdomain, conditions must be specified on such boundary surfaces (obviously, in addition to conditions on all other boundaries).

In this case, each domain is 'external' to the other, but we do not have the *a-priori* known information required to set up the necessary boundary conditions. This means that the flow problem, in terms of $p$ or $\phi$, must be solved simultaneously for both domains, as the two boundary conditions — one for each domain — will contain the two dependent variables. These conditions are

$$p|_1 = p|_2 \quad \text{or} \quad \phi|_1 = \phi|_2 \quad \text{on } B \qquad (3.4.32)$$

obtained from the first equation of (3.4.7) and for a stationary boundary

$$\frac{\mathbf{k}_1}{\mu} (\nabla p_1 + \rho g \nabla z)|_1 \cdot \mathbf{v} = \frac{\mathbf{k}_2}{\mu} (\nabla p_2 + \rho g \nabla z)|_2 \cdot \mathbf{v}$$

or, with $\rho_1 = \rho_2 = \text{const.}$, $\mu_1 = \mu_2 = \text{const.}$

$$\mathbf{K}_1 \cdot \nabla \phi|_1 \cdot \mathbf{v} = \mathbf{K}_2 \cdot \nabla \phi|_2 \cdot \mathbf{v}, \qquad (3.4.33)$$

i.e., $\mathbf{q}|_1 \cdot \mathbf{v} = \mathbf{q}|_2 \cdot \mathbf{v}$ obtained from (3.4.8).

Note that when solving a problem of this kind, it is convenient to subscript each dependent variable by a symbol (here 1 and 2) that indicates the relevant domain. We solve the appropriate balance equation within each domain, in terms of the relevant dependent variable.

The above discussion is valid for analytical solutions of mathematical models. When a model is solved numerically, there may be no need to build a separate model for each subdomain. The equivalent of the boundary conditions at a boundary of discontinuity are incorporated in the numerical code (see Chapter 10).

## 3.5. Complete Statement of Mathematical Flow Model

To solve a flow problem in a specified domain, with specified transport and storage coefficients, means to determine the spatial and temporal distributions of the values of the relevant state variable that satisfy certain partial equations at all points *within* the considered domain, as well as specified initial and boundary conditions.

We now have all the elements needed in order to construct the mathematical model of any given problem of flow in a porous medium domain. There is no need to remind the reader that the model is constructed at the (macroscopic) continuum level (Section 1.4).

Prior to the construction of any mathematical model, we should conduct investigations in order to identify the conceptual model, i.e., the set of assumptions that represent our simplified perception of the real system under consideration. Obviously, only those features that we feel are relevant to the problem at hand are included in the conceptual model. In this model we should:

(a) Identify the boundaries of the domain that we wish to model.
(b) Assume the type of flow regime (laminar, or nonlaminar) that will take place.
(c) Identify the materials contained in the flow domain (solid matrix and water in saturated flow) and those features of their behavior that are relevant to the problem on hand (e.g., fluid compressibility, solid matrix deformability). Then introduce constitutive relationships to express this behavior.
(d) Make assumptions concerning the homogeneity and isotropy of the domain with respect to the various coefficients that express transport and storage processes within the flow domain.
(e) Identify sources and sinks of water that are present within the modeled domain.
(f) Identify the behavior in the environment that is external to the considered domain and the interaction that takes place between the two domains across their common boundary.

On the basis of the conceptual model, which, we suggest, should be *explicitly stated*, the standard content of a flow model should consist of the following items:

(a) Specification of the geometrical configuration of the surface that bounds the problem domain. The boundary surface must be a closed one, but it may include segments at infinity.

(b) A list of the relevant state variables that describe the state of the system (usually pressure, $p = p(x, y, z, t)$, or piezometric head $\phi = \phi(x, y, z, t)$).
(c) Statement of the partial differential (mass balance, or continuity) equation in terms of the state variables mentioned in (b).
(d) Statement of the relevant constitutive equations, e.g., flux equation, equation of state for the water, or stress-strain relationship for the solid matrix.
(e) Specification of the numerical values of all the coefficients that appear in the constitutive relations and, hence, in the balance equations.
(f) Statement of the initial conditions of the system, i.e., values of the relevant state variables at all points within the considered domain, at some initial time, usually taken at $t = 0$. No initial conditions are required for steady flow.
(g) Statement of conditions to be satisfied at all points of the domain's boundaries specified in (a).

As we have seen in the common cases of groundwater flow, (c) and (d) are combined to yield a single partial differential equation in terms of a single state variable ($p$ or $\phi$). Then (e) refers to the coefficients appearing in that equation. In the case of more than one state variable, we should have a sufficient number of equations to enable a simultaneous solution for all of them.

As emphasized in Section 3.4, the type of boundary condition to be specified in any particular case is motivated by the physical reality of the considered flow problem and by the interaction that takes place across the boundary between the considered domain and its environment. Field investigations will provide the necessary information about these interactions. These, in turn, will dictate the boundary conditions to be employed in each case. Sometimes, because of a lack of information, various assumptions and approximations are introduced as a part of the conceptual model, before a boundary condition is stated.

In view of the approximations involved, the modeler should be careful, since not every set of conditions is acceptable. From the mathematical point of view, a *well posed boundary value problem* that corresponds to a physical reality, as is the case considered here, must satisfy the following fundamental requirements.

(a) A solution must *exist*.
(b) The solution must be *unique*, i.e., the problem as stated must have only one solution.
(c) The solution must be *stable*, i.e., it should depend in a continuous manner on the data (e.g., initial, or boundary conditions). Sufficiently small variations in the given data, should lead to arbitrarily small changes in the solution. Otherwise, we should conclude that the problem is badly formulated.

In this book we do not investigate whether a model (composed of a partial differential equation coupled with initial and boundary conditions) indeed constitutes a well posed problem. We shall implicitly assume that the model is well posed, since we base it on the physical reality, albeit with certain simplifying assumptions. This follows from the fact that the model, although stated in mathematical terms, is an attempt to describe actual physical phenomena.

## 3.6. Modeling Soil Displacement

Basically, soil deformation, or soil consolidation, is a three-dimensional phenomenon. De Josselin de Jong (1963) and Verruijt (1956, 1959, 1984), among others, present theories on three-dimensional soil consolidation. A review on land subsidence is presented by Corapcioglu (1984).

In general, three main approaches exist to the modeling of consolidation:

(a) The approach proposed by Biot (1941) which regards soil consolidation as a three-dimensional phenomenon in which the flow of water and the strain in the solid matrix are continuously interrelated through Terzaghi's (1925) concept of *effective stress* (Section 3.1). Accordingly, a *simultaneous* solution is sought for two dependent variables: pressure in the water and strain in the solid matrix. In this way, both vertical and horizontal displacement can be obtained for every point within a three-dimensional domain. Total vertical displacement in a pumped aquifer, or land subsidence can be obtained by integrating the vertical strain over the aquifer's thickness.

(b) A simplified version of the above approach, based on the assumption that soil displacements occur only in the vertical direction. Under these assumptions, the problems of water pressure distribution and solid matrix strain distribution are decoupled (Verruijt, 1969). The water pressure distribution is first determined by solving an appropriate model. Then the result is used to determine the vertical strain distribution and vertical displacement within the solid matrix.

(c) In groundwater hydrology, Jacob (1940) also assumed that only vertical displacements take place and that all stresses act only in the vertical direction. As a result, porosity becomes a function of water pressure only. The effect of the solid's velocity is neglected in determining the distribution of water pressure. Once the pressure distribution has been determined (and actually this was the sole objective in hydrology of groundwater) it can be used, as in (b), to determine vertical settlement.

The last approach was employed in developing the concept of specific storativity in Section 3.2. This approach also underlies the concept of aquifer storativity, for a confined aquifer, presented in Section 4.1.

The third approach is usually employed in the practice of soil mechanics, where it is considered to be a sufficiently good approximation of Biot's (1941) approach. Actually, in most consolidation analysis carried out in geomechanics, it is assumed

MODELING THREE-DIMENSIONAL FLOW

that both flow (when it exists) and soil deformation occur only in the vertical direction.

In this section we shall first present a three-dimensional model which is essentially based on Biot's theory. We shall then introduce Verruijt's (1969) assumption of vertical displacement only, still taking into account the effect of the solid's velocity on the water flow.

In Section 4.5, we derive a model for regional land subsidence and averaged horizontal displacements, by integrating the models presented in this section over the aquifer's thickness.

In Section 3.2, we made the assumption that $n = n(\sigma_s^*)$, assuming the effective stress to act only in the vertical direction. This led to the concept of specific storativity as defined by (3.2.15). Let us now remove these simplifying assumptions.

Our starting point is again the mass balance equation (3.3.1), noting all the assumptions underlying its development (e.g., neglecting nonadvective mass fluxes), rewritten here for convenience

$$\frac{\partial n\rho}{\partial t} + \nabla \cdot \rho \mathbf{q} = 0, \quad \mathbf{q} = n\mathbf{V}_w. \tag{3.6.1}$$

Source and sink terms may be added when necessary.

The mass balance for the solid phase can be expressed in a similar form, viz.

$$\frac{\partial (1-n)\rho_s}{\partial t} + \nabla \cdot \{\rho_s(1-n)\mathbf{V}_s\} = 0 \tag{3.6.2}$$

where $\rho_s$ and $\mathbf{V}_s$ are the solid's density and velocity, respectively, with $\mathbf{q} = \mathbf{q}_r + n\mathbf{V}_s$.

We now rewrite (3.6.1) in the form

$$n\frac{\partial \rho}{\partial t} + \rho\frac{\partial n}{\partial t} + n\mathbf{V}_w \cdot \nabla\rho + \rho\nabla \cdot (\mathbf{q}_r + n\mathbf{V}_s) = 0, \tag{3.6.3}$$

$$\rho\nabla \cdot \mathbf{q}_r + n\frac{d^w\rho}{dt} + \rho\frac{d^s n}{dt} + \rho n\nabla \cdot \mathbf{V}_s = 0 \tag{3.6.4}$$

where $d^\alpha(\ )/dt = \partial(\ )/\partial t + \mathbf{V}_\alpha \cdot \nabla(\ )$ is the *material derivative* of ( ) with respect to the moving $\alpha$ phase (here water, or solid).

Equation (3.6.2) can be rewritten in the form

$$\frac{1}{1-n}\frac{d^s(1-n)}{dt} + \frac{1}{\rho_s}\frac{d^s\rho_s}{dt} + \nabla \cdot \mathbf{V}_s = 0. \tag{3.6.5}$$

By eliminating $\nabla \cdot \mathbf{V}_s$ from (3.6.4) and (3.6.5), we obtain

$$\nabla \cdot \mathbf{q}_r + \frac{n}{\rho}\frac{d^w\rho}{dt} + \frac{1}{1-n}\frac{d^s n}{dt} - \frac{n}{\rho_s}\frac{d^s\rho_s}{dt} = 0. \tag{3.6.6}$$

We now assume that the solid phase (not the solid matrix) preserves its volume.

This means that *at the microscopic* level $d^s\rho_s/dt = 0$ and $d^s\varepsilon_s/dt \equiv \nabla \cdot \mathbf{V}_s = 0$, where $\varepsilon_s$ is the solid's volumetric strain

$$\varepsilon_s = \sum_{i=1}^{3} \varepsilon_{ii} = \nabla \cdot \mathbf{w},$$

$$\varepsilon_{ij} = \frac{1}{2}\left(\frac{\partial w_i}{\partial x_j} + \frac{\partial w_j}{\partial x_i}\right) \tag{3.6.7}$$

and $\mathbf{w}$ is the microscopic solid's displacement.

By averaging we obtain that $d^s\rho_s/dt = 0$ is valid also for the $\rho_s \equiv \overline{\rho}_s^s$, i.e., the macroscopic value of $\rho_s$, although the porous medium as a whole may undergo deformation. The deformation is then manifested by changes in porosity. In a granular porous medium, the deformation of the solid matrix is attributed to the rolling and slipping of the grains with respect to each other.

With the above assumption, (3.6.6) is now rewritten in the form

$$\nabla \cdot \rho \mathbf{q}_r + \frac{d^s\rho}{dt} + \frac{\rho}{1-n}\frac{d^s n}{dt} = 0. \tag{3.6.8}$$

With $\mathbf{w}$, $\mathbf{V}_s$ and $\varepsilon_b$ now denoting the *macroscopic* or *averaged* displacement, velocity, and volumetric strain (or *dilatation*) of the porous medium as a whole, we now have the relationships

$$\nabla \cdot \mathbf{V}_s = -\frac{1}{1-n}\frac{d^s(1-n)}{dt} \tag{3.6.9}$$

which can be obtained by inserting $d^s\rho_s/dt = 0$ in (3.6.2)

$$\mathbf{V}_s = d^s\mathbf{w}/dt \tag{3.6.10}$$

and

$$\nabla \cdot \mathbf{V}_s = \frac{d^s\varepsilon_b}{dt}, \quad \varepsilon_b = \nabla \cdot \mathbf{w}. \tag{3.6.11}$$

Employing these relationships, we now rewrite (3.6.8) in the form

$$\nabla \cdot \rho\mathbf{q}_r + n\frac{d^s\rho}{dt} + \rho\frac{d^s\varepsilon_b}{dt} = 0. \tag{3.6.12}$$

We assume that

$$|\partial\rho/\partial t| \gg |\mathbf{V}_s \cdot \nabla\rho|, \quad |\partial\varepsilon_b/\partial t| \gg |\mathbf{V}_s \cdot \nabla\varepsilon_b|.$$

Then, (3.6.12) for a compressible fluid, with a compressibility $\beta = (1/\rho)\,\partial\rho/\partial p$, reduces to

$$\nabla \cdot \rho\mathbf{q}_r + n\rho\beta\frac{\partial p}{\partial t} + \rho\frac{\partial\varepsilon_b}{\partial t} = 0. \tag{3.6.13}$$

A further approximation is obtained by assuming $|\partial \rho/\partial t| \gg |\mathbf{V}_w \cdot \nabla \rho|$. Then, (3.6.13) reduces to

$$\nabla \cdot \mathbf{q}_r + n\beta \frac{\partial p}{\partial t} + \frac{\partial \varepsilon_b}{\partial t} = 0. \qquad (3.6.14)$$

Since $\rho = \rho(p)$ and $n = n(\boldsymbol{\sigma}_s^*)$, with the effective stress, $\boldsymbol{\sigma}_s^*$, related to the pressure, $p$, by (3.1.7), Equation (3.6.13) involves two variables: $p = p(\mathbf{x}, t)$ and $\varepsilon_b = \varepsilon_b(\mathbf{x}, t)$ and we need one more equation. This role is fulfilled by an equation that expresses equilibrium of the total forces acting on a unit volume of the porous medium.

In the absence of inertial effects, the equilibrium equation takes the form

$$-\nabla \cdot \boldsymbol{\sigma} + \mathbf{f} = 0 \qquad (3.6.15)$$

where $\boldsymbol{\sigma}$ is the total stress (positive for compression) and $\mathbf{f}$ represents the total body force acting on the porous medium, per unit volume of the latter. In the case of gravity, $\mathbf{f} = \{n\rho + (1-n)\rho_s\}\mathbf{g}$, where $g$ represents gravity acceleration.

The total stress is related to the pressure in the water and to the effective stress by (3.1.7).

Following Verruijt (1969), we separate $\boldsymbol{\sigma}$, $\boldsymbol{\sigma}_s^*$, $p$ and $\mathbf{f}$ into initial steady-state values $\boldsymbol{\sigma}^0$, $\boldsymbol{\sigma}_s^{*0}$, $p^0$ and $\mathbf{f}^0$, and nonsteady, deformation producing increments $\boldsymbol{\sigma}^e$, $\boldsymbol{\sigma}_s^{*e}$, $p^e$ and $\mathbf{f}^e$, with

$$\boldsymbol{\sigma}^0 = \boldsymbol{\sigma}_s^{*0} + p^0 \mathbf{I} \quad \text{and} \quad \boldsymbol{\sigma}^e = \boldsymbol{\sigma}_s^{*e} + p^e \mathbf{I}, \qquad (3.6.16)$$

$$-\nabla \cdot \boldsymbol{\sigma}_s^{*0} + \mathbf{f}^0 - \nabla p^0 = 0, \qquad (3.6.17)$$

$$\nabla \cdot \boldsymbol{\sigma}_s^{*e} + \nabla p^e = 0 \qquad (3.6.18)$$

where we have assumed that $\mathbf{f}^e = 0$, although $n$ varies.

We now *assume* that the solid matrix is isotropic and, for the small excess stresses, $\boldsymbol{\sigma}_s^{*e}$, considered here, behaves as a *perfectly elastic body*. We further assume that the stress-strain relationship for the solid matrix (i.e., at the macroscopic level), relating average effective stress $\boldsymbol{\sigma}_s^{*e}$, to the average displacement, $\mathbf{w}$, has the form of Hooke's law of linear elasticity

$$\boldsymbol{\sigma}_s^{*e} = -\tilde{G}\{\nabla \mathbf{w} + (\nabla \mathbf{w})^T\} - \tilde{\lambda}(\nabla \cdot \mathbf{w})\mathbf{I} \qquad (3.6.19)$$

where $\tilde{G}$ and $\tilde{\lambda}$ are macroscopic constant coefficients called Lame's coefficients for a porous medium. They have to be determined experimentally for any given porous matrix.

In principle, any other stress-strain relationship, corresponding to nonelastic materials, may also be used. However, here we shall continue to demonstrate the methodology by referring to an elastic material.

The mass balance equation (3.6.14) is now also rewritten as two equations: one representing the initial steady state and the other involving the pressure increments that produce displacements. Thus, the second equation takes the form

$$\nabla \cdot \mathbf{q}_r^e + n\beta \frac{\partial p^e}{\partial t} + \frac{\partial \varepsilon_b}{\partial t} = 0 \tag{3.6.20}$$

where $\varepsilon_b^e \equiv \varepsilon_b$, since $\varepsilon_b^0 \equiv 0$. In (3.6.20), we express $\mathbf{q}_r^e$ for an isotropic porous medium by

$$\mathbf{q}_r^e = -\frac{k}{\mu}(\nabla p^e + \rho g \nabla z) \tag{3.6.21}$$

where we have made the simplification $\rho^0 \simeq \rho$, since $\rho^e \ll \rho^0$, $n \simeq n^0$, since $n^e \ll n^0$ and $k$ remains unchanged although deformation takes place.

Thus, the mathematical model involves the mass balance equation (3.6.12) or (3.6.14), the water flux equation (3.6.21), the equation of equilibrium of excessive stress, (3.6.18), the stress-strain relationship (3.6.19), the definition of $\varepsilon_b$ in (3.6.11), the equation of state for the fluid, $\rho = \rho(p)$, and an equation of state for the solid matrix' porosity, $n = n(\sigma_s^{*e})$. Altogether, these represent 16 scalar equations in the 16 variables: $\rho$, $n$, $\mathbf{q}_r^e$, $\varepsilon_b$, $p^e$, $\sigma_s^{*e}$ and $\mathbf{w}$. In principle, this is the model introduced by Biot (1941). We note that here we have a model that also yields the displacement vector, $\mathbf{w}$, which, in turn, can be used to determine soil compaction and land subsidence.

By inserting (3.6.21) into (3.6.20), we obtain

$$-\nabla \cdot \left\{ \frac{k}{\mu}(\nabla p^e + \rho g \nabla z) \right\} + n\beta \frac{\partial p^e}{\partial t} + \frac{\partial \varepsilon_b}{\partial t} = 0. \tag{3.6.22}$$

For a homogeneous porous medium, the combination of (3.6.18) and (3.6.19) can be written in the form of the three scalar equations

$$\widetilde{G}\nabla^2 w_i + (\widetilde{\lambda} + \widetilde{G})\frac{\partial \varepsilon_b}{\partial x_i} - \frac{\partial p^e}{\partial x_i} = 0, \quad i = 1, 2, 3. \tag{3.6.23}$$

By differentiating each of these equations with respect to the corresponding $x_i$, and adding the resulting three equations, we obtain the single equation (Verruijt, 1969)

$$(\widetilde{\lambda} + \widetilde{G})\nabla^2 \varepsilon_b - \nabla^2 p^e = 0 \tag{3.6.24}$$

which, together with (3.6.22), are often simplified for a homogeneous isotropic porous medium to the form

$$-\frac{k}{\mu}\nabla^2 p^e + n\beta \frac{\partial p^e}{\partial t} + \frac{\partial \varepsilon_b}{\partial t} = 0 \tag{3.6.25}$$

constitute a set of two equations is the state variables $p^e$ and $\varepsilon_b$.

Following Verruijt (1969), we integrate (3.6.24) and obtain

$$(\widetilde{\lambda} + 2\widetilde{G})\varepsilon_b = p^e + \Pi(\mathbf{x}, t) \tag{3.6.26}$$

where $\Pi$ is a function of position and time that for every value of time, $t$, satisfies

$$\nabla^2 \Pi = 0. \tag{3.6.27}$$

When $\Pi = 0$ (see below), Equation (3.6.26) reduces to

$$\varepsilon_b = \frac{p^e}{\widetilde{\lambda} + 2\widetilde{G}}. \tag{3.6.28}$$

By inserting this expression into (3.6.22), we obtain

$$\nabla \cdot \left\{ \frac{k}{\mu} (\nabla p^e + \rho g \nabla z) \right\} = \left( n\beta + \frac{1}{\widetilde{\lambda} + 2\widetilde{G}} \right) \frac{\partial p^e}{\partial t}. \tag{3.6.29}$$

By comparing (3.6.29), with (3.3.9), noting the assumptions that underlie the latter, we may conclude that the compressibility coefficient of a porous medium, $\alpha$, defined by (3.2.10), is related to $\widetilde{\lambda}$ and $\widetilde{G}$ by

$$\alpha = \frac{1}{\widetilde{\lambda} + 2\widetilde{G}}. \tag{3.6.30}$$

By inserting (3.6.21) and (3.6.26), with $\Pi = 0$, into (3.6.13) rather than into (3.6.14), we obtain an equation that is comparable to (3.3.15).

In order to identify the conditions under which $\Pi = 0$, we assume, following Verruijt (1969), that *displacements occur only in the vertical direction*, i.e., in Cartesian coordinates, $w_z \neq 0$, $w_x = w_y = 0$ and that the total stress remains unchanged, i.e., $\sigma^e = 0$, and hence $\sigma_s^{*e} = p^e \mathbf{I}$. Then (3.6.19) reduces to

$$(\sigma_s^{*e})_{xx} = (\sigma_s^{*e})_{yy} = -\widetilde{\lambda} \frac{\partial w_z}{\partial z}, \quad (\sigma_s^{*e})_{xy} = (\sigma_s^{*e})_{yx} = 0,$$

$$(\sigma_s^{*e})_{xz} = (\sigma_s^{*e})_{zx} = -\widetilde{G} \frac{\partial w_z}{\partial x},$$

$$(\sigma_s^{*e})_{zz} = -(\widetilde{\lambda} + 2\widetilde{G}) \frac{\partial w_z}{\partial z}, \quad \varepsilon_b = \frac{\partial w_z}{\partial z}. \tag{3.6.31}$$

Hence

$$-(\sigma_s^{*e})_{zz} = p^e = (\widetilde{\lambda} + 2\widetilde{G}) \frac{\partial w_z}{\partial z} = (\widetilde{\lambda} + 2\widetilde{G}) \varepsilon_b. \tag{3.6.32}$$

By comparing (3.6.28) with (3.6.32), Verruijt (1969) concludes that $\Pi$ vanishes for the simplifying assumptions leading to the latter. In other words, the balance equations (3.3.15), (3.3.16) and similar equations that involve the concept of specific storativity, are valid only when the assumption of vertical displacement only is valid (or approximately so). In most cases of groundwater flow, this situation indeed prevails.

We note that while the complete three-dimensional model for determining the soil's displacements requires a simultaneous solution of the (water) mass balance equation, say (3.6.22), and the equilibrium equation, (3.6.23), together with appropriate constitutive equations, the model based on the assumption of vertical displacement only, enables us to decouple the two equations. Thus, we may first solve (3.6.29), or any similar equation, for $p^e$, and then use (3.6.28) to determine the distribution of $\varepsilon_b$ and from it, employing the last expression in (3.6.31), the distribution of $w_z$. Then, soil compaction, or land subsidence, $\delta(x, y, t)$ can be obtained from

$$\delta(x, y, t) = \int_{(B)} \frac{\partial w_z}{\partial z} \, dz = \int_{(B)} \varepsilon_b(x, y, z, t) \, dz$$

$$= \int_{(B)} \frac{p^e(x, y, z, t)}{\widetilde{\lambda} + 2\widetilde{G}} \, dz \qquad (3.6.33)$$

where $B$ denotes the thickness of the considered layer.

In engineering practice, the two-step approach outlined above is often performed such that in the first step, the solution for the pore pressure, $p$, averaged linear compressibilities are used, whereas in the second step, the calculation of the strains in the soil, a more realistic constitutive equation is used. Thus, one can incorporate nonlinear effects such as *creep*.

CHAPTER FOUR

# Modeling Two-Dimensional Flow in Aquifers

In this chapter we shall develop models that describe groundwater flow in confined, phreatic, and leaky aquifers on the basis of the essentially horizontal flow approximation ($\equiv$ *the hydraulic approach*) discussed in Section 1.5. Obviously, such models may be used only when this approximation is indeed justified.

For a confined and a leaky aquifer, the dependent variable is the average piezometric head, $\tilde{\phi} = \tilde{\phi}(x, y, t)$, as defined by (1.5.1). However, for the sake of simplicity, we shall usually use the symbol $\phi(x, y, t)$ to indicate $\tilde{\phi}(x, y, t)$. For a phreatic aquifer, we shall employ the symbol $h = h(x, y, t)$ to indicate the elevations of points on the phreatic surface above some datum level.

For each of the three types of aquifers, the governing (balance) equation (and also the initial and boundary conditions) can be obtained by the *control box approach* (Section 3.3), stipulating the assumption of essentially horizontal flow. The control box is a column of area $\delta x\, \delta y$ and height equal to the aquifer's thickness, $B'$, or $h$. A more rigorous way to achieve the same goal is to start from the three-dimensional model and integrate it over the aquifer's thickness. In doing so, we shall notice how conditions on the top and bottom surfaces that bound the aquifer (in the three-dimensional model) become sources (or sinks) of water in the integrated balance equation. Both approaches will be presented below.

## 4.1. Aquifer Storativity

We define the *storativity of a confined aquifer*, $S$, as the volume of water, $\Delta U_w$, released from storage (or added to it) per unit horizontal area, $A$, of an aquifer and per unit decline (or rise) of piezometric head, $\phi\ (\equiv \tilde{\phi})$, i.e.,

$$S = \Delta U_w / A \Delta \phi \tag{4.1.1}$$

(Figure 4.1a) where $S$ is dimensionless. From the discussion in Section 3.2 we know that this storage is due to the elastic behavior of the water and of the solid matrix. Hence, $S$ is related to the specific storativity $S_0$ by

$$S(x, y) = \int_{b_1}^{b_2} S_0(x, y, z)\, dz \tag{4.1.2}$$

where $b_2$ and $b_1$ denote the elevations of the aquifer's top and bottom. The volume

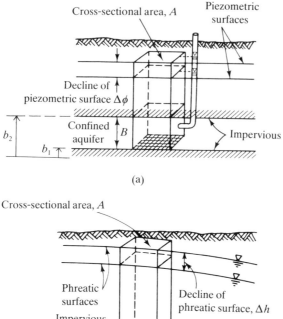

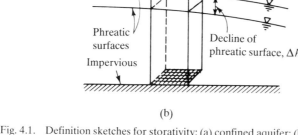

Fig. 4.1. Definition sketches for storativity: (a) confined aquifer; (b) phreatic aquifer.

of aquifer from which the volume of water, $\Delta U_w$, is released is $A \times B$ where $B$ is the aquifer's thickness (Figure 4.1a).

In (4.1.2) we have overlooked the possible change in the aquifer's thickness, $b_2 - b_1$, as $\phi$ varies. This makes (4.2.4) incomplete.

We can also define a storage coefficient for a phreatic aquifer. Consider a horizontal area, $A$, of a phreatic aquifer (Figure 4.1b). The volume of water stored in a phreatic aquifer is indicated by its water table. If, as a result of the pumping from the aquifer, a volume of water leaves this area in excess of the volume of water entering it, the water table will drop. We may define the storativity of a phreatic aquifer in the same way as the storativity of a confined aquifer was defined above, except that here the drop, $\Delta h$, is of the water table (Figure 4.1b)

$$S = \Delta U_w / A \Delta h \qquad (4.1.3)$$

In spite of the similarity in the two definitions, the storativity in each of the two types of aquifer is due to different reasons. In a confined aquifer, it is the outcome of water and matrix compressibility. In a phreatic aquifer, water is mostly drained from the pore space between the initial and final positions of the phreatic surface. The storativity of a phreatic aquifer is, therefore, sometimes referred to as *specific yield*, $S_y$; it gives the yield of an aquifer per unit area and unit drop of the water table.

Recalling that the water table is actually an approximate concept, we understand that water is actually being drained from the entire column of soil up to the ground surface. Bear (1972, p. 485) shows that when the soil is homogeneous and the fluctuating water table is sufficiently deep, the above definition for specific yield still holds (see Section 5.1).

One should be careful not to identify the specific yield with the porosity of a phreatic aquifer. As water is being drained from the interstices of the soil, the drainage is never a complete one. A certain amount of water is retained in the soil against gravity by capillary forces. After drainage has stopped, the volume of water retained in an aquifer per unit (horizontal) area and unit drop of the water table is called *specific retention*, $S_r$. Thus,

$$S_y + S_r = n \tag{4.1.4}$$

For this reason $S_y (< n)$ is sometimes called *effective porosity*. Here, again, one should note that we have been referring to the approximate concept of a water table. However, for a homogeneous soil and a sufficiently deep water table, the above definition for $S_y$ holds.

Figure 4.2 shows typical relationships between $S_y$, $S_r$, and particle size.

When drainage occurs, it takes time for the water to flow, partly under unsaturated conditions, out of the soil volume between the two positions of a water table, at $t$ and at $t + \Delta t$. This is especially true if the lowering of the water table is rapid. Under such conditions, the specific yield becomes time-dependent, gradually approaching its ultimate value. We often refer to this phenomenon as *delayed yield* (See Subsection 4.2.4). When the water level is rising or falling

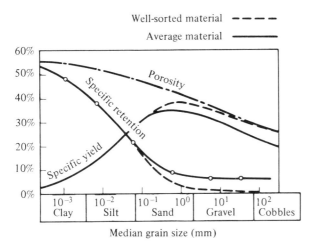

Fig. 4.2. Relationship between specific yield and grain size (*from Conkling et al., 1934, as modified by Davis and DeWiest, 1966*).

slowly, the changes in moisture distribution have time to adjust continuously and the time lag vanishes.

When the water table is lowered, the pressure drops throughout the aquifer below it. In principle, this pressure drop causes water to be released from storage in the aquifer, also due to the elastic properties of the aquifer and the water. However, when we calculate the total volume of water released from storage in the aquifer per unit area and unit decline of head: $(\Delta U_w)_1 = S_0 h$ due to the elastic storage and $(\Delta U_w)_2 = S_y$ due to the actual drainage of water from the pore space, we have $S_0 h \ll S_y$ so that $(\Delta U_w)_1$ can be neglected (see also Subsection 4.2.4).

Typical values of $S$ in a confined aquifer are of the order of $10^{-4}$–$10^{-6}$, roughly 40% of which result from the expansion of the water and 60% from the compression of the medium. In a sandy phreatic aquifer, we may have $S_0$ of the order $10^{-7}$ cm$^{-1}$, whereas $S_y$ may be 20–30%.

We shall return to the definition of aquifer storativity, both for a confined aquifer and for a phreatic one in Section 4.2, where the aquifer equations will be derived by averaging the three-dimensional flow equations along the vertical.

## 4.2. Fundamental Continuity Equations

In this section, the fundamental continuity equations for flow in confined, leaky, and phreatic aquifers are developed in two ways: first by employing the control box approach for the particular type of aquifer under consideration, and then by integrating the three-dimensional continuity equation over the thickness of the aquifer, taking into account the boundary conditions on its top and bottom surfaces.

For the sake of brevity, we shall consider first a leaky-confined aquifer and then derive the equation for a confined one as a special case. Similarly, we shall consider a leaky-phreatic aquifer, with a phreatic one as a special case.

### 4.2.1. LEAKY CONFINED AQUIFER

A leaky-confined aquifer (Subsection 1.1.3) is bounded from above and/or below by a semipervious layer through which water may leak into or out of the aquifer. We *assume that the flow in the aquifer itself is essentially horizontal*, while it is vertical in the semipervious, relatively thin bounding layers. This assumption is valid when the constrast between the permeabilities in the aquifer and in a semipervious layer (= aquitard) is at least one order of magnitude. We shall first ignore storage in the semipervious layers and then show how this storage is taken into account.

Consider the case of the inhomogeneous leaky-confined aquifer shown in Figure 4.3a. Figure 4.3b shows the control box for which the following (volumetric)

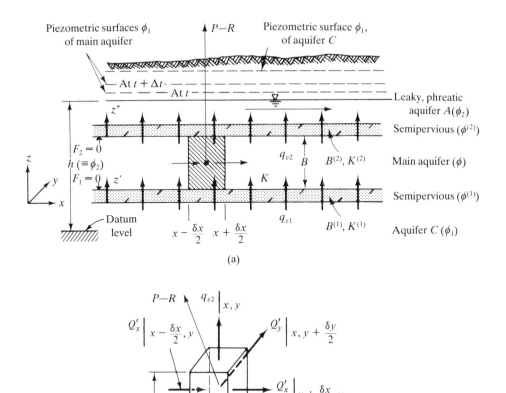

Fig. 4.3. Flow in a leaky-confined aquifer.

water balance (actually, mass balance at constant density) holds

$$\delta t \{ \delta y (Q'_x|_{x-\delta x/2, y} - Q'_x|_{x+\delta x/2, y}) + \delta x (Q'_y|_{x, y-\delta y/2} - Q'_y|_{x, y+\delta y/2}) +$$
$$+ \delta x\, \delta y (q_{v1} - q_{v2}) + \delta x\, \delta y (R - P) \}$$
$$= S\, \delta x\, \delta y (\phi|_{t+\Delta t} - \phi|_t) \quad (4.2.1)$$

where $\phi(x, y, t)$ is the piezometric head in the considered aquifer, $q_{v2}$ and $q_{v1}$ are vertical leakage rates (dims. L/T) through the top and bottom semipervious layers, respectively, and $R(x, y, t)$ and $P(x, y, t)$ are distributed rates of artificial recharge and pumping, respectively (dims. L/T). With $\phi_2(x, y, t)$ and $\phi_1(x, y, t)$ denoting the piezometric heads above the upper semipervious layer (i.e., in aquifer $A$) and below the lower one (i.e., in aquifer $C$), respectively, and assuming that the piezometric head distribution is always linear across each aquitard, the leakage rates are expressed by

$$q_{v2} = K^{(2)} \frac{\phi - \phi_2}{B^{(2)}} = \frac{\phi - \phi_2}{c^{(2)}} ; \qquad q_{v1} = K^{(1)} \frac{\phi_1 - \phi}{B^{(1)}} = \frac{\phi_1 - \phi}{c^{(1)}} \qquad (4.2.2)$$

where $c^{(1)} = B^{(1)}/K^{(1)}$ and $c^{(2)} = B^{(2)}/K^{(2)}$ are the resistances (dims. T) of the semipervious layers. Hantush (1949, 1964) calls $K^{(i)}/B^{(i)} = 1/c^{(i)}$ the *coefficient of leakage*. It is defined as the rate of flow across a unit (horizontal) area of a semipervious layer into (or out of) an aquifer under one unit of head difference across this layer. Dividing both sides of (4.2.1) by $\delta x \, \delta y$ in order to obtain a balance per unit area and per unit time, and letting $\delta x$, $\delta y$, $\delta t \to 0$, in order to obtain a balance 'at a point', i.e., in its close vicinity, we obtain

$$-\nabla' \cdot \mathbf{Q}' + q_{v1} - q_{v2} + R - P = S \frac{\partial \phi}{\partial t}. \qquad (4.2.3)$$

With $\mathbf{Q}' = -\mathbf{T} \cdot \nabla \phi$, and $q_{v1}$, $q_{v2}$ expressed by (4.2.2), Equation (4.2.3) becomes

$$\nabla' \cdot (\mathbf{T} \cdot \nabla' \phi) + \frac{\phi_1 - \phi}{c^{(1)}} - \frac{\phi - \phi_2}{c^{(2)}} + R - P = S \frac{\partial \phi}{\partial t}. \qquad (4.2.4)$$

For an inhomogeneous isotropic aquifer, (4.2.3) takes the form

$$\frac{\partial}{\partial x}\left(T \frac{\partial \phi}{\partial x}\right) + \frac{\partial}{\partial y}\left(T \frac{\partial \phi}{\partial y}\right) + \frac{\phi_1 - \phi}{c^{(1)}}$$

$$- \frac{\phi - \phi_2}{c^{(2)}} + R - P = S \frac{\partial \phi}{\partial t}. \qquad (4.2.5)$$

For a homogeneous isotropic aquifer, (4.2.5) reduces to

$$\frac{\partial^2 \phi}{\partial x^2} + \frac{\partial^2 \phi}{\partial y^2} + \frac{\phi_1 - \phi}{\lambda^{(1)2}} - \frac{\phi - \phi_2}{\lambda^{(2)2}} + R - P = S \frac{\partial \phi}{\partial t} \qquad (4.2.6)$$

where $\lambda^{(i)} = (Tc^{(i)})^{1/2}$, $i = 1, 2$, is another leaky aquifer parameter, called the *leakage factor*, that determines the areal distribution of the leakage. Equation (4.2.4) is the basic continuity equation describing groundwater flow in a leaky-confined aquifer, neglecting storage in the semipervious layers. The effect of storage in the semipervious layers will be considered in Subsection 4.2.2.

The symbols $P(x, y, t)$ and $R(xy, y, t)$ represent either distributed or point sources or sinks (e.g., recharging and pumping wells, respectively). In the latter case, we may replace these terms by $\Sigma_{(i)} P^*(x_i, y_i, t)\delta(x - x_i, y - y_i)$ and $\Sigma_{(i)} R^*(x_j, y_j, t)(x - x_j, y - y_j)$, where $P^*(x_i, y_i, t)$ and $R^*(x_j, y_j, t)$ denote pumping rates (dims. L³/T), respectively, at points $(x_i, y_i)$ and $(x_j, y_j)$, say, with $i = 1, 2, \ldots, m_p$ and $j = 1, 2, \ldots, m_R$. The symbol $\delta(x - x_i, y - y_i)$ denotes the *Dirac delta function*.

## MODELING TWO-DIMENSIONAL FLOW IN AQUIFERS

Let us now develop the same equation by integrating the three-dimensional continuity equation (3.3.13), over the aquifer's vertical thickness, with $\phi^*$ approximated by $\phi$ over the aquifer's vertical thickness. In the absence of sources and sinks (to be added later), and recalling all the assumptions underlying its development, we obtain

$$\int_{b_1(x,y)}^{b_2(x,y)} \left( \nabla \cdot \mathbf{q} - S_0 \frac{\partial \phi}{\partial t} \right) dz = 0 \qquad (4.2.7)$$

where $b_1(x, y)$ and $b_2(x, y)$ denote the elevations of the stationary bottom and top bounding surfaces, with $B(x, y) = b_2(x, y) - b_1(x, y)$. For horizontal surfaces, $b_1(x, y)$ and $b_2(x, y)$ are constants.

For the sake of simplicity, we assume that the solid's velocity is very small so that $\mathbf{q}_r \simeq \mathbf{q}$. The effect of the solid's velocity will be taken into account in Section 4.5 below. We shall make use of the Leibnitz rule (2.2.2) written in the forms

(a) for any vector $\mathbf{A}$

$$\int_{b_1(x,y,t)}^{b_2(x,y,t)} \nabla \cdot \mathbf{A} \, dz = \nabla' \cdot B\tilde{\mathbf{A}}' + \mathbf{A}|_{b_2} \cdot \nabla(z - b_2) - \mathbf{A}|_{b_1} \cdot \nabla(z - b_1), \qquad (4.2.8)$$

(b) for any scalar $\phi$

$$\int_{b_1(x,y,t)}^{b_2(x,y,t)} \frac{\partial \phi}{\partial t} dz = \frac{\partial}{\partial t} B\tilde{\phi} - \phi|_{b_2} \frac{\partial b_2}{\partial t} + \phi|_{b_1} \frac{\partial b_1}{\partial t} \qquad (4.2.9)$$

where

$$\tilde{\mathbf{A}}' = \frac{1}{B} \int_{b_1}^{b_2} \mathbf{A}' \, dz, \quad \mathbf{A}' = A_x \mathbf{1x} + A_y \mathbf{1y},$$

$$\nabla' \cdot \mathbf{A}' = \frac{\partial A_x}{\partial x} \mathbf{1x} + \frac{\partial A_y}{\partial y} \mathbf{1y}.$$

Applying these rules to (4.2.7), for the case of fixed bounding surfaces, $b_1 = b_1(x, y)$, $b_2 = b_2(x, y)$, we obtain

$$\nabla' \cdot B\tilde{\mathbf{q}}' + \mathbf{q}|_{b_2} \cdot \nabla(z - b_2) - \mathbf{q}|_{b_1} \cdot \nabla(z - b_1) = S \frac{\partial \tilde{\phi}}{\partial t} \qquad (4.2.10)$$

where $S = S_0 B$ is the aquifer's storativity. However, we may remove the constraint of fixed bounding surfaces and assume instead $\tilde{\phi} \simeq \phi|_{b_1} \simeq \phi|_{b_2}$, i.e., vertical equipotentials (essentially horizontal flow). Then

$$\tilde{\phi}\frac{\partial B}{\partial t} - \phi|_{b_2}\frac{\partial b_2}{\partial t} + \phi|_{b_1}\frac{\partial b_1}{\partial t} \simeq 0.$$

With this approximation, we also obtain (4.2.10).

The last two terms on the left-hand side of (4.2.10) denote flux components normal to the boundaries, recalling that the equations of the latter can be represented, symbolically, by

$$F_1 = F_1(x, y, z, t) = z - b_1(x, y, t) = 0,$$
$$F_2 = F_2(x, y, z, t) = z - b_2(x, y, t) = 0.$$

With these symbols, we may rewrite (4.2.10) in the form

$$-\nabla' \cdot \mathbf{Q}' - \mathbf{q}|_{F_2} \cdot \nabla F_2 + \mathbf{q}|_{F_1} \cdot \nabla F_1 = S\frac{\partial \tilde{\phi}}{\partial t} \qquad (4.2.11)$$

where $\mathbf{Q}' = B\tilde{\mathbf{q}}'$ and $S = S_0 B$.

Let $q_{l_2} = (\phi - \phi_2)/c^{(2)}$ denote the rate of leakage out of the aquifer through its fixed semipervious (not necessarily horizontal) ceiling described by $F_2(x, y, z, t) = 0$.

The condition stating continuity of flux across this boundary is (3.4.15), where $\nu_2 = \nabla F_2/|\nabla F_2|$ and $f_3(\mathbf{x}, t) = q_{l_2}(\mathbf{x}, t)$. We recall that (3.4.15) was derived for a boundary, whether stationary or moving, that is a material surface with respect to the solids. If, in addition, we *assume* $\mathbf{q} \gg n\mathbf{V}_s$, then $\mathbf{q}_r \simeq \mathbf{q}$ the continuity of flux across the boundary is expressed as

$$\mathbf{q}|_{F_2} \cdot \nabla F_2 = |\nabla F_2|q_{l_2} = |\nabla F_2|(\phi - \phi_2)/c^{(2)}.$$

A similar expression can be written for the lower boundary, $F_1(x, y, z, t) = 0$. When the two expressions are inserted into (4.2.11), we obtain

$$-\nabla' \cdot \mathbf{Q}' - |\nabla F_2|\frac{\tilde{\phi} - \phi_2}{c^{(2)}} - |\nabla F_1|\frac{\tilde{\phi} - \phi_1}{c^{(1)}} = S\frac{\partial \tilde{\phi}}{\partial t}. \qquad (4.2.12)$$

This equation is more general than (4.2.4), as it allows for a nonhorizontal bounding surface. For horizontal surfaces, with constant $b_1$ and $b_2$ and $|\nabla F_1| = |\nabla F_2| = 1$, Equation (4.2.12) reduces to (4.2.4).

To summarize, in deriving (4.2.12) we have assumed that (i) $u = 0$, or $\mathbf{u} \neq 0$, (ii) $\mathbf{u} \cdot \boldsymbol{\nu} = \mathbf{V}_s \cdot \boldsymbol{\nu}$, and (iii) $\mathbf{V}_s \simeq 0$, $\mathbf{q}_r \simeq \mathbf{q}$.

It is of interest to note that the terms expressing leakage through both $F_1 = 0$ and $F_2 = 0$, which appear as boundary conditions in a three-dimensional flow model become sink/source terms in the integrated, two-dimensional, model that appear in the balance equation. At the same time, the top and bottom surfaces do not serve anymore as boundaries of the flow domain. It is always possible to add a sink term on the left-hand side of (3.3.13) such that an integrated sink term will appear in (4.2.11) and (4.2.12). As in (4.2.3), this can take the form of a term $R - P$ added on the left-hand side of (4.2.12) to represent a net source.

## 4.2.2. EFFECT OF STORAGE IN THE SEMIPERMEABLE LAYER

In the discussion in Subsection 4.2.1 above, we assumed that the semipervious layers have zero storativity and, hence, any change in the piezometric head in the adjacent aquifers propogates instantaneously within these layers. We could, therefore, assume that a linear distribution of heads always exists in these layers. In reality, changes in water storage do take place in the semipervious layers.

In unsteady flow, as the piezometric heads in the aquifers that lie above and below each semipervious layer vary, continuous changes are also produced in the distribution of the piezometric head within each semipervious layer. Figure 4.4 shows how an instantaneous stepwise drop in the piezometric head, $\phi$, from $\phi|_{t=0^-}$ to $\phi|_{t=0^+}$ in an aquifer, even when $\phi_1$ in the aquifer underlying the semipervious layer remains unchanged, produces a gradual change in the piezometric head distribution, $\phi^{(1)}(z', t)$, within that layer. We note that the gradient $\partial \phi^{(1)}/\partial z'$ varies with both time and elevation along the aquitard's thickness, $B^{(1)}$. At the same time, the quantity of water stored (elastically) within each unit volume of porous medium along $B^{(1)}$ also varies as $\phi^{(1)} = \phi^{(1)}(z', t)$ varies. For horizontal top and bottom layers, the rates of leakage entering and leaving the aquifer shown in Figure 4.3a are expressed by

$$q_{v1}|_{F_1} = -K^{(1)} \frac{\partial \phi^{(1)}}{\partial z'}\bigg|_{F_1} ; \quad q_{v2}|_{F_2} = -K^{(2)} \frac{\partial \phi^{(2)}}{\partial z''}\bigg|_{F_2} \qquad (4.2.13)$$

where $\phi^{(2)}(z'', t)$ denotes the piezometric head in the upper aquitard and the gradients are taken at the interfaces between the main aquifer and the top and bottom aquitards.

We note in Figure 4.4 that, whereas initially the flow throughout $B^{(1)}$ is everywhere downward after some time, say $t_3$, (assuming that $\phi|_{t=0^+}$ does not vary), the flow in the lower part of the aquitard changes its direction. Eventually it will be upward throughout the entire aquitard. Employing the definition of specific

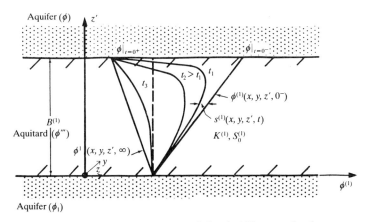

Fig. 4.4. Changes in piezometric head within an aquitard.

storativity, $S_0$ (Section 3.2), the total volume of water, $\Delta U_w$, released from storage in the aquitard (per unit horizontal area) up to time $t$ is given by

$$\Delta U_w = \int_{z'=0}^{B^{(1)}} S_0^{(1)} \{\phi^{(1)}(x, y, z', t) - \phi^{(1)}(x, y, z', 0)\} \, dz'. \tag{4.2.14}$$

Depending on the aquitard's permeability, $K^{(1)}$ (actually on the ratio $S_0^{(1)}/K^{(1)}$), it will take some time for this volume to be released from storage in the aquitard. Accordingly, the distribution of the piezometric head within an aquitard cannot respond instantaneously to head changes in the adjacent aquifers. The entire picture is more complicated when the piezometric heads in the adjacent aquifers continuously vary. We refer to this phenomenon as *delayed storage*.

With (4.2.13), Equation (4.2.11), with $\phi = \tilde{\phi}$, becomes

$$-\nabla' \cdot \mathbf{Q}' - K^{(1)} \left.\frac{\partial \phi^{(1)}}{\partial z'}\right|_{F_1} + K^{(2)} \left.\frac{\partial \phi^{(2)}}{\partial z''}\right|_{F_2} = S \frac{\partial \phi}{\partial t} \tag{4.2.15}$$

where $\mathbf{Q}' = -\mathbf{T} \cdot \nabla' \phi$. We note that in (4.2.15) we now have two additional dependent variables $\phi^{(1)}(z', t)$ and $\phi''(z'', t)$. To solve for these variables, we have to construct a model that describes the (vertical) flow within each aquitard.

To demonstrate the construction and solution of such a model and the effect of *delayed storage*, let Figure 4.4 represent the lower aquitard of Figure 4.3a. Assume that $K^{(1)} \ll K$, so that the flow in the semipervious layer is essentially vertical. Let a steady flow be established through the layer and then assume that a stepwise reduction of the head is produced in the main aquifer by pumping. After a sufficiently long time, a new steady state will be established, with a linear distribution of the head. However, during this period, the reduction of the head in the semipervious layer will lag behind that corresponding to the new steady state. The problem of determining $\phi^{(1)}(x, y, z', t)$ for $K^{(1)} = $ constant, is stated in the domain $0 \leq z' \leq B^{(1)}$, by the following partial differential equation and initial and boundary conditions

$$K^{(1)} \frac{\partial^2 s^{(1)}}{\partial z'^2} = S_0^{(1)} \frac{\partial s^{(1)}}{\partial t}, \quad s^{(1)}(x, y, z', 0) = 0, \quad t \leq 0$$

$$0 \leq z' \leq B^{(1)}, \quad s^{(1)}(x, y, 0, t) = 0, \quad t > 0$$

$$s^{(1)}(x, y, B^{(1)}, t) = \begin{cases} 0, & t \leq 0 \\ H_0, & t > 0 \end{cases} \tag{4.2.16}$$

where we have switched to the use of drawdown

$$s^{(1)}(x, y, z' \cdot t) = \phi^{(1)}(x, y, z', 0^-) - \phi^{(1)}(x, y, z', t) \quad \text{and}$$
$$H_0 = \phi|_{t=0^-} - \phi|_{t=0^+}.$$

# MODELING TWO-DIMENSIONAL FLOW IN AQUIFERS

The solution of this problem is given by (Carslaw and Jaeger, 1959, p. 310)

$$\frac{s^{(1)}(x, y, z', t)}{H_0} = \sum_{n=0}^{\infty} \left[ \text{erfc} \frac{(2n+1)B^{(1)} - z'}{2(K^{(1)}t/S_0^{(1)})^{1/2}} - \text{erfc} \frac{(2n+1)B^{(1)} + z'}{2(K^{(1)}t/S_0^{(1)})^{1/2}} \right] \quad (4.2.17)$$

Figure 4.5 (Bredehoeft and Pinder, 1970) graphically shows this solution.

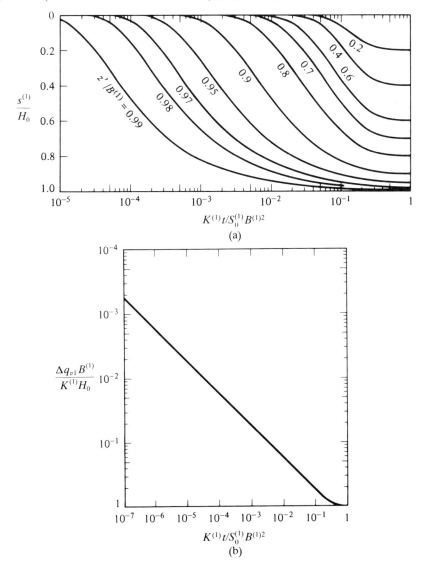

Fig. 4.5. Graphical representation of (a) Equation (4.1.17) and (b) Equation (4.1.18) (Bredehoeft and Pinder, 1970).

From $s^{(1)} = s^{(1)}(x, y, z', t)$ one can determine the increase in the rate of flow, $q_{v1}$, into the pumped aquifer produced by the stepwise reduction in head

$$\Delta q_{v1} = K^{(1)} \left. \frac{\partial s^{(1)}}{\partial z'} \right|_{z' = B^{(1)}}$$

$$= \frac{K^{(1)} H_0}{B^{(1)}(\pi K^{(1)} t/B^{(1)2} S_0^{(1)})} \left\{ 1 + 2 \sum_{n=1}^{\infty} \exp(-n^2/(K^{(1)} t/B^{(1)2} S_0^{(1)})) \right\}. \quad (4.2.18)$$

This flow is plotted in Figure 4.5b. From this figure, it follows that a long time may elapse before steady flow is re-established in the semipervious layer.

Once the distributions $\phi^{(1)}(z', t)$ and $\phi^{(2)}(z'', t)$ have been determined, the leakage terms in (4.2.15) can be calculated, leaving $\phi$ as the single variable in it.

Another approach is to reduce the set of partial differential equations (4.2.15) and (4.2.16) to a single *integro-differential equation*. Let us demonstrate this approach for the leaky-confined aquifer system shown in Figure 4.6.

The model that describes the drawdown, $s(x, y, t) = \phi(x, y, 0) - \phi(x, y, t)$ in this pumped leaky-confined aquifer is given by

$$\nabla' \cdot (\mathbf{T} \cdot \nabla s) + K^{(2)} \left. \frac{\partial s^{(2)}}{\partial z''} \right|_{F_2} - P = S \frac{\partial s}{\partial t} \quad (4.2.19)$$

where $s(x, y, 0) = 0$, and $\nabla' \cdot \mathbf{T} \cdot \nabla' \phi(x, y, 0) + K^{(2)} \partial \phi^{(2)}(x, y, 0)/\partial z''|_{F_2} = 0$, i.e., initially, without pumping, the aquifer is in a steady state. Equation (4.2.19) has to be supplemented by the appropriate initial and boundary conditions in terms of $s(x, y, t)$.

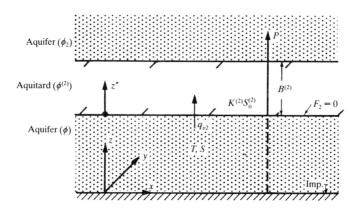

Fig. 4.6. A leaky-confined aquifer.

The model for flow in the (homogeneous) aquitard, in terms of the drawdown $s^{(1)}(x, y, z'', t)$ is given by

$$K^{(2)} \frac{\partial^2 s^{(2)}}{\partial z''^2} = S_0^{(2)} \frac{\partial s^{(2)}}{\partial t} \quad \text{in } 0 \leq z'' \leq B^{(2)},$$

$$s^{(2)}(x, y, 0, t) = s(x, y, t); \qquad s^{(2)}(x, y, z'', 0) = 0. \tag{4.2.20}$$

We shall assume $s^{(2)}(x, y, B^{(2)}, t) = 0$.

Herrera and Rodarte (1973), following Neuman and Witherspoon (1969), show that the system of Equations (4.2.19) and (4.2.20) is equivalent to the single integrodifferential equation

$$\nabla \cdot (\mathbf{T} \cdot \nabla s) - \frac{K^{(2)}}{B^{(2)}} \int_0^t \frac{\partial s(x, y, t - \tau)}{\partial t} f(K^2 \tau / S_0^{(2)} B^{(2)}) \, d\tau -$$

$$- P(x, y, t) = S \frac{\partial s}{\partial t} \tag{4.2.21}$$

where the *memory function*, $f(t')$ is expressed by

$$f(t') = 1 + 2 \sum_{m=1}^{\infty} \exp(-m^2 \pi^2 t')$$

$$= \frac{1}{(\pi t')^{1/2}} \left\{ 1 + 2 \sum_{m=1}^{\infty} \exp\left(-\frac{m^2}{t'}\right) \right\} \tag{4.2.22}$$

By solving (4.2.21), we obtain $s = s(x, y, t)$, without having to solve separately for the drop in piezometric head, $s^{(2)}(x, y, z'', t)$ along the impervious layer. We say that (4.2.21) 'has a memory' because the integral depends on the history of the drawdown, $s$, in the aquifer. This procedure does not introduce any approximation.

There is no difficulty in applying the same approach to the case of the two aquitards shown in Figure 4.3. In general, $s^{(2)}(x, y, B^{(2)}, t) \neq 0$ as it depends on what happens in the overlying aquifer. This serves to couple the two aquifers; a simultaneous solution is required.

### 4.2.3. CONFINED AQUIFER

The flow equation for a confined aquifer can easily be obtained from the one describing flow in a leaky-confined aquifer. For a confined aquifer, the two aquitards in Figure 4.3 are replaced by impervious layers. Also $q_{v1} = q_{v2} = 0$.

When the control box approach is employed, the balance represented by (4.2.1) is valid, except that $q_{v1} = q_{v2} = 0$. Hence, (4.2.4) reduces to

$$\nabla' \cdot (\mathbf{T} \cdot \nabla'\phi) + R - P = S \frac{\partial \phi}{\partial t} \qquad (4.2.23)$$

where $\phi = \phi(x, y, t) \equiv \tilde{\phi}(x, y, t)$.

This is the *basic balance equation for a confined aquifer*.

For an *inhomogeneous anisotropic aquifer*, with $T_x(x, y) \neq T_y(x, y)$, $x$, $y$ principal directions, (4.2.23) takes the form

$$\frac{\partial}{\partial x}\left(T_x \frac{\partial \phi}{\partial x}\right) + \frac{\partial}{\partial y}\left(T_y \frac{\partial \phi}{\partial y}\right) + R - P = S \frac{\partial \phi}{\partial t}. \qquad (4.2.24)$$

For a *homogeneous isotropic aquifer*, (4.2.23) takes the form

$$T\left(\frac{\partial^2 \phi}{\partial x^2} + \frac{\partial^2 \phi}{\partial y^2}\right) + R - P \equiv T\nabla^2 \phi + R - P = S \frac{\partial \phi}{\partial t}. \qquad (4.2.25)$$

For steady flow, *or* when the elastic storativity may be neglected, the right-hand side of these balance equations vanishes. We obtain

$$T\left(\frac{\partial^2 \phi}{\partial x^2} + \frac{\partial^2 \phi}{\partial y^2}\right) + R - P = 0. \qquad (4.2.25a)$$

In the absence of sources and sinks, (4.2.25a) reduces to

$$\frac{\partial^2 \phi}{\partial x^2} + \frac{\partial^2 \phi}{\partial y^2} \equiv \nabla^2 \phi = 0 \qquad (4.2.25b)$$

which is the *Laplace equation* in the $xy$ plane. We recall that in (4.2.23) through (4.2.25b), the symbol $\phi$ represents the average $\tilde{\phi} = \tilde{\phi}(x, y, t)$, or $\tilde{\phi} = \tilde{\phi}(x, y)$.

When the approach of integrating along the vertical is employed, we may start from the integrated equation (4.2.11). Along the impervious bounding surfaces, $F_1 = 0$ and $F_2 = 0$, and assuming $\mathbf{V}_s \simeq 0$, so that $\mathbf{q}_r \equiv \mathbf{q}$, by (3.4.15), the boundary condition of no flux, i.e., $f_3(\mathbf{x}, t) = 0$, takes the form

$$\begin{aligned}\mathbf{q}|_{F_1} \cdot \nabla F_1 &= 0 \quad \text{on } F_1 = 0, \\ \mathbf{q}|_{F_2} \cdot \nabla F_2 &= 0 \quad \text{on } F_2 = 0.\end{aligned} \qquad (4.2.26)$$

Hence (4.2.11), after introducing the source term, $R - P$, reduces to (4.2.23). We recall that (4.2.10) and (4.2.11) are also applicable to nonstationary top and bottom bounding surfaces if we assume that $\tilde{\phi} \simeq \phi|_{F_1} \simeq \phi|_{F_2}$, i.e. we assume vertical equipotentials.

### 4.2.4. LEAKY-PHREATIC AQUIFER

We shall first derive the balance equation by considering a control box in a leaky-phreatic aquifer (Figure 4.7). The bottom elevation, denoted by $\eta = \eta(x, y)$, is not necessarily constant. In addition to groundwater flow into and out of the box and leakage from an underlying aquifer (piezometric head $\phi$, measured from the same datum level as $h$), we also have natural replenishment (from precipitation), $N = N(x, y, t)$, artificial recharge, $R = R(x, y, t)$ and pumping, $P = P(x, y, t)$. Return flow from excess irrigation may also be added as a separate term, or incorporated in either $N$ or $R$. All these inputs and outputs may take the form of distributed sources and sinks, or of point ones (as in the comment following (4.2.6)).

Based on the Dupuit assumption of horizontal flow in the aquifer and leakage normal to the semipervious layer (Section 2.3), the balance of water volume for the control box shown in Figure 4.7, takes the form

$$\delta t \{ \delta y (Q'_x|_{x-\delta x/2, y} - Q'_x|_{x+\delta x/2, y}) +$$
$$+ \delta x (Q'_y|_{x, y-\delta y/2} - Q'_y|_{x, y+\delta y/2}) +$$
$$+ (R + N - P)\, \delta x\, \delta y + q_v\, \delta x\, \delta y \}$$
$$= S_y\, \delta x\, \delta y (h|_{t+\Delta t} - h|_t) \qquad (4.2.27)$$

where we have neglected the elastic storativity, as the storativity due to drainage from the pore space is much larger than that resulting from the elasticity of the water and the solid matrix: $S_y \gg S_0(h - \eta)$ (Bear, 1972, p. 376). In (4.2.27), $S_y$ is the specific yield of the phreatic aquifer. Expressing $\mathbf{Q}'$ by (2.3.12), and following the usual procedure of dividing both sides of (4.2.27) by $\delta x\, \delta y\, \delta t$ and letting

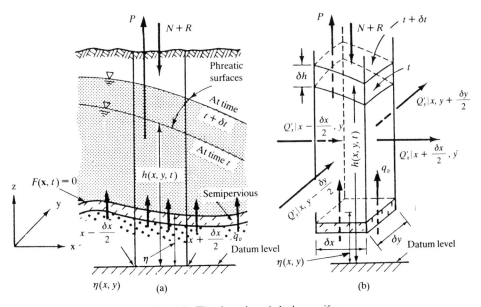

Fig. 4.7. Flow in a phreatic leaky-aquifer.

$\delta x$, $\delta y$, $\delta t \to 0$, we obtain for an inhomogeneous isotropic aquifer, with $K = K(x, y)$

$$\frac{\partial}{\partial x}\left\{K(h-\eta)\frac{\partial h}{\partial x}\right\} + \frac{\partial}{\partial y}\left\{K(h-\eta)\frac{\partial h}{\partial y}\right\} + R + N - p + q_v$$

$$= S_y \frac{\partial h}{\partial t}. \tag{4.2.28}$$

The rate of vertical leakage, $q_v$, can be expressed by employing (4.2.2), i.e., $q_v = (\phi - h)/c^{(1)}$, where $c^{(1)} =$ is the coefficient of leakage of the semipervious layer. Accordingly, (4.2.28) can be rewritten in the form

$$\frac{\partial}{\partial x}\left\{K(h-\eta)\frac{\partial h}{\partial x}\right\} + \frac{\partial}{\partial y}\left\{K(h-\eta)\frac{\partial h}{\partial y}\right\} + R + N - p + \frac{\phi - h}{c^{(1)}}$$

$$= S_y \frac{\partial h}{\partial t}. \tag{4.2.29}$$

This is the *basic continuity equation for water flow in a leaky-phreatic aquifer*. It is often called the *Boussinesq equation*.

For a homogeneous aquifer, $K = $ const, we obtain

$$K\left\{\frac{\partial}{\partial x}(h-\eta)\frac{\partial h}{\partial x} + \frac{\partial}{\partial y}(h-\eta)\frac{\partial h}{\partial y}\right\} + R + N - P + \frac{\phi - h}{c^{(1)}}$$

$$= S_y \frac{\partial h}{\partial t}. \tag{4.2.30}$$

As in the case of a leaky-confined aquifer, in order to develop the balance equation for a leaky-phreatic aquifer by integration, we start by integrating the three-dimensional balance equation (3.3.13) over the vertical

$$\int_{\eta(x,y)}^{h(x,y,t)} \left(\nabla \cdot \mathbf{q} + S_0 \frac{\partial \phi}{\partial t}\right) dz = 0. \tag{4.2.31}$$

By employing the modified forms of the Leibnitz rule, (4.2.8) and (4.2.9), we obtain for $\mathbf{q} \simeq \mathbf{q}_r$

$$\nabla' \cdot (h-\eta)\tilde{\mathbf{q}}' + \mathbf{q}|_h \cdot \nabla(z-h) - \mathbf{q}|_\eta \cdot \nabla(z-\eta) +$$

$$+ S_0(h-\eta)\frac{\partial h}{\partial t} = 0 \tag{4.2.32}$$

where we have assumed

$$\tilde{\phi} \simeq \phi|_\eta \simeq \phi|_h \equiv h. \tag{4.2.33}$$

For the third term on the right-hand side of (4.2.32), that expresses the upward leakage normal to the semipervious layer, we use an expression similar to the third term in (4.2.12), i.e.,

$$\mathbf{q}|_\eta \cdot \nabla(z - \eta) = \mathbf{q}|_\eta \cdot \nabla F = |\nabla F|(\phi - h)/c^{(1)}. \tag{4.2.34}$$

The second term on the left-hand side of (4.2.32) expresses the rate of accretion (from all sources) at the phreatic surface $F = z - h(x, y, t) = 0$. To obtain this expression, we refer to the phreatic surface boundary condition (3.4.25), from which, replacing $\mathbf{N}$ by $\mathbf{R} + \mathbf{N} - \mathbf{P}$, we obtain

$$\mathbf{q}|_h \cdot \nabla F = (\mathbf{R} + \mathbf{N} - \mathbf{P}) \cdot \nabla F - S_y \frac{\partial F}{\partial t} \tag{4.2.35}$$

where $\mathbf{R} + \mathbf{N} - \mathbf{P} = -(R + N - P)\nabla z$, i.e., vertical accretion. By inserting (4.2.34) and (4.2.35) into (4.2.32), we obtain

$$\nabla' \cdot (h - \eta)\tilde{\mathbf{q}}' - (R + N - P) - |\nabla F|(\phi - h)/c^{(1)} +$$

$$+ \{S_0(h - \eta) + S_y\} \frac{\partial h}{\partial t} = 0. \tag{4.2.36}$$

Finally, since $S_y \gg S_0(h - \eta)$ and introducing $(h - \eta)\tilde{\mathbf{q}}' = \mathbf{Q}' = -K(h - \eta)\nabla'h$, we obtain

$$\nabla' \cdot \{K(h - \eta)\nabla'h\} + R + N - P + |\nabla F|(\phi - h)/c^{(1)} = S_y \frac{\partial h}{\partial t} \tag{4.2.37}$$

which is identical to (4.2.29), except that it allows for a nonhorizontal semipervious bottom. In performing the integration, we note again how conditions on the top and bottom bounding surfaces, here the semipervious bottom and the phreatic surface, become sink/source terms in the two-dimensional balance equation.

In (4.2.29), (4.2.30) and (4.2.37), the product $K(h - \eta)$ represents the local instantaneous transmissivity $T = T(x, y, t)$ of the aquifer (see Chapter 10).

In the development presented above, we have neglected changes of storage within the semipervious layer. There is no difficulty in applying the discussion presented in Subsection 4.2.2 above, also to the development of the balance equation in this subsection.

When the bottom of the aquifer is impervious, the third term in (4.2.37) vanishes.

### 4.2.5. LINEARIZATION

Equations (4.2.29) and (4.2.37) are *nonlinear*, as they contain the product $h\nabla'h$, in addition to terms that are linear in $h$ (e.g., $S_y \, \partial h/\partial t$). We may regard the product $K(h - \eta)$ in these equations as the transmissivity of the phreatic aquifer. However, unlike the transmissivity of a confined aquifer, the product $K(h - \eta)$ may also vary in time because $h = h(x, y, t)$.

There are only minor difficulties in numerically solving the nonlinear balance equation. Nevertheless, the equations are often first approximated by linearization and only then solved numerically. In the method of linearization that is commonly employed, we replace $h(x, y, t)$ in the product $K(h - \eta)$ by some mean value, $\bar{h}(x, y)$, assuming that $|h - \bar{h}| \ll \bar{h}$. Equation (4.2.37) then becomes

$$\nabla' \cdot \{K(\bar{h} - \eta) \nabla' h\} + R + N - P + |\nabla F|(\phi - h)/c^{(1)} = S_y \frac{\partial h}{\partial t} \quad (4.2.38)$$

which is *linear* in $h = h(x, y, t)$. The introduction of an average thickness $(\bar{h} - \eta)$ is justified whenever fluctuations in the water table elevations are much smaller than the thickness $(h - \eta)$ of a phreatic aquifer. In a numerical solution scheme, the average thickness may be updated gradually as time progresses.

## 4.3. Initial and Boundary Conditions

To describe a specific problem, the partial differential (balance) equation that describes flow in an aquifer must be supplemented by appropriate initial and boundary conditions. We recall that the state variable is $\phi = \phi(x, y, t)$, or $h = h(x, y, t)$ for a phreatic aquifer, and that the model is based on the assumption of *essentially horizontal flow*. For the sake of simplicity, we shall refer here only to $\phi$.

The (possibly moving) boundary itself is described by $F(x, y, t) = 0$ in the horizontal $xy$-plane.

Initial conditions take the form

$$\phi = \phi(x, y, 0) = g(x, y) \quad \text{in } (D) \quad (4.3.1)$$

where $g = g(x, y)$ is a known function.

Several types of boundary conditions may be encountered.

(a) *Boundary of prescribed piezometric head*. In this case

$$\phi = f_1(x, y, t) \quad \text{on } B \quad (4.3.2)$$

where the boundary, $B$, is a straight line or a curve in the $xy$-plane, and $f_1(x, y, t)$ is a known function.

A special case is the *equipotential boundary*, i.e., $f_1(x, y, t) = \text{const}$ or $f_1(x, y, t) = f_1^*(t)$, where $f_1^*$ is a known function.

A known potential boundary is encountered whenever the aquifer is in direct hydraulic contact with a river or a lake in which the water level is known.

Another special case of this kind of boundary is a *spring* through which groundwater emerges to the ground surface. The outlet threshold is at fixed elevation. Water emerges from the aquifer into the atmosphere ($p = 0$) at that fixed elevation and, hence, this is a boundary of a fixed piezometric head. Sometimes a water layer exists above the threshold which may vary with the rate of flow. However, when the piezometric heads in the aquifer in the vicinity of the spring are lower than this threshold, the spring dries up and ceases to serve as a boundary to the flow domain. It is thus a boundary of fixed potential only as long as the water heads in the vicinity are above the spring's outlet: they drop toward the spring (= loss of head in the converging flow in the aquifer).

When a spring drains into a lake, the specified head is given by the water level in the lake.

(b) *Boundary of prescribed flux.* Along such a boundary

$$Q'_n = f_3(x, y, t) \tag{4.3.3}$$

where $f_3$ is a known function. For an isotropic medium, this condition may be expressed as

$$\partial \phi / \partial \nu = f_4(x, y, t) \quad \text{on } B \tag{4.3.4}$$

where $\nu$ is the distance measured as normal to the boundary; $f_4$ is a known function. In an anisotropic aquifer, we modify (2.2.8) to express $\mathbf{Q}'$ and, hence, the boundary condition is

$$Q'_n = \mathbf{Q}' \cdot \mathbf{\nu} = -(\mathbf{T} \cdot \nabla' \phi) \cdot \mathbf{\nu} = f_5(x, y, t) \quad \text{on } B. \tag{4.3.5}$$

Or, in view of (3.4.5)

$$(\mathbf{T} \cdot \nabla' \phi) \cdot \nabla F = f_6(x, y, t) \quad \text{on } B \tag{4.3.6}$$

where $f_5$ is a known function, and $f_6 = |\nabla F| f_3$.

For $x, y$ that are principal directions, (4.3.6) may be rewritten in the form

$$T_{xx} \frac{\partial \phi}{\partial x} \frac{\partial F}{\partial x} + T_{yy} \frac{\partial \phi}{\partial y} \frac{\partial F}{\partial y} = f_6(x, y, t) \quad \text{on } B. \tag{4.3.7}$$

For an impervious boundary, $f_4 = 0$ in (4.3.4) and $f_5 = 0$ in (4.3.5). We recall that a *streamline* and a *water-divide* behave as an impervious boundary.

(c) *Semipervious boundary.* This boundary condition is encountered when a partly clogged river bed (e.g., by a thin layer of silt or clay) serves as a boundary of the flow domain. Because of the resistance to the flow offered by the semipervious layer, the water level (or piezometric head) in the river (point $A$ of Figure 4.8) differs from that at point $P$ in the aquifer, on the other side of the semipervious boundary. Since the flow is assumed horizontal, continuity of flux through the entire thickness of an anisotropic aquifer requires that

$$Q'_n = -(\mathbf{T} \cdot \nabla' \phi) \cdot \mathbf{\nu} = K^{(1)} \frac{H(x, y, t) - \phi}{b^{(1)}} \quad B \tag{4.3.8}$$

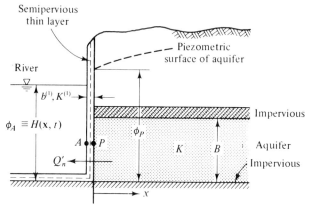

Fig. 4.8. A partly clogged river bed serving as a semipervious boundary.

where $H(x, y, t)$ is a known function. Or, making use of (3.4.5)

$$-(\mathbf{T} \cdot \nabla'\phi) \cdot \nabla F = \frac{H(x, y, t) - \phi}{c^{(1)}} B |\nabla F| \qquad (4.3.9)$$

where $c^{(1)} = b^{(1)}/K^{(1)}$.

## 4.4. Complete Statement of Aquifer Flow Model

Following the discussion in Section 3.5, and recalling that flow in the aquifer models considered here takes place only in the horizontal, say $xy$-plane, the standard content of such a model should include the following items.

(a) Specification of the geometrical configuration of the closed curve that bounds the problem area.
(b) Specification of the state variable, usually $\phi$ ($\equiv \tilde{\phi}$) $= \phi(x, y, t)$ for a confined, or leaky confined aquifer and $h = h(x, y, t)$ for a phreatic, or leaky-phreatic aquifer, for which a solution is sought.
(c) Statement of the partial differential (balance) equation in terms of the state variables stated in (b).
(d) Specification of the numerical values of the (transport and storage) coefficients that appear in (c).
(e) Statement of the numerical values of the various source and sink terms that appear in (c).
(f) Statement of initial conditions that the state variable of (b) satisfies at $t = 0$ within the considered domain. No initial conditions are required for steady flow.
(g) Statement of boundary conditions in terms of the state variables specified in (b).

The regions under investigation need not extend to the natural boundaries of a considered aquifer, such as a river or a lake in contact with the aquifer, or an impervious fault. Many reasons (sometimes merely economical, or political ones) may dictate boundaries other than natural ones. Since boundary conditions introduce the effect of the environment on the considered groundwater system, *modeling any portion of an aquifer is permitted, provided we specify the appropriate conditions along its boundaries.*

In the case of a system of leaky aquifers, we have first to completely model each aquifer separately, in terms of its own state variable, and then to solve all models simultaneously.

The discussion in Subsection 3.4.9 is applicable also when discontinuities are encountered in an aquifer, whether in the coefficients, or in the type of aquifer, e.g., when in the same formation the groundwater changes from confined to phreatic conditions, or when a horizontal impervious layer splits an aquifer into two. Accordingly, the aquifer system is split into subdomains along such lines of discontinuity. A complete model is then constructed for each subdomain, with its own state variable, recalling the required conditions along a discontinuity (Subsection 3.4.9). The entire set of models is then solved simultaneously. The comment at the end of Subsection 3.4.9 is also valid here.

## 4.5. Regional Model for Land Subsidence

When water is pumped from an aquifer, the pressure prevailing in the aquifer is reduced. With the nomenclature of Section 3.6, we have $p^e \neq 0$. Changes in excess stress, $\sigma^e$, at points within an aquifer may also be produced by changing the overburden load, e.g., by excavation or construction at the ground surface. From (3.6.16) it follows that any reduction in water pressure, whether also accompanied by a change in the total stress, or not, produces an increase in the effective stress in the solid matrix. The latter produces solid matrix deformation that manifests itself as *compaction* and *horizontal displacement*. The former leads to observable *land subsidence*. Land subsidence may serve as one of the constraints in determining water withdrawal from an aquifer. Hence, the need for a model that will provide information on land subsidence in response to planned pumping. The San Joaquin Valley (California), Mexico City (Mexico), Bangkok (Thailand) and Venice (Italy), are often mentioned as areas with very large land subsidence attributed to pumping. In most cases, the actual (irreversible) compaction is of layers or lenses of soft material (e.g., clay) within the aquifer, whereas the aquifer material lends itself to very small deformation only.

In this section, following Bear and Corapcioglu (1981a, b), we shall present a regional, two-dimensional mathematical model of land subsidence and averaged horizontal displacement due to pumping, i.e., assuming $\sigma^e = 0$. There is no difficulty in extending the discussion to the case where $\sigma^e \neq 0$.

Although, as mentioned above, the deformation is primarily that of the softer portion of an aquifer, we shall *assume* that the aquifer as a whole behaves as some average, fictitious elastically deformable material, with averaged elastic coefficients.

In Section 3.6, we constructed a mathematical model of soil deformation in a three-dimensional space. As in Sections 4.1 through 4.4, in this section we shall develop a model that averages the three-dimensional behavior over the aquifer's thickness. In such a model, the variables for which a solution is sought are the integrated vertical compaction of the aquifer, that manifests itself as land subsidence, and the averaged horizontal displacements. All these variables are functions of the horizontal coordinates and of time only.

For the sake of simplicity, we shall limit the discussion only to a confined aquifer. Corapcioglu and Bear (1983) also consider a phreatic one.

### 4.5.1. THE INTEGRATED WATER BALANCE EQUATION

Our starting point is the three-dimensional mass balance equation (3.6.14) for flow of a compressible fluid in a deformable porous medium, noting all the assumptions underlying its development. Rewritten in terms of Hubbert's potential, $\phi^*$, defined by (2.1.22), and adding a term $P'(x, y, z, t)$ to symbolically represent water withdrawal, this equation takes the form

$$\nabla \cdot \mathbf{q}_r + \rho g n \beta \frac{\partial \phi^*}{\partial t} + \frac{\partial \varepsilon_b}{\partial t} + P'(x, y, z, t) = 0. \tag{4.5.1}$$

We recall that $\mathbf{q}_r$ represents the specific discharge of the fluid relative to the solid matrix.

Following the methodology presented in Section 4.2, we now integrate this equation over the thickness, $B = B(x, y, t)$, of a confined aquifer, obtaining

$$\int_{b_1(x,y,t)}^{b_2(x,y,t)} \left( \nabla \cdot \mathbf{q}_r + \rho g n \beta \frac{\partial \phi^*}{\partial t} + \frac{\partial \varepsilon_b}{\partial t} + P' \right) dz$$

$$= \nabla' \cdot B\widetilde{\mathbf{q}}_r + \mathbf{q}_r|_{F_2} \cdot \nabla F_2 - \mathbf{q}_r|_{F_1} \cdot \nabla F_1 + B \frac{\widetilde{\partial \varepsilon_b}}{\partial t} + \widetilde{\rho g n} \beta \frac{\widetilde{\partial \phi^*}}{\partial t} +$$

$$+ \widetilde{\rho g n} \beta \left\{ \widetilde{\phi^*} \frac{\partial B}{\partial t} + \phi^*|_{F_2} \frac{\partial F_2}{\partial t} - \phi^*|_{F_1} \frac{\partial F_1}{\partial t} \right\} + BP' = 0 \tag{4.5.2}$$

where $F_i = F_i(x, y, t) = z - b_i(x, y, t) = 0$ for $i = 1$ and $2$, describe the bottom and top surfaces bounding the aquifer, and

$$\widetilde{(\ )} = \frac{1}{B} \int_{(B)} (\ ) \, dz.$$

In writing (4.5.2), we have already introduced the approximation

$$\int_{(B)} \rho g n \beta \frac{\partial \phi^*}{\partial t} dz \simeq \tilde{\rho} g \tilde{n} \beta \int_{(B)} \frac{\partial \phi^*}{\partial t} dz.$$

We recall that the prime symbol over a vector or an operator indicates that the vector or the operator are in the $xy$-plane only.

For the impervious bottom and top bounding surfaces considered here, we use boundary condition (3.4.15), with $f_3(x, t) = 0$, written in the forms

$$\mathbf{q}_r|_{F_1} \cdot \nabla F_1 = 0, \qquad \mathbf{q}_r|_{F_2} \cdot \nabla F_2 = 0. \tag{4.5.3}$$

By assuming also that equipotentials (i.e., surfaces of $\phi^* = $ constant) are essentially vertical), i.e., $\phi^*|_{F_1} \simeq \phi^*|_{F_2} \simeq \tilde{\phi}^*$, Equation (4.5.2) reduces to

$$\nabla \cdot B\widetilde{\mathbf{q}'} + B \frac{\partial \widetilde{\varepsilon_b}}{\partial t} + \tilde{\rho}\tilde{n}\beta B \frac{\partial \widetilde{\phi^*}}{\partial t} + P(x, y, t) = 0 \tag{4.5.4}$$

where $P = BP'$ represents the volume of water withdrawn from the aquifer per unit horizontal area, per unit time, and (Section 2.2) $B\widetilde{\mathbf{q}'_r} = -B\widetilde{\mathbf{K}'} \cdot \nabla \tilde{\phi}^*$.

We note that in (4.5.4)

$$\frac{\partial \widetilde{\phi^*}}{\partial t} \simeq \frac{1}{\tilde{\rho}g} \frac{\partial \tilde{p}}{\partial t} + \frac{\partial \tilde{z}}{\partial t} \tag{4.5.5}$$

where $\tilde{z} = (b_1 + b_2)/2$ is the elevation of the midpoint of the aquifer. Also

$$\nabla' \tilde{\phi}^* \simeq \frac{1}{\tilde{\rho}g} \nabla' \tilde{p} + \nabla' \tilde{z}. \tag{4.5.6}$$

With $|\partial \varepsilon_b/\partial t| \gg |\mathbf{V}_z \cdot \nabla \varepsilon_b|$, the total derivative, $d^s\varepsilon_b/dt$, can be replaced by the partial one, $\partial \varepsilon_b/\partial t$, leading to

$$B\frac{\partial \widetilde{\varepsilon_b}}{\partial t} = B\widetilde{\nabla \cdot \mathbf{V}_s} = \nabla' \cdot B\widetilde{\mathbf{V}'_s} + \mathbf{V}_s|_{F_2} \cdot \nabla F_2 - \mathbf{V}_s|_{F_1} \cdot \nabla F_1. \tag{4.5.7}$$

Since the top and bottom surfaces of the aquifer are assumed to be material surfaces with respect to the solid matrix, following (3.4.10), we have on them

$$(\mathbf{V}_s - \mathbf{u})|_{F_1} \cdot \nabla F_1 = 0, \qquad (\mathbf{V}_s - \mathbf{u})|_{F_2} \cdot \nabla F_2 = 0. \tag{4.5.8}$$

Or, in view of (3.4.4),

$$\mathbf{V}_s|_{F_1} \cdot \nabla F_1 = -\partial F_1/\partial t, \qquad \mathbf{V}_s|_{F_2} \cdot \nabla F_2 = -\frac{\partial F_2}{\partial t}. \tag{4.5.9}$$

Hence, (4.5.7) becomes

$$B\frac{\partial \widetilde{\varepsilon_b}}{\partial t} = \nabla' \cdot B\widetilde{\mathbf{V}'_s} - \frac{\partial(F_2 - F_1)}{\partial t} = \nabla' \cdot B\widetilde{\mathbf{V}'_s} + \partial B/\partial t. \tag{4.5.10}$$

With the solid's velocity, $\mathbf{V}_s$, related to the displacement, $\mathbf{w}$, by (3.6.10), and introducing the approximation $d^s\mathbf{w}/dt \simeq \partial\mathbf{w}/\partial t$, we obtain

$$B\widetilde{\mathbf{V}'_s} = \int_{(B)} \mathbf{V}'_s\, dz = \int_{(B)} \frac{\partial \mathbf{w}}{\partial t}\, dz = \frac{\partial}{\partial t}(B\widetilde{\mathbf{w}'}) +$$

$$+ \mathbf{w}'|_{F_2} \frac{\partial F_2}{\partial t} - \mathbf{w}'|_{F_1} \frac{\partial F_1}{\partial t}$$

$$= B\frac{\partial \widetilde{\mathbf{w}'}}{\partial t} + \left( \widetilde{\mathbf{w}'}\frac{\partial B}{\partial t} + \mathbf{w}'|_{F_2}\frac{\partial F_2}{\partial t} - \mathbf{w}'|_{F_1}\frac{\partial F_1}{\partial t} \right). \quad (4.5.11)$$

At this point, we need information on the displacements $\mathbf{w}'|_{F_1}$ and $\mathbf{w}'|_{F_2}$ which are the displacement boundary conditions on $F_1$ and $F_2$, respectively. This information is not available. To circumvent this difficulty we introduce a simplifying assumption, namely that, practically, horizontal displacements are the same along the vertical, i.e.,

$$\mathbf{w}'|_{F_1} \simeq \mathbf{w}'|_{F_2} \simeq \widetilde{\mathbf{w}'}. \quad (4.5.12)$$

Then, by combining (4.5.10) through (4.5.12), we obtain

$$B\frac{\widetilde{\partial \varepsilon_b}}{\partial t} = \nabla' \cdot B \frac{\partial \widetilde{\mathbf{w}'}}{\partial t} + \frac{\partial B}{\partial t}.$$

Following the discussion in Section 3.6, we separate the specific discharge and piezometric head distributions into initial steady-state values and ones that express excess above the latter.

$$\widetilde{\phi^*}(x, y, t) = \widetilde{\phi^{*0}}(x, y) + \widetilde{\phi^{*e}}(x, y, t),$$
$$\widetilde{\mathbf{q}'_r}(x, y, t) = \widetilde{\mathbf{q}'^0_r}(x, y) + \widetilde{\mathbf{q}'^e_r}(x, y, t), \quad (4.5.13)$$
$$P(x, y, t) = P^0(x, y) + P^e(x, y, t).$$

In terms of these variables, (4.5.4) is separated into two equations: a steady-state balance equation

$$\nabla' \cdot B\widetilde{\mathbf{q}'^0_r} + P^0 = 0 \quad (4.5.14)$$

and an unsteady one

$$\nabla' \cdot B\widetilde{\mathbf{q}'^e_r} + \nabla' \cdot B\frac{\partial \widetilde{\mathbf{w}'}}{\partial t} + \frac{\partial B}{\partial t} + \widetilde{\rho}g\widetilde{n}\beta B\frac{\partial \widetilde{\phi^{*e}}}{\partial t} + P^e = 0. \quad (4.5.15)$$

Equation (4.5.15) can be linearized by introducing

$$B(x, y, t) \equiv b_2(x, y, t) - b_1(x, y, t) = b_2^0(x, y) + w_z|_{F_2} - $$
$$- (b_1^0(x, y) + w_z|_{F_1}) = B^0(x, y) + \Delta_z,$$
$$\Delta_z = (w_z|_{F_2} - w_z|_{F_1}) \ll B^0.$$

Thus, neglecting the effect of consolidation on permeability, we may write

$$B\widetilde{\mathbf{q}_r^{\prime e}} = -B^0 \widetilde{\mathbf{K}'} \cdot \nabla' \widetilde{\phi^{*e}} \qquad (4.5.16)$$

### 4.5.2. THE INTEGRATED EQUILIBRIUM EQUATION

The total stress tensor, $\sigma$, at a point within an aquifer satisfies the equilibrium equation (3.6.15) rewritten here for convenience in the form

$$-\nabla \cdot \boldsymbol{\sigma} + \mathbf{f} = 0 \qquad (4.5.17)$$

where $\mathbf{f} = -\{\rho_w n + \rho_s(1-n)\} g \nabla z$ represents the body force. We recall that the inertial effects have been neglected in writing (4.5.17). We may now replace (4.5.17) by (3.6.17) and (3.6.18), assuming, as a good approximation, that the body force remains unchanged.

For the sake of simplicity in presenting the methodology of constructing a regional land subsidence model, let us assume that the solid matrix of the aquifer behaves like an isotropic and (for the relatively small displacements considered here) perfectly elastic body, for which the stress-strain relationship (3.6.19) is valid.

By integrating the equation for the $x$ direction in (3.6.18), we obtain

$$\int_{(B)} (\nabla \cdot \boldsymbol{\sigma}_s^{*e} + \nabla p^e) \, dz$$

$$= \nabla' \cdot B(\widetilde{\boldsymbol{\sigma}_s^{*e}})' - (\boldsymbol{\sigma}_s^{*e} + p^e \mathbf{I})|_{F_2} \cdot \nabla F_2 +$$
$$+ (\boldsymbol{\sigma}_s^{*e} + p^e \mathbf{I})|_{F_1} \cdot \nabla F_1 - \nabla' B \widetilde{p^e} = 0. \qquad (4.5.18)$$

To derive the stress boundary conditions on $F_1$ and $F_2$, to be inserted in (4.5.18), let us consider the surface $F_2$ bounding the aquifer from above. With subscripts $u$ and $l$ denoting the upper and lower sides of $F_2$, equilibrium requires that

$$\boldsymbol{\sigma}|_u \cdot \nabla F_2 = \boldsymbol{\sigma}|_l \cdot \nabla F_2. \qquad (4.5.19)$$

Similar to the flow, we assume here that a certain initial stress distribution prevails in the aquifer, and that the pumping introduces an excess stress distribution. Hence, we rewrite (4.5.19) in the form of the two equations

$$\boldsymbol{\sigma}^e|_u \cdot \nabla F_2 = \boldsymbol{\sigma}^e|_l \cdot \nabla F_2, \qquad \boldsymbol{\sigma}^0|_u \cdot \nabla F_2 = \boldsymbol{\sigma}^0|_l \cdot \nabla F_2 \qquad (4.5.20)$$

or, in terms of effective stress and pressure.

$$(\boldsymbol{\sigma}_s^{*e} + p^e \mathbf{I})|_u \cdot \nabla F_2 = (\boldsymbol{\sigma}_s^{*e} + p^e \mathbf{I})|_l \cdot \nabla F_2 = 0 \tag{4.5.21}$$

and a similar equation for the initial steady state.

When the excess stress and pressure in an aquifer are due only to pumping from the latter and not to changes in the overburden load, the total stress on the upper side of $F_2$ remains unchanged, i.e., $\boldsymbol{\sigma}^e|_u \cdot \nabla F_2 = 0$. Hence

$$(\boldsymbol{\sigma}_s^{*e} + p^e \mathbf{I})|_l \cdot \nabla F_2 = 0. \tag{4.5.22}$$

Similar considerations lead to

$$(\boldsymbol{\sigma}_s^{*e} + \boldsymbol{\rho}^e \mathbf{I})|_u \cdot \nabla F_1 = 0. \tag{4.5.23}$$

There is no difficulty in also extending the above discussion to the case $\boldsymbol{\sigma}^e \neq 0$.

By inserting (4.5.22) and (4.5.23) into (4.5.18), we obtain

$$\nabla' \cdot B(\widetilde{\boldsymbol{\sigma}_s^{*e}})' + \nabla' B\widetilde{p}^e = 0 \tag{4.5.24}$$

in which averaged values are functions of $x$, $y$ and $t$ only.

In indicial notion, (4.5.24) is rewritten as

$$\frac{\partial}{\partial x} B\widetilde{(\sigma_s^{*e})}_{xx} + \frac{\partial}{\partial y} B\widetilde{(\sigma_s^{*e})}_{xy} + \frac{\partial}{\partial x} B\widetilde{p}^e = 0, \tag{4.5.25}$$

$$\frac{\partial}{\partial y} B\widetilde{(\sigma_s^{*e})}_{yx} + \frac{\partial}{\partial y} B\widetilde{(\sigma_s^{*e})}_{yy} + \frac{\partial}{\partial y} B\widetilde{p}^e = 0, \tag{4.5.26}$$

$$\frac{\partial}{\partial x} B\widetilde{(\sigma_s^{*e})}_{zx} + \frac{\partial}{\partial y} B\widetilde{(\sigma_s^{*e})}_{zy} = 0. \tag{4.5.27}$$

We now express the averaged excess effective stress tensor in terms of the averaged displacement, $\widetilde{\mathbf{w}}$, making use of the stress-strain relationship (3.6.19). Obviously, other relationships, describing other types of soils, can be introduced at this point instead of (3.6.19). We also make use of the assumption introduced by (4.5.12). We obtain

$$\widetilde{\varepsilon}_b = \widetilde{(\varepsilon_b)}_{xx} + \widetilde{(\varepsilon_b)}_{yy} + \widetilde{(\varepsilon_b)}_{zz}$$

$$= \frac{\partial \widetilde{w}_x}{\partial x} + \frac{\partial \widetilde{w}_y}{\partial y} + \frac{\partial \widetilde{w}_z}{\partial z} = \frac{\partial \widetilde{w}_x}{\partial x} + \frac{\partial \widetilde{w}_y}{\partial y} + \frac{\Delta_z}{B}, \tag{4.5.28}$$

$$-\widetilde{(\sigma_s^{*e})}_{xx} = (\widetilde{\lambda} + 2\widetilde{G}) \frac{\partial \widetilde{w}_x}{\partial x} + \widetilde{\lambda} \left( \frac{\partial \widetilde{w}_y}{\partial y} + \frac{\Delta_z}{B} \right), \tag{4.5.29}$$

$$-\widetilde{(\sigma_s^{*e})}_{yy} = \widetilde{\lambda} \frac{\partial \widetilde{w}_x}{\partial x} + (\widetilde{\lambda} + 2\widetilde{G}) \frac{\partial \widetilde{w}_y}{\partial y} + \widetilde{\lambda} \frac{\Delta_z}{B}, \tag{4.5.30}$$

$$-\widehat{(\sigma_s^{*e})}_{xy} = -\widehat{(\sigma_s^{*e})}_{yx} = \widetilde{G}\left(\frac{\widehat{\partial w_y}}{\partial x} + \frac{\widehat{\partial w_x}}{\partial y}\right), \qquad (4.5.31)$$

$$-\widehat{(\sigma_s^{*e})}_{xz} = -\widehat{(\sigma_s^{*e})}_{zx} = \widetilde{G}\left(\frac{\widehat{\partial w_z}}{\partial x} + \frac{\widehat{\partial w_x}}{\partial z}\right)$$

$$= \widetilde{G}\frac{\widehat{\partial w_z}}{\partial x} + \frac{\widetilde{G}}{B}\left\{\widetilde{w_z}\frac{\partial B}{\partial x} + w_z|_{F_2}\frac{\partial F_2}{\partial x} - w_z|_{F_1}\frac{\partial F_1}{\partial x}\right\}, \qquad (4.5.32)$$

$$-\widehat{(\sigma_s^{*e})}_{yz} = -\widehat{(\sigma_s^{*e})}_{zy}$$

$$= \widetilde{G}\frac{\widehat{\partial w_z}}{\partial y} + \frac{\widetilde{G}}{B}\left\{\widetilde{w_z}\frac{\partial B}{\partial y} + w_z|_{F_2}\frac{\partial F_2}{\partial y} - w_z|_{F_1}\frac{\partial F_1}{\partial y}\right\}, \qquad (4.5.33)$$

$$-\widehat{(\sigma_s^{*e})}_{zz} = \widetilde{\lambda}\left(\frac{\widehat{\partial w_x}}{\partial x} + \frac{\widehat{\partial w_y}}{\partial y}\right) + (\widetilde{\lambda} + 2\widetilde{G})\frac{\Delta_z}{B}. \qquad (4.5.34)$$

By inserting these expressions into (4.5.25) through (4.5.27), we obtain three equations in the four averaged variables $\widetilde{p^e}$, $\widetilde{w_x}$, $\widetilde{w_y}$ and $\widetilde{w_z}$, all functions of $x$, $y$ and $t$ only

$$\frac{\partial}{\partial x}\left\{B\left[(\widetilde{\lambda} + 2\widetilde{G})\frac{\widehat{\partial w_x}}{\partial x} + \widetilde{\lambda}\left(\frac{\widehat{\partial w_y}}{\partial y} + \frac{\Delta_z}{B}\right)\right]\right\} +$$

$$+ \frac{\partial}{\partial y}\left\{B\widetilde{G}\left(\frac{\widehat{\partial w_x}}{\partial y} + \frac{\widehat{\partial w_y}}{\partial x}\right)\right\} - \frac{\partial}{\partial x}B\widetilde{p^e} = 0, \qquad (4.5.35)$$

$$\frac{\partial}{\partial x}\left\{B\widetilde{G}\left(\frac{\widehat{\partial w_x}}{\partial x} + \frac{\widehat{\partial w_y}}{\partial y}\right)\right\} +$$

$$+ \frac{\partial}{\partial y}\left\{B\left[\widetilde{\lambda}\frac{\widehat{\partial w_x}}{\partial x} + (\widetilde{\lambda} + 2\widetilde{G})\frac{\widehat{\partial w_y}}{\partial y} + \widetilde{\lambda}\frac{\Delta_z}{B}\right]\right\} - \frac{\partial}{\partial y}B\widetilde{p^e} = 0, \qquad (4.5.36)$$

$$\frac{\partial}{\partial x}\left\{B\widetilde{G}\frac{\widehat{\partial w_x}}{\partial x} + \widetilde{G}\left[\widetilde{w_z}\frac{\partial B}{\partial x} + w_z|_{F_2}\frac{\partial F_2}{\partial x} - w_z|_{F_1}\frac{\partial F_1}{\partial x}\right]\right\} +$$

$$+ \frac{\partial}{\partial y}\left\{B\widetilde{G}\frac{\widehat{\partial w_z}}{\partial y} + \widetilde{G}\left(\widetilde{w_z}\frac{\partial B}{\partial y} + w_z|_{F_2}\frac{\partial F_2}{\partial y} - w_z|_{F_1}\frac{\partial F_1}{\partial y}\right)\right\} = 0. \qquad (4.5.37)$$

For constant $\tilde{\lambda}$ and $\tilde{G}$, we assume $\Delta_z \ll B^0$ to linearize (4.5.35) and (4.5.36), obtaining

$$\tilde{G}\nabla'^2 w_x + (\tilde{\lambda} + \tilde{G})\frac{\partial \tilde{\varepsilon}_b}{\partial x} - \tilde{G}\frac{\partial(\Delta_z/B^0)}{\partial x} - \frac{\partial \tilde{p}^e}{\partial x} = 0, \qquad (4.5.38)$$

$$\tilde{G}\nabla'^2 w_y + (\tilde{\lambda} + \tilde{G})\frac{\partial \tilde{\varepsilon}_b}{\partial y} - \tilde{G}\frac{\partial(\Delta_z/B^0)}{\partial y} - \frac{\partial \tilde{p}^e}{\partial y} = 0 \qquad (4.5.39)$$

where $\tilde{\varepsilon}_b$ is expressed in terms of $\widetilde{w_x}$, $\widetilde{w_y}$ and $\widetilde{w_z}$ by (4.5.28).

The same linearization, together with $|\mathbf{V}'_s \cdot \nabla' B| \ll |\partial B/\partial t|$, leads to

$$\nabla' \cdot B\frac{\partial \widetilde{\mathbf{w}'}}{\partial t} + \frac{\partial B}{\partial t} \simeq B^0 \frac{\partial \tilde{\varepsilon}_b}{\partial t} \qquad (4.5.40)$$

and the mass balance equation (4.5.15) becomes

$$\nabla' \cdot B^0 \widetilde{\mathbf{q}'^e_r} + B^0 \frac{\partial \tilde{\varepsilon}_b}{\partial t} + \tilde{\rho} g \tilde{n} \beta B \frac{\partial \widetilde{\phi^{*e}}}{\partial t} + P^e = 0. \qquad (4.5.41)$$

In principle, (4.5.23), (4.5.37), (4.5.38), (4.5.39), and (4.5.41) constitute four equations in the five dependent variables: $\tilde{\varepsilon}_b$, $\widetilde{w_x}$, $\widetilde{w_y}$, $\Delta_z$ and $\tilde{p}^e$. However, in these equations we have the terms $w_z|_{F_2}$, $w_z|_{F_1}$ and $(\Delta_z = w_z|_{F_2} - w_z|_{F_1})$ that constitute conditions on the boundaries $F_1 = 0$ and $F_2 = 0$. No information is available for these boundary conditions. In fact, the land subsidence, expressed by, or related to $w_z|_{F_2}$ is the very variable for which a solution is sought in most subsidence problems.

At this point we may continue by introducing further simplifying assumptions to replace the missing information. For example, we may assume that the bottom of the aquifer is fixed, i.e., $w_z|_{F_1} = 0$ and that $w_z$ varies linearly with $z$, i.e., $\widetilde{w_z} = \frac{1}{2}w_z|_{F_2} = -\delta/2$, where $\delta$ is the extent of land subsidence (positive downward). We end up with four equations for $\widetilde{w_x}$, $\widetilde{w_y}$, $\delta$ and $\widetilde{p^e}$ (or $\widetilde{\phi^{*e}}$).

Another approach, suggested by Verruijt (1969, p. 347), is to assume that *consolidation occurs under conditions of plannar incremental total stress*. This means

$$\sigma^e_{zz} = 0, \qquad \sigma^e_{xz} = \sigma^e_{zx} = 0, \qquad \sigma^e_{yz} = \sigma^e_{zy} = 0. \qquad (4.5.42)$$

Verruijt indicates that this assumption is justified when the aquifer is bounded from above and below by soft confining layers (e.g., clay) which cannot resist shear stress. Furthermore, this assumption also justifies (4.5.12), since in a relatively thin aquifer, as implied by the plane stress assumption, lateral deformation is, more or less, uniform along the relatively small thickness of the aquifer.

From (4.5.42) it follows that (3.6.18) reduces to

$$\nabla' \cdot (\boldsymbol{\sigma}_s^{*e})' + \nabla' p^e = 0 \quad \text{or} \quad \frac{\partial(\sigma_s^{*e})_{ij}}{\partial x_j} + \frac{\partial p^e}{\partial x_i} = 0, \quad i,j = x, y. \qquad (4.5.43)$$

The boundary conditions (4.5.22) and (4.5.23) are also written in the $xy$ coordinates only. Following the integration procedure which led above to (4.5.25) through (4.5.27), we now obtain only (4.5.25) and (4.5.26), as (4.5.27) drops out.

Accordingly, (4.5.37) also drops out, leaving only (4.5.35) and (4.5.36), or (4.5.38) and (4.5.39). We now have to solve (4.5.41), (4.5.38) and (4.5.39) for $\widetilde{p^e}$, $\widetilde{w_x}$, $\widetilde{w_y}$ and $\Delta_z$. The necessary fourth equation is obtained from the first condition in (4.5.42), viz.

$$\widetilde{(\sigma_s^{*e})_{zz}} = -\widetilde{p^e} \tag{4.5.44}$$

From the expression for $\widetilde{(\sigma_s^{*e})_{zz}}$ in (4.5.34), we obtain the fourth equation in the form of

$$\widetilde{p^e} = \tilde{\lambda}\left(\frac{\partial \widetilde{w_x}}{\partial x} + \frac{\partial \widetilde{w_y}}{\partial y}\right) + (\tilde{\lambda} + 2\tilde{G})\frac{\Delta_z}{B} = \tilde{\lambda}\widetilde{\varepsilon_b} + 2\tilde{G}\frac{\Delta_z}{B}. \tag{4.5.45}$$

This completes the formulation of the problem in terms of $\widetilde{p^e}$, $\widetilde{w_x}$, $\widetilde{w_y}$ and $\Delta_z$ as the sought variables. With $w_z|_{F_1} = 0$, the land subsidence is given by $\delta(x, y, t) = -w_z|_{F_2} = -\Delta_z$.

Initially, the values of these variables is zero.

As boundary conditions on the external boundaries of an aquifer, or at sufficiently large distances from a pumping well field, we shall usually assume vanishing values of all these variables.

### 5.4.3. COMPARISON BETWEEN BIOT'S AND JACOB'S APPROACHES

In Sections 3.2 and 3.3 we developed the concept of specific storativity and the mass balance equation, making use of Jacob's (1940) approach of *vertical displacement only*. In Subsection 4.5.1 we have developed the fundamental mass balance equation, making use of Biot's three-dimensional displacement theory. Let us compare these two approaches.

By differentiating (4.5.38) with respect to $x$, and (4.5.39) with respect to $y$, linearizing them, adding them and integrating the resulting equation with constant $\tilde{\lambda}$, $\tilde{G}$ and $B^0$, we obtain (Verruijt, 1969)

$$(\tilde{\lambda} + 2\tilde{G})\left(\frac{\partial \widetilde{w_x}}{\partial x} + \frac{\partial \widetilde{w_y}}{\partial y}\right) + \tilde{\lambda}\frac{\Delta_z}{B^0}$$

$$= (\tilde{\lambda} + 2\tilde{G})\widetilde{\varepsilon_b} - 2\tilde{G}\frac{\Delta_z}{B^0}$$

$$= \widetilde{p^e} + \widetilde{\Pi}(x, y, t) \tag{4.5.46}$$

where $\widetilde{\Pi}$ satisfies the Laplace equation, $\nabla'^2\widetilde{\Pi} = 0$, in the $xy$ plane for every time $t$.

By comparing (4.5.45) with (4.5.46), obtained by introducing the plane stress assumption, we find that

$$\widetilde{\Pi} = 2\widetilde{G}\left(\widetilde{\varepsilon}_b - 2\frac{\Delta_z}{B}\right). \qquad (4.5.47)$$

If, following Jacob (1940), we assume no horizontal displacements, i.e., $\mathbf{w}' = 0$, the mass balance equation (4.5.15) for $p^e = 0$, reduces to

$$\nabla' \cdot B\widetilde{\mathbf{q}_r'^e} + \frac{\partial B}{\partial t} + \widetilde{\rho}g\widetilde{n}\beta B \frac{\partial \widetilde{\phi^{*e}}}{\partial t} = 0. \qquad (4.5.48)$$

Under the same conditions, we obtain from (4.5.45)

$$\widetilde{p}^e \simeq (\widetilde{\lambda} + 2\widetilde{G})\frac{\Delta_z}{B}. \qquad (4.5.49)$$

With $\Delta_z \ll B$, Equations (4.5.48) and (4.5.49) can be combined to the single equation =

$$\nabla' \cdot B^0\widetilde{\mathbf{q}_r'^e} + B^0\left\{\frac{1}{\widetilde{\lambda} + 2\widetilde{G}} + \widetilde{n}\beta\right\}\frac{\partial \widetilde{p}^e}{\partial t} = 0. \qquad (4.5.50)$$

By comparing (4.5.46) with $\widetilde{\mathbf{w}}' = 0$, with (4.5.45), obtained by the horizontal plane stress assumption and vertical displacement only, we conclude that

$$\widetilde{\Pi} = -2\widetilde{G}\frac{\Delta_z}{B}. \qquad (4.5.51)$$

By comparing (4.5.50) with (3.6.28) and (3.3.9), we conclude again that, $\alpha = 1/(\widetilde{\lambda} + 2\widetilde{G})$, as in (3.6.30).

As pointed out by Verruijt (1969), the function $\Pi$ defined by (3.6.26) and (3.6.27), describes the deviation of the simplified Terzaghi—Jacob theory from Biot's one, where the former assumes vertical consolidation only, while the latter takes into account the three-dimensional nature of consolidation. Similarly, here $\widetilde{\Pi}$ expresses the deviation when integrated aquifer consolidation equations are employed. It is of interest to note that here $\widetilde{\Pi}$ does not vanish, while Verruijt (1969) shows that $\Pi$ vanished in the three-dimensional model.

## 4.6. Streamlines and Stream Function

So far, in this chapter, we have described the flow in an aquifer in terms of piezometric head, with the gradient in the latter serving as a driving force to the flow. Information on the piezometric head is readily obtainable by using observation wells, or piezometers. In what follows, we shall introduce the concepts of the *streamline* and the *stream function* which can be used as an alternative description of the flow. In Chapter 11 we shall discuss a case in which the description of the flow in terms of streamlines has a definite advantage.

### 4.6.1. PATHLINES

In the macroscopic approach to flow in porous media, we can define the average movement of a *fluid particle*. For example, we may label the fluid within an REV (Section 1.4), by some tracer, or we may continuously inject tracer-labeled fluid at a point in a moving fluid. In laminar flow, in spite of the spreading of the tracer that will occur (Chapter 6), it is possible to define the average path of the tracer-labeled fluid. In what follows, we shall use the term *fluid particle* to indicate an ensemble of molecules contained in a small volume, e.g., an REV. However, as shown in Section 1.4, the size of the latter is not a single constant value, as it may vary within a certain range (Figure 1.5). The lower limit of this range is determined by the heterogeneity of the configuration of the void space at the microscopic level (e.g., pore size), while its upper limit is determined by the heterogeneity in macroscopic coefficients that characterize the same configuration (e.g., porosity and permeability).

Let us consider a single-species fluid and label the fluid molecules initially within the REV around a point. As this cloud moves, the labelled molecules comprising it spread out (by diffusion and dispersion — see Chapter 6) and occupy a growing volume around the centroid of the cloud. Because of the upper bound imposed on the REV, after a certain time interval, a 'new' particle has to be defined and labelled around the centroid of the cloud. The motion of the new particle, i.e., the movement of its centroid, can then be traced for an additional period. By repeating this procedure, i.e., by labeling the molecules whenever the volume occupied by them becomes too large, with the end of a 'former' particle serving as the centroid for a new one, a continuous path of a fluid particle is obtained. We thus speak of tracing the *path of a fluid particle*, although the molecules comprising it continuously change.

With the above background, a *pathline* of a fluid particle is the locus of its positions in space as time passes. It is thus the trajectory (in the continuum sense) of a particle of fixed identity. It is a *Lagrangian* concept (to be compared with the Eulerian point of view (end of Subsection 3.3.1) of watching *different* particles passing through a *fixed* point). As the particle moves at a velocity $\mathbf{V}(=\mathbf{q}/n)$ from location $(x, y, z)$ at time $t$ to location $(x + dx, y + dy, z + dz)$ at time $t + dt$, we have

$$dx = V_x\, dt, \qquad dy = V_y\, dt, \qquad dz = V_z\, dt$$

or

$$\frac{dx}{V_x(x, y, z, t)} = \frac{dy}{V_y(x, y, z, t)} = \frac{dz}{V_z(x, y, z, t)} = dt. \tag{4.6.1}$$

Equation (4.6.1) contains three equations. Their solution can be written in the parametric form

$$\lambda = \lambda(x, y, z, t), \qquad \chi = \chi(x, y, z, t), \qquad \omega = \omega(x, y, z, t)$$

where $\lambda$ = const, $\chi$ = const and $\omega$ = const describe three surfaces in space. Each set of three such surfaces defines (at their point of intersection) the location of a particle in space at time $t$. Together, they describe the pathlines of different particles.

In somewhat simpler terms, let us consider a porous medium domain in which flow takes place, and identify a fluid particle at some location at time $t$. If we know the velocity at that location, say by solving the flow model for the piezometric head, $\phi$, and using Darcy's law to determine the velocity $\mathbf{V}$ ($= \mathbf{q}/n$), we can use (4.6.1) to determine the location of the considered particle at $t + \mathrm{d}t$.

In Section 6.7 we shall return to this subject in order to consider the movement of polluted fluid particles.

### 4.6.2. STREAMLINES

At any instant of time, a velocity (or specified discharge) vector can be defined at every point within a considered flow domain. The *instantaneous* curves that at every point are tangent to the velocity vector at that point are called *streamlines*. In steady flow, since velocities do not change with time, pathlines (Subsection 4.6.1) and streamlines are identical.

The mathematical expression for a streamline, as defined above, is $\mathbf{V} \times \mathrm{d}\mathbf{s} = 0$, where $\times$ denotes the vector product and $\mathrm{d}\mathbf{s}$ is an element of arc along the streamline, or

$$\frac{\mathrm{d}x}{V_x(x, y, z, t_0)} = \frac{\mathrm{d}y}{V_y(x, y, z, t_0)} = \frac{\mathrm{d}z}{V_z(x, y, z, t_0)} \tag{4.6.2}$$

(Figure 4.9), where $t_0$ indicates a certain instant of time. Equation (4.6.2) states

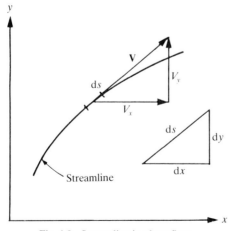

Fig. 4.9. Streamline in plane flow.

# MODELING TWO-DIMENSIONAL FLOW IN AQUIFERS

that the vectors **V** and d**s** are parallel. We could also say that it states that the velocity component normal to a streamline always vanishes, i.e., that no flow crosses it.

To complete the picture, let us mention two additional definitions that are often encountered in the literature.

A *streakline* represents the locus of locations within a flow-domain, occupied, or to be occupied, by all the fluid particles that at some earlier time have passed though a certain fixed point within the domain (or on its boundary). A plume of pollutants originating from a point source may serve as an example.

A *front*, or *interface* is a surface that always consists of the same particles. Let the fluid particles forming a continuous surface within the flow domain at some initial time, $\tau = 0$, be labelled by some tracer. It is then possible to determine the subsequent positions of this surface by tracing the individual fluid particles comprising it. The phreatic surface without accretion (Subsection 3.4.5) and the interface in a coastal aquifer (Chapter 7) may serve as examples.

### 4.6.3. THE STREAM FUNCTION IN PLANE FLOW

In two-dimensional flow, (4.6.2) rewritten in terms of the specific discharge, **q**, becomes

$$\frac{dx}{q_x} = \frac{dy}{q_y} \quad \text{or} \quad q_y\, dx - q_x\, dy = 0. \tag{4.6.3}$$

The solution of (4.6.3) is

$$\Psi = \Psi(x, y) = \text{const} \quad \text{or} \quad \Psi = \Psi(x, y, t) \tag{4.6.4}$$

which describes the geometry (or, in unsteady flow, the instantaneous geometry) of the streamlines. The condition for (4.6.3) to be an exact differential of some function $\Psi(x, y)$ is $\partial q_x/\partial x + \partial q_y/\partial y = 0$, which is nothing but the mass conservation equation (3.3.2) for two-dimensional steady flow and constant density fluid (or for nondeformable medium and constant density fluid). Thus the function $\Psi$ as defined here is valid only for the flow of a constant density fluid in a nondeformable porous medium for which div **q** = 0. Such flow is referred to as *macroscopically isochoric flow*.

By using (4.6.3), it is possible to define at any point $(x, y)$ in the plane, an angle $\alpha = \tan^{-1}(q_y/q_x)$ with respect to the $+x$ axis, which is tangent to the integral curve (4.6.4).

Equation (4.6.4) describes a family of curves, for various values of the constant. Since $\Psi$ is an exact differential along any streamline, we have

$$d\Psi = \frac{\partial \Psi}{\partial x}\, dx + \frac{\partial \Psi}{\partial y}\, dy = q_y\, dx - q_x\, dy \tag{4.6.5}$$

from which we obtain the relationships

$$q_x = -\frac{\partial \Psi}{\partial y}, \qquad q_y = \frac{\partial \Psi}{\partial x}. \tag{4.6.6}$$

The function $\Psi = \Psi(x, y)$, which is *constant along streamlines* (i.e., $d\Psi = 0$), is called the *stream function of two-dimensional flow* (dims. $L^2/T$).

The physical interpretation of the stream function, $\Psi$, may be obtained as follows. Figure 4.10 shows some streamlines labeled, $\Psi_A$, $\Psi_A + \Delta\Psi$, $\Psi_A + 2\Delta\Psi$, etc. Let d**A** be an element of area of unit width (in the $z$ direction) and length d$s$, where d$s$ is an arbitrary path connecting points $A$ and $B$ on the streamlines $\Psi_A$ and $\Psi_B$, respectively. The direction of the vector d**A** is perpendicular to d**s**, with d**A** $=$ (d$A$)**1n**, d**s** $=$ (d$s$)**1s** and the unit vector **1n** is obtained from the unit vector **1s** by a counterclockwise rotation. We therefore have d**A** $=$ **1z** $\times$ d**s**. Consider the integral

$$Q_{AB} = \int_A^B \mathbf{q} \cdot d\mathbf{A} = \int_A^B \mathbf{q} \cdot (\mathbf{1z} \times d\mathbf{s}) = -\int_A^B q_x\, dy - q_y\, dx$$

$$= \int_A^B d\Psi = \Psi_A - \Psi_B. \tag{4.6.7}$$

Thus, the total discharge between two streamlines (actually two stream surfaces and the planes $z = 0$ and $z = 1$) is given by the difference between the values of the stream function corresponding to these lines. The dimensions of $\Psi$ are, therefore, volume per unit time per unit width.

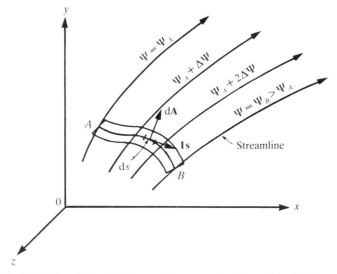

Fig. 4.10. The relationship between the stream function and the discharge.

MODELING TWO-DIMENSIONAL FLOW IN AQUIFERS 119

From (4.6.7) it follows that for $\Psi$ to be constant along a streamline, we should require that no sources or sinks should be present in the flow field.

Stream functions can also be defined for three-dimensional flow (e.g., Bear, 1972).

It may be of interest to note the relationship between the stream function and the piezometric head, $\phi$ in two-dimensional flow.

From (4.6.6) and Darcy's law *for a homogeneous isotropic porous medium*, we obtain

$$q_x = -\frac{\partial \Psi}{\partial y} = -\frac{\partial K\phi}{\partial x} \quad \text{or} \quad \frac{\partial \Psi}{\partial y} = \frac{\partial \Phi}{\partial x},$$

$$q_y = \frac{\partial \Psi}{\partial x} = -\frac{\partial K\phi}{\partial y} \quad \text{or} \quad \frac{\partial \Psi}{\partial x} = -\frac{\partial \Phi}{\partial y} \quad (4.6.8)$$

where $\Phi$ is often referred to as the *specific discharge potential* (noting that it is *valid only for a homogeneous, isotropic porous medium!*).

From (4.6.8) it follows that the curves $\Phi$ = constant and those of $\Psi$ = const are orthogonal to each other.

For the more general case of a nonhomogeneous anisotropic porous medium, we obtain

$$K_x \frac{\partial \phi}{\partial x} = \frac{\partial \Psi}{\partial y}; \quad K_y \frac{\partial \phi}{\partial y} = \frac{\partial \Psi}{\partial x} \quad (4.6.9)$$

where $x$ and $y$ were taken along the principal directions of **K**. The relationships between $\Phi$ and $\Psi$ in (4.6.8) and between $\phi$ and $\Psi$ in (4.6.9), are known as *Cauchy–Riemann conditions*.

In a homogenous isotropic porous media, it is also possible to define another stream function, $\psi$ defined by $\psi = \Psi/K$. Then the Cauchy–Riemann conditions (4.6.8) reduce to the form

$$\frac{\partial \psi}{\partial y} = \frac{\partial \phi}{\partial x}, \quad \frac{\partial \psi}{\partial x} = -\frac{\partial \phi}{\partial y}. \quad (4.6.10)$$

In this case, the families of curves $\phi$ = const and $\psi$ = const are orthogonal to each other.

### 4.6.4. PLANE FLOW MODELS IN TERMS OF $\Psi$

Consider the steady two-dimensional flow of a constant density fluid in a homogeneous isotropic porous medium. In such flow, div $\mathbf{q} = 0$ and $\mathbf{q} = -\text{grad } \Phi$, $\Phi = K\phi$, and hence curl $\mathbf{q} = 0$ (*potential flow*). From (curl $q$)$_z$ = 0, we obtain

$$\nabla^2 \Psi \equiv \frac{\partial^2 \Psi}{\partial x^2} + \frac{\partial^2 \Psi}{\partial y^2} = 0 \quad (4.6.11)$$

which is the partial differential equation that describes the flow in terms of $\Psi$.

For an inhomogeneous, yet isotropic, porous medium, where $K = K(x, y)$, we obtain from the requirement that $(\text{curl } \mathbf{J})_z \equiv \{\text{curl}(\mathbf{q}/K)\}_z = 0$

$$K\nabla^2 \Psi - \left( \frac{\partial K}{\partial x} \frac{\partial \Psi}{\partial x} + \frac{\partial K}{\partial y} \frac{\partial \Psi}{\partial y} \right) = 0. \tag{4.6.12}$$

When the porous medium is homogeneous, but anisotropic, the equation is

$$K_x \frac{\partial^2 \Psi}{\partial x^2} + K_y \frac{\partial^2 \Psi}{\partial y^2} = 0. \tag{4.6.13}$$

For flow in a confined aquifer, we define another stream function, $\Psi$, such that $\Delta\Psi$ gives the flow *through the entire thickness of an aquifer*, between adjacent streamlines. Then, for an isotropic aquifer

$$Q'_x = -\frac{\partial \Psi}{\partial y} = -T \frac{\partial \phi}{\partial x}, \qquad Q'_y = \frac{\partial \Psi}{\partial x} = -T \frac{\partial \phi}{\partial y} \tag{4.6.14}$$

In the absence of sources and sinks, and for steady flow, or neglecting the aquifer's storativity (i.e., $S(\partial\phi/\partial t) \approx 0$), Equation (4.2.23) reduces to

$$\nabla' \cdot \mathbf{Q}' \equiv \nabla' \cdot (T\nabla'\phi) = 0.$$

Then, with $\mathbf{J} = -\nabla'\phi$ and curl $\mathbf{J} = 0$, we obtain

$$(\text{curl } \mathbf{J})_z \equiv \{\text{curl}(\mathbf{Q}'/T)\}_z = 0$$

or

$$\frac{\partial J_x}{\partial y} - \frac{\partial J_y}{\partial x} \equiv \frac{\partial}{\partial x}\left(\frac{Q'_y}{T}\right) - \frac{\partial}{\partial y}\left(\frac{Q'_x}{T}\right) = 0$$

whence we obtain

$$\frac{\partial}{\partial x}\left(\frac{1}{T} \frac{\partial \Psi}{\partial x}\right) + \frac{\partial}{\partial y}\left(\frac{1}{T} \frac{\partial \psi}{\partial y}\right) = 0. \tag{4.6.15}$$

When we wish to express the flow model in terms of $\Psi$, we have also to provide appropriate initial and boundary conditions in terms of this variable. For example, an impervious boundary is also a streamline along which $\Psi = $ const. Along a boundary of specified flow, we express the flow in terms of increments in $\Psi$.

### 4.6.5. THE FLOWNET

Together, the two families of curves: equipotentials $\phi = $ const and streamlines $\Psi = $ const are called a *flownet*. In an isotropic porous medium, the two families are orthogonal to each other. In a homogeneous isotropic porous medium, we may

## MODELING TWO-DIMENSIONAL FLOW IN AQUIFERS

also draw a $\Phi - \Psi$ network, or a $\phi - \psi$ one, with the two families of curves being orthogonal to each other.

Since for $\Psi$ to be constant along streamlines, we require the absence of sources or sinks, whenever they are present, we exclude them from the flow domain and make them part of the latter's boundary.

Figure 4.11 shows a portion of a $\phi - \psi$ flownet in a homogeneous isotropic aquifer. It shows three streamlines and three equipotentials. It is customary to draw the flownet such that the increment $\Delta\phi$ between any two adjacent equipotentials is constant. For each *streamtube* (i.e., the space between adjacent streamlines), we have

$$\Delta Q' = T \, \Delta n_1 \, \frac{\Delta\phi}{\Delta s_1} = T \, \Delta n_2 \, \frac{\Delta\phi}{\Delta s_2} \tag{4.6.16}$$

where $Q'$ denotes the discharge through the entire thickness of the aquifer.

From (4.6.16) we obtain

$$\frac{\Delta n_1}{\Delta s_1} = \frac{\Delta n_2}{\Delta s_2} \tag{4.6.17}$$

that is, in a homogeneous medium the ratio $\Delta n / \Delta s$ must remain constant throughout the flownet.

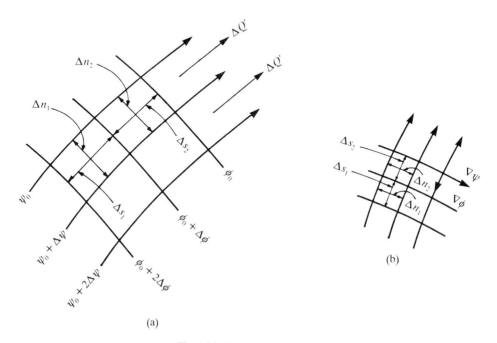

Fig. 4.11. Portion of a flownet.

For an inhomogeneous aquifer

$$T_1 \, \Delta n_1 \frac{\Delta \phi}{\Delta s_1} = T_2 \, \Delta n_2 \frac{\Delta \phi}{\Delta s_2}, \qquad T_1 \frac{\Delta s_1}{\Delta n_1} = T_2 \frac{\Delta s_2}{\Delta n_2} \qquad (4.6.18)$$

that is, the ration $\Delta s/\Delta n$ varies. When streamlines are approximately parallel (that is, $\Delta n_1 \approx \Delta n_2$), we have $(T_1/T_2) \simeq (\Delta \phi_2/\Delta s_2)/(\Delta \phi_1/\Delta s_1)$ that is, the transmissivity is inversely proportional to the hydraulic gradient; equipotentials will be closely spaced in regions of low transmissivity.

It is convenient to draw the flownet for a homogeneous isotropic aquifer so that approximate curvilinear squares are formed (Figure 4.11b). For this case $\Delta s = \Delta n$ and $Q = T \, \Delta \phi$. In certain cases, however, it is more convenient to draw the flownet for a given domain such that we have $m$ streamtubes, each carrying the same discharge $\Delta Q' = Q'_{\text{total}}/m$, and $n$ equal drops in piezometric head, $\Delta \phi$ $(= \phi_{\max} - \phi_{\min}/n)$. Then

$$Q'_{\text{total}} = m \, \Delta Q' = mT \, \Delta n \frac{\Delta \phi}{\Delta s} = mT \, \Delta n \frac{\phi_{\max} - \phi_{\min}}{n \, \Delta s}$$

$$= \frac{m}{n} \frac{\Delta n}{\Delta s} T(\phi_{\max} - \phi_{\min}) \qquad (4.6.19)$$

Note that in Figure 4.11, the stream function is such that $\Delta \Psi$ gives the discharge through the entire aquifer thickness, between adjacent streamlines.

CHAPTER FIVE

# Modeling Flow in the Unsaturated Zone

In order to reach a phreatic aquifer, water from precipitation, from irrigation, or from an influent river, infiltrates through the ground surface and percolates downward through the unsaturated zone. The same is true for pollutants carried with the water. These pollutants may be already present in the water reaching the ground surface, or they may be added to the water by processes of leaching, dissolution, and desorption along its path, from the ground surface to an underlying aquifer. Solid waste in landfills, septic tanks, fertilizers, pesticides and herbicides, applied over extended areas and dissolved in the water applied to the ground surface, may serve as examples of sources of pollutants that travel through the unsaturated zone.

Hence, the understanding of, and consequently the ability to calculate and predict the movement of water in the unsaturated zone, is essential when we wish to determine the (total) replenishment of a phreatic aquifer as part of our groundwater flow model. Information on the movement of water is also needed in order to forecast the movement and accumulation of pollutants in the unsaturated zone and the rate and concentration at which pollutants reach the water table.

In the following subsections, a brief review is presented on the motion and continuity equations of unsaturated flow. Only these concepts which are directly related to the modeling of the movement of water are reviewed. The movement of pollutants is considered in Chapter 6.

## 5.1. Capillarity and Retention Curves

### 5.1.1. MOISTURE CONTENT AND SATURATION

In unsaturated flow, the void space is partly filled by air and partly by water. Two state variables may be used to define the relative quantity of water at a certain time at a point in a porous medium domain (i.e., in an REV for which this point is a centroid)

$$\theta_w = \frac{\text{Volume of water in REV}}{\text{Volume of REV}} ; \quad 0 \leqslant \theta_w \leqslant n, \tag{5.1.1}$$

$$S_w = \frac{\text{Volume of water in REV}}{\text{Volume of voids in REV}} ; \quad 0 \leqslant S_w \leqslant 1. \tag{5.1.2}$$

Here $\theta_w$ is called the *water* (or *moisture*) *content*; $S_w$ is called the *water saturation*. Obviously, the two definitions are related to each other by

$$\theta_w = nS_w \tag{5.1.3}$$

where $n$ is the porosity at the considered point.

### 5.1.2. CONTACT ANGLE AND WETTABILITY

When a liquid is in contact with another substance (another liquid immiscible with the first, a gas, or a solid), a *free interfacial energy* exists between them. The interfacial energy arises from the difference between the resultant attraction of all the molecules surrounding a molecule located in the interior of each phase and that of a molecule located on the surface separating the phases. In the former case, due to the spherical symmetry in the type and spatial distribution of the molecules, the resultant force is zero. On the other hand, for a molecule on the surface of separation, a nonzero resultant force arises. Since a surface possessing free energy contracts, if it can do so, the free interfacial energy manifests itself as an *interfacial tension*. The interface behaves *as if it were a thin membrane* under tension that tends to reduce its surface area, if it can do so. Thus, the interfacial tension, $\sigma_{ik}$, for a pair of substances $i$ and $k$ is defined as the amount of work that must be performed in order to increase their common contact area by one unit. For air and water, $\sigma_{ik} = 72.5$ erg/cm$^2$ (or 72.5 dyne/cm) at 20°C. The interfacial tension between a substance and its own vapor is sometimes called *surface tension*. However, the term surface tension is often also used to indicate the interfacial tension between two phases. The interfacial tension and the surface tension are temperature dependent.

Figure 5.1a shows two immiscible fluids in contact with a solid plane surface. The angle $\theta$, called the *contact angle*, denotes the angle between the solid surface and the liquid-gas, or liquid-liquid interface, measured through the denser fluid. Equilibrium of forces requires that

$$\sigma_{LG} \cos \theta + \sigma_{SL} = \sigma_{GS} \quad \text{or} \quad \cos \theta = \frac{\sigma_{GS} - \sigma_{SL}}{\sigma_{LG}}. \tag{5.1.4}$$

Equation (5.4), called *Young's equation*, states that $\cos \theta$ is defined as the ratio of

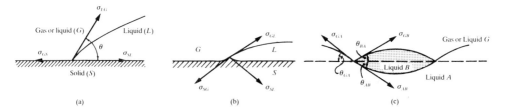

Fig. 5.1. Interfacial tension.

the energy released in forming a unit area of an interface between a solid, S, and a liquid, L, instead of between a solid, S, and a fluid, G, to the energy required to form a unit area between the liquid L and the fluid G. An attempt to write a balance of forces along the normal to the plane will show that, in principle, Figure 5.1a is incorrect and that the situation shown in Figure 5.1b is probably closer to reality, as is the case when three fluids are in contact (Figure 5.1c).

Sometimes, a factor is introduced in (5.1.4) to account for the roughness of the solid. From (5.1.4) it follows that no equilibrium is possible if $(\sigma_{GS} - \sigma_{SL})/\sigma_{LG} > 1$. In such case, the liquid, L, (Figure 5.1a) will spread indefinitely over the surface. This leads to the *concept of wettability* of a solid by a liquid.

The product $\sigma_{LG} \cos \theta$, called *adhesion tension*, determines which of the two fluids (L or G) will preferentially wet the solid, i.e., adhere to it and tend to spread over it.

When $\theta < 90°$, the fluid (e.g., L in Figure 5.1a) is said to wet the solid and is called a *wetting fluid*. When $\theta > 90°$, the fluid (G in Figure 5.1a) is called a *nonwetting fluid*. In any system similar to that shown in Figure 5.1a, it is possible to have either a fluid L-wet or a fluid G-wet solid surface, depending on the chemical composition of the two fluids and the solid. In the unsaturated (air-water) zone in a soil, water is the wetting phase, while air is the nonwetting one. Additives (called *surface active agents*) in a liquid alter surface tension and, hence, also the contact angle.

Interfacial tension and wettability may be different when a fluid-fluid interface (e.g., an air-water interface) is advancing or receding on a solid surface. This phenomena is called *hysteresis* (see Subsection 5.1.4).

With the concept of wettability as defined above, and for the air-water system considered here, we may distinguish three ranges of water saturation between the limits of 0 and 100%. Figure 5.2 shows water in a water wet granular soil (e.g., sand). At a very low saturation (Figure 5.2a), water forms rings, called *pendular rings*, around grain contact points. The air-water interface has the shape of a 'saddle'.

At this low saturation, the rings are usually isolated and do not form a continuous water phase. Although a very thin film of water, several molecules thick (which does not behave as ordinary liquid water, due to the very strong forces of attraction between the water and solid molecules) does remain on the solid surface, practically no pressure can be transmitted through it from one ring to the next. Figure 5.2b shows a pendular ring between two spheres. For this idealized case, it is possible to relate the volume of the ring to the radius of curvature of the air-water interface; the latter, in turn is related, as we shall see below, to the difference in pressures in the air and the water across it.

As water saturation increases, the pendular rings expand until a continuous water phase is formed. The saturation at which this occurs is called *equilibrium water saturation*. Above this critical saturation, the saturation is called *funicular* and flow of water is possible (Figure 5.2c). Both the water and the air phases are

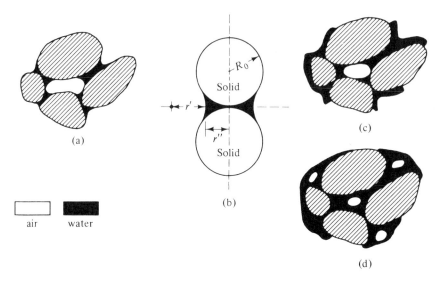

Fig. 5.2. Water and air saturation states. (a) Pendular saturation. (b) Pendular ring between two spheres. (c) Funicular saturation. (d) Insular air saturation.

continuous. As the water saturation increases, a situation develops in which the air (= nonwetting phase) is no longer a continuous phase; it breaks into individual bubbles lodged in the larger pores (Figure 5.2d). The air is then said to be in a state of *insular saturation*. A bubble of air can move only if a pressure difference, sufficient to squeeze it through a capillary size restriction, is applied across it in the water. Obviously, if all the air can escape from the void space (or be dissolved in the water) we have complete water saturation.

Sometimes the term *adsorbed stage* is used for water present in the pore space at a very low saturation, such that it forms continuous, or discontinuous, films of one or more molecular layers on adsorption sites on the solid.

5.1.3. CAPILLARY PRESSURE

When two immiscible fluids (here, air and water) are in contact, a discontinuity in pressure exists across the interface separating them. This is a consequence of the interfacial tension which exists between the two phases in contact. The magnitude of the pressure difference depends on the curvature of the interface at that point (which, in turn, depends on the saturation). Here 'point' is a microscopic point on the air-water interface inside the void space. The difference in pressure, $p_c = p_{air} - p_{water}$, is called *capillary pressure*, where the pressures are taken in the two phases as the interface is approached from their respective sides. In a general two-phase system, $p_c$ is the difference between the pressure on the nonwetting fluid side of the interface and that on the wetting one.

In order to determine the relationship between the curvature of the interface

## MODELING FLOW IN THE UNSATURATED ZONE

between air and water (or between any two phases), the interfacial tension and the pressures, in the two fluids separated by the interface, the latter is considered as a (two-dimensional) material body (actually, surface) which has rheological properties of its own. Its behavior is similar to that of a stretched membrane under tension in contact with the adjacent two fluids. In fact, with this assumption, the consideration of equilibrium surface tension leads to the conclusion that the normal component of fluid stress, or pressure, must be discontinuous at a curved interface. Scriven (1960) and Slattery (1967) present detailed analyses of interface behavior. However, for our purpose, a much simpler approach will suffice.

Figure 5.3 shows an infinitesimal element of a curved air-water interface. By writing a balance of force components along the normal to the interface, with a constant interfacial tension, $\sigma_{aw}$, we obtain

$$p_c = p_a - p_w = \sigma_{aw}\left(\frac{1}{r'} + \frac{1}{r''}\right) = \frac{2\sigma_{aw}}{r^*} \tag{5.1.5}$$

where $p_w$ and $p_a$ denote the pressure in the water and in the air, respectively, $\sigma_{aw}$ denotes the air-water interfacial tension, $r'$ and $r''$ denote the principal radii of curvature, and $r^*$ is the mean radius of curvature defined by $2/r^* = 1/r' + 1/r''$. Equation (5.1.5) is known as the *Laplace formula for capillary pressure*. The capillary pressure is thus a measure of the tendency of the partially saturated porous medium to suck in water, or to repel air. In soil science, the negative value of the capillary pressure is often called *suction*, or *tension*.

If we assume the air in the void space to be everywhere at atmospheric pressure, then the water in the void space is at a pressure, $p_w$, less than

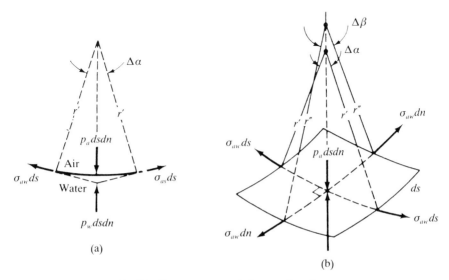

Fig. 5.3. Forces at a curved air-water interface.

atmospheric. In the more general case, $p_a \neq 0$. A simple model that explains what happens in the void space is the water in a capillary tube (simulating the narrow opening between grains) shown in Figure 5.4. For $p_a = 0$, the pressure, $p_w$, in the water just below the meniscus is $p_w = -p_c$.

We recall that once we understand the phenomena that take place at the microscopic level, i.e., at points within each pore, we average the microscopic values of state variables in order to obtain the corresponding macroscopic ones. Here, the pressure in the water ($= p_w$) and that in the air ($= p_a$) are the state variables under consideration. At every instant of time, the water is distributed within the void space, with meniscii separating it from the air. The pressures in the water are distributed within the entire void space, such that the water is under dynamic equilibrium (for a moving water phase, see Subsection 5.1.4 and Figure 5.7). Similarly, there exists a certain air pressure distribution within the air phase. Hence, for every point within the unsaturated domain, we employ (1.4.3) to define averaged water and air pressures, $\overline{p}_w^w$ and $\overline{p}_a^a$. Then the averaged capillary pressure (denoted here by $p_c$) at a point within a porous medium domain is defined by

$$p_c = \overline{p}_a^a - \overline{p}_w^w. \tag{5.1.6}$$

Figures 5.5 and 5.6 show how the negative (i.e., less than atmospheric) pressure in the water can be determined. In Figure 5.5a, the unsaturated soil sample is placed on a *porous membrane* (or *porous plate*) which has very small openings, such that air cannot be sucked through them into the manometer, even through the largest openings (recall that $p_c = 2\sigma_{aw}/r^*$) at the range of suctions planned for the experiment. Suction is applied by draining water through the stopcock. After some time, equilibrium is reached between the water in the soil and in the manometer. The manometer reads an average pressure over the area of contact between the water in the soil and in the manometer (see Bear et al., 1968, p. 46). Porous plates with smaller openings (e.g., unglazed earthenware or porcelain) are used for higher

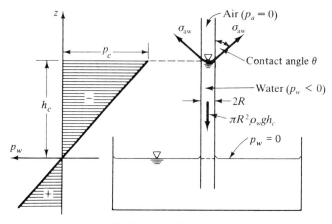

Fig. 5.4. A capillary tube.

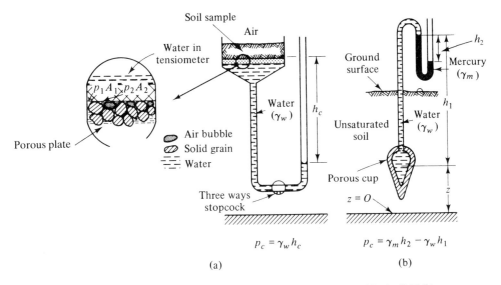

Fig. 5.5. Measurement of capillary pressure in the laboratory (a) and in the field (b).

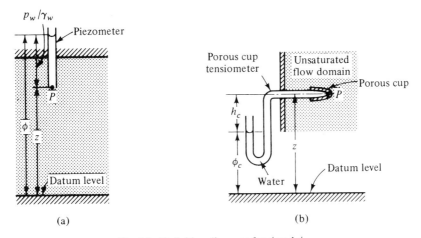

Fig. 5.6. Definition diagrams for $\phi$ and $\phi_c$.

suctions. The porous plate, which is thus permeable to water but impermeable to air, is necessary in order to establish hydraulic contact between the water in the soil and that in the manometer, without air being sucked into the latter.

The instrument used for measuring the capillary pressure in an unsaturated soil is called a *tensiometer* (a name introduced by Richards and Gardner, 1936). The contact between the water in the tensiometer and that in the soil is established through a porous cup.

In using a tensiometer, one has to make sure that equilibrium has been reached, as sometimes this may take a very long time. Also, that suction is such that air is not drawn through the porous plate at any point. The pressure at which air will enter is called *bubbling pressure* (or *air entry pressure*, or *threshold pressure*). Membrane materials are available with bubbling pressures of 20—50 atmospheres. We note that all measurements of averaged pressures by tensiometers, or piezometers, are actually areal averaged ones, rather than volumetric. In using the former to replace the latter in our (macroscopic) mathematical models, we implicitly assume that both averages are the same.

By analyzing the forces acting on the water column in the capillary tube of Figure 5.4, we find

$$h_c \pi R^2 \rho_w g = 2\pi R \sigma_{aw} \cos \theta, \qquad h_c = 2\sigma_{aw} \cos \theta / R \rho_w g \qquad (5.1.7)$$

where $\sigma_{aw}$ denotes surface tension and $\rho_w$ is the water density. With the average radius of curvature of the meniscus, $r^*$, equal to $R/\cos \theta$, we obtain from (5.1.5)

$$h_c = p_c/\rho_w g = -p_w/\rho_w g, \qquad p_a = 0 \qquad (5.1.8)$$

where $h_c$ is called the *capillary pressure head*. By analogy, the same definition and (5.1.8) are employed for a soil, with $h_c$, $p_c$ and $\rho_w$ representing average values.

As in a saturated flow of water, also in an unsaturated flow, we may define a piezometric head, $\phi = z + p_w/\gamma_w$, at every point of the flow domain. Often, the term *capillary head* (symbol $\phi_c$) is used to denote the piezometric head in unsaturated flow (Figure 5.6b)

$$\phi_c = z + p_w/\gamma_w = z - p_c/\gamma_w = z - h_c = z - \psi, \quad p_c = -p_w > 0, p_a = 0 \quad (5.1.9)$$

The term *suction* (symbol $\psi$) is used here for the negative of the pressure head so that $p_w < 0$, but $\psi \equiv -p_w/\gamma_w > 0$. As in saturated flow, the concepts of head ($\phi_c$, $\psi$, and $h_c$) should be used only when $\rho_w = $ const. Some authors use $\psi = p_w/\gamma_w$.

### 5.1.4. MOISTURE RETENTION

Let water be drained by lowering the manometer limb or through the stopcock from an initially saturated sample placed on the porous plate of the tensiometer shown in Figure 5.5a. Figure 5.7 (after Childs, 1969) shows the distribution of water within the pore space at several successive stages of drainage. The initial stage is denoted by 1. As water is drained, interfaces (meniscii) are formed (2). The radii of curvature of the interfaces depend on the magnitude of the suction. As water is drained and the interfaces are drawn further down, the curvature becomes larger (i.e., small radius of curvature) and the suction increases. At every stage, the greatest suction that can be maintained by the interface corresponds to the largest curvature that can be accommodated in the channel through which the interface is being withdrawn and the largest curvature occurs at the narrowest part (e.g., interface 3 in Figure 5.7). As drainage progresses, the interface retreats into channels which support a curvature of a greater radius (e.g., interface 4). However, since this means reduced suction, this is a nonequilibrium stage and the water will

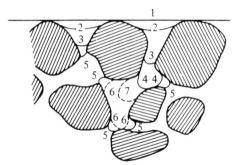

Fig. 5.7. Stages of water drainage (1—5) and re-entry (6—7) (*after Childs, 1969*).

continue to retreat until the interfaces have taken up a position of equilibrium in channels which are sufficiently narrow to support the interfaces of a larger curvature. Obviously, if all channels are equal and large, at a certain suction equilibrium cannot be maintained any more and a sudden, almost complete, withdrawal of the water from the soil will be observed. We say 'almost' because some water will still remain as isolated pendular rings. Within such rings, which are completely isolated from each other, the pressure is independent of the pressure in the remaining, continuous water body in the void space (except for dependence through the water vapor phase).

In general, the pores have different dimensions and, therefore, will not empty at the same suction. The large pores (or those with large channels of entry) will empty at low suctions, while those with narrow channels of entry, supporting interfaces of a larger curvature, will empty at higher suctions.

Let us now reverse the process, and rather than increase the suction in order to empty more pores, reduce it in an attempt to refill the pore space. The transition is now from stage 5 to stages 6 and 7 in Figure 5.7. The interface curvature becomes progressively smaller.

Figure 5.8 shows typical examples of curves $p_c = p_c(S_w)$, or $h_c = h_c(S_w)$, during drainage. In soil science, these curves are called *retention curves*, as they show how much water is retained in the soil by capillary forces against gravity. Some authors refer to a drainage retention curve as a *desorption curve* and to an imbibition curve as a *sorption curve*. Point $A$ in Figure 5.8 is the *critical capillary head, $h_{cc}$*. If we start from a saturated sample, say, in the apparatus shown in Figure 5.5a, and produce a small capillary head $h_c$, almost no water will leave the sample (i.e., no air will penetrate the sample) until the critical capillary head is reached. When expressed in terms of pressure, the critical value (point $A$ of Figure 5.8) is called the *bubbling pressure*. As the value of $h_c$ is increased, an initial small reduction in $\theta_w$, associated with the retreat of the air-water meniscii into the pores at the external surface of the sample, is observed. Then, at the critical value $h_{cc}$, the larger pores begin to drain.

The shape of the retention curve and, hence, also the threshold pressure, depends on the pore-size distribution and pore shapes of the porous medium.

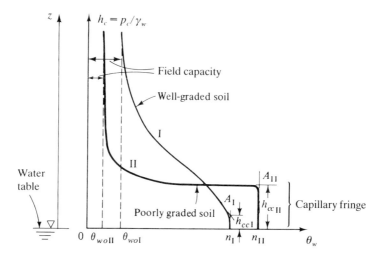

Fig. 5.8. Schematic retention curves during drainage.

As drainage progresses, we observe that a certain quantity of water remains (in the form of isolated pendular rings and immobile thin films) in the sample even at very high capillary pressures. This value of $\theta_w$, denoted by $\theta_{w0}$, is called *irreducible water content*. In terms of saturation, it is denoted by $S_{w0}$ ($= \theta_{w0}/n$) and called *irreducible water saturation* (Figure 5.8).

Upon rewetting (or imbibition), we observe that the retention curve $h_c = h_c(\theta_w)$ differs from that obtained during drainage. We have here the phenomenon of *hysteresis* resulting from two phenomena. The first, called the *ink-bottle effect* (Figure 5.9a), results from the fact that as water re-enters narrow channels, a local

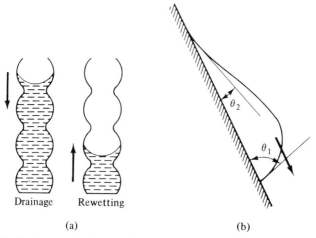

Fig. 5.9. Factors causing hysteresis in retention curve. (a) The ink-bottle effect. (b) The raindrop effect.

MODELING FLOW IN THE UNSATURATED ZONE    133

increase of suction is required. In the soil (Figure 5.7) at this stage we have instability and the interface cannot advance until a neighboring pore is filled. Equilibrium at a given suction may be obtained with somewhat different $\theta_w$. The second effect, sometimes called the *raindrop effect* (Figure 5.9b), is due to the fact that the contact angle at an advancing interface differs from that at a receding one. Entrapped air is another factor causing hysteresis.

Figure 5.10 shows hysteresis in the relationship $h_c = h_c(\theta_w)$. The drainage and imbibition curves form a closed loop. In fine-grained soil, the effect indicated as caused by entrapped air, is also cauesd by subsidence or shrinkage. Entrapped air may be removed with time, e.g., by flow and dissolution of the air in the water. It is possible to start the imbibition process from any point on the drainage curve, or to start the drainage process from any point on the imbibition curve, leading to the dashed lines, called *drying* and *wetting scanning curves*. In this way, the relationship between capillary pressure and saturation (expressed by the retention curve) also depends on the wetting-drying history of the particular sample under consideration. For a given capillary pressure, a higher saturation is obtained when a sample is being drained than during imbibition. As long as the soil remains stable (i.e., no consolidation), the hysteresis loop can be repeatedly traced.

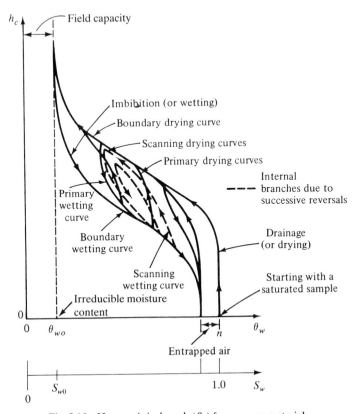

Fig. 5.10. Hysteresis in $h_c = h_c(\theta_w)$ for a coarse material.

For a given soil, the retention curve $p_c = p_c(S_w)$, or $h_c = h_c(S_w)$, can be obtained in the laboratory using an apparatus similar to that described in Figure 5.5a.

At equilibrium, with no flow taking place, the piezometric head, $\phi$, is the same for all points of an unsaturated zone. Consider points 1 and 2 with $p_{c1}$, $\phi_1$, and $p_{c2}$, $\phi_2$, respectively. We have

$$\phi_1 = z_1 + \frac{p_{w1}}{\gamma_w}, \quad \phi_2 = z_2 + \frac{p_{w2}}{\gamma_w}, \quad \gamma_w = \text{const.} \tag{5.1.10}$$

From $\phi_1 = \phi_2$, and since the air phase is taken at $p_a = 0$ (i.e., atmospheric pressure), it follows that

$$z_1 - z_2 = (p_{w2} - p_{w1})/\gamma_w = (p_{c1} - p_{c2})/\gamma_w. \tag{5.1.11}$$

If $z_2 = 0$ is chosen as a point on the phreatic surface, where $p_{w2} = 0$, then $p_{c2} = 0$. Denoting $z_1 \equiv z$, $p_{c1} \equiv p_c$, we obtain

$$z = p_c(S_w) \quad \text{or} \quad z = z(S_w) = h_c. \tag{5.1.12}$$

From (5.1.12) it follows that the curves $z = z(S_w)$ and the retention curve, $h_c = h_c(S_w)$ are identical. This means that for a homogeneous soil, with a sufficiently deep water table, the retention curve can be used to describe the moisture distribution between the phreatic surface and the ground surface at equilibrium.

Accordingly, immediately above a water table ($p_w = 0$) we have a zone that is saturated with water, or nearly so, because a certain suction must be reached before any substantial reduction in water content can be produced. Then, above this zone there is a marked drop in the water content with a relatively small rise in the capillary pressure. This zone contains most of the water present in the zone of aeration. From Figure 5.8 it is clear that this statement better describes the situation for poorly graded or coarse-textured soil (sand, gravel, etc.), but is also valid for fine-textured, or well-graded, soils when the water table is sufficiently deep below the ground surface. As this phenomenon is analogous to the rise in a capillary tube, where the water rises to a certain height above the free water surface, with a fully saturated tube below the meniscus and zero saturation above it, the nearly saturated zone above the phreatic surface, when it occurs, is called the *capillary fringe*, or *capillary rise* (see Figure 1.3). Thus, $h_c$ in Figure 5.8 is the capillary rise for a poorly graded soil.

The capillary fringe is thus an approximate practical concept that is very useful and greatly simplifies the treatment of phreatic flows when we wish to take into account the fact that a certain saturated zone (or nearly so) is present above a phreatic surface (and up to the water table; see Figure 1.3).

The following formulas may be used to estimate the thickness of the capillary fringe $h_{cc}$. Mavis and Tsui (1939):

$$h_{cc} = \frac{2.2}{d_H}\left(\frac{1-n}{n}\right)^{3/2}, \quad d_H \text{ and } h_{cc} \text{ in inches,} \tag{5.1.13}$$

where $d$ is the mean grain diameter, and Polubarinova-Kochina (1952):

$$h_{cc} = \frac{0.45}{d_{10}} \frac{1-n}{n}, \quad h_{cc} \text{ and } d_{10} \text{ in cm,} \qquad (5.1.14)$$

where $d_{10}$ is the effective particle diameter. Silin-Bekchurin (1958) suggested a capillary rise of 2–5 cm in coarse sand, 12–35 cm in sand, 35–70 cm in fine sand, 70–150 cm in silt, and more than 2m in clay.

### 5.1.5. FIELD CAPACITY

*Field capacity* is usually defined as that value of water content remaining in a unit volume of soil after downward gravity drainage has ceased, or materially done so, say, after a period of rain, or excess irrigation. A difficulty inherent in this definition is that no quantitative specification of what is meant by 'materially ceased' is given. Although, according to this definition, field capacity is a property of a unit volume of soil (depending on the soil structure, grain-size distribution, etc.), it is obvious from any of the curves describing the moisture distribution above the water table (e.g., Figure 5.8, with $h_c$ replaced by the elevation, $z$, above the water table) that the amount of water retained in a unit volume of soil at equilibrium under field conditions, depends on the elevation of this unit volume above the water table. In addition, in the soil-water zone adjacent to the ground surface (Figure 1.3), equilibrium is seldom reached, as water in this zone constantly moves up or down and the water content is also being reduced by plant uptake. From these observations, it follows that the above definition of field capacity should be supplemented by the constraint that the soil sample should be at a point sufficiently high above the water table. Returning to the relationship $h_c = h_c(S_w)$, the notion of field capacity of unsaturated flow is identical to the notion of the irreducible moisture content in Figures 5.8 and 5.10. The field capacity, $\theta_{w0}$, is shown in the figures. The complement of the field capacity, i.e., volume of water drained by gravity from a unit volume of saturated soil, is called effective porosity and is denoted by $n_e$ ($= n - \theta_{w0}$). At any elevation, once gravity drainage has materially ceased, a certain amount of moisture is retained in the soil. The moisture distribution is shown by the retention curve, with $z$ instead of $h_c$ as ordinate.

### 5.1.6. SPECIFIC YIELD

*Specific yield* is another unsaturated flow concept employed in investigations of drainage of agricultural lands and in groundwater hydrology. It is defined as the volume of water drained from a soil column of the unit horizontal area extending from the water table to the ground surface, per unit lowering of the water table. The corresponding amount of water retained in the soil against gravity when the water table is lowered, is called *specific retention*. When expressed in terms of moisture content, we obtain for every instant

$$\theta_{wy} + \theta_{wr} = n \qquad (5.1.15)$$

where $\theta_{wy}$ denotes specific yield and $\theta_{wr}$ denotes specific retention. By dividing (5.1.15) by porosity, we obtain the same relationship in terms of saturation

$$S_{wy} + S_{wr} = 1 \qquad (5.1.16)$$

We note that in Figure 4.2, in (4.1.4), and, in general, in the definition of specific yield as equivalent to the storativity of a phreatic aquifer, the specific yield (and also the specific retention) are in terms of moisture content. Thus, (5.1.15) is actually identical to (4.1.4), with $S_y \equiv \theta_{wy} (= nS_{wy})$.

Thus, specific retention is a field concept obtained by averaging what actually happens in the zone of aeration when the water table is lowered. In Section 4.1, $S_y$ is defined as the storativity, or specific yield, of a phreatic aquifer. Figure 5.11 shows the effect of depth on the specific yield. With the nomenclature of this figure, we have per unit area

$$S_y(d', d''; t', t'') = \frac{\text{Volume of water drained between } t' \text{ and } t''}{(d'' - d')}$$

$$= \frac{1}{d'' - d'} \left\{ n(d'' - d') + \int_{z'=0}^{d'} \theta'_w(z', t') \, dz' - \int_{z''=0}^{d''} \theta''_w(z'', t'') \, dz'' \right\} \qquad (5.1.17)$$

where $S_y \equiv \theta_{wy} = nS_{wy}$. The volume of water drained is indicated by the shaded area in Figure 5.11. For a homogeneous isotropic soil, the two curves $\theta'_w = \theta'_w(z', t')$ and $\theta''_w = \theta''_w(z'', t'')$ are identical in shape. If both water table positions are sufficiently deep below the ground surface, the two curves will merge at $\theta_w = \theta_{w0}$. Hence, we have for very large $d'$ and $d''$

$$S_{y\infty} \equiv S_y|_{d \to \infty} = n - \theta_{w0} = (1 - S_{w0})n. \qquad (5.1.18)$$

It is thus apparent that for a homogeneous isotropic soil and very deep water table, the specific retention is identical to the field capacity. For such conditions, Figure 4.2 shows the relationship between specific yield and specific retention for various soils. However, when the soil is inhomogeneous (e.g., composed of layers), or when the water table is at a shallow depth, the moisture distribution curves, corresponding to the two water table positions, are no longer parallel, and (5.1.18) is no longer valid; we must distinguish between field capacity and specific retention.

## MODELING FLOW IN THE UNSATURATED ZONE

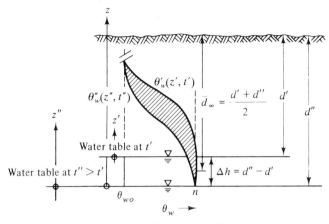

(a) Deep water table.

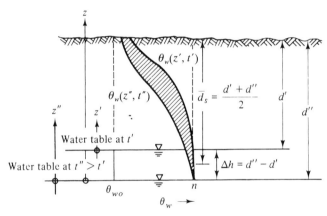

(b) Shallow water table.

Fig. 5.11. Effect of depth on specific yields.

Since it takes time for drainage to be completed, we obtain a specific yield that is time-dependent and that approaches asymptotically the values corresponding to the depths considered. When the water table is lowered instantaneously (or relatively fast), say, as a result of drainage, the corresponding changes in the moisture distribution lag behind and reach a new equilibrium (or practically so) only after a certain time interval that depends on the type of soil. A time lag will also take place when infiltration causes the water table to rise. When the water table is rising or falling slowly, the changes in moisture distribution have sufficient time to adjust continuously and the time lag practically vanishes.

### 5.1.6. CONDITIONS BETWEEN TWO POROUS MEDIA

At an interface between two porous media (say, a coarse sand and a fine one), we require that the pressure, $p_w$, in the water (actually also that in the air, $p_a$) be the same as the surface of separation is approached from both sides. Denoting the two media by subscripts 1 and 2, this means that since we assume $p_{a1} = p_{a2} = 0$, we have

$$p_{w1} = p_{w2} \quad \text{or} \quad p_{c1} = p_{c2}. \tag{5.1.19}$$

Figure 5.12 shows, schematically, the two retention curves. It is obvious that *we have a jump in the water saturation across the surface separating the two media.*

## 5.2. Motion Equations

In principle, both the water and the air move simultaneously in the void space. Movement of water vapor affected by variations in temperature and in the concentration of dissolved solids, may also take place. Nevertheless, in what follows we shall ignore the movement of water in the vapor phase. We shall also ignore movement due to pressure differences resulting from variations in salt concentration (*osmotic effect*) and movement due to temperature variations (*thermo-osmotic effect*). We shall assume that the solid matrix is rigid and stable (i.e., with no consolidation or subsidence). Some of the above phenomena, however, may be important in certain situations and should not be ignored. Thus, we shall consider flow resulting only from variations in water pressure; the density, $\rho_w$, may vary.

### 5.2.1. MOTION EQUATIONS FOR WATER AND AIR

Many investigators conclude from experiments that when two immiscible fluids flow simultaneously through a porous medium, each fluid establishes its own

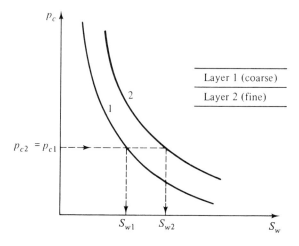

Fig. 5.12. Saturation discontinuity across the surface separating two porous media.

tortuous paths through the void space. They assume that a unique (or nearly so) set of channels corresponds to every degree of saturation. From the discussion in Section 5.1, it follows that as the degree of saturation of a nonwetting fluid (here, the air) is reduced, the channels of that fluid tend to break down until only isolated regions of it remain at *residual nonwetting fluid saturation*. Similarly, when the saturation of the wetting fluid, here the water, is reduced, it becomes discontinuous at the *irreducible wetting fluid saturation*. When any of these fluids becomes discontinuous, no flow of that fluid can take place.

With these ideas in mind, it seems natural to employ the concept of permeability established for the saturated flow of a single fluid, modifying its value due to the presence of the second phase which occupies part of the void space. In fact, as in the case of saturated flow, the motion equations for unsaturated flow can be derived as an averaged momentum balance equation for each phase, taking into account that the latter occupies only part of the void space. Accordingly, we may now use (2.1.34), with $nS_w V_{wi} = q_i$, for an anisotropic porous medium, to describe, separately, the flow of each of the two phases — the wetting one (here, water) and the nonwetting one (here, air). The difference, however, is that in this case the *permeability for each of the phases is a function of the degree of saturation*. In what follows, we shall neglect the solid's velocity (if such exists), assuming $|\mathbf{q}_w| \gg |\theta_w \mathbf{V}_s|$, and $|\mathbf{q}_a| \gg |\theta_a \mathbf{V}_s|$ so that $\mathbf{q}_w \approx \mathbf{q}_{wr}$ and $\mathbf{q}_a \approx \mathbf{q}_{ar}$.

Accordingly, the motion equations for the air and water in unsaturated flow in an anisotropic porous medium are

$$q_{wi} = -\frac{k_{wij}}{\mu_w}\left\{\frac{\partial p_w}{\partial x_j} + \rho_w g \frac{\partial z}{\partial x_j}\right\}, \qquad \mathbf{q}_w = -\frac{\mathbf{k}_w}{\mu_w}(\nabla p_w + \rho_w g \nabla z),$$

$$q_{ai} = -\frac{k_{aij}}{\mu_a}\left\{\frac{\partial p_a}{\partial x_j} + \rho_a g \frac{\partial z}{\partial x_j}\right\}, \qquad \mathbf{q}_a = -\frac{\mathbf{k}_a}{\mu_a}(\nabla p_a + \rho_a g \nabla z) \quad (5.2.1)$$

where the subscripts *a* and *w* denote air and water, respectively, and the summation convention is employed.

The unsaturated permeabilities, $\mathbf{k}_w$ for the water and $\mathbf{k}_a$ for the air, are referred to as *effective permeabilities* of the respective fluids (see Subsection 5.2.2).

For constant densities, $\rho_w$ = constant, $\rho_a$ = constant, (5.2.1) can be rewritten in terms of the piezometric heads $\phi_w = z + p_w/\gamma_w$ and $\phi_a = z + p_a/\gamma_a$, in the form

$$\mathbf{q}_w = -\frac{\mathbf{k}_w(S_w)\gamma_w}{\mu_w} \cdot \nabla \phi_w = -\mathbf{K}_w(S_w) \cdot \nabla \phi_w,$$

$$\mathbf{q}_a = -\frac{\mathbf{k}_a(S_a)\gamma_a}{\mu_a} \cdot \nabla \phi_a = -\mathbf{K}_a(S_a) \cdot \nabla \phi_a \quad (5.2.2)$$

where $\mathbf{K}_w$ and $\mathbf{K}_a$ are the *effective hydraulic conductivities* of the water and of the air, respectively.

If we assume that the air is stationary, or ignore its movement, i.e., assuming $\phi_w$ = const, (5.2.2) reduces to a single equation — that of the water.

We recall that $p_a$ and $p_w$ in (5.2.1) are not independent of each other, as their difference defines the *capillary pressure* which is a function of the degree of saturation, say, $S_w$.

Other forms of the water motion equation for the case $p_a = 0$, are

$$\mathbf{q}_w = -\mathbf{K}_w(S_w) \cdot \nabla \phi_w = -\frac{\mathbf{K}_w(S_w)}{\gamma_w} \cdot \nabla p_w - \mathbf{K}_w(S_w) \cdot \nabla z, \tag{5.2.3}$$

$$\mathbf{q}_w = \mathbf{K}_w(S_w) \cdot \nabla \psi - \mathbf{K}_w(S_w) \cdot \nabla z \tag{5.2.4}$$

where the negative pressure head $\psi = -p_w/\gamma_w > 0$ is called *suction*. Since in this case $-p_w = p_c$, $\psi = p_c/\gamma_w \equiv h_c$, i.e., the *capillary head*. Because $\psi = \psi(S_w)$, and $p_c = p_c(S_w)$, we could replace $\mathbf{K}_w(S_w)$ by $\mathbf{K}_w(\psi)$ in (5.2.4). All equations could also be written in terms of $\theta_w$ rather than in terms of $S_w$. However, when the porosity, $n$, varies in space, it is advisable to use the saturation, $S_w$, as a variable, with a possibility that the retention curve will vary from one point to the next. In a consolidating porous medium, the porosity may also vary with time. We wish to emphasize again that motion equations can be written in terms of $\phi$ or $\psi$ only when water and air densities are assumed constant.

It is important to recall that the relationship $p_c = p_c(S_w)$ is not a unique one because of hysteresis: hence, the history of wetting and drying may play an important role in the analysis of flow problems.

Nevertheless, assuming that the relationship $p_c = p_c(S_w)$ is a unique one (as when dealing with a problem of only drainage), (5.2.4) written for an isotropic medium and $\gamma_w$ = const, becomes

$$\mathbf{q}_w = K_w(\theta_w) \frac{\mathrm{d}\psi}{\mathrm{d}\theta_w} \nabla \theta_w - K_w(\theta_w) \nabla z$$

$$= \left\{ \frac{K_w(\theta_w)}{\mathrm{d}\theta_w/\mathrm{d}\psi} \right\} \nabla \theta_w - K_w(\theta_w) \nabla z \tag{5.2.5}$$

where we have replaced $S_w$ by $\theta_w$ and $\psi = p_c/\gamma_w = \psi(\theta_w)$ is also assumed unique. Klute (1952) calls the group $D_w(\theta_w) = -K_w(\theta_w)\,\mathrm{d}\psi/\mathrm{d}\theta_w = -K_w(\theta_w)/(\mathrm{d}\theta_w/\mathrm{d}\psi) = K_w(\theta_w)/\gamma_w(\mathrm{d}\theta_w/\mathrm{d}p_w)$ *coefficient of moisture diffusivity*, or *capillary diffusivity* (dims. $L^2T^{-1}$). Then, (5.2.5) becomes

$$\mathbf{q}_w = -D_w(\theta_w)\nabla \theta_w - K(\theta_w)\nabla z. \tag{5.2.6}$$

For horizontal, two-dimensional water flow in the $xy$-plane

$$\mathbf{q}'_w = -D_w(\theta_w)\nabla' \theta_w, \quad \nabla'(\ ) = \frac{\partial(\ )}{\partial x}\mathbf{1x} + \frac{\partial(\ )}{\partial y}\mathbf{1y}. \tag{5.2.7}$$

For vertical water flow

$$q_{wz} = -D_w(\theta_w)\,\partial\theta_w/\partial z - K_w(\theta_w). \qquad (5.2.8)$$

Sometimes the definitions $C_w = d\theta_w/dp_w$ (= *water capacity*) and $D_w = -K_w/(d\theta_w/dp_w)$ are introduced. The similarity between (5.2.7) and Fick's law of diffusion explains why the term *diffusivity* is used here.

The second term on the right-hand side of (5.2.6) gives the effect of gravity. If the model describes the simultaneous flow of air and water in the unsaturated zone, equations similar to (5.2.5) through (5.2.8) may also be written for the air.

The dependence of $D$ on $\theta_w$, or of $K$ on $\psi$, introduces a nonlinearity into the equations of motion presented above and, hence, also into the continuity equation (Section 5.3). The gravity term in the motion equation also makes the continuity equation a rather difficult one for exact solution by analytical methods. Without the assumptions of uniqueness, $K(\theta_w)$, $p_c(\theta_w)$ and $D_w(\theta)$ are subject to hysteresis.

One should note that although the relationship $p_c = p_c(\theta_w)$ is usually obtained from a static test (i.e., in the absence of flow), we use it in the motion equation. We *assume* that the relationship remains unchanged also under dynamic conditions. Another interesting observation is that in the motion equations presented above, say (5.2.1), we have *assumed* no momentum transfer between the two phases across their common microscopic interfaces inside the pore space.

Had we taken such transfer into account, there would be coupling between the flow of the two phases: a pressure gradient in one phase would also produce flow in the adjacent phase. As is commonly done in water flow in the unsaturated zone in the soil, in what follows this possibility will be ignored.

In all the motion equations presented above, it was assumed that the solid is stationary and nondeformable (or practically so). Otherwise, similar to the situation that occurs in the case of saturated flow, as expressed by (2.1.34), the motion equations will take the forms

$$V_{wi} - V_{si} = -\frac{k_{wij}(S_w)}{\theta_w \mu_w}\left(\frac{\partial p_w}{\partial x_j} + \rho_w g \frac{\partial z}{\partial x_j}\right), \qquad (5.2.9)$$

$$V_{ai} - V_{si} = -\frac{k_{aij}(S_a)}{\theta_a \mu_a}\left(\frac{\partial p_a}{\partial x_j} + \rho_a g \frac{\partial z}{\partial x_j}\right) \qquad (5.2.10)$$

### 5.2.2. PERMEABILITY

In (5.2.2), $\mathbf{k}_w(\theta_w)$, or $\mathbf{k}_w(S_w)$ is the *effective water permeability* of the unsaturated soil, while $\mathbf{k}_a(\theta_a)$, or $\mathbf{k}_a(S_a)$ is the *effective air permeability*. Similarly, $\mathbf{K}_w(\theta_w)$ and $\mathbf{K}_a(\theta_a)$ are the *effective water and air hydraulic conductivities*, respectively.

In a deformable porous medium, the permeability is also a function of the porosity which varies as the soil is compacted. Thus, in principle, we have $\mathbf{k}_w = \mathbf{k}_w(S_w, n)$.

Similar to the case of saturated flow, for an anisotropic porous medium, effective permeabilities, to water and to air flow, are second-rank symmetric tensors. Accordingly, assuming that the soil is anisotropic, the effective permeability, as well as the corresponding effective hydraulic conductivity, appear in the motion equations as second rank tensors (with components $k_{wij}$, $k_{aij}$, $K_{wij}$, $K_{aij}$). For an isotropic porous medium, the effective permeabilities and the corresponding effective hydraulic conductivities reduce to scalars.

In principle, since momentum is transferred across the microscopic interfaces within the void space from one fluid to the other, the effective permeability should depend not only on the saturation, but, at least to some extent, also on the interfacial tension between the fluids, the preferential wettability of the solid to the two fluids, the rate of flow and the fluids' viscosities. However, most investigations seem to indicate that momentum transfer between the two fluids may be ignored and, hence, the *effective permeabilities are functions of the saturation only.*

For an isotropic porous medium, it is often convenient to employ the concept of *relative permeability*, defined as the ratio of effective permeability to the corresponding permeability at saturation, viz.,

$$k_{wr} = \frac{k_w(S_w)}{k_w|_{S_w=1}}, \qquad k_{ar} = \frac{k_a(S_a)}{k_a|_{S_a=1}}. \tag{5.2.11}$$

However, *the concept of relative permeability should not be applied to an anisotropic porous medium.* If we do employ the concept for an anisotropic medium, for each component $k_{wij}(S_w)$ we have to define a relative permeability $(k_{wr})_{ij}(S_w) = k_{wij}(S_w)/k_w|_{S_w=1}$, with a different functional relationship $(k_{wr})_{ij}(S_w)$ for each $ij$-component. Moreover, the nine terms $(k_{wr})_{ij}(S_w)$ thus defined, will not constitute components of a second-rank tensor (Bear, Braester and Menier, 1987).

Investigations, including visual studies, over the years have led to the conclusion that when immiscible fluids (here, air and water) flow simultaneously through a porous medium, each fluid establishes its own tortuous path through a certain network of channels within the porous medium. These channels are very stable, with a different set of channels corresponding to every degree of saturation. As the saturation of the nonwetting fluid is reduced, the channels for that fluid tend to break down until only isolated bubbles of the nonwetting fluid remain. At this saturation (called *residual nonwetting fluid saturation*, $S_{nw0}$), pressure can no more be transmitted through the nonwetting fluid and its permeability vanishes. These bubbles can be forced to move only by applying very high pressure gradients in the wetting fluid.

Similarly, as the wetting fluid saturation decreases, its flow channels tend to break down and become discontinuous. Permeability vanishes at the *irreducible wetting fluid saturation*, $S_{w0}$.

Figure 5.13 shows the variations of relative permeability to the water, $k_{wr}$, with saturation $S_w$, for an isotropic porous medium, according to experiments by Wyckoff and Botset (1936). As the saturation decreases, the large pores drain first

MODELING FLOW IN THE UNSATURATED ZONE    143

so that the flow takes place through the smaller ones. This causes both a reduction in the cross-sectional area available for the flow and an increase in tortuosity of the flow paths. The combined effect causes a rather rapid reduction in the permeability as the moisture content decreases. Point $A$ in Figure 5.13 indicates the irreducible water saturation, $S_{w0}$. At this point, the discontinuous water phase exists only in the very small pores, as a very thin film on the solid in the larger pors and as isolated pendular rings. We note that due to the steepness of the curve at high saturation, when saturation is reduced, say by entrapped air, even by 10 or 20%, the relative permeability is appreciably reduced.

Several authors suggest relationships between the effective hydraulic conductivity, $K_w$, and saturation, $S_w$ (or water content, $\theta_w$). Childs and Collis-George (1950) assume

$$K_w(\theta_w) = B\theta_w^3/M^2 \qquad (5.2.12)$$

where $M$ is the specific surface area of the solid phase and $B$ is a constant.

Irmay (1954) derives a similar relationship, assuming that the resistance to flow offered by the solid matrix is proportional to the solid-liquid interfacial area. The effective hydraulic conductivity, $K_w$, then becomes proportional to the hydraulic radius (= volume of voids divided by wetted area of solid). This leads to the relationship

$$K_w(S_e) = K_0 S_e^3 \qquad (5.2.13)$$

where $S_e = (S_w - S_{w0})/(1 - S_{w0}^{..})$ is called the *effective saturation*, and $K_0$ is the hydraulic conductivity at saturation. The experimental curve of Figure 5.13 fits such a cubic parabola. Experiments by several authors with soils of uniform grain size seem to agree with the relationship (5.2.13).

In general, we have also to consider the permeability to the nonwetting phase

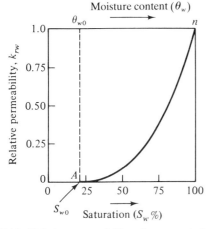

Fig. 5.13. Relative permeability of unsaturated sand according to experiments by Wyckoff and Botset (1936) and theoretical analysis by Irmay (1954).

(= air). Figure 5.14a shows typical $k_{wr}(S_w)$ and $k_{nwr}(S_{nw})$ curves; the dashed portions of the curves correspond to the case when we start from complete saturation of the considered phase. Otherwise, we cannot obtain saturations $S_w > (100 - S_{nw0})$ and $S_{nw0} > (100 - S_{w0})$.

We have seen above that, due to hysteresis, we may have different flow channels at the same saturation during wetting and drying. This leads to a certain amount of hysteresis also in the relationship $k_{wr}(S_w)$ and $k_{nwr}(S_{nw})$. Figure 5.14b shows a typical example of this phenomenon.

Effective permeability may also be presented as a function of $p_c$ or of the capillary pressure head, $\psi$. However, the relationship $k_{wr}(\psi)$ shows much hysteresis, probably because of the large hysteresis in the function $\psi(\theta_w)$. When an initially saturated soil is drained and then rewet, full saturation cannot be reached due to entrapped air. Under such conditions, although the soil seems to be almost saturated, its permeability is much smaller, say 50–60% of its permeability at full saturation.

Corey (1957) finds that for many consolidated rocks, $K_{wr}$ is proportional to $S_e^4$, while $K_{nwr}$ is proportional to $(1 - S_e)^2(1 - S_e^2)$.

Brooks and Corey (1964) have generalized these relationships by proposing

$$k_{wr} = (p_{cb}/p_c)^{(2+3\lambda)/\lambda}, \quad p_c \geq p_{cb},$$
$$k_{nwr} = (1 - p_{cb}/p_c)^2 \{1 - (p_{cb}/p_c)^{(2+\lambda)/\lambda}\} \quad (5.2.14)$$

where $p_{cb}$ is the *bubbling pressure* and $\lambda$ is a pore size distribution factor.

Gardner (1958) suggests the empirical relationship

$$K_w(\psi) = a/(b + \psi^m) \quad (5.2.15)$$

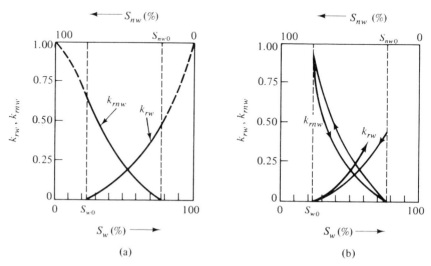

Fig. 5.14. Typical relative permeability curves (a) and the effect of hysteresis (b).

where $a$, $b$ and $m$ are constants, with $m \approx 2$ for heavy soil and $m \approx 4$ for sand, or

$$K_w(\psi) = K_0 \exp(-a\psi) \tag{5.2.16}$$

which does not fit experimental data too well, but is convenient both for analytical and numerical models.

Burdine (1953), related the relative permeabilities to the retention curve, $p_c(S_w)$

$$k_{wr}(S_e) = S_e^2 \int_0^{S_e} \frac{dS_w}{p_c^2(S_w)} \bigg/ \int_0^{1.0} \frac{dS_w}{p_c^2(S_w)},$$

$$k_{nwr}(S_e) = (1 - S_e)^2 \int_{S_e}^{1.0} \frac{dS_w}{p_c^2(S_w)} \bigg/ \int_0^{1.0} \frac{dS_w}{p_c^2(S_w)}. \tag{5.2.17}$$

Equations (5.2.17) are known as *Burdine's equations* (see Brooks and Corey, 1964).

Mualem (1976) suggested

$$k_{wr} = S_e^{1/2} \int_0^{S_e} \frac{dS_w}{p_c(S_w)} \bigg/ \int_0^{1.0} \frac{dS_w}{p_c(S_w)}. \tag{5.2.18}$$

Both (5.2.17) and (5.2.18) require information about the retention curve, $p_c = p_c(S_w)$.

## 5.3. Balance Equations

We start from the mass balance equation (3.3.2) which, since the mass of water per unit volume of porous medium is expressed by $nS_w\rho_w$, takes the form

$$\frac{\partial(\rho_w n S_w)}{\partial t} + \nabla \cdot \rho_w \mathbf{q}_w + \rho_w P = 0 \tag{5.3.1}$$

where $P = P(\mathbf{x}, t)$ represents a sink (dims. $L^3/L^3/T$), e.g., due to water uptake by roots. The symbol $P$ may also represent water lost by a phase change from water to vapor (or vice-versa). However, in this book this possibility is disregarded.

We recall that in writing the (macroscopic) balance equation (3.3.1), and, hence, also in (5.3.1), the nonadvective (i.e., dispersive plus diffusive) fluxes of the total mass have been neglected. For the sake of simplicity, let us henceforth omit the source term; when necessary, it can always be added at a later stage.

A similar equation can be written for the air.

By inserting the motion equation for $\mathbf{q}_w$, (5.2.1), assuming $|n\mathbf{V}_s| \ll |\mathbf{q}_w|$ and therefore $\mathbf{q}_w \equiv \mathbf{q}_{wr}$, into (5.3.1), we obtain

$$\frac{\partial(\rho_w n S_w)}{\partial t} - \nabla \cdot \left\{ \rho_w \frac{\mathbf{k}(S_w)}{\mu_w} (\nabla p_w + \rho_w g \nabla z) \right\} = 0. \tag{5.3.2}$$

For $\rho_w =$ const, Equation (5.3.2) is further reduced to one of the following forms

$$\frac{\partial(n S_w)}{\partial t} - \nabla \cdot \left\{ \frac{\mathbf{k}_w(S_w)}{\mu_w} (\nabla p_w + \rho_w g \nabla z) \right\} = 0, \tag{5.3.3}$$

$$\frac{\partial(n S_w)}{\partial t} - \nabla \cdot \{\mathbf{K}_w(S_w) \cdot \nabla \phi\} = 0, \tag{5.3.4}$$

$$\frac{\partial \theta_w}{\partial t} - \nabla \cdot \{\mathbf{K}_w(\theta_w) \cdot \nabla \phi\} = 0, \tag{5.3.5}$$

$$\frac{\partial \theta_w}{\partial t} - \nabla \cdot \{\mathbf{D}_w(\theta_w) \cdot \nabla \theta_w\} - \partial K(\theta_w)/\partial z = 0, \tag{5.3.6}$$

where the use of $\theta_w$ is recommended only for cases of $n =$ const. Another example, for an isotropic porous medium, is

$$\frac{\partial \psi}{\partial t} + \frac{1}{d\theta_w/d\psi} \left\{ \nabla \cdot K(\psi)\nabla \psi - \frac{\partial K(\psi)}{\partial z} \right\} = 0. \tag{5.3.7}$$

Because most of the applications are in connection with one-dimensional vertical flow, let us give the balance equations for such flows

$$\frac{\partial \theta_w}{\partial t} - \frac{\partial}{\partial z}\left( K_w(\theta_w) \frac{\partial \phi}{\partial z} \right) = 0, \tag{5.3.8}$$

$$\frac{\partial \theta_w}{\partial t} - \frac{\partial}{\partial z}\left( D_w(\theta_w) \frac{\partial \theta_w}{\partial z} \right) - \frac{\partial K_w(\theta_w)}{\partial z} = 0. \tag{5.3.9}$$

Often the gravity term in (5.3.9) is neglected.

We note that because of the dependence of the permeability on saturation, all mass balance equations for the unsaturated zone are nonlinear and require special attention in their solution, whether by analytical or numerical methods.

For the intermediate range of moisture contents for isotropic soils, Gardner and Mayhugh (1958) suggested the expression

$$D_w(\theta_w) = D_{w0} \exp\{A(\theta_w - \theta_{w0})\} \tag{5.3.10}$$

where $D_{w0}$ and $A$ are empirical coefficients.

Let us now take into account the compressibility of the water and of the porous medium. In the unsaturated zone, (3.1.7) is replaced by $\sigma = \sigma_s^* - p_{av}$, where $p_{av}$ is the average pressure in the fluids filling the void space. For example, $p_{av} = S_w p_w + S_a p_a$. Bishop et al. (1960) suggested the relationship $\sigma = \sigma_s^* - p_a + S_w(p_a - p_w)$. For comparison, Bishop et al. (1960) suggested the relationship $\sigma = \sigma_s^* - p_a + \chi(S_w)(p_a - p_w)$, where $\chi$ ($= 1$ for saturated soil and 0 for dry soil) is an empirical coefficient that represents the fraction of the soil's cross-sectional area occupied by water. For $p_a = 0$ and $d\sigma = 0$, we obtain $d\sigma_s^* = d\{\chi(S_w)p_w\}$.

By developing the first (change of storage) term on the left-hand side of (5.3.1), we obtain

$$\frac{\partial(\rho_w n S_w)}{\partial t} = nS_w \frac{\partial \rho_w}{\partial t} + \rho_w n \frac{\partial S_w}{\partial t} + \rho_w S_w \frac{\partial n}{\partial t}$$

$$= \rho_w \left\{ C_w + nS_w \beta + S_w(1-n)\alpha \left( \chi + p_w \frac{\delta \chi}{\delta p_w} \right) \right\} \frac{\partial p_w}{\partial t}$$

$$= \rho_w \{ C_w + S_{0p}(S_w) \} \frac{\partial p_w}{\partial t} \tag{5.3.11}$$

where $C_w = d\theta_w/dp_w \simeq n\, dS_w/dp_w$ (due to possible variations in $n$) is the *water capacity*, and $S_{0p}(S_w)$ is defined by (5.3.11). In (5.3.11), we note the changes of storage due to solid matrix compressibility, to water compressibility, and to moisture retention in the void space. Usually, the effect of the first two is negligible with respect to the third in the unsaturated zone. The last effect vanishes in the saturated zone. In (5.3.11) we have assumed that $p_w = p_w(S_w)$ is known.

The second term on the left-hand side of (5.3.1) may either be left as it is, or, similar to what was done in saturated flow (Section 3.3), simplified by assuming that $|\mathbf{q}_w \cdot \nabla \rho_w| \ll |n\, \partial \rho_w/\partial t|$, so that $\nabla \cdot \rho_w \mathbf{q}_w \simeq \rho_w \nabla \cdot \mathbf{q}_w$. In the latter case, (5.3.1) reduces to

$$\{C_w + S_{0p}(S_w)\} \frac{\partial p_w}{\partial t} + \nabla \cdot \mathbf{q}_w + P = 0. \tag{5.3.12}$$

We recall that $\mathbf{q}_w = \mathbf{q}_{wr} - n\mathbf{V}_s$, where $\mathbf{q}_{wr}$ is the specific discharge relative to the (possibly moving) solids, expressible by the motion equation (5.2.9). In order to take into account the difference between $\mathbf{q}_w$ and $\mathbf{q}_{wr}$, due to $\mathbf{V}_s \neq 0$ in a deformable porous medium, we follow the development of (3.6.14). Then (5.3.1), without the sink term, becomes

$$\rho_w S_w \alpha \left( \chi + p_w \frac{\partial \chi}{\partial p_w} \right) \frac{d^s p_{av}}{dt} + \rho_w (C_w + n S_w \beta) \frac{d^s p_w}{dt} +$$
$$+ \rho_w \nabla \cdot \mathbf{q}_{wr} + \rho_w \beta \mathbf{q}_{wr} \cdot \nabla p_w = 0 \qquad (5.3.13)$$

By making the following simplifying assumptions

$$|\partial(n\rho_w S_w)/\partial t| \gg |\mathbf{V}_s \cdot \nabla(n\rho_w S_w)|$$
$$|\mathbf{q}_w \cdot \nabla \rho_w| \ll |n \partial \rho_w / \partial t|, \quad p_a = 0, p_{av} = \chi(p_w) p_w,$$

Equation (5.3.13) reduces to

$$\nabla \cdot \mathbf{q}_{wr} + \left( C_w + S_w \left\{ \alpha \left( \chi + p_w \frac{\partial \chi}{\partial p_w} \right) + n\beta \right\} \right) \frac{\partial p_w}{\partial t}$$
$$= (C_w + S'_{0p}(S_w)) \frac{\partial p_w}{\partial t} = 0 \qquad (5.3.14)$$

to be compared with (3.3.17); $S'_{0p}$ is defined by (5.3.14).

Obviously, we could have followed the discussion in Section 3.6 in order to obtain a more general model for soil deformation under unsaturated flow conditions.

When solving for the flow of water only, neglecting any flow of air, we assume that the entire air phase is stationary at atmospheric pressure, taken as $p_a = 0$. However, it is possible (e.g., Noblanc and Morel-Seytoux, 1972; Morel-Seytoux, 1973) to consider the simultaneous flow of both the water and the air in the unsaturated zone. In the latter case, we have to solve the continuity equations for air and water simultaneously

$$\frac{\partial(n\rho_w S_w)}{\partial t} + \nabla \cdot \rho_w \mathbf{q}_w = 0,$$

$$\frac{\partial(n\rho_a S_a)}{\partial t} + \nabla \cdot \rho_a \mathbf{q}_a = 0. \qquad (5.3.15)$$

Altogether, we have here nine variables: $n$, $S_w$, $S_a$, $p_w$, $p_a$, $\rho_w$, $\rho_a$, $\mathbf{q}_w$ and $\mathbf{q}_a$. In order to determine them, we have: the two mass balance equations, the relationship $S_a + S_w = 1$, the equations of state $\rho_w = \rho_w(p_w)$, $\rho_a = \rho_a(p_a)$, $n = n(p_w)$ or $n = n(p_w, p_a)$, the capillary pressure curve $p_c = p_a - p_w = p_c(S_w)$ and the two motion equations (5.2.1). Altogether — nine equations.

In studies of flow in large unsaturated domains in the field, the effect of air flow is generally neglected. However, as water infiltrates into a soil, air must escape by flow towards the boundaries. This air flow may affect the flow of water (Morel-Seytoux, 1973). The possibility that pressure in the air phase may differ significantly from atmospheric pressure should not be overlooked (Vachaud et al., 1974). This is especially true when air cannot escape freely from the system.

In regional groundwater problems, we are interested mainly in the flow in the saturated zone. The unsaturated zone is treated separately, primarily as vertical downward infiltration, leading to the values of natural replenishment which, in turn, serve as a source term in the horizontal, two-dimensional model of flow in the aquifer. The same is true for the movement of pollutants carried with the infiltrating water.

In some localized, three-dimensional problems, however, e.g., flow through an earth embankment or contamination problems, we may wish to treat the flow in both the unsaturated zone and the saturated zone underlying it as flow in a single domain.

Although such a problem sounds more complicated, in most cases it is easier for a numerical solution, as we avoid the need to define and follow the phreatic surface that serves as a common boundary to both zones. A single set of equations serves to describe the flow in both zones. For example, assuming the flow of water only, we have:

*Saturated zone*  
$p_w > 0, \quad S_w = 1$  
$\mathbf{k}_w = \mathbf{k}_w(\mathbf{x})$

*Unsaturated zone*  
$p_w < 0;\ S_w = S_w(p_w), \quad S_{w0} \leq S_w < 1$  
$\mathbf{k}_w = \mathbf{k}_w(S_w(\mathbf{x}, t)), \quad S_w > S_{w0}$  
$k_w = 0, \quad 0 < S_w \leq S_{w0}.$

The mass balance equation which is common to both zones is, for example,

$$\nabla \cdot \left\{ \frac{\mathbf{k}_w}{\mu_w} (\nabla p_w + \rho_w g \nabla z) \right\} = (S'_{0p} + C_w) \frac{\partial p_w}{\partial t} \tag{5.3.16}$$

or, depending on the underlying assumptions, one of the other water mass balance equations considered in this section. Computer codes are available for solving this system.

Altogether, usually the state variables for a problem of flow in the unsaturated zone are $p_w$, $S_w$ (and $p_a$, $S_a$ if air flow is not neglected) for which we have one balance equation, say (5.3.16), and the relationship $p_w = p_w(S_w)$.

We assume that the retention curve $\psi = \psi(S)$, or $p_c = p_c(S_w)$, is known and so is the relationship $\mathbf{k}_w(S_w)$. In principle, in a deformable soil, $\mathbf{k}_w = \mathbf{k}_w(S_w, n)$, where the porosity varies continuously as consolidation, or compaction, takes place. However, because of hysteresis, in both relationships, we have to specify whether a drying or a wetting process is taking place.

## 5.4. Initial and Boundary Conditions

As with the solution of flow problems in the saturated zone, the solution of the partial-differential equations of unsaturated flow requires the specification of initial and boundary conditions in terms of the relevant state variable, usually $p_w$ (or $\psi$), or $S_w$ (or $\theta_w$). However, unlike the case of saturated flow, it is also necessary to state whether a drying or wetting process is taking place along the boundary because $K_w(\theta_w)$ and $\psi(\theta_w)$ are subject to hysteresis.

Only water flow is being considered in this section. The extension to simultaneous air-water flow is obvious and requires no further discussion.

*Initial conditions* include the specification of the considered state variables ($S_w$, $\theta_w$, $p_w$, or $\psi$) at every point inside the considered flow domain.

Boundary conditions may be of several types:

(a) *Prescribed water content*, $\theta_w$, (or piezometric head $\phi_w$, pressure $p_w$, or suction $\psi$) at all points of the boundary. The condition of prescribed $p_w$ occurs when we have ponded water on the soil surface, dictating there a certain water pressure. In a limit situation, we may have a thin sheet of water over the surface so that practically $p_w = 0$. Instead, we can always specify the $\theta_w$ at saturation, corresponding to $p_w = 0$. We shall do so when the partial differential equation is given in terms of $\theta_w$. We note that except for the example given above, usually in the practice, we do not know the values of water pressure and moisture content along a boundary.

This is a *boundary condition of the first kind*, or (*Dirichlet boundary condition*).

(b) *Prescribed flux of water at the boundary*. This case occurs, for example, when water (rainfall or irrigation by sprinklers), at a known rate reaches the ground surface, which serves as a boundary to the unsaturated flow domain. For a rate of accretion denoted by the vector **N** we have

$$\mathbf{N} \cdot \boldsymbol{\nu} = \mathbf{q}_w \cdot \boldsymbol{\nu} \tag{5.4.1}$$

where $\boldsymbol{\nu}$ denotes the outward normal to the boundary surface and $\mathbf{q}_w$ denotes the specific discharge on the soil side of the boundary. For vertically downward accretion at a rate $N$, we replace **N** in (5.4.1) by $-N\nabla z$. For evaporation at a rate $E$, we replace $\mathbf{N} \cdot \boldsymbol{\nu}$ by the evaporation rate, $E\nabla z$. For $\mathbf{q}_w$, we may use any of the flux equations given in Section 5.2. For example, with (5.2.4), Equation (5.4.1) in terms of $\psi$, for a horizontal ground surface, $\boldsymbol{\nu} \equiv \nabla z$, and vertically downward accretion, becomes,

$$N = -\left\{ K_w(\psi) \frac{\partial \psi}{\partial z} - K_w(\psi) \right\}. \tag{5.4.2}$$

Or, with (5.2.6), in terms of $\theta_w$

$$N = D_w(\theta_w) \frac{\partial \theta_w}{\partial z} - K_w(\theta_w). \tag{5.4.3}$$

For an impervious boundary, we set $\mathbf{N} \cdot \boldsymbol{\nu} = 0$ in (5.4.1).

The boundary condition of prescribed flux is thus either a *third-kind boundary condition*, or, in the absence of a gravity term, a *second kind* one.

In the case of accretion, there is a limit to the capacity of a soil to take in water.

If the rate of accretion, $N$, say, on a horizontal surface, exceeds a certain value, ponding will occur. This happens when $N = K_0$ (i.e., $K$ at saturation). At that time,

$\theta_w$ reaches saturation at the surface, $\psi = 0$, $\partial\psi/\partial z = 0$, and the rate (= specific discharge) of downward flow is equal to $K_0$. If $N > K_0$, ponding, or surface runoff removing part of $N$, will take place and we have to switch to a first-type boundary condition.

In the case of evaporation, the flux leaving the soil surface is dictated by the energy supplied to the soil (overlooking the possibility of soil heating, phase change etc., in the domain). Like in the case of accretion, the actual flux across the soil surface is constrained by the ability of the soil to transmit water from the soil's interior to the ground surface. The actual transmission (which may be only a fraction of the potential evapotranspiration of the soil) is governed by the soil's permeability which, in turn, depends on the moisture content, and by the moisture gradient, both at the soil's surface.

Let $E$ denote the actual rate of evaporation at the soil's surface, with $\mathbf{N} = E\nabla z$ in (5.4.1), or $N = -E$ in (5.4.3). At low values of $E$, the boundary condition, say (5.4.3), is of the third kind. As $E$ is increased, a point is reached where the moisture content is reduced to the irreducible one, $\theta_w = \theta_{w0}$ at which point, $K(\theta_w) = 0$ and $\nabla\theta_w \to \infty$.

To some extent, imposing isothermal conditions as we have done here, with no soil heating, no phase change (except at the soil surface), and no vapor flux, is unrealistic.

We may summarize that in both cases (time-dependent boundary conditions cannot be assigned *a-priori* due to the limiting capacity of the soil to transmit the water), a limiting situation may develop at some unknown time. This situation is obtained by maximizing the absolute value of the water flux, maintaining the correct sign.

Similar to saturated flow, a boundary condition of the third type also occurs when the soil (e.g., the bottom of an artificial recharge pond) is covered by a very thin semipervious layer through which flow takes place.

Regarding the semipervious layer (Figure 5.15) as a very thin, saturated, membrane having a resistance $c^{(1)}$ to the flow, the flux continuity condition requires that

$$\mathbf{q}_w \cdot \boldsymbol{\nu} = \frac{\phi - \phi_0}{c^{(1)}} \qquad (5.4.4)$$

where $\phi_0$ and $\phi$ are the piezometric heads on the top and bottom of the semipervious layer, respectively

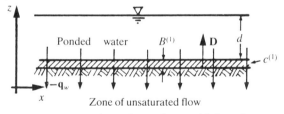

Fig. 5.15. Nomenclature for semipermeable boundary.

For a horizontal boundary and isotropic soil, the boundary condition (5.4.4) takes on the form

$$K_w(\psi) \frac{\partial \psi}{\partial z} - K_w(\psi) + \frac{\psi}{c^{(1)}} = -\frac{B^{(1)} + d}{c^{(1)}} \qquad (5.4.5)$$

This is *a third-kind* (= *Cauchy*) *boundary condition*.

When the flow domain is made up of regions of different (homogeneous) porous media (e.g., a layered soil), we require that at points of the boundary between two media, both the normal flux component and the pressure be equal, i.e.,

$$p_{w1} = p_{w2}; \; q_{w1n} = q_{w2n}$$

We have seen above (Subsection 5.1.6) that the requirement of pressure continuity means a discontinuity in water content.

If we wish to consider the flow of both air and water in the unsaturated zone, we also have to state boundary conditions for the air phase, similar to those described above for the water phase.

## 5.5. Complete Statement of Unsaturated Flow Model

The discussion on the statement of mathematical models for unsaturated flow is similar to that presented in Section 3.5 for saturated flow, and need not be repeated here. Before constructing any model, we should decide, within the framework of the conceptual model, whether we wish to model the flow of water only, neglecting any air movement, or to model the simultaneous flow of both phases. Once this has been decided, the model should include:

(a) Specification of the flow domain.
(b) Specification of the relevant state variable, or variables.
(c) Statement of the partial differential equations that express mass balances of the water and air.
(d) Statement of the relevant motion equation, or equations.
(e) Statement of the constitutive equations for the water, air, and solid phases.
(f) Information on the retention curve and effective permeability curves, as well as on the numerical values of all coefficients that appear in the constitutive equations.
(g) Statement of initial conditions of the relevant state variable, or variables.
(h) Statement of boundary conditions in terms of the relevant state variables.

CHAPTER SIX

# Modeling Groundwater Pollution

So far, we have discussed only the movement and storage of water in various types of aquifers, overlooking a major problem which is of interest in any development and management of a water resources system, namely that of water quality. In fact, with the increased demand for water in most parts of the world, and with the intensification of water utilization, the quality problem becomes the limiting factor in the development and use of water resources. Although in some regions, the quality of both surface and groundwater resources deteriorates, special attention should be devoted to the pollution of groundwater in aquifers due to the very slow velocity of the water and to the possibility of an interaction of the pollutants with the solid matrix. Although it may seem that groundwater is more protected than surface water, it is still subject to pollution, and when the latter occurs, the restoration to the original, nonpolluted state, is usually more difficult and lengthy.

The problem of groundwater pollution need not be associated only with water supply for domestic, industrial, or agricultural purposes. Serious environmental problems arise when polluted groundwater emerges at ground surface, or discharges into rivers and lakes.

The term 'quality' usually refers either to energy — in the form of heat or nuclear radiation — or to materials contained in the water. Many materials dissolve in water, whereas others may be carried with the water in suspension. Given the very large number of polluting constituents — and new materials are coming onto the market every day — groundwater quality can be defined in terms of hundreds of parameters. The relevance of any of these materials depends on the water use that is being considered. For example, salinity may be important if the water is intended for drinking, for irrigation, or for certain industries, but less important for recreation. Radioactive substances released into groundwater, in connection with nuclear power productions, by accidents, or from nuclear waste repositories, pose a pollution problem that requires special attention. Standards have been issued by national and international health authorities with respect to the various constituents, according to the origin of the water and the type of consumer.

When we speak of 'water pollution', rather than of 'water quality', we usually have in mind a situation in which the quality of the water has been deteriorating towards the point of being hazardous to the consumer. However, even under undisturbed conditions, and without man's intervention, groundwater already contains a certain amount of dissolved matter, sometimes reaching levels which render the water unsuitable for certain usages. With this in mind we shall, henceforth, use the term 'pollutant' to denote dissolved matter carried with the water and accumulating in the aquifer, without inferring that concentrations have necessarily reached dangerous levels.

Groundwater pollution may usually be traced back to four sources:

(i) *Environmental.* This type of pollution is due to the environment through which the flow of groundwater takes place. For example, in flow through carbonate rocks, water dissolves small, yet sometimes significant, amounts of the rock. Sea water intrusion, or pollution of good quality aquifers by invading brackish groundwater from adjacent aquifers as a result of disturbing an equilibrium that existed between the two bodies of water, may also serve as examples of environmental pollution.

(ii) *Domestic.* Domestic pollution may be caused by accidental breaking of sewers, by percolation from septic tanks, by rain infiltrating through sanitary landfills, or by artificial recharge of aquifers by sewage water, after being treated to different levels. Biological contaminants (e.g., bacteria and viruses) are usually related to this source.

(iii) *Industrial.* In many cases, a single sewage disposal system serves both industrial and residential areas. In this case, one cannot separate between industrial and domestic pollution, although their compositions — and, hence, the type of treatment they require and the pollution they cause — are completely different. Heavy metals, for example, constitute a major problem in industrial waste. Industrial waste may also contain radioactive materials and various non-deteriorating, highly toxic compounds.

(iv) *Agricultural.* This source is due to irrigation water and rain water dissolving and carrying fertilizers, salts, herbicides, pesticides, etc., as they infiltrate through the ground surface, travel through the unsaturated zone, and replenish the aquifer. Irrigation with reclaimed sewage water may also serve as a source of pollution for an underlying phreatic aquifer.

We are dealing with the transport of mass, where the considered 'mass' is that of some substance, e.g., a solute, that moves with the water in the interstices of a porous medium, both in the saturated and unsaturated zones. The mechanisms affecting the transport of a pollutant in a porous medium are: advective, dispersive, and diffusive fluxes, solid-solute interactions and various chemical reactions and decay phenomena, which may be regarded as source-sink phenomena for the solute.

Our objective in this chapter is to present and discuss the laws governing the movement and accumulation of pollutants in groundwater flow, and, as in the case of groundwater flow discussed in the previous chapters, to construct models that enable the engineer and planner to predict future pollutants' distributions in an aquifer. We shall consider the general case of three-dimensional flow. However, the procedure of averaging along the vertical, employed in Chapter 4, will also be used here to derive a model based on the assumption of essentially horizontal flow and pollution transport in aquifers.

Obviously, one should be careful to employ the *essentially horizontal flow concept only when justified*. Whereas equipotentials in an aquifer are more or less vertical, even when an aquifer is stratified (i.e., consists of several layers of different hydraulic conductivities), velocities in the different strata may vary appreciably, resulting in a marked difference in the rates of advance and spreading of a pollutant in the different strata. Situations may arise, where the average concentration along the vertical is meaningless and one should take into account the stratification in water quality.

Under certain conditions, the transition zone between two bodies of groundwater of different qualities may be approximated as an abrupt front. Water of one quality injected into an aquifer containing water of another quality may serve as an example. This approximate approach is treated in Chapter 11.

## 6.1. Hydrodynamic Dispersion

Consider saturated flow through a porous medium, and let a portion of the flow domain contain a certain mass of solute. This solute will be referred to as a *tracer*. The tracer, which is a labeled portion of the same liquid, may be identified by its density, concentration of some pollutant, color, electrical conductivity, etc.

In Section 2.1, e.g., (2.1.9), we defined the water's velocity. With this definition in mind, let us conduct two field experiments.

Figure 6.1a shows an (assumed) abrupt front in an aquifer, at $t = 0$. Let the abrupt front separate a porous medium domain occupied by tracer labeled water ($C = 1$) from one occupied by unlabeled water ($C = 0$). If uniform flow (normal to the initial front) at a velocity $V$ takes place in the aquifer, Darcy's law enables us to calculate the new position of the (assumed) abrupt front; its new position is at $x = Vt$. On the basis of Darcy's law alone, the two types of water should continue to occupy domains separated by an abrupt front. However, if we measure concentrations in a number of observation wells scattered in the aquifer, we note that no such front exists. Instead, we shall observe a gradual transition from a domain containing water at $C = 1$, to a domain containing water at $C = 0$. Experience shows that as flow continues, the width of the transition zone increases. This spreading of the tracer-labeled water, beyond the zone it is supposed to occupy according to the description of water movement by Darcy's law, cannot be explained by the averaged movement of the water.

As a second experiment, consider the injection of a certain quantity of tracer-labeled water at point $x = 0$ at some initial time $t = 0$. Making use of the (averaged) velocity as calculated by Darcy's law, we should expect the tracer-labeled water to move as a volume of fixed shape, reaching point $x = Vt$ at time $t$. Again, field observations (shown in Figure 6.1b) reveal a completely different picture. We note the spreading of the tracer-labeled water, not only in the direction of the uniform (averaged) flow, but also normal to it. The area occupied by the tracer-labeled water, which has the shape of an ellipse in a horizontal two-

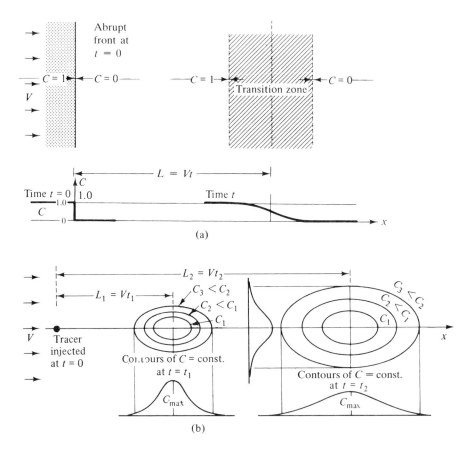

Fig. 6.1. Longitudinal and transversal spreading of a tracer. (a) Longitudinal spreading of an initially sharp front. (b) Spreading of a point injection.

dimensional flow field, will continue to grow, both longitudinally, i.e., in the direction of the uniform flow, and transversally, i.e., normal to it. Curves (in two-dimensional flow) of equal concentration have the shape of confocal ellipses. Again, this spreading cannot be explained by the averaged flow alone (especially noting that we have spreading perpendicular to the direction of the uniform flow).

The spreading phenomenon described above in a porous medium is called *hydrodynamic dispersion* (or *miscible displacement*). It is a nonsteady *irreversible process* (in the sense that the initial tracer distribution cannot be obtained by reversing the direction of the uniform flow) in which the tracer mass mixes with the nonlabeled portion of the water.

One of the earliest observations of this phenomenon is reported by Slichter (1905), who used an electrolyte as a tracer in studying the movement of groundwater. Slichter observed that at an observation well downstream of a (continuous) injection point, the tracer's concentration increases gradually, and that even in a uniform (average) flow field, the tracer advances in the direction of the flow in a pear-like shape that becomes longer and wider as it advances.

The dispersion phenomenon may also be demonstrated by a simple laboratory experiment. Consider steady flow in a cyclindrical column of homogeneous sand, saturated with water. At a certain instant, $t = 0$, a tracer-marked water (e.g., water with NaCl at a low concentration, so that the effect of density variations on the flow pattern is negligible) starts to displace the original unlabeled water in the column. Let the tracer concentration, $C = C(t)$ be measured at the end of the column and presented in a graphic form, called a *breakthrough curve*, as a relationship between the relative tracer concentration and time, or volume of effluent, $U$.

In the absence of dispersion, the breakthrough curve should have taken the form of the broken line shown in Figure 6.2, where $U_0$ is the pore volume of the column, and $Q$ is the constant discharge. Actually, owing to hydrodynamic dispersion, it will take the form of the S-shaped curve shown in full line in Figure 6.2.

We cannot explain all the above observations on the basis of the average water flow. We must refer to what happens at the microscopic level, viz., inside the pore. There we have velocity varying in both magnitude and direction across any pore cross-section. We usually assume zero fluid velocity on the solid surface, with a maximum velocity at some internal point (compare with the parabolic velocity distribution in a straight capillary tube). The maximum velocity itself varies according to the size of the pore. Because of the shape of the interconnected pore space, the (microscopic) streamlines fluctuate in space with respect to the mean direction of flow (Figures 6.3a and b). This phenomenon causes the spreading of any initially close group of tracer particles; as flow continues, they will occupy an ever increasing volume of the flow domain. The two basic factors that produce this kind of spreading are, therefore, flow and the presence of a pore system through which flow takes place.

Although this spreading is in both the longitudinal direction, namely that of the average flow, and in the direction transversal to the average flow, it is primarily in the former direction. Very little spreading can be caused in a direction perpendicular to the average flow by velocity variations alone. Also, such velocity variations alone cannot explain the ever-growing width of the zone occupied by dispersed tracer particles normal to the direction of flow. In order to explain the

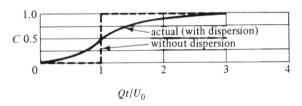

Fig. 6.2. Breakthrough curve in one-dimensional flow in a sand column.

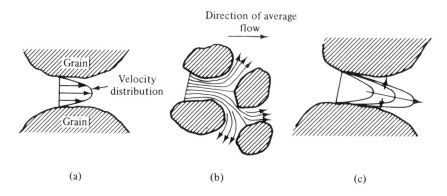

Fig. 6.3. Spreading due to mechanical dispersion (a, b) and molecular diffusion (c).

latter observed spreading, we have to refer to an additional phenomenon that takes place in the void space, viz., *molecular diffusion*.

Molecular diffusion, caused by the random movement of molecules in a fluid, produces an additional flux of tracer particles (at the microscopic level) from regions of higher tracer concentrations to those of lower ones. This means, for example, that as the marked particles spread *along* each microscopic streamtube, as a result of velocity variations, a concentration gradient of these particles is produced, which, in turn, produces a flux of tracer by the mechanism of molecular diffusion. The latter phenomenon tends to equalize the concentrations along the steamtube. Relatively, this is a minor effect. However, at the same time, a tracer concentration gradient will also be produced between adjacent streamlines, causing *lateral molecular diffusion across streamtubes* (Figure 6.3c), tending to equalize the concentration across pores. It is this phenomenon that explains the observed transversal dispersion.

In addition to the role played at the microscopic level by molecular diffusion in enhancing the transversal component of mechanical dispersion, it produces a macroscopic flux of its own. This is easily demonstrated by letting the velocity vanish. Then the tracer is transported by (macroscopic) molecular diffusion only.

We shall refer to the spreading caused by the velocity variations at the microscopic level, enhanced by molecular diffusion, especially in the direction transversal to the average flow, as *mechanical dispersion*.

We use the term *hydrodynamic dispersion* to denote the spreading (at the microscopic level) resulting from both mechanical dispersion and molecular diffusion. Actually, the separation between the two processes is rather artificial, as they are inseparable. However, molecular diffusion alone does also take place in the absence of motion (both in a porous medium and in a fluid continuum). Because molecular diffusion depends on time, its effect on the overall dispersion is more significant at low velocities. It is molecular diffusion which makes the phenomenon of hydrodynamic disperison in purely laminar flow irreversible.

In addition to inhomogeneity on a microscopic scale (i.e., presence of pores and grains), we may also have inhomogeneity on a macroscopic scale, due to variations in permeability from one portion of the flow domain to the next. This inhomogeneity also produces dispersion of marked particles, but on a much larger scale (see Section 6.7).

Dispersion may take place both in a laminar flow regime, where the liquid moves along definite paths that may be averaged to yield streamlines, and in a turbulent regime, where the turbulence may cause yet an additional mixing. In what flows, we shall focus our attention only on flow of the first type.

In addition to advection (at average velocity), mechanical dispersion, and molecular diffusion, several other phenomena may affect the concentration distribution of a tracer as it moves through a porous medium. The tracer (say, a solute) may interact with the solid surface of the porous matrix in the form of adsorption of tracer particles *on the solid surface*, deposition, solution of the solid matrix, ion exchange, etc. All these phenomena cause changes in the concentration of a tracer in a flowing liquid. Radioactive decay and chemical reactions *within the liquid* also cause tracer concentration changes.

In general, variations in tracer concentration cause changes in the liquid's density and viscosity. These, in turn, affect the flow regime (i.e., velocity distribution) that depends on these properties. We use the term *ideal tracer* when the concentration of the latter does not affect the liquid's density and viscosity. At relatively low concentrations, the ideal tracer approximation is sufficient for most practical purposes. However, in certain cases, for example in the problem of sea water intrusion, the density may vary appreciably, and the ideal tracer approximation should not be used.

## 6.2. Advective, Dispersive, and Diffusive Fluxes

As explained above, at every (microscopic) point within a porous medium domain, we have a velocity $\mathbf{V}$ and a concentration, $c$, of some considered substance; $c$ expresses the mass of the substance per unit volume of the liquid. Figure 6.4 shows a point $\mathbf{x}'$ belonging to a *Representative Elementary Volume* (REV) centered at point $\mathbf{x}$. The product $c\mathbf{V}$ at $\mathbf{x}'$ denotes the local *flux* (= quantity of the considered substance per unit area of liquid) vector at that point. However, we already know that we cannot predict values of $\mathbf{V}$ and $c$ at this microscopic level, and that, instead, we should aim at predicting the average concentration, $\overline{c}^w$, and the average tracer flux $\overline{c\mathbf{V}}^w$ at the macroscopic level.

6.2.1. ADVECTIVE AND DISPERSIVE FLUXES

To achieve this goal, without going into the details of the continuum approach to transport in porous media, let the liquid's velocity at an arbitrary point, $\mathbf{x}'$, within the liquid that completely occupies the pore space, be denoted by $\mathbf{V}(\mathbf{x}', t; \mathbf{x})$. The

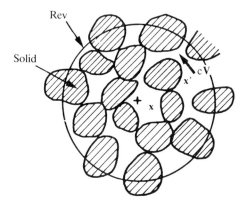

Fig. 6.4. Nomenclature for the dispersive flux.

symbol **x** in this parenthesis indicates that point **x'** belongs to a Representative Elementary Volume (REV) centered at **x** (Figure 6.4). The velocity **V**, can be decomposed into two parts: the average velocity $\overline{\mathbf{V}}^w$, of the liquid within the REV, and a deviation, $\mathring{\mathbf{V}}$, from that average. Thus

$$\mathbf{V}(\mathbf{x}', t; \mathbf{x}) = \overline{\mathbf{V}}^w(\mathbf{x}, t) + \mathring{\mathbf{V}}(\mathbf{x}', t; \mathbf{x}), \tag{6.2.1}$$

$$c(\mathbf{x}', t, \mathbf{x}) = \overline{c}^w(\mathbf{x}, t) + \mathring{c}(\mathbf{x}', t; \mathbf{x}). \tag{6.2.2}$$

In both cases, the average has the meaning of an intrinsic phase average as defined by (1.4.3).

To obtain the average flux, we write

$$\overline{c\mathbf{V}}^w = \overline{(\overline{c}^w + \mathring{c})(\overline{\mathbf{V}}^w + \mathring{\mathbf{V}})}^w = \overline{\overline{c}^w \overline{\mathbf{V}}^w}^w + \overline{\overline{c}^w \mathring{\mathbf{V}}}^w + \overline{\mathring{c} \overline{\mathbf{V}}^w}^w + \overline{\mathring{c}\mathring{\mathbf{V}}}^w. \tag{6.2.3}$$

However, in view of (1.4.3), $\overline{\mathring{c}\overline{\mathbf{V}}^w}^w \equiv 0$ and $\overline{\overline{c}^w\mathring{\mathbf{V}}}^w = 0$. Hence

$$\overline{c\mathbf{V}}^w = \overline{c}^w \overline{\mathbf{V}}^w + \overline{\mathring{c}\mathring{\mathbf{V}}}^w, \tag{6.2.4}$$

i.e., the average flux of the considered substance is equal to the sum of two macroscopic fluxes:

(a) An *advective flux*, $\overline{c}^w \overline{\mathbf{V}}^w$, expressing the flux carried by the water at the latter's average velocity, $\overline{\mathbf{V}}^w$, as determined by Darcy's law (or any of the motion equations presented in Section 2.1).

(b) A flux $\overline{\mathring{c}\mathring{\mathbf{V}}}^w \equiv \overline{c\mathring{\mathbf{V}}}^w$ expressing an additional flux resulting from the fluctuating velocity in the vicinity (i.e., within the REV) of the considered point. Recalling the discussion in the previous section, this is the flux that produces the spreading, or dispersion. We refer to it as the *dispersive flux*. It is a macroscopic flux that expresses the effect of the microscopic variations of the velocity in the vicinity of a considered point. We note that this flux is created by the averaging procedure. It does not exist at the microscopic level. In employing this flux, we are

MODELING GROUNDWATER POLLUTION                                                                161

losing the information about the behavior at the microscopic level (which we do not have anyway).

Our next objective is to express the dispersive flux in terms of averaged (and measurable) quantities, such as averaged velocity and averaged concentration. Investigations over a period of about two decades, starting from the mid-50s (see review, for example, in Bear, 1972), have led to the *working assumption* that the dispersive flux can be expressed as a Fickian type law, viz. in the form

$$\overline{c\overset{\circ}{\mathbf{V}}}^w = -\mathbf{D} \cdot \nabla \overline{c}^w; \qquad \overline{c\overset{\circ}{\mathbf{V}}}_i^w = -D_{ij}\frac{\partial \overline{c}^w}{\partial x_j} \qquad (6.2.5)$$

where $\mathbf{D}$ is a second rank symmetric tensor called the *coefficient of (mechanical) dispersion*. We recall that $\overline{c}^w$ denotes the mass of the dispersing substance per unit volume of water, and $\overline{c\overset{\circ}{\mathbf{V}}}^w$ represents a flux per unit area of the water. Equation (6.2.5) indicates that the dispersive flux is linearly proportional to the gradient of the average concentration and that this flux takes place from high concentrations to lower ones.

6.2.2. COEFFICIENT OF DISPERSION

Several authors (e.g., Nikolaevskii, 1959; Bear, 1961; Scheidegger, 1961; Bear and Bachmat, 1967) derived the following expression for the relationship between the coefficient $\mathbf{D}$ and microscopic porous matrix configuration, flow velocity, and molecular diffusion

$$D_{ij} = a_{ijkm}\frac{\overline{V}_k^w \overline{V}_m^w}{\overline{V}^w} f(\text{Pe}, \delta) \qquad (6.2.6)$$

where $\overline{V}^w = |\overline{\mathbf{V}}^w|$ is the average velocity, Pe is the Peclet number defined as Pe $= L\overline{V}^w/D_d$, $L$ being some characteristic length of the pores, $D_d$ is the coefficient of molecular diffusion of the solute in the liquid phase, $\delta$ = the ratio of the length characterizing the individual pores of a porous medium to the length characterizing their cross-section, and $f(\text{Pe}, \delta)$ is a function which introduces the effect of tracer transfer by molecular diffusion between adjacent streamlines at the microscopic level. In this way, molecular diffusion affects mechanical dispersion. One should not identify this effect with the macroscopic flux due to molecular diffusion (see below), but with the transfer between streamtubes at the microscopic level, as explained in the definition of mechanical dispersion in Section 6.1. Bear and Bachmat (1967) suggested the relationship $f(\text{Pe}, \delta) = \text{Pe}/(\text{Pe} + 2 + 4\delta^2)$. In most cases, we assume that $f(\text{Pe}, \delta) \simeq 1$. Henceforth, we shall also make this assumption.

The coefficient $a_{ijkm}$, (dims. L) called the *dispersivity of the porous medium*, is a fourth-rank tensor which expresses the microscopic configuration of the solid-liquid interface. Bear and Bachmat (1967) and Bear (1972, p. 614) expresses $a_{ijkm}$ by

$$a_{ijkm} = [\overline{B\overset{\circ}{T}{}^*_{ij} B\overset{\circ}{T}{}^*_{jp}} / \overline{BT^*_{lk} BT^*_{pm}}] \bar{L}, \qquad (6.2.7)$$

where $B$ is the conductance of an elementary medium channel, $BT^*_{ij}$ is an oriented conductance of a channel, $\overline{T^*_{ij}}$ is the medium's *tortuosity*, $n\overline{BT^*_{ij}} = k_{ij}$ is the medium's permeability and $L$ is a characteristic length of the medium. Thus, the medium's dispersivity is related to the variance of $BT^*_{ij}$, while its permeability is related to the average, $\overline{BT^*_{ij}}$, of $BT^*_{ij}$.

A fourth rank tensor has 81 components in a three-dimensional space (and 16 in a two-dimensional one) Scheidegger (1961) and Bear (1961) showed that $a_{ijkm}$ has a number of symmetries that reduce the number of nonzero components of the dispersivity tensor, in a three-dimensional space, to only 36.

For an *isotropic porous medium*, the number of nonzero components is further reduced to 21. Furthermore, these 21 components are related to two parameters: $a_L$ (dim. L) called the *longitudinal dispersivity* of the isotropic porous medium, and $a_T$ (dim. L) called the *transversal dispersivity*. In the theoretical developments mentioned above, it is shown that $a_L$ expresses the heterogeneity of the porous medium at the microscopic scale, i.e., due to the presence of pores and solids. Hence, in laboratory experiments in homogeneous sand columns it was found that $a_L$ is of the order of magnitude of the average sand grain. The transversal dispersivity is estimated as 10 to 20 times smaller than $a_L$.

With $a_L$ and $a_T$, the components of the dispersivity for an isotropic porous medium can be expressed in the form

$$a_{ijkm} = a_T \delta_{ij} \delta_{km} + \frac{a_L + a_T}{2} (\delta_{ik} \delta_{jm} + \delta_{im} \delta_{jk}) \qquad (6.2.8)$$

where $\delta_{ij}$ denotes the Kronecker delta (with $\delta_{ij} = 0$ for $i \neq j$ and $\delta_{ij} = 1$ for $i = j$). For an isotropic porous medium, the components $a_{ijkm}$ do not change with the rotation of the coordinate system.

For an *anisotropic porous medium with axial symmetry*, e.g., a medium made up of a large number of thin layers normal to the axis of symmetry, the dispersivity can be expressed in the form

$$\begin{aligned} a_{ijkm} = &\, a_\mathrm{I} \delta_{ij} \delta_{km} + a_\mathrm{II}(\delta_{ik} \delta_{jm} + \delta_{im} \delta_{jk}) + \\ &+ a_\mathrm{III}(\delta_{ij} h_k h_m + \delta_{km} h_i h_j) + \\ &+ a_\mathrm{IV}(\delta_{ik} h_j h_m + \delta_{jk} h_i h_m + \delta_{im} h_j h_k + \delta_{jm} h_i h_k) + \\ &+ a_\mathrm{V} h_i h_j h_k h_m \end{aligned} \qquad (6.2.9)$$

where $a_\mathrm{I}$, $a_\mathrm{II}$, $a_\mathrm{III}$, $a_\mathrm{IV}$ and $a_\mathrm{V}$ are five independent parameters and **h** is a unit vector directed along the axis of symmetry. Similar expressions can be written for other types of anisotropy.

By combining (6.2.8) with (6.2.6) for $f(\text{Pe}, \delta) = 1$, we obtain

$$D_{ij} = a_T V \delta_{ij} + (a_L - a_T) V_i V_j / V \qquad (6.2.10)$$

where here, and henceforth, we have omitted the symbol $\overline{(\ )}^w$ that indicates that the velocity is an average one.

The permeability, $k_{ij}$, of a porous medium is also a second-rank symmetric tensor. However, there is a basic difference between tensors $k_{ij}$ and $D_{ij}$. In an isotropic porous medium, any three mutually orthogonal directions in space may serve as principal directions. However, due to the effect of the velocity pattern, the principal axes of the dispersion coefficient, $D_{ij}$ *at a point* are always in the direction of the tangent to the streamline passing through that point and in the directions of the two principal normals to that direction. Figure 6.5 shows these directions. The unit vectors **N**, **T** and **B** are called the *principal normal*, the *tangent*, and the *binormal* to the curve (see any text on differential geometry).

Thus, although the porous medium is isotropic, we have a distinct set of principal directions at every point of a flow domain. As the velocity varies from point to point, so do the principal axes of the dispersion. Furthermore, at every point, these directions may vary continuously as the flow pattern varies. This dependence of the dispersion coefficient on the velocity introduces a major

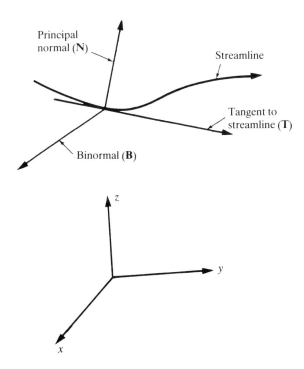

Fig. 6.5. Principal axes of the coefficient of dispersion.

difficulty in the solution of pollution problems, especially under unsteady flow conditions and when the velocity is density (and hence, concentration) dependent.

In Cartesian coordinates, and velocity components $V_x$, $V_y$, $V_z$, we obtain from (6.2.10)

$$D_{xx} = a_T V + (a_L - a_T)V_x^2/V = [a_T(V_y^2 + V_z^2) + a_L V_x^2]/V,$$
$$D_{xy} = (a_L - a_T)V_x V_y/V = D_{yx},$$
$$D_{xz} = (a_L - a_T)V_x V_z/V = D_{zx},$$
$$D_{yy} = a_T V + (a_L - a_T)V_y^2/V = [a_T(V_x^2 + V_z^2) + a_L V_y^2]/V,$$
$$D_{yz} = (a_L - a_T)V_y V_z/V = D_{zy},$$
$$D_{zz} = a_T V + (a_L - a_T)V_z^2/V = [a_T(V_x^2 + V_y^2) + a_L V_z^2]/V. \qquad (6.2.11)$$

If we choose a Cartesian coordinate system at a point, such that one of its axes, say $x_1$, coincides with the direction of the average uniform velocity $\mathbf{V}$, then at that point (6.2.10) reduces to

$$D_{11} = a_L V, \qquad D_{22} = a_T V, \qquad D_{33} = a_T V, \qquad D_{ij} = 0, \quad \text{for } i \neq j \qquad (6.2.12)$$

which can be written in the matrix form

$$[D_{ij}] = \begin{bmatrix} a_L V & 0 & 0 \\ 0 & a_T V & 0 \\ 0 & 0 & a_T V \end{bmatrix} \qquad (6.2.13)$$

We note that lateral dispersion, in the directions of $x_2$ and $x_3$, can still take place in such uniform flow. The axes of the coordinate system in which $D_{ij}$ is expressed by (6.2.13) — namely, in the direction of the flow at a point and perpendicular to it — are the *principal axes of the dispersion*. The coefficients $D_{11}$, $D_{22}$, and $D_{33}$ are the *principal values of the coefficient of mechanical dispersion*. In this case, $D_{11}$ is called the *coefficient of longitudinal dispersion*, while $D_{22}$ and $D_{33}$ are called *coefficients of transversal dispersion*.

6.2.3. MOLECULAR DIFFUSION

We shall simplify the presentation by assuming a *binary system*, i.e., a single solute and a solvent. The discussion can be extended to multicomponent systems.

At the microscopic level, the flux vector, $\mathbf{J}^{(d)}$, due to molecular diffusion is expressed by Fick's law

$$\mathbf{J}^{(d)} = -D_d \cdot \nabla c; \qquad J_i^{(d)} = -D_d \frac{\partial c}{\partial x_i} \qquad (6.2.14)$$

where $D_d$ is the coefficient of molecular diffusion in a fluid continuum (equals about $10^{-5}$ cm$^2$/sec in dilute systems). By averaging (6.2.14) over the REV, and

MODELING GROUNDWATER POLLUTION

introducing certain simplifying assumptions, Bear and Bachmat (1984, 1986) derived an expression for the macroscopic flux in the form

$$\overline{\mathbf{J}^{(d)}}^w = -D_d \mathbf{T}^* \cdot \nabla \overline{c}^w = -\mathbf{D}_d^* \cdot \nabla \overline{c}^w \qquad (6.2.15)$$

where $\mathbf{D}_d^* = \mathbf{T}^* D_d$ is the *coefficient of molecular diffusion in a porous medium* and $\mathbf{T}^*$ is a second-rank symmetric tensor that expresses the effect of the configuration of the water occupied portion of the REV. We used the averaging symbol $\overline{(\ )}^w$ in (6.2.15) in order to emphasize the difference between this equation and (6.2.14).

The coefficient $\mathbf{T}^*$, often referred to as a *tortuosity*, is defined by (Bear and Bachmat, 1984, 1986)

$$T_{ij}^* = \frac{1}{U_{0w}} \int_{(S_{ww})} (x_j - x_{0j}) \nu_i \, dS \qquad (6.2.16)$$

where $S_{ww}$ denotes the water-water portion of the bounding surface of the REV, $\mathbf{x}_0$ is the centroid of the REV, $\boldsymbol{\nu}$ is the outwardly directed normal to the surface $S_{ww}$, and $U_{0w}$ denotes the volume occupied by water within the REV.

For an isotropic porous medium $T_{ij}^*$ reduces to

$$T_{ij}^* = \frac{\theta_w^s}{\theta_w} \delta_{ij} \qquad (6.2.17)$$

where $\theta_w^s = S_{ww}/S_0$, $\theta_w = U_{0w}/U_0$, and $\delta_{ij}$ is the Kroenecker delta.

### 6.2.4. COEFFICIENT OF HYDRODYNAMIC DISPERSION

By adding the dispersive flux, expressed by (6.2.5), and the diffusive flux, expressed by (6.2.15), we obtain

$$\overline{\overset{\circ}{c}\mathbf{V}}^w + \overline{\mathbf{J}^{(d)}}^w = -(\mathbf{D} + \mathbf{D}_d^*) \cdot \nabla \overline{c}^w = -\mathbf{D}_h \cdot \nabla \overline{c}^w \qquad (6.2.18)$$

where the coefficient $\mathbf{D}_h = \mathbf{D} + \mathbf{D}_d^*$ is called the *coefficient of hydrodynamic dispersion*.

The total flux, $\mathbf{q}_{c,\,total}$ of a pollutant, by advection, dispersion, and diffusion can now be written in the form

$$\mathbf{q}_{c,\,total} = \theta_w(\overline{c}^w \overline{\mathbf{V}}^w - \mathbf{D}_h \cdot \nabla \overline{c}^w) \qquad (6.2.19)$$

This is the amount of the pollutant passing through a unit area of porous medium.

In (6.2.6), deleting the function $f(\text{Pe}, \delta)$, we have a linear relationship between the coefficient of mechanical dispersion, $\mathbf{D}$, and the average velocity $V$. However, $f(\text{Pe}, \delta)$ introduces a nonlinear effect of the velocity (as $\text{Pe} = LV/D_d$). Many experiments and some analytical studies seem to indicate that the coefficient of dispersion is not exactly a linear function of the velocity. Often expressions of the form

$$D_{11} = a_L V(\text{Pe})^{m_1}, \qquad D_{22} = a_T V(\text{Pe})^{m_2} \qquad (6.2.20)$$

where $m_1$ and $m_2$ are constants, are suggested instead of the linear ones given by (6.2.12).

Figure 6.6 gives a schematic representation of results of a large number of one-dimensional flow experiments for determining the coefficient of longitudinal dispersion, $D_{hL}$. Practically all the experiments were conducted in unconsolidated porous media.

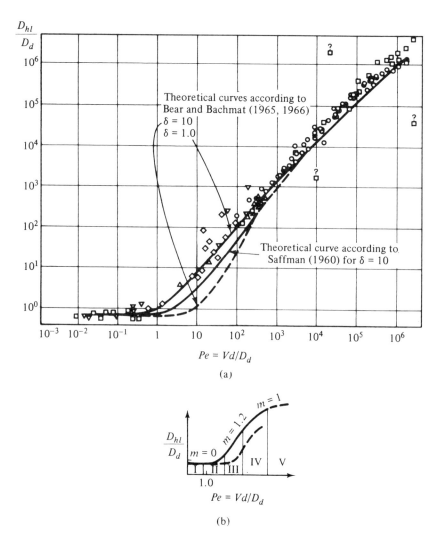

Fig. 6.6. Relationship between molecular diffusion and longitudinal hydrodynamic dispersion (after Pfannkuch, 1963; Saffman, 1960).

Figures 6.6a and b may be divided into several zones:

*Zone I*: In this zone, molecular diffusion predominates, as the average flow velocity is very small ($a_L V \ll D_d T^*$).

*Zone II*: Corresponds approximately to Pe between 0.4 and 5. In this zone, the effects of mechanical dispersion and molecular diffusion are of the same order of magnitude.

*Zone III*: Here spreading is caused mainly by mechanical dispersion. In this zone

$$D_{hL}/D_d = \alpha(\text{Pe})^m, \quad \alpha \approx 0.5, \quad 1 < m < 1.2. \tag{6.2.21}$$

*Zone IV*: This is the region in which mechanical dispersion predominates (as long as we stay in the range of validity of Darcy's law). The effect of molecular diffusion is negligible except as a factor that governs the transversal disperison. In the diagram, we obtain a straight line at 45 degrees

$$D_{hL}/D_d = \beta \text{Pe}, \quad \beta \approx 1.8. \tag{6.2.22}$$

In practice, in both Zones III and IV, the coefficient $D_{hL}$ is taken as proportional to the velocity.

*Zone V*: This is another zone of pure mechanical dispersion, but beyond the range of Darcy's law, so that the effects of inertia and turbulence can no longer be neglected.

Much less information is available on transversal dispersion. Ratios of $a_L/a_T$ of 5:1 to 24:1 and even up to 100:1 have been reported in the literature. A relationship for $D_{hT}/D_d$ similar to that given by (6.2.21) is often used, but with different values for $\alpha$ and $m$ (e.g., $\alpha = 1/40$ and $m = 1.1$ for Zone III).

#### 6.2.5. COMMENT ON UNSATURATED FLOW

The entire discussion presented for saturated flow in Subsections 6.2.1 through 6.2.4 can also be extended to an unsaturated one, where the polluted water occupies only part of the pore space. The phenomena of velocity variations and concentration gradients within the water-occupied portion will lead, when averaged, to mechanical dispersion and molecular diffusion at the macroscopic level. The entire presentation is therefore valid, except that the dispersivity that expresses the effect of the configuration of the water within the REV at a point, will now depend on the saturation at that point. Similarly, the coefficient of molecular diffusion in the porous medium will also be a function of the saturation.

## 6.3. Balance Equation for a Pollutant

As in the case of water flow, each of the various flux equations presented in Section 6.2 involves two variables: the flux and the concentration. Additional information is contained in the mass balance of the considered pollutant. This macroscopic balance takes the form of a second-order partial-differential equation that expresses the balance at (i.e., in the close vicinity of) a point inside a porous medium domain.

We shall consider the general case of unsaturated flow, with $\theta$ denoting the moisture content. For saturated flow, $\theta$ is replaced by the porosity, $n$. We recall that in unsaturated flow, the dispersivity is a function of the moisture content, and so is the coefficient of molecular diffusion in a porous medium.

When there is no danger of ambiguity, $c$ will denote the concentration of a pollutant at the macroscopic level ($\equiv \overline{c}^w$).

### 6.3.1. THE FUNDAMENTAL BALANCE EQUATION

Five components should be taken into account in the construction of a balance equation for a constituent.

(i) The quantity of the pollutant entering and leaving a control volume around a considered point by advection dispersion and diffusion, or the total flux, $q_{c,\text{total}}$, expressed by (6.2.19).

We recall that in Section 3.3, using a parallelpiped control box, we have shown that *minus the divergence of a flux (of any extensive quantity) represents the excess of inflow (of that quantity) over outflow, per unit volume of porous medium, per unit time*. Hence, here $-\text{div } \mathbf{q}_{c,\text{total}}$ represents the excess of inflow of a considered pollutant over outflow, per unit volume of porous medium, per unit time.

(ii) Pollutant leaving the fluid phase through the water-solid interface as a result of chemical or electrical interactions between the pollutant and the solid surface. Phenomena of *ion exchange* and *adsorption* may serve as examples. Let $f$ denote the quantity of pollutant that leaves the water by such mechanisms, per unit volume of porous medium, per unit time (see Subsection 6.3.2).

(iii) Pollutant added to the water (or leaving it) as a result of chemical interactions among species *inside* the water, or by various decay phenomena. Let $\Gamma$ denote the rate at which the mass of a pollutant is added to the water per unit mass of fluid (so that $\theta \rho \Gamma$ denotes the mass added by such phenomena, per unit volume of porous medium per unit time).

(iv) Pollutant may be added by injecting polluted water into a porous medium domain, e.g., as part of artificial recharge or waste disposal operations. Pollutant may be removed from a porous medium domain by withdrawing (polluted) water, e.g., by pumping. With $P(\mathbf{x}, t)$ and $R(\mathbf{x}, t)$ denoting the rates of water withdrawn or added, respectively, per unit volume of porous medium per unit time, and $c(\mathbf{x}, t)$ and $c_R(\mathbf{x}, t)$ denoting the pollutant's concentration in the water present in the porous medium and in the water added by injection, respectively, the total quantity of pollutant added per unit volume of porous medium per unit time is expressed by $Rc_R - Pc$.

# MODELING GROUNDWATER POLLUTION

(v) As a result of the above components, the quantity of the pollutant is increased within a control box. With $\theta c$ denoting the mass of a pollutant per unit volume of porous medium, $\partial \theta c / \partial t$ denotes the rate at which this quantity increases.

Combining the above components, we obtain

$$\frac{\partial \theta c}{\partial t} = -\nabla \cdot \mathbf{q}_{c,\text{total}} - f + \theta \rho \Gamma - Pc + Rc_R \qquad (6.3.1)$$

or, using (6.2.19) to express $\mathbf{q}_{c,\text{total}}$

$$\frac{\partial \theta c}{\partial t} = -\nabla \cdot (c\mathbf{q} - \theta \mathbf{D} \cdot \nabla c - \theta \mathbf{D}_d^* \cdot \nabla c) - f + \theta \rho \Gamma - Pc + Rc_R. \qquad (6.3.2)$$

Equation (6.3.2) is the (macroscopic) mass balance equation of a pollutant, expressed in terms of $c = c(\mathbf{x}, t)$. It is often called the *equation of hydrodynamic dispersion*, or the *advectivion — dispersion equation*.

As in Subsection 3.3.1, $P$ and $R$ are merely symbols that indicate pumping and recharge rates. The sinks and sources represented by these symbols may be of the distributed type, or point sources and sinks. In the latter case we employ the *Dirac delta function*, as in (3.3.4).

We note that (6.3.2), as well as all other equations derived from it in this section, are written from the *Eulerian point of view*. For $\theta = n = $ const, combined with (5.3.1) for $\rho_w = 0$ and $P$ replaced by $P - R$, (6.3.2) may also be written as

$$\frac{d^w c}{dt} = \nabla \cdot (\mathbf{D}_h \cdot \nabla c) + \left\{ \frac{n\rho \Gamma - f - R(c_R - c)}{n} \right\}$$

where $d^w c/dt$ ($\equiv \partial c / \partial t + \mathbf{V}_w \cdot \nabla c$) is the material derivative of the concentration $c$. We note that in this formulation, the left-hand side is written as a *Lagrangian formulation*, following a fixed particle of concentration $c$. The right-hand side is written as an *Eulerian formulation*, observing what happens at a point in space. Some numerical methods are based on this mixed *Eulerian–Lagrangian formulation*. In Section 6.6 we consider the case of $d^w c/dt = 0$ as a basis for numerical models presented in Chapter 11.

Another way of deriving (6.3.2) is to start from the microscopic balance equation

$$\frac{\partial c}{\partial t} = -\nabla \cdot (c\mathbf{V} - \mathbf{D} \cdot \nabla c) + \rho \Gamma \qquad (6.3.3)$$

and average it over the fluid phase contained within the REV. Note that in (6.3.3), $c$ is the concentration of the pollutant at the microscopic level. The resulting equation, into which we insert the expression (6.2.5) for the dispersive flux and (6.2.15) for the averaged diffusive flux, is identical to (6.3.2), except that $f$ is represent by $(1/U_0) \int_{(S_{fs})} \mathbf{J}_d \cdot \boldsymbol{\nu} \, dS$, in which $S_{fs}$ is the surface area of the solid-fluid interface and $\mathbf{J}_d$ is the microscopic flux due to molecular diffusion.

In order to obtain an expression for $f$, we now turn to the equation of balance of the same polluting component *on the solid phase*. Let $F$ denote the mass of the pollutant on the solid per unit mass of solid. With $\rho_s$ denoting the solid's density and $\theta_s \, (= (1 - n))$ denoting the solid's volumetric fraction, it is easy to show that the pollutant mass balance on the solid surface, reduces to

$$\frac{\partial(\theta_s \rho_s F)}{\partial t} = f + \theta_s \rho_s \Gamma_s \tag{6.3.4}$$

where $\Gamma_s$ is the rate of production of the pollutant per unit mass of solid. We note that in (6.3.4), $f$ has the same meaning as in (6.3.2) and that we have neglected any advective, dispersive, or diffusive flux of the pollutant present on the solid.

By eliminating $f$ from (6.3.3) and (6.3.4), and expressing the advective flux, $c\mathbf{q}$, by $\theta c \mathbf{V}$, we obtain

$$\frac{\partial(\theta c)}{\partial t} = -\nabla \cdot \theta(c\mathbf{V} - \mathbf{D} \cdot \nabla c - \mathbf{D}_d^* \cdot \nabla c) -$$

$$-\frac{\partial \theta_s \rho_s F}{\partial t} + \theta_s \rho_s \Gamma_s + \theta \rho \Gamma - Pc + Rc_R. \tag{6.3.5}$$

Equation (6.3.5) is a single equation in the two state variables, $c$ and $F$ (as $\Gamma$ and $\Gamma_s$ may also be functions of $c$ and $F$). We need an additional relationship between $c$ and $F$. This relationship is discussed in Subsection 6.3.2.

### 6.3.2. SOURCES AND SINKS ON THE SOLID SURFACE

*Adsorption* is the phenomenon of increase in the mass of a substance (e.g., a pollutant) on the solid at a fluid-solid interface. The component's affinity to the solid surface is due to electrical attraction, van der Waals attraction, i.e., intermolecular forces of attraction between molecules of the solid and adsorbed components, and chemisorption, i.e., chemical interaction between the solid and the adsorbed substances. Hence, the main factors affecting the adsorption and desorption of chemicals to or from the solid are the physical and chemical characteristics of the considered substance and of the solid's surface. Additional factors are temperature and the presence of other components in the fluid phase (e.g., through the pH that results from their concentrations in the fluid phase).

*Ion exchange* is a process of exchange between ions in the solution and ions present at sites on the solid's surface.

An *adsorption isotherm* is an expression that relates the quantity of an adsorbed component to its quantity (expressed as concentration) in the fluid phase, at constant temperature (i.e., under isothermal conditions, hence the term *isotherm*).

Thus, the isotherm relates $F$ to $c$. Different adsorbate-adsorbent pairs have different isotherms. However, in general, we may distinguish two classes of isotherms.

(i) *Equilibrium isotherms* that are based on the assumption (substantiated, of course, by experiments) that the quantities of the component on the solid and in the adjacent solution are continuously at equilibrium. Any change in the concentration of one of them produces an *instantaneous* change in the other.

(ii) *Nonequilibrium isotherms*, which assume that equilibrium is not achieved instantaneously, but rather that it is approached at a certain rate which, in general, depends on both $F$ and $c$.

Following are a number of examples of the more commonly encountered isotherms.

(a) Fruendlich (1926), suggested the nonlinear equilibrium isotherm

$$F = bc^m \tag{6.3.6}$$

where $b$ and $m$ are constant coefficients.

(b) For $m = 1$, and replacing the symbol $b$ by the more commonly used symbol $K_d$, Equation (6.3.6) reduces to

$$F = K_d c \tag{6.3.7}$$

known as the *linear equilibrium isotherm*. It assumes that adsorption is instantaneous, reversible, and linear. The coefficient $K_d$ is called the *distribution coefficient*, or *partitioning coefficient*. From (6.3.7), it follows that $K_d$ ($\equiv F/c$) gives, at every instant, the mass of the component on the solid, per unit mass of the latter, per unit concentration of the component in the fluid phase. It describes the partitioning of the total amount of the component, say in a unit volume of porous medium, between the part adsorbed on the solid walls and the part remaining in the fluid phase. Sometimes, $K_d$ for the adsorption process differs from that of the desorption one. This means that the process is not completely reversible. Another observation is that there is often a limit to the adsorptive capacity of the solid walls. This requires a modification of the isotherm (6.3.7).

In unsaturated flow, water occupies only part of the void space, at the volumetric fraction $\theta_w$, or saturation $S_w$ ($= \theta_w/n$), overlooking the presence of a thin stagnant liquid film that covers the solid in the air occupied zone. Then, only part of the total area of the solid is exposed to adsorption, or ion-exchange phenomena. The portion of the total surface of the solid that is in contact with the liquid phase, depends on $\theta_w$. Let us *assume* (as one of many possibilities to be verified by experiments for a particular porous medium) that the ratio of the area of the solid-liquid interface to the total area of the solid is equal to the ratio of active solid mass (i.e., solid mass participating in the surface phenomena) to the total mass of the solid, and that each of these ratios, in turn, is equal to the ratio of the liquid occupied portion of the void space to the total void space volume, i.e., equal to $S_w$. Then, the linear equilibrium isotherm (6.3.7) reduces to

$$F = S_w K_d c \tag{6.3.8}$$

where $K_d$ has the same definition as in (6.3.7). In other models, $S_w$ in (6.3.8) may be replaced by some known function of $S_w$.

Obviously, it is possible to assume that due to the presence of the film and to its ability to transport a solute through it by molecular diffusion, the entire surface of the solid is available for adsorption. In this case, the isotherm (6.3.7) remains valid also for $S_w < 1$.

Considerations similar to those discussed above are also applicable to the isotherms considered below.

(c) A more general form of the linear equilibrium isotherm is

$$F = k_1 c + k_2 \tag{6.3.9}$$

where $k_1$ and $k_2$ are constant coefficients.

(d) Langmuir (1916, 1918) suggested *the nonlinear equilibrium isotherm*

$$F = \frac{k_3 c}{1 + k_4 c} \tag{6.3.10}$$

where $k_3$ and $k_4$ are constant coefficients.

(e) Lindstrom *et al.* (1971) and Van Genuchten (1974) mention the *nonlinear isotherm*

$$F = k_5 c \exp(-2 k_6 F) \tag{6.3.11}$$

where $k_5$ and $k_6$ are constant coefficients.

(f) the simplest *nonequilibrium isotherm* for an irreversible system (Langmuir, in Adamson, 1967) is

$$\frac{\partial F}{\partial t} = k_7 c \tag{6.3.12}$$

where $k_7$ is a constant coefficient.

(g) Lapidus and Amundson (1952) proposed the nonlinear isotherm

$$\frac{\partial F}{\partial t} = k_r (k_8 c + k_9 - F) \tag{6.3.13}$$

where $k_8$ and $k_9$ are constant coefficients and $k_r$ is a *kinetic rate coefficient*.

(h) A *nonequilibrium Langmuir isotherm* is (e.g., Hendricks, 1972)

$$\frac{\partial F}{\partial t} = k_r \left( \frac{k_{10} c}{1 + k_{11} c} - F \right) \tag{6.3.14}$$

(i) Van Genuchten et al. (1974) also mention the *nonequilibrium Freundlich isotherm*

$$\frac{\partial F}{\partial t} = k_r(k_{12}c^{k_{13}} - F). \quad (6.3.15)$$

Only one of the above isotherms would apply for each particular case. The selection of the appropriate isotherm and the determination of the value of the various coefficients appearing in it, should be based on the study of the thermodynamics of the interacting components and on experiments with the particular soil under consideration.

When the species adsorbed on the solid undergoes radioactive, or any other type of decay, the sink term $\Gamma_s$ takes the form

$$\Gamma_s = -\lambda F \quad \text{or} \quad \Gamma_s = -k_s F \quad (6.3.16)$$

where $\lambda = 1/T$, $T$ is the half-life, and $k_s$ is a *degradation rate constant* of the decaying species.

### 6.3.3. SOURCES AND SINKS WITHIN THE LIQUID PHASE

Sources and sinks of the pollutant, expressed by the term $\theta\rho\Gamma$, result from various processes, e.g., chemical reactions among components within the liquid, radioactive decay and biodegradation, and growth due to bacterial activities.

When the pollutant present in the water is a radioactive species, or any other decaying species, then

$$\theta\rho\Gamma = -\theta\lambda c \quad \text{or} \quad \theta\rho\Gamma = -\theta k_f c \quad (6.3.17)$$

where $k_f$ is a degradation rate constant in the water.

When the considered $\gamma$-component participates in chemical reactions which cause its quantity to increase, we may express the $\gamma$-source by

$$\theta\rho\Gamma = \sum_{(j)} R_{\gamma j} \quad (6.3.18)$$

where $R_{\gamma j}$ is the rate of production of the mass of the $\gamma$-component by the $j$th reaction, per unit volume of porous medium. We could also express the rate of production per unit volume of water, or per unit mass of it. In general, $R_{\gamma j} = R_{\gamma j}(c_{\gamma 1}, c_{\gamma 2}, c_{\gamma 3}, \ldots)$, i.e., a function of the concentrations of the various components that are present in the water. Often

$$R_{\gamma j} = k_m(c_\gamma)^m \quad (6.3.19)$$

where $c_\gamma$ is the concentration of the $\gamma$-component in moles per unit volume of water and $m$ indicates the 'order' of the reaction. For a first-order reaction, $m = 1$, and $k_m$ has the dimension of reciprocal time.

The general chemical reaction can be described by the stoichiometric equation

$$eE + fF \underset{k_r}{\overset{k_f}{\rightleftarrows}} gG + hH. \tag{6.3.20}$$

At equilibrium, with the forward rate of reaction (described by $k_f$) equalling the reverse one (described by $k_r$), the law of mass action

$$K = \frac{[a_G]^g [a_H]^h}{[a_E]^e [a_F]^f} \tag{6.3.21}$$

has to be satisfied. In (6.3.21), the symbol $K$ represents a (known) *thermodynamic equilibrium constant*, that depends on the temperature, and the square brackets denote thermodynamical concentrations, or activities, $a_\beta$, $\beta = E, F, G, H$. These activities can be related to molar concentrations, $c_\beta$ (i.e., expressed in moles per liter) by $a_\beta = \gamma_\beta c_\beta$, where $\gamma_\beta$ is the activity coefficient of $\beta$. For dilute solutions, $\gamma_\beta$ approaches unity and $a_\beta = c_\beta$.

It is important to emphasize that (6.3.21) describes the relationship among the reacting species when the reaction is at equilibrium. In many groundwater situations, equilibrium may not be reached for a long, sometimes, very long, time. In order to deal with the rates at which reactions occur, we need information about the kinetics of the chemical process involved, e.g., in the form of (6.3.19).

### 6.3.4. MASS BALANCE EQUATION WITH ADSORPTION AND DECAY

Let the considered polluting component be one that is adsorbed to the solid surfaces and, in addition, undergoes degradation, but at rates that are different for the component on the solids and within the fluid. In addition, point sources and sinks of the component exist within the considered domain due to the recharge and withdrawal of water. Expressing the sink, $\Gamma_s$, in (6.3.5), due to the degradation of the component on the solid, by (6.3.16), the sink, $\Gamma$, in the water, by (6.3.17) the sources, $Rc_R$, due to artificial recharge at rates $R^{(m)}$ at points $\mathbf{x}$, by

$$\sum_{(m)} R^{(m)}(\mathbf{x}^{(m)}, t) \, \delta(\mathbf{x} - \mathbf{x}^{(m)}) c_R^{(m)}(\mathbf{x}^{(m)}, t)$$

and the sink, $Pc$, due to pumping at rates $P^{(r)}(\mathbf{x}^{(r)}, t)$ at points $\mathbf{x}^{(r)}$, by

$$\sum_{(r)} P^{(r)}(\mathbf{x}^{(r)}, t) \, \delta(\mathbf{x} - \mathbf{x}^{(r)}) c(\mathbf{x}^{(r)}, t),$$

we obtain the component's mass balance in the form

$$\frac{\partial \theta c}{\partial t} = -\nabla \cdot \theta(c\mathbf{V} - \mathbf{D} \cdot \nabla c - \mathbf{D}_d^* \cdot \nabla c) - \frac{\partial \theta_s \rho_s F}{\partial t} -$$

$$- \theta_s \rho_s k_s F - \theta k_f c + \sum_{(m)} R^{(m)}(\mathbf{x}^{(m)}, t)\, \delta(\mathbf{x} - \mathbf{x}^{(m)}) c_R^{(m)}(\mathbf{x}^{(m)}, t) -$$

$$- \sum_{(r)} P^{(r)}(\mathbf{x}^{(r)}, t)\, \delta(\mathbf{x} - \mathbf{x}^{(r)}) c(\mathbf{x}^{(r)}, t). \tag{6.3.22}$$

For saturated flow, $\theta$ should be replaced by $n$ and $\theta_s$ by $(1 - n)$. For unsaturated flow, $\mathbf{D} = \mathbf{D}(\theta)$ and $\mathbf{D}_d^* = \mathbf{D}_d^*(\theta)$.

Equation (6.3.22) contains the two variables: $c(\mathbf{x}, t)$ and $F(\mathbf{x}, t)$. Hence, we have to supplement this equation by the appropriate isotherm that expresses the relationship between them. For example, with (6.3.8), in which $S_w$ is replaced by a more general term $f_a(\theta)$, we obtain from (6.3.22)

$$\frac{\partial}{\partial t}\{\theta + \theta_s \rho_s f_a(\theta) K_d\} c$$

$$= -\nabla \cdot \theta(c\mathbf{V} - \mathbf{D} \cdot \nabla c - \mathbf{D}_d^* \cdot \nabla c) -$$

$$-(\theta k_f + \theta_s \rho_s f_a(\theta) k_s K_d) c +$$

$$+ \sum_{(m)} R^{(m)}(\mathbf{x}^{(m)}, t)\, \delta(\mathbf{x} + \mathbf{x}^{(m)}) c_R^{(m)}(\mathbf{x}^{(m)}, t) -$$

$$- \sum_{(r)} P^{(r)}(\mathbf{x}^{(r)}, t)\, \delta(\mathbf{x} - \mathbf{x}^{(r)}) c(\mathbf{x}^{(r)}, t) \tag{6.3.23}$$

which now involves only the single variable $c(\mathbf{x}, t)$ expressing the spatial concentration distribution of the considered $\gamma$-component. We recall the comments that in unsaturated flow, $\mathbf{D}$ and $\mathbf{D}_d^*$ depend on the saturation (Subsection 6.2.5).

We note that in order to solve (6.3.23), we need information on the velocity distribution $\mathbf{V}(\mathbf{x}, t)$, as well as on $\theta(\mathbf{x}, t)$. In saturated flow, we replace $\theta$ by $n$, and we need information on $n(\mathbf{x}, t)$.

By combining (6.3.22) with the mass balance equation (5.3.1), to which we add a term expressing artificial recharge, we obtain, for $\rho_w = \text{const}$

$$\theta \frac{\partial c}{\partial t} = \nabla \cdot (\theta \mathbf{D} h \cdot \nabla c) - \theta \mathbf{V} \cdot \nabla c - \frac{\partial \theta_s \rho_s F}{\partial t} - \theta_s \rho_s k_s F -$$

$$- \theta k_f c + \sum_{(m)} R^{(m)}(\mathbf{x}^{(m)}, t)\, \delta(\mathbf{x} - \mathbf{x}^{(m)})(c_R^{(m)} - c). \tag{6.3.24}$$

Consider the special case of saturated flow with no external sources and sinks, i.e., $R^{(m)}(\mathbf{x}, t) \equiv 0$ and $P^{(r)}(\mathbf{x}, t) \equiv 0$, and with $\partial \rho_s/\partial t = 0$ and $\partial n/\partial t = 0$. We shall further assume that adsorption takes place, obeying the linear equilibrium isotherm (6.3.7) with $\partial K_d/\partial t = 0$. Under these conditions, (6.3.23) reduces to

$$nR_d \frac{\partial c}{\partial t} = -\nabla \cdot n(c\mathbf{V} - \mathbf{D} \cdot \nabla c - \mathbf{D}_d^* \cdot \nabla c) -$$

$$- \{nk_f + (1-n)\rho_s k_s K_d\} c \qquad (6.3.25)$$

where

$$R_d = 1 + \frac{(1-n)\rho_s K_d}{n} \quad (> 1)$$

is called the *coefficient of retardation*, or the *retardation factor*. To understand the significance of $R_d$ and the reason for calling it a 'retardation factor', let us further simplify (6.3.25) by assuming $k_s = k_f$, and a homogeneous porous medium. Then, $R_d$ = const and (6.3.25) may be rewritten as

$$n \frac{\partial c}{\partial t} = -\nabla \cdot n \left( c \frac{\mathbf{V}}{R_d} - \frac{\mathbf{D}}{R_d} \cdot \nabla c - \frac{\mathbf{D}_d^*}{R_d} \cdot \nabla c \right) - nk_f c. \qquad (6.3.26)$$

For comparison, let us rewrite (6.3.23) in which $\theta$ is replaced by $n$, $K_d = 0$ and the external source and sink terms have been deleted, leaving

$$n \frac{\partial c}{\partial t} = -\nabla \cdot n(c\mathbf{V} - \mathbf{D} \cdot \nabla c - \mathbf{D}_d^* \cdot \nabla c) - nk_f c. \qquad (6.3.27)$$

We note that (6.3.26) and (6.3.27) are similar, except that in the former the average water velocity carrying the component *seems to be* $\mathbf{V}/R_d$ and the coefficient of hydrodynamic dispersion is reduced to $\mathbf{D}_h/R_d$. Thus, since $R_d > 1$, the effect of adsorption and similar activities is to *retard* the advance of the considered component (as part of it is adsorbed to the solid surface, rather than advance with the water moving at the average velocity, $\mathbf{V}$). At the same time, the coefficient of advective dispersion, $\mathbf{D}$, which is shown in (6.2.6) to be proportional to the average velocity, is also reduced by the factor $R_d$. The coefficient of molecular diffusion in a porous medium, $\mathbf{D}_d^*$, is also reduced by the factor $R_d$.

Although we have reduced (6.3.23) to the simpler form (6.3.26) in order to explain the phenomenon of retardation, this phenomenon obviously also exists in the more general case of unsaturated flow, expressed by (6.3.23)

$$\frac{\partial}{\partial t}\{\theta R_d(\theta)c\}$$
$$= -\nabla \cdot \theta(c\mathbf{V} - \mathbf{D}_h \cdot \nabla c) - \theta k_f \left(1 + \frac{\theta_s \rho_s f_a(\theta) k_s K_d}{\theta k_f}\right) c +$$
$$+ \sum_{(m)} R^{(m)}(\mathbf{x}^{(m)}, t) \, \delta(\mathbf{x} - \mathbf{x}^{(m)}) c_R^{(m)}(\mathbf{x}^{(m)}, t) -$$
$$- \sum_{(r)} P^{(r)}(\mathbf{x}^{(r)}, t) \, \delta(\mathbf{x} - \mathbf{x}^{(r)}) c(\mathbf{x}^{(r)}, t) \qquad (6.3.28)$$

where

$$R_d(\theta) = 1 + \frac{\theta_s \rho_s f_a(\theta) K_d}{\theta}$$

now depends on $\theta(\mathbf{x}, t)$.

The phenomenon of retardation also exists when $F$ in (6.3.23) is expressed by any other isotherm, not necessarily by the linear equilibrium isotherm (6.3.7). The structure of $R_d$ will then depend on the selected isotherm.

### 6.3.5. THE EFFECT OF IMMOBILE WATER

Another phenomenon that affects the movement of pollutants is that of *immobile water* (or almost so), often encountered in both saturated and unsaturated zones. In a saturated flow domain, *immobile*, or *stagnant water* is the water occupying *dead-end pores*. These are pores that, although being part of the general interconnected void space, have very narrow connections with the latter, so that the water in them is almost stagnant. However, stagnant water may also be due to local zones of very low permeability. In unsaturated flow, immobile water may also occur in pendular rings of drained pores. Although (almost) immobile, the water in the immobile zones is part of the *continuous* water phase.

Due to its very low (or zero) velocity, it is common to assume that no advection of a pollutant, or hydrodynamic dispersion, can take place in a body of immobile water. However, these water bodies can exchange a pollutant with the water surrounding them by molecular diffusion. Thus, the behavior of this portion of the void space is equivalent to that of sources or sinks for the pollutant. The considered pollutant will always diffuse from the portion of the water where the concentration is higher to that where it is lower.

The changes that take place in the concentration of the pollutant in the immobile water, $c_{im}$, can be described by a continuum model, similar to that expressed by (6.3.4) for the adsorbed pollutant. With $\theta_{im}$ ($= S_{im}n$) denoting the fractional volume of the porous medium occupied by immobile water, and $c_{im}$ denoting the concentration of the latter, the pollutant's mass balance takes the form

$$\frac{\partial(\theta_{im}c_{im})}{\partial t} = -f_{im} + \rho\theta_{im}\Gamma \tag{6.3.29}$$

where $f_{im}$ denotes the net rate at which the pollutant leaves the immobile water (with $c_{im}$ and $\theta_{im}$ denoting its concentration and volumetric fraction; $\theta = \theta_m + \theta_{im}$) per unit volume of porous medium.

For the mobile water, the pollutant's balance equation is

$$\frac{\partial(\theta_m c_m)}{\partial t} = -\nabla \cdot \theta_m(c_m \mathbf{V} - \mathbf{D}_h \cdot \nabla c_m) + f_{im} + \rho\theta_m\Gamma. \tag{6.3.30}$$

Often, the net rate of exchange, $f_{im}$, is expressed by

$$f_{im} = \alpha_d^*(c_{im} - c_m) \tag{6.3.31}$$

where $\alpha_d^*$ is a transfer coefficient that depends on the coefficient of molecular diffusion, $D_d$, and on the geometry of the immobile water's contact area with the mobile water.

When adsorption occurs, it does so both on the solid-mobile water and the solid-immobile water contact areas. One possible model is to assume that fractions $p$ and $(1 - p)$ of the total solid-water contact area (which in itself is a function of $\theta$) constitute solid-mobile water and solid-immobile water contact areas, respectively. The corresponding isotherms will be

$$F_m = pK_d c_m; \qquad F_{im} = (1-p)K_d c_{im} \tag{6.3.32}$$

where $F_m$ and $F_{im}$ denote the mass of adsorbed pollutant per unit *total* mass of solid.

The two balance equations for a radioactively decaying pollutant in the mobile and in the immobile water, are

$$\frac{\partial(\theta_m c_m R_{dm})}{\partial t}$$
$$= -\nabla \cdot \theta_m(c_m \mathbf{V} - \mathbf{D}_h(\theta_m) \cdot \nabla c_m) + \alpha_d^*(c_{im} - c_m) -$$
$$- \theta_m R_{dm}\lambda c_m, \quad R_{dm} = 1 + \frac{\theta_s \rho_s p K_d}{\theta_m}, \tag{6.3.33}$$

$$\frac{\partial(\theta_{im}c_{im}R_{dim})}{\partial t} = \alpha_d^*(c_m - c_{im}) - \theta_{im}R_{dim}\lambda c_{im},$$

$$R_{dim} = 1 + \frac{\theta_s \rho_s (1-p) K_d}{\theta_{im}}. \tag{6.3.34}$$

We note here the possibility of $\partial\theta_{im}/\partial t \neq 0$. It is possible to use some average

# MODELING GROUNDWATER POLLUTION

retardation factor $R_d = 1 + \theta_s \rho_s K_d / \theta$. For saturated flow, $\partial \theta_{im}/\partial t = 0$ and $\theta_m = n - \theta_m$.

### 6.3.6. RADIONUCLIDE AND OTHER DECAY CHAINS

Consider the case of a decay chain of elements

$$A_1 \to A_2 \to A_3 \to \ldots A_N$$

such that

$$A_2 = A_1 \exp(-\lambda_1 t); \ A_3 = A_2 \exp(-\lambda_2 t) \ldots$$
$$A_N = A_{N-1} \exp(-\lambda_{N-1} t) \tag{6.3.35}$$

with $c_1, c_2, \ldots, c_N$ denoting the concentrations of the respective $N$ elements of the chain.

If the various components can also be adsorbed to the solid matrix, let us denote the corresponding distribution coefficients by $K_{d1}, K_{d2}, \ldots, K_{dN}$. Some of these may be zero.

The source terms are

$$\rho \Gamma_i = -\lambda_i c_i + \lambda_{i-1} c_{i-1},$$
$$\Gamma_{si} = -K_{di} \lambda_i c_i + K_{d(i-1)} \lambda_{i-1} c_{i-1},$$
$$i = 1, 2, \ldots, N, \ c_0 = 0.$$

The balance equation for the $i$th component becomes

$$\theta R_{di} \frac{\partial c_i}{\partial t} = -\nabla \cdot (c_i \mathbf{q} - \theta \mathbf{D}_h \cdot \nabla c_i) -$$
$$- \theta(\lambda_i R_{di} c_i - \lambda_{i-1} R_{d(i-1)} c_{i-1}) \tag{6.3.36}$$

where

$$R_{di} = 1 + \frac{1-n}{\theta} \rho_s K_{di}.$$

For saturated flow, we replace $\theta$ in (6.3.36) by $n$. Usually, $\lambda_N = 0$, i.e., a nondecaying component. Thus, (6.3.36) represents $N$ equations which should be solved simultaneously for the $N$ concentration, $c_i$.

## 6.4. Initial and Boundary Conditions

As in the cases of saturated and unsaturated flows considered in Chapters 3 through 5, here also the partial differential equation expressing the balance of a polluting constituent has to be supplemented by appropriate initial and boundary conditions, in order to yield a solution for a particular studied case. These conditions should, therefore, be obtained from actual observations (and/or future anticipated values) of these conditions in the particular studied case.

The discussions in Subsections 3.4.2 and 3.4.3 are also applicable here. We shall continue to use $F = F(x, y, z, t)$ as the equation describing the boundary surface, with $\boldsymbol{\nu} = \nabla F/|\nabla F|$. We shall consider saturated as well as unsaturated flows. As always, $\theta = n$ for saturated flow.

*Initial conditions* include information on the concentration distribution at $t = 0$ at all points within the considered region, $R$.

$$c = c(\mathbf{x}, 0) \quad \text{in } R. \tag{6.4.1}$$

### 6.4.1. THE GENERAL BOUNDARY CONDITION

We shall use (3.4.6) as a starting point; it expresses the condition of flux continuity of the considered polluting species across the boundary, assuming that no sources or sinks of that species exist on the latter. Since in this case, $S_{\alpha\beta}$, which denotes the combined solid and air portions the boundary, is a material surface with respect to the considered species, the general condition of no-jump in total flux, takes the form

$$[\theta c(\mathbf{V} - \mathbf{u}) - \theta \mathbf{D}_h \cdot \nabla c]_{1,2} \cdot \boldsymbol{\nu} = 0 \quad \text{on } S_i \tag{6.4.2}$$

where $S_i$ denotes the $i$th segment of the boundary, and we have used (6.2.18) to express the flux of hydrodynamic dispersion. Note that since $(\mathbf{V}_s - \mathbf{u}) \cdot \boldsymbol{\nu} = 0$, we can always replace $(\mathbf{V} - \mathbf{u}) \cdot \boldsymbol{\nu}$ by $\mathbf{q}_r \cdot \boldsymbol{\nu}$.

Let us consider several cases of special interest.

### 6.4.2. BOUNDARY OF PRESCRIBED CONCENTRATION

When the concentration, $c = c(\mathbf{x}, t)$, can be specified as a known function, say $g_1(\mathbf{x}, t)$, at all points of a given boundary (or boundary segment), $S_1$, due to phenomena occurring in the domain's environment, independent of what happens within $R$, we make use of the assumption leading to (3.4.7b), writing the boundary condition in the form of

$$c(\mathbf{x}, t) = g_1(\mathbf{x}, t) \quad \text{on } S_1 \tag{6.4.3}$$

where $c$ and $g$ refer to the domain's side and to the external side of $S_1$, respectively.

This is a *first kind*, or *Dirichlet boundary condition*.

### 6.4.3. BOUNDARY OF PRESCRIBED FLUX

When phenomena occurring in the environment *impose a known flux*, say $g_2(x, t)$, at all points of a boundary segment, $S_2$, independent of what happens within the considered domain, we use (6.4.2), combined with (3.4.7b), to write the condition

$$\{c\mathbf{q}_r - \theta_w \mathbf{D}_h \cdot \nabla c\} \cdot \boldsymbol{\nu} = g_2(\mathbf{x}, t) \quad \text{on } S_2. \tag{6.4.4}$$

Since both $c$ and $\nabla c$ are involved in (6.4.4), this is a *Cauchy boundary condition*.

A boundary of special interest is the *impervious boundary*. Then, since $g_2(\mathbf{x}, t) = 0$, and by (3.4.5), $\mathbf{q}_r \cdot \boldsymbol{\nu} = 0$, Equation (6.4.4) reduces to

$$\mathbf{D}_h \cdot \nabla c = 0 \quad \text{on } S_2 \tag{6.4.5}$$

which is a *Neumann boundary condition*

### 6.4.4. BOUNDARY BETWEEN TWO POROUS MEDIA

In this case, we assume the existence of discontinuities in all porous matrix properties, such as porosity, permeability, and dispersivity. Neither the concentration nor the flux are known *a-priori* on the boundary, $S_3$. However, both (3.4.7b) and (6.4.4) must be satisfied, i.e.,

$$c(\mathbf{x}, t)|_1 = c(\mathbf{x}, t)|_2 \quad \text{on } S_3, \tag{6.4.6}$$

$$\{c\mathbf{q}_r - \theta \mathbf{D}_h \cdot \nabla c\}|_1 \cdot \boldsymbol{\nu}$$
$$= \{c\mathbf{q}_r - \theta \mathbf{D}_h \cdot \nabla c\}|_2 \cdot \boldsymbol{\nu} \quad \text{on } S_3. \tag{6.4.7}$$

We note that in view of (3.4.10), we have used

$$\theta(\mathbf{V} - \mathbf{u}) \cdot \boldsymbol{\nu} = (\mathbf{q} - \theta \mathbf{\dot{u}}) \cdot \boldsymbol{\nu} = (\mathbf{q} - \theta \mathbf{V}_s) \cdot \boldsymbol{\nu}$$
$$= \mathbf{q}_r \cdot \boldsymbol{\nu}.$$

We note here that since neither $c|_1$ nor $c|_2$ are known on $S_3$, we need two conditions to solve, simultaneously, for the concentrations on both sides of $S_3$.

### 6.4.5. BOUNDARY WITH A 'WELL MIXED ZONE'

Here we consider the boundary (say, $S_4$) between a porous medium domain and a body of water (a river, lake, or sea) assumed to be a *well-mixed domain*, in which the concentration of the considered species, $c$, is maintained constant in space, but not necessarily in time, say $c_0$. This assumption is often introduced to circumvent the need to solve explicitly for the concentration distribution in the body of water. It is, obviously, an approximation of reality.

For such boundary, the no-jump condition in the flux of the considered species takes the form

$$\{c_0(\mathbf{V} - \mathbf{u})\}|_{wb} \cdot \boldsymbol{\nu} = \{c\mathbf{q}_r - \theta \mathbf{D}_h \cdot \nabla c\}|_{pm} \quad \text{on } S_4 \tag{6.4.8}$$

where subscripts *wb* and *pm* denote that the boundary is approached from the water body or the porous medium sides, respectively. We note that on the water body side, because $\nabla c_0 = 0$, we have advective flux only, while on the porous medium side we have advection, dispersion, and diffusion.

As a simplification, let us consider a stationary boundary ($\mathbf{u} = 0$), so that (6.4.8) reduces to

$$(c_0 \mathbf{V})|_{wb} \cdot \boldsymbol{\nu} = (c\mathbf{q} - \theta \mathbf{D}_h \cdot \nabla c)|_{pm} \cdot \boldsymbol{\nu} \quad \text{on } S_4. \tag{6.4.9}$$

Consequently, when no advection takes place across the boundary, i.e., $\mathbf{V}|_{wb} \cdot \boldsymbol{\nu} = \mathbf{q}|_{pm} \cdot \boldsymbol{\nu} = 0$, Equation (6.4.9) yields

$$(\mathbf{D}_d^* \cdot \nabla c)|_{pm} \cdot \boldsymbol{\nu} = 0 \quad \text{on } S_4. \tag{6.4.10}$$

This implies that for a stationary boundary without advection, there is no transport of the considered species to or from the porous medium domain, even when $c_0 \neq c|_{pm}$. This is obviously *unacceptable*, as under such conditions we would intuitively expect transport by molecular diffusion between the water body and the porous medium. Obviously, there is some flaw in the above development.

The error stems from the assumptions that (a) a well-mixed zone exists, and (b) that such zone, combined with the sharp boundary approximation, yields no flux of the considered species by molecular diffusion and, perhaps, by dispersion, across the boundary to the adjacent porous medium domain. In order to 'reinstate' this diffusive-dispersive flux, that we know should take place at the boundary, we introduce the concept of a *buffer zone*, or *transition zone* of width $\Delta$ (Figure 6.7), across which the concentration varies from $c|_{wb} = c_0$ to $c|_{pm}$. We assume that the abrupt boundary passes through the middle of this zone. By writing a balance of the advective, dispersive and diffusive fluxes entering and leaving this zone, and assuming no storage of the considered species within it, we obtain

$$c_0 \mathbf{V}|_{wb} \cdot \boldsymbol{\nu} - \alpha(c|_{pm} - c_0) = (c\mathbf{q} - \theta \mathbf{D}_h \cdot \nabla c)|_{pm} \cdot \boldsymbol{\nu} \tag{6.4.11}$$

where $\alpha$ is a coefficient that is related to both diffusion and dispersion. In (6.4.11),

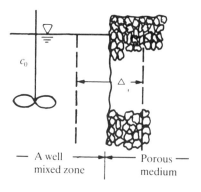

Fig. 6.7. A transition zone ($\Delta$) adjacent to an open water body.

the diffusive-dispersive flux entering (or leaving) the transition zone from the fluid domain, is approximated by

$$-(\mathbf{D}_h \cdot \nabla c) \cdot \boldsymbol{\nu} = -(D_d^* + a_L \mathbf{V} \cdot \boldsymbol{\nu}) \frac{c|_{pm} - c_0}{\Delta} = \alpha(c|_{pm} - c_0). \quad (6.4.12)$$

Thus $\alpha = (D_d^* + a_L \mathbf{V} \cdot \boldsymbol{\nu})/\Delta$. Since $\mathbf{V}|_{wb} \cdot \boldsymbol{\nu} = \mathbf{q}_{pm} \cdot \boldsymbol{\nu}$, Equation (6.4.11) reduces to

$$(c_0 - c|_{pm})(\mathbf{q} \cdot \boldsymbol{\nu} + \alpha) + (\theta \mathbf{D}_h \cdot \nabla c)|_{pm} \cdot \boldsymbol{\nu} = 0 \quad \text{on } S_4. \quad (6.4.13)$$

If we now return to a case of no advective flux, (6.4.13) reduces to

$$\alpha(c_0 - c|_{pm}) + (\theta \mathbf{D}_h \cdot \nabla c)|_{pm} \cdot \boldsymbol{\nu} = 0 \quad \text{on } S_4 \quad (6.4.14)$$

where $\alpha = D_d^*/\Delta$. This is consistent with our intuition that a diffusive flux should exist across the boundary, even in the absence of advection. We note that in this case, due to our conceptual model of a well-mixed zone, we accept the condition $c_0 \neq c|_{pm}$.

When $|c_0 \mathbf{V} \cdot \boldsymbol{\nu}| \gg |\alpha(c|_{pm} - c_0)|$, Equation (6.4.14) reduces to

$$(c_0 - c|_{pm})\mathbf{q} \cdot \boldsymbol{\nu} + (\theta \mathbf{D}_h \cdot \nabla c)|_{pm} \cdot \boldsymbol{\nu} = 0 \quad \text{on } S_4. \quad (6.4.15)$$

### 6.4.6. BOUNDARY OF EXIT TO THE ATMOSPHERE

Here the polluted water is drained into the atmospheric environment under saturated conditions. The seepage face may serve as an example. We shall assume that on the external side of the boundary, the concentration of the considered pollutant remains the same as on the internal side. Hence

$$\{nc(\mathbf{V} - \mathbf{u}) - n\mathbf{D}_h \cdot \nabla c\}|_{pm} \cdot \boldsymbol{\nu} = \{c(\mathbf{V} - \mathbf{u})\}|_{\text{atm.}} \cdot \boldsymbol{\nu}. \quad (6.4.16)$$

Since $n(\mathbf{V} - \mathbf{u})|_{pm} \cdot \boldsymbol{\nu} = (\mathbf{V} - \mathbf{u})|_{\text{atm.}} \cdot \boldsymbol{\nu}$ and $c|_{pm} = c|_{\text{atm}}$, Equation (6.4.16) reduces to

$$(\mathbf{D}_h \cdot \nabla c)|_{pm} \cdot \boldsymbol{\nu} = 0 \quad \text{on } S_5, \quad (6.4.17)$$

i.e., a *second kind*, or a *Neumann boundary condition*.

### 6.4.7. PHREATIC SURFACE

Finally, we consider the boundary condition for the transport of a polluting species across a phreatic surface (Subsection 3.4.6). Let $c'$ denote the concentration of the infiltrating water. We shall assume that the mass density of the water is unaffected by changes in the concentration, $c$, of the polluting species. The boundary condition is derived from the requirement of flux continuity normal to the phreatic surface. The resulting boundary condition is

$$(c\mathbf{q} - c'\mathbf{N}) \cdot \nabla F + (nc - \theta_{w0}c') \frac{\partial F}{\partial t} - n\mathbf{D}_h \cdot \nabla c|_{\text{sat}} \cdot \nabla F = 0 \quad (6.4.18)$$

where $\theta_{w0}$ is the irreducible moisture content, assumed to exist everywhere above the phreatic surface.

If we combine this condition with (3.4.25), we obtain

$$(c - c') \left( \mathbf{N} \cdot \nabla F + \theta_{w0} \frac{\partial F}{\partial t} \right) - n\mathbf{D}_h \cdot \nabla c \cdot \nabla F = 0. \tag{6.4.19}$$

We note that in this case $c \neq c'$ on the phreatic surface. Thus, the zone above the phreatic surface represents a well-mixed zone in the sense discussed in Subsection 6.4.6.

## 6.5. Complete Statement of Pollution Model

### 6.5.1. THE GENERAL STATEMENT

Similar to the models of saturated (Section 3.5) and unsaturated (Section 5.5) flow problems, the complete model of a pollution problem consists of the following items:

(i) Specification of the geometrical configuration of the closed surface that bounds the problem area, with possible segments at infinity.
(ii) Specification of the dependent variable(s) of the pollution problem, i.e., the concentration, $c(\mathbf{x}, t)$, of the specific constituent or constituents under consideration.

In the case of interacting constituents, the concentration of each of them is a state variable and we need information on how they interact with each other.

Recalling that in all dispersion equations, the velocity, $\mathbf{V}(\mathbf{x}, t)$ appears in both the advective flux and as a building block of the dispersion coefficient, $\mathbf{D}$, we must have information on $\mathbf{V}(\mathbf{x}, t)$. This information can either be provided as part of the input to the pollution problem, or we may construct a model in which the velocity is another state variable for which a solution is sought. If changes in concentration affect the water's density, $\rho(\mathbf{x}, t)$, the latter becomes another state variable to be solved for, and we need information on the relationship $\rho = \rho(c)$. In unsaturated flow, we need information on $\theta(\mathbf{x}, t)$, or we regard $\theta(\mathbf{x}, t)$ as another dependent variable of the problem.

(iii) Statement of a partial differential (balance) equation, for every relevant species. Balance equations, in terms of the various state variables of the problem, as listed in (ii) above, are also required for every extensive quantity that is relevant to the problem.
(iv) Specification of the numerical values of the (transport and storage) coefficients that appear in (iii). Of special interest here is the information on the dispersivity and on the coefficient of molecular diffusion in the porous medium under consideration.

(v) Statement of the numerical values of the various source and sink terms that appear in (iii).

(vi) Statement of initial and boundary conditions that the state variables appearing in (ii) have to satisfy within the considered domain at $t = 0$, and on its boundaries at $t > 0$, respectively.

6.5.2. THE MODEL FOR A CONCENTRATION DEPENDENT DENSITY

When the fluid's density, $\rho$, is unaffected by changes in concentration, the core of the model includes only the equation of hydrodynamic dispersion, say (6.3.23) or (6.3.25).

Information on $\mathbf{V} = \mathbf{V}(\mathbf{x}, t)$ should be provided. It can be obtained by solving a flow model separately.

When $\rho = \rho(c)$ or $\rho = \rho(p, c)$, and $\theta = \theta(\mathbf{x}, t)$, the flow and the pollution models have to be solved simultaneously. For saturated flow, the model then consists of the following set of equations

$$nR_d \frac{\partial c}{\partial t} = -\nabla \cdot n(c\mathbf{V} - \mathbf{D}_h \cdot \nabla c),$$

$$\mathbf{V} = -\frac{k}{n\mu}(\nabla p + \rho g \nabla z),$$

$$n\frac{\partial \rho}{\partial t} + \nabla \cdot n\rho\mathbf{V} = 0, \qquad \rho = \rho(p, c) \qquad (6.5.1)$$

where, for the purpose of the demonstration, we have assumed no sources, no sinks, no decay of the considered pollutant, and an isotropic nondeformable porous medium. The set of four equations in (6.5.1) involves four variables: $c(\mathbf{x}, t)$, $\mathbf{V}(\mathbf{x}, t)$, $p(\mathbf{x}, t)$ and $\rho(\mathbf{x}, t)$. To solve it, we need information on $R_d$ (i.e., on $K_d$, $\rho_s$ and $n(\mathbf{x}, t)$), $k(\mathbf{x}, t)$, $\mu$, $a_L$, $a_T$, and the details of $\rho = \rho(p, c)$. Examples of forms of $\rho = \rho(p, c)$ are given in Section 3.2.

Obviously, the above equations have to be supplemented by appropriate initial and boundary conditions.

In unsaturated flow, the model consists of the equations

$$nR_d \frac{\partial S_w c}{\partial t} = -\nabla \cdot nS_w(c\mathbf{V} - \mathbf{D}_h \cdot \nabla c),$$

$$\mathbf{V} = -\frac{k}{nS_w\mu}(\nabla p + \rho g \nabla z),$$

$$n\frac{\partial S_w \rho}{\partial t} = -\nabla \cdot nS_w\rho\mathbf{V},$$

$$\rho = \rho(p, c); \qquad p_w = p_w(S_w) \qquad (6.5.2)$$

where the underlying assumptions should be obvious from the selected equations. The variables are: $c$, $\mathbf{V}$, $\theta$ ($= nS_w$), $p$ and $\rho$.

## 6.6. Pollution Transport by Advection Only

So far in the present chapter, pollutants in groundwater were shown to be transported simultaneously by advection, dispersion, and diffusion. The total flux was given by (6.2.19). However, in many cases, or as a first approximation, the fluxes due to hydrodynamic dispersion are much smaller than those due to advection, i.e., for saturated flow

$$|\mathbf{q}c| \gg |n\mathbf{D}_h \cdot \nabla c|. \tag{6.6.1}$$

Under such conditions, polluted groundwater bodies move in the aquifer along the pathlines of the water itself, at the velocity of the latter. This is the topic considered in the present section.

### 6.6.1. ABRUPT FRONT APPROXIMATION

Sometimes the transition zone that develops by hydrodynamic dispersion between two zones, one occupied by polluted water only and the other occupied by nonpolluted water, is narrow relative to the dimensions of the individual zones. Within each zone, the concentration of the pollutant is constant, or approximately so. Under such conditions, we may replace the real situation by an *approximation* in which the transition zone is replaced by an assumed *abrupt front*, as defined in Subsection 4.6.2. This front, while moving in the flow domain according to the velocity distribution in the latter, continuously separates the two zones, each occupied by water at a different pollutant concentration. For example, when a large volume of water of one concentration (of some component) is injected into an aquifer in which the water has a different concentration of the same component, we sometimes assume, at least as a first approximation, that the effect of hydrodynamic dispersion may be neglected so that a moving abrupt front continuously separates the injected water from the aquifer water.

In most pollution cases, unlike the case of the interface in a coastal aquifer considered in Chapter 7, the density and viscosity of the two kinds of water are assumed to be identical. This simpler case will be considered throughout this section.

Figure 6.8 shows an abrupt front that separates two domains, each occupied by water of a different concentration of some polluting species. Let $F(x, y, z, t) = 0$ define the configuration of the front as it is being displaced (see Subsection 3.4.2). The front divides the flow domain into two subdomains $R_1$ and $R_2$. Since $\rho_1 = \rho_2$ and $\mu_1 = \mu_2$, we also have $K_1 = K_2 = K$. In Subsection 7.2.1, we discuss the case of different densities and viscosities.

From the discussion on boundary conditions in Subsection 3.4.3, it follows that

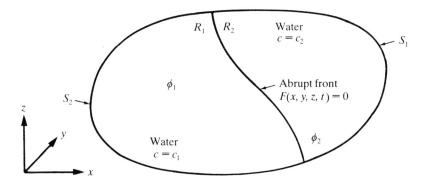

Fig. 6.8. Nomenclature for an abrupt front between two fluids of different concentrations.

since no fluid crosses the front, we have along the latter

$$(\mathbf{V} - \mathbf{u})|_1 \cdot \boldsymbol{\nu} = (\mathbf{V} - \mathbf{u})|_2 \cdot \boldsymbol{\nu} = 0$$

or

$$\mathbf{V}|_1 \cdot \boldsymbol{\nu} = \mathbf{V}|_2 \cdot \boldsymbol{\nu} = \mathbf{u} \cdot \boldsymbol{\nu} \tag{6.6.2}$$

where $\mathbf{u}$ is the speed of displacement of the front, $\boldsymbol{\nu} = \nabla F / |\nabla F|$ is the unit vector normal to the front and $\mathbf{V}|_1$ and $\mathbf{V}|_2$ are the water velocities on sides 1 and 2 of the front, respectively. Hence, from (3.4.2) it follows that as the front is being approached from within $R_1$ or $R_2$, we have to satisfy the conditions

$$\frac{\partial F}{\partial t} + \mathbf{V}_1 \cdot \nabla F = 0, \qquad \frac{\partial F}{\partial t} + \mathbf{V}_2 \cdot \nabla F = 0 \tag{6.6.3}$$

where $\mathbf{V}_1$ and $\mathbf{V}_2$ are obtained from Darcy's law, e.g.,

$$\mathbf{V}_1 = -\frac{K}{n} \nabla \phi_1 \quad \text{in } R_1, \qquad \mathbf{V}_2 = -\frac{K}{n} \nabla \phi_2 \quad \text{in } R_2 \tag{6.6.4}$$

where $\phi_1$ and $\phi_2$ are the piezometric head distributions within $R_1$ and $R_2$, respectively.

When the entire domain is homogeneous and isotropic, and because the changes in storage due to front movement are much larger than those associated with elastic storativity, the problem of determining the location of the moving front, $F(x, y, z, t) = 0$, can, in principle, be stated as follows:

Determine $\phi_1$ in $R_1$ and $\phi_2$ in $R_2$, such that (a)

$$\nabla^2 \phi_1 = 0 \quad \text{in } R_1, \qquad \nabla^2 \phi_2 = 0 \quad \text{in } R_2$$

or appropriate equations when sources and sinks are present; (b) $\phi_1 = \phi_2$ on $F = 0$; (c)

$$\nabla \phi_1 \cdot \nabla F = \nabla \phi_2 \cdot \nabla F \quad \text{on } F = 0$$

obtained from condition (6.6.2) and (6.6.4), and (d) appropriate conditions for $\phi_1$ on $S_1$ and for $\phi_2$ on $S_2$, where $S_1$ and $S_2$ are external boundaries of $R_1$ and $R_2$, respectively. As in the case of the phreatic surface (which is also an abrupt front) in Subsection 3.4.6, we have an inherent difficulty in solving the above-stated problem, as the (continuously changing) location of the boundary, $F = 0$, is *a-priori* unknown. In fact, the shape of the surface $F = 0$ is the sought solution. Once we solve for $\phi_1(x, y, z, t) = 0$ in $R_1$, and for $\phi_2(x, y, z, t) = 0$ in $R_2$, the shape of the interface is given by

$$F(x, y, z, t) = \phi_1 - \phi_2 = 0 \tag{6.6.5}$$

Actually, since $\rho$ and $\mu$ were assumed to be constant throughout $R_1 + R_2$, the problem discussed above can be solved in a much simpler way. We first solve the usual problem of determining $\phi$ in the entire flow domain, $R_1 + R_2$; the presence of a front, $F = 0$, has no effect on this solution. Then we use the known $\phi$-distribution and Darcy's law to calculate the velocity and corresponding displacement at points on the front, and move the latter within the domain, making use of the fact that $\mathbf{V} \cdot \mathbf{\nu} = \mathbf{u} \cdot \mathbf{\nu}$.

Muskat (1937) presents a number of analytical solutions for the movement of the front that delineates labeled fluid injected through a well and moving towards a pumping well in an infinite reservoir. Bear and Jacobs (1965) (see also Bear, 1979) present a soluton for the shape of advancing fronts in the case of an injection well in two-dimensional uniform flow. Altogether, analytical solutions for the problem of front movement, as described above, are possible only for a small number of relatively simple cases. Numerical solutions are discussed in Chapter 11. In Subsection 7.2.1, we return to the problem of a moving front, in connection with the interface between fresh water and salt water in a coastal aquifer.

6.6.2. ADVECTION OF POLLUTED WATER PARTICLES

When the situation justifies the assumption expressed by (6.6.1), the pollutant balance equation, say (6.3.2) for saturated flow ($\theta = n$), and in the absence of adsorption, sources and sinks, reduces to

$$\frac{\partial nc}{\partial t} = -\nabla \cdot c\mathbf{q}, \quad \mathbf{q} = n\mathbf{V}. \tag{6.6.6}$$

For the simple case of a homogeneous nondeformable porous medium ($\nabla n = 0$ and $\partial n/\partial t = 0$), Equation (6.6.6) reduces to

$$\frac{\partial c}{\partial t} = -\mathbf{V} \cdot \nabla c - c\nabla \cdot \mathbf{V}. \tag{6.6.7}$$

For steady flow of an incompressible fluid in a homogeneous nondeformable porous medium, (3.3.2) reduces to $\nabla \cdot \mathbf{V} = 0$. Under such conditions, (6.6.7)

MODELING GROUNDWATER POLLUTION

reduces to

$$\frac{\partial c}{\partial t} = -\mathbf{V} \cdot \nabla c. \tag{6.6.8}$$

In two-dimensional flow in the $xy$-plane, (6.6.8) takes the form

$$\frac{\partial c}{\partial t} = -V_x \frac{\partial c}{\partial x} - V_y \frac{\partial c}{\partial y}. \tag{6.6.9}$$

In order to solve (6.6.9) for $c = c(x, y, t)$, we need information on $V_x = V_x(x, y, t)$ and $V_y = V_y(x, y, t)$. In the case considered here, this information is obtained by solving the flow model $\nabla \cdot \mathbf{V} = 0$. We note that quite a number of simplifying assumptions underlie the development of (6.6.9). As mentioned above, in spite of these simplifications, analytic solutions of (6.6.9) can be obtained only for a rather small number of elementary cases. A numerical model and computer program for solving (6.6.9) are presented in Chapter 11.

We note that the solution of (6.6.9), which is a *linear hyperbolic partial-differential equation* (while (6.3.2) is a *parabolic* one) in the three independent variables $x, y, t$, rewritten in the form

$$\frac{dc}{dt} \equiv \frac{\partial c}{\partial t} + V_x \frac{\partial c}{\partial x} + V_y \frac{\partial c}{\partial y} = 0 \tag{6.6.10}$$

can be represented, at least in principle, by lines of constant concentration, called *characteristics*. Along such a characteristic, the variation of $c$ vanishes, i.e.,

$$dc = \frac{\partial c}{\partial t} dt + \frac{\partial c}{\partial x} dx + \frac{\partial c}{\partial y} dy = 0. \tag{6.6.11}$$

Equation (6.6.10) describes the variations of $c$ from the Lagrangian point of view. The statement $dc/dt = 0$ means that the concentration of an observed *fixed* particle does not change with time as it travels in the considered domain (see the comment following (6.3.2)).

By comparing (6.6.10) with (6.6.11), we conclude that the direction of the characteristic is defined by the relationship

$$1 : V_x : V_y = dt : dx : dy.$$

Hence,

$$dx = V_x \, dt, \qquad dy = V_y \, dt. \tag{6.6.12}$$

This means that the direction of the characteristic coincides with that of the flow, namely that of the streamline. Actually, this should have been expected in view of the definition and discussion on streamlines in Subsection 4.6.2. We may compare, for example, (4.6.1) with (6.6.12), noting that we consider here the transport of a polluting component by advection only, which means that the water and the component carried by it move together, with no relative motion in respect to each other.

Thus, the concentration of particles remains constant along streamlines The problem that remains to be solved is to trace the advance of the particles along the streamlines in accordance with (6.6.12). Semi-analytical and fully numerical methods for achieving this goal are presented in Chapter 11.

## 6.7. Macrodispersion

The equation of (three-dimensional) hydrodynamic dispersion at a point in an aquifer is discussed in Section 6.3. For example, for the case of saturated flow, we obtain from (6.3.2)

$$\frac{\partial(nc)}{\partial t} + \nabla \cdot \{c\mathbf{q} + \mathbf{J}_c^*\} - \tau_c^* = 0 \qquad (6.7.1)$$

where $\mathbf{J}_c^* = -n\mathbf{D}_h \cdot \nabla c$ is the sum of the dispersive and diffusive fluxes, $\mathbf{D}_h = \mathbf{D} + \mathbf{D}_d^*$, and $\tau_c^* = -f + n\rho\Gamma - Pc + Rc_R$ denotes to total source function, due to surface phenomena at the solid-water interface, to chemical reaction and decay phenomena in the water, to external injection and production of water.

In (6.7.1), we have $c = c(x, y, z, t)$ and $\mathbf{q} = \mathbf{q}(x, y, z, t)$. However, under certain conditions, the 'hydraulic approach' of treating the dispersion problem as one of essentially two-dimensional flow in the horizontal $xy$ plane is justified. Figure 6.9 shows, schematically, several cases of aquifer pollution. In Case 1 we have a continuous surface source, say leachate from a landfill, contaminating a phreatic aquifer with accretion. In the absence of dispersion, the pollutant will advance within a well-defined streamtube. Dispersion would cause both longitudinal and transversal spreading, the latter beyond the surface bounding the streamtube. At a sufficiently large distance from the source (say 10—15 times the thickness of the flow domain), the plume of contaminant will occupy most of the thickness of the aquifer. Beyond such distance, the two-dimensional approach seems justified. In the absence of accretion, the contaminant will remain close to the surface, with lateral dispersion causing spreading in the downward direction. The lower layers may remain uncontaminated for a rather large distance.

Case 2 describes a fully (or partially) penetrating well injecting water of a different quality into a stratified (that is, $K = K(z)$) confined aquifer. Again, because of transversal dispersion, beyond some distance from the injection well, the average quality of the mixed water may be considered as depending on $x$ and $y$ only. One should remember, that a three-dimensional approach requires also measurements at points in space. On the other hand, a pumping well, even a partially penetrating one, performs an averaging, or mixing, of the water quality along the different elevations (or along different streamlines terminating in the well).

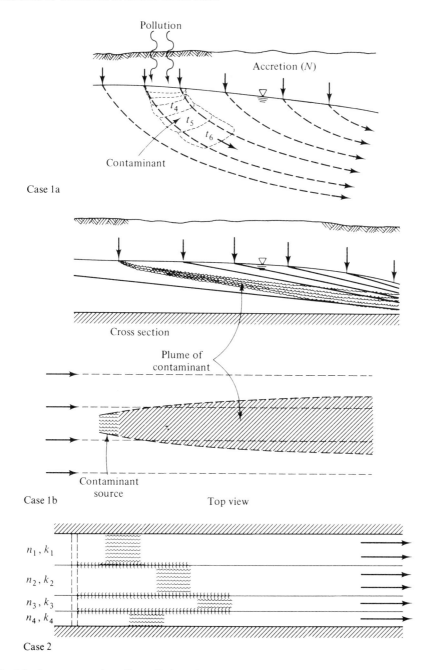

Fig. 6.9. Some cases of aquifer pollution. *Case 1a*: Pollution of a homogeneous water table aquifer with accretion; local pollution source; short time and travel distance; phenomenon is three-dimensional; some contamination beyond streamtube by lateral dispersion. *Case 1b*: Same as Case 1a, except that time is long and travel distance is large; may be considered two-dimensional phenomenon (in horizontal plane). *Case 2*: Uniform flow in a stratified confined aquifer; fully penetrating pollution source; longitudinal spreading within layers and lateral dispersion between layers.

Thus, under certain conditions, a model based on the assumption of essentially horizontal flow and pollution transport in an aquifer, may be justified. The averaged, or integrated equation is then obtained by integrating (6.7.1) along the vertical height, $B$, of the aquifer.

For a *confined aquifer*, making use of (4.2.8) and (4.2.9) in order to integrate (6.7.1) from the aquifer's bottom, at $z = b_1(x, y)$ to its top, at $z = b_2(x, y)$, with $B = b_2 - b_1$, we obtain

$$\int_{b_1}^{b_2} \left( \frac{\partial nc}{\partial t} + \nabla \cdot (c\mathbf{q} + \mathbf{J}_c^*) - \tau_c^* \right) dz$$

$$= \frac{\partial}{\partial t} B\widetilde{nc} + \nabla' \cdot B(\widetilde{c\mathbf{q}'} + \widetilde{\mathbf{J}_c^{*'}}) - B\widetilde{\tau_c^*} +$$

$$+ (c\mathbf{q} + \mathbf{J}_c^*)|_{b_2} \cdot \nabla(z - b_2) -$$

$$- (c\mathbf{q} + \mathbf{J}_c^*)|_{b_1} \cdot \nabla(z - b_1) = 0 \qquad (6.7.2)$$

where the symbol $\widetilde{(\,)}$ denotes an average over the vertical as defined by (2.2.5). From the discussion in Subsection 6.4.1, it follows that the last two terms on the left-hand side of (6.7.2) express the flux of the considered component normal to the upper and lower bounding surfaces, respectively. For a confined aquifer, both these terms vanish and we obtain an average equation in the form

$$\frac{\partial}{\partial t} B\widetilde{nc} + \nabla' \cdot B(\widetilde{c\mathbf{q}'} + \widetilde{\mathbf{J}_c^{*'}}) - B\widetilde{\tau_c^*} = 0. \qquad (6.7.3)$$

Employing the same procedure, with appropriate boundary conditions on the top and bottom surfaces, Bear (1979, p. 256) derives averaged balance equations for a phreatic aquifer and for a leaky one.

It is of interest to note that, similar to the integrated continuity equations discussed in Section 4.2, the integrated solute balance equations, e.g., (6.7.2), can also be derived by assuming only that we have essentially horizontal flow in the aquifer.

In (6.7.2)

$B\widetilde{nc}$ = total mass of solute per unit (horizontal) area of aquifer,
$B\widetilde{c\mathbf{q}'}$ = total horizontal flux of solute by convection through the entire thickness of the aquifer,
$B\widetilde{\mathbf{J}_c^{*'}}$ = total horizontal flux of solute by hydrodynamic dispersion through the entire thickness of the aquifer,
$B\widetilde{\tau_c^*}$ = total source function for the entire thickness of the aquifer (per unit area and unit time).

All these variables are now *functions of $x$ and $y$ and $t$ only.*

Let us now discuss the two fluxes $B\widetilde{c\mathbf{q}'}$ and $B\widetilde{\mathbf{J}_c^{*'}}$ mentioned above. For the first, $\widetilde{c\mathbf{q}'} \equiv \widetilde{(nc)\mathbf{V}'}$, we can write

$$\mathbf{q}' = \widetilde{\mathbf{q}'} + \hat{\mathbf{q}}', \qquad c = \tilde{c} + \hat{c}, \qquad \widetilde{c\mathbf{q}'} = \tilde{c}\widetilde{\mathbf{q}'} + \widetilde{\hat{c}\hat{\mathbf{q}}'} \tag{6.7.4}$$

where the hat (^) symbol indicates deviation over the vertical.

As in the passage from the microscopic level to the macroscopic one (e.g., (6.2.4)), the total solute flux here is also made up of two fluxes, an advective flux, $\tilde{c}\widetilde{\mathbf{q}'}$, and a dispersive one, $\widetilde{\hat{c}\hat{\mathbf{q}}'}$. The latter, referred to as *macrodispersion*, results from fluctuations in $\mathbf{q}'$ along the *vertical*. This dispersive flux is distinct from $\overline{c'\mathbf{V}}$ of (6.2.4), which results from fluctuations in the microscopic velocity. Nevertheless, one should regard both dispersive fluxes as produced basically by the same mechanisms, except that each involves fluctuations and averaging at a different scale. In mechanical dispersion, we have velocity fluctuations at the microscopic level and the average is taken over an REV, the size of which is related to the microscopic inhomogeneity (i.e., the presence of pores and solids). In macrodispersion, the fluctuations, $\hat{\mathbf{q}}'$, are in the specific discharge ($\mathbf{q}'$) and the averaging is performed over the entire thickness of the aquifer.

In a stratified aquifer, the hydraulic conductivity varies only along the vertical, i.e., $K = K(z)$, and the flow is essentially horizontal. We may then write

$$\begin{aligned}\mathbf{q}'(z) &= -\mathbf{K}'(z) \cdot \nabla \tilde{\phi}, \\ \mathbf{K}' &= \widetilde{\mathbf{K}'} + \hat{\mathbf{K}}', \qquad \mathbf{q}' = \widetilde{\mathbf{q}'} + \hat{\mathbf{q}}' = -\widetilde{\mathbf{K}'} \cdot \nabla \tilde{\phi} - \hat{\mathbf{K}}' \cdot \nabla \tilde{\phi}.\end{aligned} \tag{6.7.5}$$

The macrodispersive flux is, therefore

$$\widetilde{\hat{c}\hat{\mathbf{q}}'} = -\widetilde{\hat{c}\hat{\mathbf{K}}'} \cdot \nabla \tilde{\phi} \tag{6.7.6}$$

which is thus related to the variability in hydraulic conductivity.

We may continue the analogy between dispersion and macrodispersion and define, as a *working hypothesis*, a coefficient of macrodispersion $\widetilde{\mathbf{D}}_{ij}$ ($i, j = 1, 2$) and a *macrodispersivity* $A_{ijkm}$ ($i, j, k, m = 1, 2$) related to *longitudinal macrodispersivity*, $A_L$, and *transversal macrodispersivity*, $A_T$, of the aquifer. With these, the macrodispersive flux will be given by

$$\widetilde{\hat{c}\hat{\mathbf{q}}'} = -\widetilde{\mathbf{D}} \cdot \nabla \tilde{c}, \qquad \widetilde{D}_{ij} = \frac{\widetilde{\hat{q}'_i \hat{q}'_j}}{\widetilde{q}'} \tilde{L} = A_{ijkm} \frac{\widetilde{q}'_k \widetilde{q}'_m}{\widetilde{q}'} f(\mathrm{Pe}^*, \delta^*),$$

$$A_{ijkm} = \frac{\widetilde{\hat{K}'_{in}\hat{K}'_{jl}}}{\widetilde{K}'_{nk}\widetilde{K}'_{lm}} \tilde{L}, \qquad i, j, k, m = 1, 2 (\equiv x, y) \tag{6.7.7}$$

(to be compared with (6.2.5) and (6.2.6)), where Pe* is a *Peclet number* related to the dispersivity of the medium, and especially to transversal dispersivity of the medium, $\delta^*$ is some dimensionless parameter describing the thickness of the layers relative to the thickness of the aquifer, and $\tilde{L}$ is a length characterizing the inhomogeneity of the aquifer due to the stratification. Here the transversal dispersivity plays the same role as that played by molecular diffusion inside an individual channel of a porous medium in producing mixing between adjacent streamlines and making the reference to the average concentration meaningful.

Gelhar (1976) analyzed the dependence of macrodispersion on permeability variations. For horizontal flow in a confined aquifer he suggested

$$A_L = \frac{1}{3} \frac{L_1^2 \sigma_{\ln k}^2}{a_T} \qquad (6.7.8)$$

where $L_1$ is a correlation distance (= distance along which permeabilities are still correlated) and $\sigma_{\ln k}$ is the standard deviation of $\ln k$ (= natural logarithm of the permeability, $k$).

Following the same line of thinking, one may extend the above ideas to inhomogeneity, say in permeability, in a general three-dimensional flow domain. In the absence of detailed information about the spatial variation of permeability (say in the form of the surfaces that separate subdomains of different permeabilities), we may attempt to obtain some average description of concentration variations. The permeability inhomogeneity will produce a dispersion, or spreading, phenomenon with respect to the average flow, similar to that occurring in a layered aquifer, where under the same head gradient flow is faster in the layer with larger $K/n$ value.

Accordingly, (6.7.7) may intuitively be extended to three-dimensional flow with $i, j, k, m = 1, 2, 3$ (or: $x, y, z$) and $\mathbf{q}$ replacing $\mathbf{q}'$.

Let us now go back to (6.7.3) and examine the averaged dispersive flux, $\mathbf{J}_c^{*\prime}$

$$\widetilde{\mathbf{J}_c^{*\prime}} = \frac{1}{B} \int_{b_1}^{b_2} \mathbf{J}_c^{*\prime} \, dz = -\frac{1}{B} \int_{b_1}^{b_2} \widetilde{n\mathbf{D}_h'} \cdot \widetilde{\nabla' c} \, dz$$

$$= -\widetilde{n\mathbf{D}_h'} \cdot \widetilde{\nabla' c} - \widetilde{(n\mathbf{D}_n') \cdot (\nabla' c)}. \qquad (6.7.9)$$

Neglecting the second term on the right-hand side of (6.7.9), we obtain

$$\widetilde{\mathbf{J}_c^{*\prime}} = -\widetilde{n\mathbf{D}_h'} \cdot \widetilde{\nabla' c} = -\widetilde{n\mathbf{D}_h'} \cdot \frac{1}{B} \int_{b_1}^{b_2} \nabla' c \, dz$$

$$= -\widetilde{n\mathbf{D}_h'} \cdot \frac{1}{B} \left[ \nabla' \int_{b_1}^{b_2} c \, dz - c|_{b_2} \nabla b_2 + c|_{b_1} \nabla b_1 \right]$$

$$= -\widetilde{n\mathbf{D}_h'} \cdot \left[ \nabla' \tilde{c} + \frac{1}{B} \left\{ \tilde{c} \nabla B - c|_{b_2} \nabla b_2 + c|_{b_1} \nabla b_1 \right\} \right]$$

$$\approx -\widetilde{n\mathbf{D}_h'} \cdot \nabla' \tilde{c} \qquad (6.7.10)$$

where we have assumed $\tilde{c} \approx c|_{b_2} \approx c|_{b_1}$. From the discussion above, it seems reasonable to assume that $\widetilde{\mathbf{D}} \gg \widetilde{n\mathbf{D}_h'}$. This explains why the dispersivity as derived

# MODELING GROUNDWATER POLLUTION

from field experiments of solute spreading is several orders of magnitude that of the dispersivity related to size of individual grains or openings as obtained in laboratory experiments with a homogeneous medium.

In conclusion, the movement and accumulation of a contaminant in a confined aquifer in which we can justify an averaged approach (i.e., the hydraulic approach with respect to concentration), are governed by

$$\frac{\partial}{\partial t} B\widetilde{nc} + \nabla' \cdot B\{\widetilde{c}\widetilde{\mathbf{q}}' - \widetilde{\mathbf{D}} \cdot \nabla'\widetilde{c}\} - B\widetilde{\tau_c^*} = 0. \tag{6.7.11}$$

The solute is carried by advection and macrodispersion and we have neglected averaged hydrodynamic dispersion. Similar integrated equations can be written for a leaky aquifer and a phreatic one.

The extension to macrodispersion in three-dimensional flow requires the definition of a new, larger REV and averaging over it.

CHAPTER SEVEN

# Modeling Seawater Intrusion

Coastal aquifers constitute important sources for water. Many coastal areas are also heavily urbanized, a fact which makes the need for fresh water even more acute. However, the proximity of the sea, with the contact between freshwater and seawater in a coastal aquifer, requires special attention and special management techniques.

The objective of this chapter is to present models that enable the prediction of seawater intrusion into coastal aquifers, in response to variations in the components of the freshwater balance of the latter. This information is required for the management of coastal aquifers.

The conceptual and mathematical models are presented in this chapter. Chapter 13 develops and presents the numerical models for seawater intrusion.

## 7.1. The Interface in a Coastal Aquifer

In coastal aquifers, a hydraulic gradient generally exists toward the sea that serves as a recipient for the excess of their fresh water (replenishment minus pumpage). Owing to the presence of sea water in the aquifer formation under the sea bottom, a zone of contact is formed between the lighter freshwater (specific weight $\gamma_f$) flowing to the sea and the heavier, underlying, sea water (specific weight $\gamma_s > \gamma_f$). Typical cross-sections showing interfaces under various natural conditions are shown in Figure 7.1, while Figure 7.2 shows a typical cross-section of a coastal phreatic aquifer with groundwater exploitation. In all cases, a body of seawater, often in the form of a wedge, exists underneath the freshwater. One should note that, like most figures describing flow in aquifers, these are also *highly distorted figures*, not drawn to scale.

Freshwater and seawater are actually miscible fluids and therefore the zone of contact between them takes the form of a transition zone caused by hydrodynamic dispersion (Chapter 6). Across this zone, the density of the mixed water varies from that of freshwater to that of seawater. However, under certain conditions, the width of this zone is small, relative to the thickness of the aquifer, so that the zone of gradual transition from freshwater to seawater may be approximated as a sharp interface (see Section 6.6).

In a way, this is similar to the introduction of a phreatic surface as an approximation of the gradual moisture variation in the soil. For example, observations (Jacobs and Schmorak 1960; Schmorak 1967) along the coast of Israel indicate that indeed this assumption of an abrupt interface is justified. On the other hand,

# MODELING SEAWATER INTRUSION

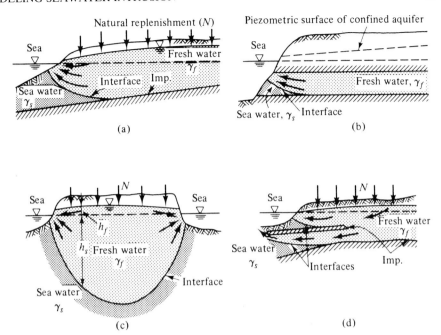

Fig. 7.1. Examples of interfaces in coastal aquifers.

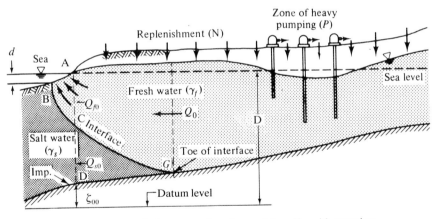

Fig. 7.2. A typical cross-section of a coastal aquifer with pumping.

Cooper (1959) describes a case where the transition zone is very wide so that the interface approximation is no longer valid.

In this chapter, we shall assume that a sharp interface always separates the regions occupied by the two fluids.

Under natural undisturbed conditions, a state of equilibrium is maintained, in a coastal aquifer, with a *stationary interface* and a freshwater flow to the sea above it. At every point on this interface, the elevation and the slope are determined by the freshwater potential and gradient (or by the flow velocity). The continuous change of slope results from the fact that as the sea is approached, the specific discharge of freshwater tangent to the interface increases. By pumping from a coastal aquifer in excess of replenishment, the water table (or the piezometric surface in a confined aquifer) in the vicinity of the coast is lowered to the extent that the piezometric head in the freshwater body becomes less than in the adjacent seawater wedge, and the interface starts to advance landward until a new equilibrium is reached. This phenomenon is called *seawater intrusion* (or *encroachment*). As the interface advances, the transition zone widens; nevertheless, we shall assume that the abrupt interface approximation remains valid. When the advancing interface reaches inland pumping wells, the latter become contaminated. When pumping takes place in a well located above the interface, the latter upcones towards the pumping well. Unless the rate of pumping is carefully controlled, seawater will eventually enter the pumped well. Actually the real situation is even more dangerous in view of the presence of a transition zone, rather than an abrupt interface. Because of this transition zone, an increase in salinity in a pumping well serves as a warning of advancing salinization of the aquifer.

Rather than treat the problem as one with an abrupt interface, we can always consider it as one with a continuous variation of salt concentration and density, using the material presented in Chapter 6 on hydrodynamic dispersion. However, the assumption of an abrupt interface, especially when certain assumptions related to horizontal flow are also introduced, greatly simplifies the model in most cases of practical interest.

As we shall see below, there exists a relationship between the rate of freshwater discharge to the sea and the extent of seawater intrusion. This makes seawater encroachment a *management problem*, as the freshwater discharge to the sea is the difference between the rate of natural and artificial recharge and that of pumping.

## 7.2. Modeling Seawater Intrusion in a Vertical Plane

### 7.2.1. EXACT MATHEMATICAL STATEMENT OF THE PROBLEM

Let us state the mathematical model that describes the flow problem of two liquids, assuming that an abrupt interface (always) separates them, such that each liquid occupies a separate part of the entire flow domain (here, the aquifer). In general, the interface is not stationary. Sources and sinks of liquid (i.e., pumping and artificial recharge) may exist in both regions.

In each of the two regions, we may define a piezometric head. Assuming that both freshwater and saltwater are compressible, we obtain

$$\phi_f^* = z + \int_{p_0}^{p_f} \frac{dp_f}{\gamma_f(p_f)} \quad \text{in the freshwater region,}$$

$$\phi_s^* = z + \int_{p_0}^{p_s} \frac{dp_s}{\gamma_s(p_s)} \quad \text{in the saltwater region.}$$

# MODELING SEAWATER INTRUSION

For the sake of simplicity, we shall henceforth assume that

$$\phi_f \simeq \phi_f^*, \qquad \phi_s \simeq \phi_s^*.$$

In spite of the introduction and use of the piezometric heads, $\phi_f$ and $\phi_s$, in the present subsection, it may sometimes be found more convenient to use pressure as a single-state variable for both freshwater and saltwater domains. There is also no jump in pressure as an interface is crossed. We shall return to this point in Subsection 7.3.1.

Let Figure 7.3 represent the two regions, $R_1$ and $R_2$, occupied by freshwater and saltwater, respectively. Then the model describing the flow in both regions can be stated in the following way.

Determine $\phi_f$ in $R_1$ and $\phi_s$ in $R_2$ such that

$$\nabla \cdot (\mathbf{K}_f \cdot \nabla \phi_f) = S_0 \frac{\partial \phi_f}{\partial t} \quad \text{in } R_1,$$

$$\nabla \cdot (\mathbf{K}_s \cdot \nabla \phi_s) = S_0 \frac{\partial \phi_s}{\partial t} \quad \text{in } R_2 \qquad (7.2.1)$$

where $\mathbf{K}_f = k\gamma_f/\mu_f$; $\mathbf{K}_s = k\gamma_s/\mu_s$ and the specific storativity, $S_0$, is assumed the same for $R_1$ and $R_2$. Obviously, other equations can be used when necessary, depending on the assumptions we make with respect to the medium and the fluids. Often, since a change in storage due to a moving interface is involved here, we neglect the elastic storage, i.e., $S_0 \simeq 0$ in (7.2.1).

In addition, we have to specify initial conditions for $\phi_f$ in $R_1$ and for $\phi_s$ in $R_2$. Boundary conditions for $\phi_f$ on $B_1$ and $\phi_s$ on $B_2$ are the usual ones encountered in the flow of a single fluid (Section 3.4). However, the boundary condition on the interface requires special attention. Moreover, as in the case of a phreatic surface, the location of the interface is unknown until the problem is solved. In fact, the location and shape of the interface, say expressed in the form of

$$F(x, y, z, t) = 0 \qquad (7.2.2)$$

is what we are looking for.

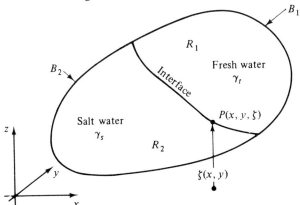

Fig. 7.3. An abrupt interface between regions occupied by fresh water and by salt water.

Denoting the elevation, z, of points on the interface by $\zeta = \zeta(x, y, t)$, the relationship for $F$ becomes

$$z = \zeta(x, y, t) \quad \text{or} \quad F \equiv z - \zeta(x, y, t) = 0. \tag{7.2.3}$$

The pressure at a point on the interface is the same when the point is approached from both sides. Hence, from the definitions of $\phi_f$ and $\phi_s$, we have

$$\gamma_f(\phi_f - \zeta) = \gamma_s(\phi_s - \zeta)$$

or

$$\zeta(x, y, t) = \phi_s \frac{\gamma_s}{\gamma_s - \gamma_f} - \phi_f \frac{\gamma_f}{\gamma_s - \gamma_f}$$

$$= \phi_s(1 + \delta) - \phi_f \delta, \quad \delta = \gamma_f/(\gamma_s - \gamma_f). \tag{7.2.4}$$

For example, for $\gamma_f = 1.00$ gr/cm$^3$, $\gamma_s = 1.03$ gr/cm$^3$, $\delta \approx 33$.

Once we solve the model and obtain the distributions $\phi_f = \phi_f(x, y, z, t)$ and $\phi_s = \phi_s(x, y, z, t)$, Equation (7.2.4) becomes the sought interface equation

$$F \equiv z - \phi_s(1 + \delta) + \phi_f \delta = 0. \tag{7.2.5}$$

The boundary conditions on the interface are as follows.

Same specific discharge on both sides: $(q_n)_f = (q_n)_s$ on $F$.
Same pressures on both sides: $\gamma_f(\phi_f - \zeta) = \gamma_s(\phi_s - \zeta)$ on $F$.

Since the interface is a material surface, with the same fluid particles always remaining on it, we have

$$\frac{dF}{dt} \equiv \frac{\partial F}{\partial t} + \mathbf{V}_f \cdot \nabla F = 0, \quad \frac{\partial F}{\partial t} + \mathbf{V}_s \cdot \nabla F = 0 \tag{7.2.6}$$

where, assuming a rigid solid, $n\mathbf{V}_f = -\mathbf{K}_f \cdot \nabla \phi_f$, $n\mathbf{V}_s = -\mathbf{K}_s \cdot \nabla \phi_s$. Note that here subscripts denotes saltwater and not solid. By combining (7.2.5) with (7.2.6) we obtain

$$n\delta \frac{\partial \phi_f}{\partial t} - n(1 + \delta) \frac{\partial \phi_s}{\partial t} -$$

$$- \mathbf{K}_f \cdot \{\nabla z - (1 + \delta)\nabla \phi_s + \delta \nabla \phi_f\} \cdot \nabla \phi_f = 0, \tag{7.2.7}$$

$$n\delta \frac{\partial \phi_f}{\partial t} - n(1 + \delta) \frac{\partial \phi_s}{\partial t} -$$

$$- \mathbf{K}_s \cdot \{\nabla z - (1 + \delta)\nabla \phi_s + \delta \nabla \phi_f\} \cdot \nabla \phi_s = 0. \tag{7.2.8}$$

Thus, the boundary conditions on an interface take the form of two *nonlinear partial differential equations* in terms of $\phi_f$ and $\phi_s$. This is the reason why the

# MODELING SEAWATER INTRUSION

derivation of the shape and position of an interface by solving the partial differential equations (7.2.1) subject to the boundary conditions (7.2.7) and (7.2.8) on the surface defined by (7.2.5), is practically impossible. In Section 7.3, the *hydraulic approach* is employed to reduce the interface problem essentially to one of flow in a plane, thus eliminating the boundary conditions on the interface.

It is of interest to determine the slope at a point on a stationary interface. Figure 7.4 shows an element of an interface in two-dimensional flow in the vertical $xz$ plane. The components of the specific discharge tangential to the interface in the two regions are expressed by

$$(q_f)_\xi = -\frac{k\gamma_f}{\mu_f}\frac{\partial \phi_f}{\partial \xi} = -\frac{k}{\mu_f}\left(\frac{\partial p}{\partial \xi} + \gamma_f \frac{\partial z}{\partial \xi}\right), \tag{7.2.9}$$

$$(q_s)_\xi = -\frac{k\gamma_s}{\mu_s}\frac{\partial \phi_s}{\partial \xi} = -\frac{k}{\mu_s}\left(\frac{\partial p}{\partial \xi} + \gamma_s \frac{\partial z}{\partial \xi}\right) \tag{7.2.10}$$

where $k = $ const. By eliminating $\partial p/\partial \xi$ from both equations, we obtain

$$\sin \theta = \frac{\partial z}{\partial \xi} = \frac{(q_f)_\xi \mu_f - (q_s)_\xi \mu_s}{k(\gamma_s - \gamma_f)} \tag{7.2.11}$$

where $\theta$ is the angle that the interface makes with the $+x$ direction. For $(q_s)_\xi = 0$ (i.e., immobile sea water), we obtain

$$\sin \theta = \frac{(q_f)_\xi \mu_f}{k(\gamma_s - \gamma_f)}, \qquad (q_f)_\xi = -\frac{K_f}{\delta}\frac{\partial z}{\partial \xi}. \tag{7.2.12}$$

For the sake of comparison, it is interesting to note that along a phreatic surface, $(q_f)_\xi = -K_f \partial z/\partial \xi$.

The shape of a stationary interface as a coast is approached follows from the

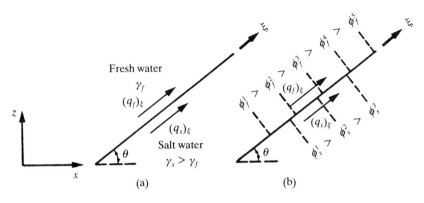

Fig. 7.4. Dynamic equilibrium conditions at a stationary interface.

observation that $(q_f)_\xi$ increases as the coast is approached and, hence, $\theta$ also increases.

When $(q_f)_\xi = 0$

$$\sin \theta = -\frac{(q_s)_\xi \mu_s}{k(\gamma_s - \gamma_f)}, \quad (q_s)_\xi = -\frac{K_s}{\delta}\frac{\partial z}{\partial s} \qquad (7.2.13)$$

and the interface will tilt down in the direction of the flow. At most, we may have $\theta = -\pi/2$. Then $(\partial \phi_s/\partial \xi)_{\max} = -1/\delta$.

Bear and Shapiro (1984) discuss the shape of a moving interface intersecting a surface of discontinuity in permeability.

### 7.2.2. THE GHYBEN–HERZBERG APPROXIMATION

Beginning with Badon–Ghyben (1888) and Herzberg (1901), investigations of the interface in a coastal aquifer have aimed at determining the relationship between its shape and position, and the various components of the groundwater balance in the region near the coast.

Figure 7.5 shows the idealized Ghyben–Herzberg model of an interface in a coastal phreatic aquifer. Essentially, Ghyben and Herzberg assume *static equilibrium*, with a hydrostatic pressure distribution in the freshwater region and with stationary seawater. Instead, we may assume *dynamic equilibrium*, i.e., steady flow, but with *horizontal velocities in the freshwater region*. This means that equipotentials are vertical lines or surfaces, *identical to the Dupuit assumption*. With the nomenclature of Figure 7.5, we have under these conditions

$$h_s = \frac{\gamma_f}{\gamma_s - \gamma_f} h_f \equiv \delta h_f. \qquad (7.2.14)$$

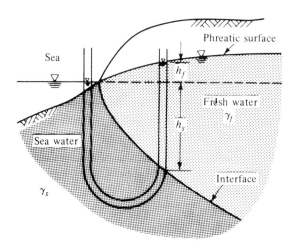

Fig. 7.5. Ghyben–Herzberg's interface model.

For example, for $\gamma_s = 1.025$ gr/cm³, $\gamma_f = 1.000$ gr/cm³, $\delta = 40$, and $h_s = 40 h_f$, that is, at any distance from the sea, the depth of a *stationary interface* below sea level is 40 times the height of the freshwater table above it. For every meter of freshwater stored in the soil above mean sea level, 40 m of freshwater are stored below it, down to the interface. The situation is similar to that of a floating iceberg, where most of the ice is below sea level. Obviously, as the sea is approached the assumption of horizontal flow is no longer valid, because vertical flow components can no longer be neglected. Moreover, in Figure 7.5, no outlet is left for the freshwater flow to the sea. Figure 7.6 shows the actual flow conditions near the coast. Point $A$ on the interface indicates the actual depth of the interface at that distance from the coast. Point $B$ is located at the intersection of the interface and the freshwater equipotential $\phi_f = h_f$.

Accordingly, point $B$ is at a depth equal to $\delta h_f$ which is the depth predicted by the Ghybenen–Herzberg relationship for the interface corresponding to $h_f$. The actual depth (point $A$) is thus greater than that predicted by the Ghyben–Herzberg relationship.

If we set $\phi_s = \text{const} = 0$ (i.e., immobile sea water) in (7.2.4), we see that the difference between the Ghyben–Herzberg approximation (7.2.14) and the exact expression (7.2.4) stems from the difference between $h_f$ and $\phi_f$ (that is, between the assumed vertical equipotential with $\phi_f = h_f$ and the actual curved one). In a confined aquifer, $h_s$ in (7.2.14) is the depth of a point on the interface below sea level, whereas $h_f$ is the (freshwater) piezometric head.

From (7.2.4) it follows that when the interface is in motion, $\phi_s$, which is no more a constant, also affects the shape of the interface.

Bear and Dagan (1964a) investigated the validity of the Ghyben–Herzberg

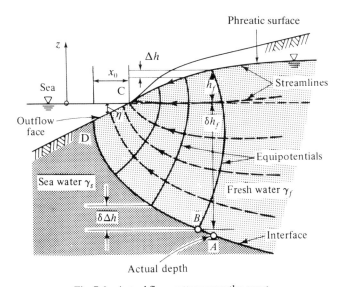

Fig. 7.6. Actual flow pattern near the coast.

relationship. They used the *hodograph method* to derive an exact solution that shows that in steady flow in a horizontal confined aquifer of constant thickness, $B$, the approximation is good, within an error of 5%, for determining the depth of the interface toe (point $G$ in Figure 7.7) and, hence, also the length of the intruding seawater wedge, provided $\pi KB/Q_0\delta > 8$, where $Q_0$ is the freshwater discharge to the sea. Figure 7.7 shows this relationship.

As the coast is approached, the depth of the interface is greater than that predicted by the Ghyben–Herzberg relationship (Figure 7.6). In the case of a phreatic aquifer, a seepage face is also present above sea level.

One should also note that for a horizontal or a downward sloping flat sea bottom, it follows from the exact solution that the interface always terminates at the sea bottom with a tangent in the vertical direction. The interface always terminates on the sea bottom at some distance from the coast (Figure 7.6).

7.2.3. THE STATIONARY INTERFACE IN A VERTICAL PLANE

To demonstrate the shape of the interface and the relationship which exists between the extent of seawater intrusion and the flow of freshwater to the sea, consider a stationary interface, i.e., steady flow, in a case where the flow is everywhere perpendicular to the coast line. Figure 7.2 shows such a cross-section in a phreatic aquifer. For the sake of simplicity, we shall assume a horizontal bottom (Figure 7.8). Let the origin $x = 0$ be located at the interface toe (point $G$). The seaward freshwater flow above this point is $Q_0$. It is the difference between

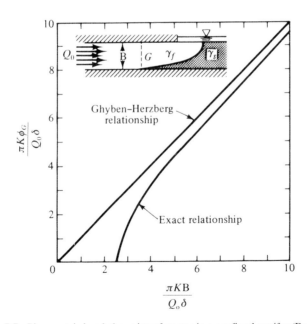

Fig. 7.7. Piezometric head above interface toe in a confined aquifer (Bear and Dagan, 1964a).

# MODELING SEAWATER INTRUSION

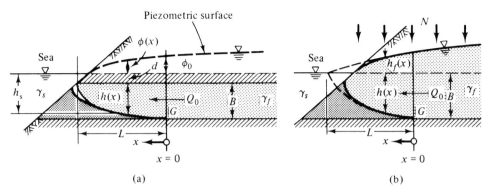

Fig. 7.8. The shape of a stationary interface by the Dupuit–Ghyben–Herzberg approximation.

the total replenishment of the aquifer and the total withdrawal in the coastal strip to the right of point $G$.

Consider first the confined aquifer shown in Figure 7.8a. Using Dupuit's assumption of horizontal flow of fresh water (and vertical equipotentials), continuity leads to

$$Q_0 \equiv Q|_{x=0} = -Kh(x)\frac{\partial \phi(x)}{\partial x},$$

$$K \equiv K_f = \frac{k_f \gamma_f}{\mu_f}, \quad \phi \equiv \phi_f. \tag{7.2.15}$$

Since

$$h_s = d + h(x) = \phi \delta, \qquad \phi_0 \delta = d + B,$$

Equation (7.2.15) becomes

$$Q_0 = -\frac{Kh}{\delta}\frac{dh(x)}{dx}, \quad \text{or} \quad Q_0 = -K\{\delta\phi(x) - d\}\frac{d\phi(x)}{dx}. \tag{7.2.16}$$

Because the flow is steady and assumed horizontal, we have used the Ghyben–Herzberg relationship (7.2.14).

By integrating (7.2.16) with $x = 0$, $\phi = \phi_0$ (or $h = B$), we obtain

$$Q_0 x = \frac{K\{B^2 - h^2(x)\}}{2\delta}, \qquad Q_0 x = \frac{K\delta(\phi_0^2 - \phi^2)}{2} - Kd(\phi_0 - \phi) \tag{7.2.17}$$

which shows that the interface has the form of a *parabola*.

At $x = L$ we set $h = 0$, $\phi = d/\delta$. Then, with $\phi_0 = (B + d)/\delta$, we obtain

$$Q_0 L = \frac{K\phi_0}{2}(\delta\phi_0 - 2d) + \frac{Kd^2}{2\delta} = \frac{KB^2}{2\delta}. \tag{7.2.18}$$

The relationship among the length of seawater intrusion, $L$, the discharge to the sea, $Q_0$, and the piezometric head above the toe, $\phi_0$, is clearly expressed by (7.2.18).

From (7.2.18) it follows that as $Q_0$ increases, $L$ decreases. This means that the extent of *seawater intrusion is a decision variable in the management of a coastal aquifer*; it is controlled by controlling $Q_0$, or, alternatively, by controlling the recharge and/or pumping in the coastal aquifer strip.

Figure 7.8b shows a phreatic aquifer with uniform natural replenishment from precipitation, $N$. Again, assuming that the steady flow in the aquifer is essentially horizontal, and that $h(x) = \delta h_f(x)$, continuity of fresh water discharge leads to

$$Q_0 + Nx = -K(h + h_f)\frac{\partial h_f}{\partial x} = -K(1 + \delta)h_f\frac{\partial h_f}{\partial x}. \tag{7.2.19}$$

Integrating (7.2.19) from $x = 0$, $h_f = \phi_0$, $h = B$, to any $x$ and $h_f$, leads to

$$\phi_0^2 - h_f^2 = \frac{2Q_0 x + Nx^2}{K(1 + \delta)}. \tag{7.2.20}$$

The rate of freshwater flowing to the sea from a phreatic aquifer with constant accretion, $N$, is $Q_0 + NL$. Bear and Dagan (1963, 1964b and 1966) and Strack (1973) studied the possibility of intercepting part of the freshwater flowing to the sea in a coastal aquifer. The field technique, developed by Water Planning for Israel, Ltd., Tel Aviv, in the early 60s is called the *coastal collector*. According to this technique, an array of shallow wells is placed along a line parallel to the coast, and not far from it, in order to intercept part of the freshwater flow to the sea, without causing the toe of the interface to advance farther inland. The method is implemented in the coastal aquifer in Israel.

At $x = L$, $h_f = 0$ and, hence, (7.2.20) can be rewritten in the form

$$\phi_0^2 = \frac{2Q_0 L + NL^2}{K(1+\delta)} \quad \text{or} \quad Q_0 = \frac{KB^2}{2L}\frac{1+\delta}{\delta^2} - \frac{NL}{2}, \quad \phi_0 = \frac{B}{\delta} \tag{7.2.21}$$

For $N = 0$

$$\phi_0^2 = \frac{2Q_0 L}{K(1+\delta)} = \frac{B^2}{\delta^2} \tag{7.2.22}$$

Again, the interface has a parabolic shape and we have here a relationship between $L$ and $Q_0$. Equation (7.2.22) relates $\phi_0$ (= piezometric head above the toe) to $L$. By controlling $\phi_0$ (say, by means of artificial recharge), the water table may be lowered both landward and seaward of the toe, without causing any additional seawater intrusion. Landward of the toe, water levels may fluctuate as a result of some management scheme. When pumpage takes place seaward of the toe, the interface there will rise and may contaminate wells if their screened portion is not at a sufficient distance above it.

It may be of interest to note that when $Q_0 = 0$, that is, no seaward fresh water

## MODELING SEAWATER INTRUSION

flow takes place above the toe, $L = (B/\delta)[K(1+\delta)/N]^{1/2}$. This case corresponds to the lowest value of freshwater flow to the sea ($= NL$), with seaward freshwater flow everywhere above the interface.

Instead of $\phi = 0$ at $x = L$, we could use another boundary condition for the confined aquifer, viz.

$$x = L, \quad h = \frac{\beta \delta Q}{K}. \tag{7.2.23}$$

For $\beta = 1$, this is the exact solution for the depth of the interface as derived by Glover (1959) for the case in which the interface has the shape of the parabola

$$y^2 = \frac{2Q\delta}{K} x + \left(\frac{Q\delta}{K}\right)^2 \tag{7.2.24}$$

(Bear, 1972, p. 552). Henry (1959) obtained (7.2.23) with $\beta = 0.741$, using slightly different conditions at the outflow boundary.

If $N = N(x) \neq \text{const}$, we have to start from a continuity equation in the form

$$-\frac{\partial Q}{\partial x} + N(x) = 0, \quad Q = K(h + h_f)\frac{\partial h_f}{\partial x} \tag{7.2.25}$$

and solve it for the given $N = N(x)$ and appropriate boundary conditions at $x = 0$ and $x = L$.

### 7.2.4. MOVING INTERFACE IN THE VERTICAL PLANE

Consider the following simplifying changes introduced in the model represented by (7.2.1), where $R_1$ and $R_2$ are the fresh- and salt-water regions above and below an interface, respectively:

— Because changes in storage due to a moving interface are involved, we may neglect the elastic storage, i.e., $S_0 \simeq 0$. Equivalently, we assume $\alpha = \beta = 0$ in the definition (3.3.11) of $S_0$. Thus, we assume $\rho = \rho(c)$ only.
— The flow domain is two-dimensional in the $xz$-vertical plane.

Then, the two equations in (7.2.1) reduce to a single equation

$$\frac{\partial}{\partial x}\left(\frac{k}{\mu}\frac{\partial p}{\partial x}\right) + \frac{\partial}{\partial z}\left(\frac{k}{\mu}\frac{\partial p}{\partial z}\right) = -\frac{\partial}{\partial z}\left(\frac{k\rho g}{\mu}\right) \tag{7.2.26}$$

where $\rho = \rho(c)$ and $c = c(\mathbf{x}, t)$. This equation is the same as (3.3.26).

Equation (7.2.26) is valid for all point in $(R_1 + R_2)$. However, a question arises as to the meaning of the derivative $\partial(\rho g)/\partial z$ along a sharp interface. We shall return to this point in Section 13.1, where a numerical program is presented for the solution of (7.2.26).

## 7.3. Modeling Regional Seawater Intrusion

### 7.3.1. INTEGRATED FRESHWATER AND SEAWATER BALANCE EQUATIONS

At the end of Subsection 7.2.1, we reached the conclusion that attempting to derive a solution for the shape and position of an interface in a coastal aquifer by solving the three-dimensional model with the nonlinear interface boundary conditions is practically impossible. Instead, similar to the derivation of the equations which govern the flow in a phreatic aquifer (Section 4.2), we may employ the *hydraulic approach* and average the three-dimensional balance equations (7.2.1), separately for each region, over the vertical. In this way, we obtain a model composed of two equations: one for essentially horizontal flow of freshwater and a similar one for the saltwater. A numerical model and a program for the solution of this model are presented in Chapter 13. We shall use the nomenclature of Figure 7.9, which shows an interface in a phreatic aquifer.

For the freshwater region, bounded from below by an interface at elevations $\zeta_1(x, y, t)$ and from above by a phreatic surface with accretion at elevations $\zeta_2(x, y, t)$, and the seawater region bounded from above by the interface at $\zeta_1(x, y, t)$ and from below by an impervious bottom at elevations $\zeta_0(x, y)$, we rewrite (7.2.1) in the form

$$\nabla \cdot \mathbf{q}_f + S_{0f} \frac{\partial \phi_f}{\partial t} = 0, \qquad \nabla \cdot \mathbf{q}_s + S_{0s} \frac{\partial \phi_s}{\partial t} = 0 \qquad (7.3.1)$$

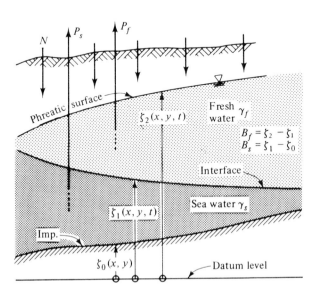

Fig. 7.9. Nomenclature for an interface in a phreatic coastal aquifer.

# MODELING SEAWATER INTRUSION

where, neglecting solid velocity

$$\mathbf{q}_f = -\mathbf{K}_f \cdot \nabla \phi_f, \qquad \mathbf{q}_s = -\mathbf{K}_s \cdot \nabla \phi_s. \tag{7.3.2}$$

By integrating the second equation in (7.3.1) along the vertical, making use of Leibnitz rule in the form of (4.2.8) and (4.2.9), we obtain for the seawater region

$$\int_{\zeta_0}^{\zeta_1} \left( \nabla \cdot \mathbf{q}_s + S_{0s} \frac{\partial \phi_s}{\partial t} \right) dz$$

$$= \nabla' \cdot \int_{\zeta_0}^{\zeta_1} \mathbf{q}'_s \, dz - \mathbf{q}'_s|_{\zeta_1} \cdot \nabla' \zeta_1 + \mathbf{q}'_s|_{\zeta_0} \cdot \nabla' \zeta_0 +$$

$$+ \int_{\zeta_0}^{\zeta_1} \frac{\partial q_{sz}}{\partial z} dz + S_{0s} \left( \frac{\partial}{\partial t} \int_{\zeta_0}^{\zeta_1} \phi_s \, dz - \phi_s|_{\zeta_1} \frac{\partial \zeta_1}{\partial t} \right) = 0. \tag{7.3.3}$$

where

$$\nabla'(\ ) \equiv \frac{\partial(\ )}{\partial x} \mathbf{1}\mathbf{x} + \frac{\partial(\ )}{\partial y} \mathbf{1}\mathbf{y}, \qquad \mathbf{q}' = q_x \mathbf{1}\mathbf{x} + q_y \mathbf{1}\mathbf{y}.$$

By (4.2.9)

$$\frac{\partial}{\partial t} \int_{\zeta_0}^{\zeta_1} \phi_s \, dz = \frac{\partial}{\partial t} \{\tilde{\phi}_s B_s\} = \tilde{\phi}_s \frac{\partial B_s}{\partial t} + B_s \frac{\partial \tilde{\phi}_s}{\partial t} = \tilde{\phi}_s \frac{\partial \zeta_1}{\partial t} + B_s \frac{\partial \tilde{\phi}_s}{\partial t}$$

where

$$\tilde{\phi}_s = \frac{1}{B_s} \int_{\zeta_0}^{\zeta_1} \phi_s \, dz, \qquad B_s = \zeta_1 - \zeta_0.$$

Assuming that saltwater equipotentials are vertical within the saltwater zone, i.e., $\tilde{\phi}_s \approx \phi_s|_{\zeta_1}$, we obtain from (7.3.3)

$$\nabla' \cdot (\widetilde{B_s \mathbf{q}'_s}) - \mathbf{q}'_s|_{\zeta_1} \cdot \nabla' \zeta_1 + \mathbf{q}'_s|_{\zeta_0} \cdot \nabla' \zeta_0 +$$

$$+ q_{sz}|_{\zeta_1} - q_{sz}|_{\zeta_0} + S_{0s} B_s \frac{\partial \tilde{\phi}_s}{\partial t} = 0 \tag{7.3.4}$$

where

$$\widetilde{B_s \mathbf{q}'_s} = \int_{\zeta_0}^{\zeta_1} \mathbf{q}'_s \, dz \equiv \mathbf{Q}'_s$$

is the flow of seawater per unit width of aquifer.

For the fresh water zone, we obtain in a similar way

$$\int_{\zeta_1}^{\zeta_2} \left( \nabla \cdot \mathbf{q}_f + S_{0f} \frac{\partial \phi_f}{\partial t} \right) dz$$

$$= \nabla' \cdot (B_f \widetilde{\mathbf{q}_f}) - \mathbf{q}'_f|_{\zeta_2} \cdot \nabla' \zeta_2 + \mathbf{q}'_f|_{\zeta_1} \cdot \nabla' \zeta_1 + q_{fz}|_{\zeta_2} -$$

$$- q_{fz}|_{\zeta_1} + S_{0f} B_f \frac{\partial \tilde{\phi}_f}{\partial t} = 0 \qquad (7.3.5)$$

where $B_f = \zeta_2 - \zeta_1$, and we have assumed vertical freshwater equipotentials in the freshwater zone, i.e., $\tilde{\phi}_f = \phi_f|_{\zeta_2} = \phi_f|_{\zeta_1}$. This is equivalent to the *Dupuit assumption* (Section 2.3). With this approximation, we also have from (7.2.4)

$$\zeta_1 = (1 + \delta)\tilde{\phi}_s - \delta\tilde{\phi}_f. \qquad (7.3.6)$$

The shape of the interface surface, $F = F(x, y, z, t)$, is described by

$$F \equiv z - \zeta_1 = z - (1 + \delta)\tilde{\phi}_s + \delta\tilde{\phi}_f = 0. \qquad (7.3.7)$$

Since $F(x, y, z, t) = 0$ is maintained unchanged as the interface moves, the material derivative of $F$ vanishes, i.e., $dF/dt = 0$. This statement can be written as the interface is approached from the freshwater domain and from the saltwater one, in the forms

$$n\frac{\partial F}{\partial t} + \mathbf{q}_f \cdot \nabla F = 0, \qquad n\frac{\partial F}{\partial t} + \mathbf{q}_s \cdot \nabla F = 0, \qquad (7.3.8)$$

respectively, or

$$n(1+\delta)\frac{\partial \tilde{\phi}_s}{\partial t} - n\delta \frac{\partial \tilde{\phi}_f}{\partial t}$$

$$= \mathbf{q}_f|_{\zeta_1} \cdot \nabla(z - \zeta_1) = q_{fz}|_{\zeta_1} - \mathbf{q}'_f|_{\zeta_1} \cdot \nabla' \zeta_1, \qquad (7.3.9)$$

$$n(1+\delta)\frac{\partial \tilde{\phi}_s}{\partial t} - n\delta \frac{\partial \tilde{\phi}_f}{\partial t}$$

$$= \mathbf{q}_s|_{\zeta_1} \cdot \nabla(z - \zeta_1) = q_{sz}|_{\zeta_1} - \mathbf{q}'_s|_{\zeta_1} \cdot \nabla' \zeta_1. \qquad (7.3.10)$$

By combining (7.3.4) and (7.3.10), we obtain for the saltwater zone

$$\nabla' \cdot (B_s \widetilde{\mathbf{q}_s}) + [S_{0s} B_s + n(1+\delta)] \frac{\partial \tilde{\phi}_s}{\partial t} - n\delta \frac{\partial \tilde{\phi}_f}{\partial t} +$$

$$+ \mathbf{q}'_s|_{\zeta_0} \cdot \nabla' \zeta_0 - q_{sz}|_{\zeta_0} = 0. \qquad (7.3.11)$$

Then, we derive an expression for $\tilde{\mathbf{q}}_s$ in the form

$$\tilde{\mathbf{q}}_s = \frac{1}{B_s}\int_{\zeta_0}^{\zeta_1} \mathbf{q}_s \, dz = -\frac{1}{B_s}\int_{\zeta_0}^{\zeta_1} \mathbf{K}_s \cdot \nabla \phi_s \, dz$$

$$= -\frac{\mathbf{K}_s}{B_s} \cdot \left[\nabla' \int_{\zeta_0}^{\zeta_1} \phi_s \, dz - \phi_s|_{\zeta_1}\nabla'\zeta_1 + \right.$$

$$\left. + \phi_s|_{\zeta_0}\nabla'\zeta_0 + (\phi_s|_{\zeta_1} - \phi_s|_{\zeta_0})\mathbf{1z}\right]$$

$$\simeq -\mathbf{K}'_s \cdot \nabla'\tilde{\phi}_s \qquad (7.3.12)$$

where $\mathbf{K}'_s$ includes only the components in the $xy$ plane, and we have assumed that $\mathbf{K}'_s = \mathbf{K}'_s(x, y)$, independent of $z$, and $\phi_s|_{\zeta_1} \simeq \phi_s|_{\zeta_0} \simeq \tilde{\phi}_s$. From (7.3.11) and (7.3.12), we obtain for the saltwater zone

$$\nabla' \cdot (B_s\mathbf{K}'_s \cdot \nabla'\tilde{\phi}_s) - \mathbf{q}'_s|_{\zeta_0} \cdot \nabla'\zeta_0 + q_{sz}|_{\zeta_0} -$$
$$- [S_{0s}B_s + n(1 + \delta)]\frac{\partial \tilde{\phi}_s}{\partial t} + n\delta\frac{\partial \tilde{\phi}_f}{\partial t} = 0. \qquad (7.3.13)$$

For an impervious bottom, the sum of the second and third terms on the left-hand side of (7.3.13) vanishes, and we obtain for the saltwater zone

$$\nabla' \cdot (B_s\mathbf{K}'_s \cdot \nabla'\tilde{\phi}_s) - [S_{0s}B_s + n(1 + \delta)]\frac{\partial \tilde{\phi}_s}{\partial t} + n\delta\frac{\partial \tilde{\phi}_f}{\partial t} = 0. \qquad (7.3.14)$$

Usually, $S_{0s}B_s \ll n$, so that $S_{0s}B_s$ may be neglected in (7.3.14).

If sinks (e.g., wells) are located in the seawater zone, we add the term $-P_s(\mathbf{x}, t)$, expressing distributed pumping, or

$$-\sum_{(m)} P_s^{(m)}(\mathbf{x}^m, t)\delta(\mathbf{x} - \mathbf{x}^{(m)}),$$

expressing localized pumping at point $\mathbf{x}^{(m)}$, on the left-hand side of (7.3.13), or (7.3.14).

For the phreatic surface, it follows from (3.4.25), with $n - \theta_{w0}$ replaced by $n_e$, and (7.3.7) that

$$n_e\frac{\partial F}{\partial t} + (\mathbf{q}_f - \mathbf{N}) \cdot \nabla F = 0, \quad \mathbf{N} = -N\mathbf{1z},$$

$$F = z - \zeta_2 = z - \phi_f|_{\zeta_2} \simeq z - \tilde{\phi}_f,$$

$$n_e\frac{\partial \tilde{\phi}_f}{\partial t} = (\mathbf{q}_f - \mathbf{N}) \cdot \nabla(z - \zeta_2)$$

$$= q_{fz}|_{\zeta_2} + N - \mathbf{q}'_f|_{\zeta_2} \cdot \nabla'\zeta_2. \qquad (7.3.15)$$

For $\tilde{\mathbf{q}}_f$ we obtain

$$\tilde{\mathbf{q}}_f = \frac{1}{B_f} \int_{\zeta_1}^{\zeta_2} \mathbf{q}_f \, dz = -\frac{1}{B_f} \int_{\zeta_1}^{\zeta_2} \mathbf{K}_f \cdot \nabla \phi_f \, dz$$

$$= -\frac{\mathbf{K}_f'}{B_f} \cdot \left[ \nabla' \int_{\zeta_1}^{\zeta_2} \phi_f \, dz - \right.$$

$$\left. - \phi_f|_{\zeta_2} \nabla' \zeta_2 + \phi_f|_{\zeta_1} \nabla' \zeta_1 + \phi_f|_{\zeta_2} - \phi_f|_{\zeta_1} \right]$$

$$\approx -\mathbf{K}_f' \cdot \nabla \tilde{\phi}_f. \tag{7.3.16}$$

By combining (7.3.5), (7.3.9), and (7.3.15), we obtain for the freshwater region

$$\nabla' \cdot (B_f \mathbf{K}_f' \cdot \nabla' \tilde{\phi}_f) + n(1+\delta) \frac{\partial \tilde{\phi}_s}{\partial t} -$$

$$- [n_e + n\delta + S_{0f} B_f] \frac{\partial \tilde{\phi}_f}{\partial t} + N = 0. \tag{7.3.17}$$

If sinks (e.g., wells) are located in the freshwater region, we add the term $-P_f(\mathbf{x}, t)$ or $-\Sigma_{(m)} P_f^{(m)}(\mathbf{x}^{(m)}, t) \delta(\mathbf{x} - \mathbf{x}^{(m)})$ on the left-hand side of (7.3.17).

In general, $S_{0f} B_f \ll n_e$, so that the elastic storativity expressed by $S_{0f} B_f$ may be neglected in (7.3.17).

For a confined aquifer, $\zeta_2 = \zeta_2(x, y)$ is a known function and the freshwater equation reduces to

$$\nabla' \cdot (B_f \mathbf{K}_f' \cdot \nabla' \tilde{\phi}_f) + \mathbf{q}_f'|_{\zeta_2} \cdot \nabla' \zeta_2 - q_{fz}|_{\zeta_2} +$$

$$+ n(1+\delta) \frac{\partial \tilde{\phi}_s}{\partial t} - (n\delta + S_{0f} B_f) \frac{\partial \tilde{\phi}_f}{\partial t} = 0 \tag{7.3.18}$$

in which we have left out $N$ that represents sources of water in the freshwater zone.

Since at an impervious bundary

$$-\mathbf{q}'|_{\zeta_2} \cdot \nabla' \zeta_2 + q_z|_{\zeta_2} \equiv \mathbf{q}|_{\zeta_2} \cdot \nabla(z - \zeta_2) = 0,$$

Equation (7.3.18) reduces to

$$\nabla'(B_f \mathbf{K}' \cdot \nabla' \tilde{\phi}_f) + n(1+\delta) \frac{\partial \tilde{\phi}_s}{\partial t} - (n\delta + S_{0f} B_f) \frac{\partial \tilde{\phi}_f}{\partial t} = 0. \tag{7.3.19}$$

Essentially, we have assumed here horizontal flow in both the freshwater and the seawater regions. With this assumption, and with $S_{0f} B_f \ll n$, $S_{0s} B_s \ll n$, we

MODELING SEAWATER INTRUSION

could have written the two water balances for a phreatic aquifer, with $n \simeq n_e$, as

$$-\nabla' \cdot \mathbf{Q}'_f + N - n \frac{\partial(\zeta_2 - \zeta_1)}{\partial t} = 0, \quad -\nabla' \cdot \mathbf{Q}'_s - n \frac{\partial \zeta_1}{\partial t} = 0 \quad (7.3.20)$$

in which

$$\mathbf{Q}'_f = -B_f \mathbf{K}'_f \cdot \nabla' \tilde{\phi}_f, \quad \mathbf{Q}'_s = -B_s \mathbf{K}'_s \cdot \nabla' \tilde{\phi}_s,$$
$$\zeta_2 = \tilde{\phi}_f, \quad \zeta_1 = (1+\delta)\tilde{\phi}_s - \delta\tilde{\phi}_f. \quad (7.3.21)$$

### 7.3.2. INITIAL AND BOUNDARY CONDITIONS

We have to specify the initial and boundary conditions for both the freshwater and seawater regions. In view of (7.3.6), they are not independent.

When we specify initial freshwater levels $\tilde{\phi}_f = \tilde{\phi}_f(x, y, 0) \equiv \zeta_2(x, y, 0)$ and seawater heads $\tilde{\phi}_s = \tilde{\phi}_s(x, y, 0)$, then the interface elevation $\zeta_1(x, y, 0)$ is dictated by (7.3.6).

With respect to the boundary conditions along the coast, we have first to decide where to locate this boundary. As mentioned above, we have always an *outflow face* on the bottom of the sea, through which the freshwater leaves the aquifer (*AB* in Figure 7.2 and *CD* in Figure 7.6). From the definition of $\phi_f$, it follows that along the bottom of the sea

$$\phi_f = z + \frac{p}{\gamma_f} = -\eta + \frac{\eta \gamma_s}{\gamma_f} = \eta \frac{\gamma_s - \gamma_f}{\gamma_f} = \frac{\eta}{\delta} > 0 \quad (7.3.22)$$

where $\eta(x)$ is the depth of the sea bottom below sea level (Figure 7.6). Hence, $\phi_f|_C < \phi_f|_D$.

Often, the vertical surface through point *C* in Figure 7.6 is taken as the boundary of both the freshwater and saltwater regions. The length $x_0$ (Figures 7.6 and 7.10) of the outflow face can be estimated from the parabolic shape of a steady interface in an infinitely thick aquifer (Figure 7.10; Glover, 1959)

$$\zeta_2^2 = \frac{2Q\delta}{K} x + \left(\frac{Q\delta}{K}\right)^2, \quad \delta = \gamma_f/(\gamma_s - \gamma_f),$$

$$x = 0, \quad \zeta_2|_{x=0} = \frac{\delta Q}{K}, \quad \zeta_2 = 0, \quad x = -x_0 = -\frac{\delta Q}{2K}. \quad (7.3.23)$$

For a steady flow and a vertical outflow face (i.e., assuming that $x = 0$ is the sea boundary), Henry (1959) obtained

$$\zeta_2|_{x=0} = 0.741 \frac{\delta Q}{K}.$$

Because of the approximation involved, there is a no unique way of expressing

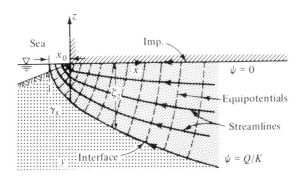

Fig. 7.10. The flow net near the coast (Glover, 1959).

the boundary conditions for the freshwater and seawater flow domains along $AC$ and $CD$ in Figure 7.2, respectively. The following considerations may be useful:

(1) For the fresh water, we may assume $\phi_f|_{AC} = 0$ (measured with respect to sea level). However, as we know that $\phi_f|_C > \phi_f|_B > \phi_f|_A \,(= 0)$, that is, equipotentials are not vertical in the domain $ABC$, a certain error is introduced. We may reduce it by taking some higher value for $\phi_f|_{AC}$.

(2) If we know the freshwater discharge rate, $Q_{f0}$ (Figure 7.2), to the sea through $AC$, (e.g., in a two-dimensional case), we can estimate $\overline{AC} = \delta Q_{f0}/K$. Then, knowing $\phi_f|_{AC}$ and $AC$, we can determine $\phi_s|_{CD}$ from (7.3.6).

(3) In general, the seaward freshwater and saltwater discharge rates, through $AC$ and $CD$, $Q_{f0}$ and $Q_{s0}$ (per unit length of coast) vary along the boundary and are unknown. If we *assume* $\phi_f|_{AB} \simeq d/2\delta \simeq 0$, then

$$Q_{f0} = \frac{\phi_f|_{AC}}{R}, \qquad Q_{f0} = -K_f \overline{AC} \left. \frac{\partial \phi_f}{\partial x} \right|_{AC} \tag{7.3.24}$$

where $R$ is the resistance to the freshwater flow to the sea through $ACB$. One may estimate $R \simeq \bar{L}/K\bar{A} = \alpha/K$ where $\bar{L}$ is an average length of flow, $\bar{A}$ is an average flow cross-section, and $\alpha \,(= \bar{L}/\bar{A})$ is a dimensionless coefficient which has to be estimated. With these considerations, (7.3.24) becomes

$$\frac{\phi_f}{\alpha} + \overline{AC}\frac{\partial \phi_f}{\partial x} = 0 \quad \text{on } AC \tag{7.3.25}$$

which is a *third-type boundary condition*.

A similar expression, with a coefficient $\beta$, can be written for the seawater boundary $CD$, i.e.,

$$\frac{\phi_s}{\beta} + \overline{CD}\frac{\partial \phi_s}{\partial x} = 0 \quad \text{on } CD. \tag{7.3.26}$$

In addition, we have from (7.3.6) the conditions

$$z|_C = -\overline{AC} = \phi_s(1+\delta) - \phi_f\delta, \quad \overline{CD} + \overline{AC} = D|_{x=0} - \zeta_{00} \tag{7.3.27}$$

where $D$ and $\zeta_{00}$ are shown in Figure 7.2.

With these conditions, (7.3.25) and (7.3.26) become

$$\frac{\phi_f}{\alpha} + [\phi_s(1+\delta) - \phi_f\delta]\frac{\partial \phi_f}{\partial x} = 0, \tag{7.3.28}$$

$$\frac{\phi_s}{\beta} + [D - \zeta_{00} + \phi_s(1+\delta) - \phi_f\delta]\frac{\partial \phi_s}{\partial x} = 0 \tag{7.3.29}$$

which are the boundary conditions in terms of $\phi_f$ and $\phi_s$. The coefficients $\alpha$ and $\beta$ have to be estimated (possibly as part of the procedure of parameter estimation) and the depth of the interface (point $C$) will adjust itself such that (7.3.6) is satisfied.

At no point in the development of the equations and boundary conditions presented above, was use made of the *Ghyben–Herzberg relationship* (7.2.14), which corresponds to steady flow with a stationary interface and $\phi_s$ = const, $\phi_f = h_f$. Instead, we have used the relationship (7.2.4), based only on the identity of pressure as the interface is approached from both sides.

Equations (7.3.14) and (7.3.17) or (7.3.19), are two equations in the variables $\tilde{\phi}_f(x, y, t)$ and $\tilde{\phi}_s(x, y, t)$ which have to be solved simultaneously, subject to appropriate boundary conditions on $\tilde{\phi}_f$ and $\tilde{\phi}_s$ in the $xy$ plane. For most problems of practical interest, the solution will have to be derived numerically. Once $\tilde{\phi}_f$ and $\tilde{\phi}_s$ are known, the position and shape of the interface are obtained from (7.3.6). However, one should note that in (7.3.13) and (7.3.17), we have $B_f = \zeta_2 - \zeta_1$ and $B_s = \zeta_1 - \zeta_0$, where $\zeta_0$ is the known bottom elevation, $\zeta_2 = \tilde{\phi}_f$, while $\zeta_1$ is related to $\tilde{\phi}_f$ and $\tilde{\phi}_s$ by (7.3.6). In an iterative numerical solution scheme, this poses no special difficulty. Also, special attention should be given in the solution scheme to the possibility that the seawater region terminates at some distance from the coast (Figure 7.2), where the interface intersects the impervious bottom of the aquifer. In fact, in most cases, the location of this boundary is the sought solution.

CHAPTER EIGHT

# Introduction to Numerical Methods

In the previous chapters, the phenomena of transport in porous media have been described by mathematical models. The complete description was made up of a partial differential equation, or a system of several partial differential equations, together with initial conditions and boundary conditions. In order to solve a given groundwater problem, this system of equations must be solved, for the specific data of that problem. This can be done by using analytical methods, or numerical techniques. For most problems of practical interest, because of the irregular shape of the boundaries, the spatial variability of the coefficients appearing in the equations and in the boundary conditions, the nonuniformity of the initial conditions, and the nonanalytic form of the various source and sink terms, analytical solutions are virtually impossible, except for relatively simple problems. Solutions of most problems can be obtained only by numerical methods. Hence, in this book numerical methods of solution are used almost exclusively. They provide a most powerful and general tool for solving problems encountered in practice. In this chapter, some general aspects will be discussed. In Chapters 9—13 specific problems will be solved, and actual numerical models will be presented.

## 8.1. Analytical versus Numerical Solutions

Before the advent of digital computers, the standard method for the analysis of groundwater problems involved the analytical solution of a partial differential equation, with the corresponding initial and boundary conditions. This means that the appropriate method of solution has to be determined for each particular problem. For example, for problems of steady flow in two dimensions, the most powerful method appears to be the method of complex variables, which involves such concepts as the velocity hodograph and conformal mapping. Nonsteady problems, in general, include a time derivative of the dependent variable, usually the groundwater head $\phi$. A powerful method to deal with this complication is the Laplace transform technique. The complete solution of a problem often requires the application of integral transformation techniques (Fourier transformation, Hankel transformation, etc.). A typical analytical solution may then be in the form of an infinite series of algebraic terms, or even a double infinite series, or an infinite series of definite integrals. Only in some special cases, such as flow in a region of very simple form (a circular region or an infinite strip), can the solution be expressed in the form of a single analytical formula.

Many of the analytical solutions that have been obtained for groundwater flow problems require a considerable amount of computational effort in order to yield numerical values for a specific set of input data. This can be done by hand. However, since there is a proliferation of large and small computers, this task is usually performed by a computer program. A great advantage of an analytical solution, apart from its immediate availability, is that it can give a good insight into the dependence of the soluton on various physical parameters, such as transmissivity, storativity, or geometrical dimensions. This advantage may be partially or completely lost, however, when the form of the solution is very complicated, and the evaluation of data requires a computer program. In such cases, it may well be that a numerical solution is just as convenient.

The main limitation of analytical methods of solution is that they are available only for relatively simple problems. The geometry of the problem must be regular, e.g., a circular aquifer, or a rectangular one. The properties of the soil in the region considered must be homogeneous, or at least homogeneous in sub-regions. With the advent of large digital computers, numerical methods have received a large impulse, to such the extent that a variety of powerful general techniques have become available to the scientific and professional community. In recent years, the introduction of cheap, smaller computers (*micro-computers* or *personal computers*) has been a new stimulant for the application of numerical methods. As these small computers rapidly develop into powerful machines, the solution of engineering problems by numerical models is within the reach of every scientist and engineer.

## 8.2. Survey of Numerical Methods

In this section, a brief survey of the most popular and powerful methods for the numerical solution of groundwater flow and pollution problems is presented. Some of these are further elaborated in Chapters 9 to 13. In general, these numerical methods consist of certain procedures as a result of which the partial differential equation is replaced by an algebraic equation, or a system of algebraic equations. Usually these constitute a system of *linear equations*. The final solution then requires the solution of this system of equations. Many powerful methods are available for this purpose. The various numerical methods differ mainly in the way the system of equations is derived, and sometimes also in the basic approach to the problem.

### 8.2.1. FINITE DIFFERENCES

Probably the oldest numerical method is the *method of finite differences* (Southwell, 1940; Forsythe and Wasow, 1960; Fox, 1962; Kantorovich and Krylov, 1964). In this method, the partial derivatives appearing in the basic differential equation are replaced by an algebraic equivalent, with a quotient of two finite differences of the dependent and an independent variable replacing the differential quotient. The basic idea is that the derivative $df/dt$ of a function $f(t)$ is defined as

$$\frac{df}{dt} = \lim_{\Delta t \to 0} \frac{f(t + \Delta t) - f(t)}{\Delta t} \qquad (8.2.1)$$

and that an obvious approximation of the derivative can be obtained by simply omitting the limiting process, $\Delta t \to 0$.

As a simple example, let us consider the differential equation

$$\frac{df}{dt} = -f \qquad (8.2.2)$$

subject to the initial condition $f(0) = 1$. If the derivative is now approximated by the finite difference suggested by omitting the limit operation in (8.2.1), the differential equation is replaced by the equation

$$\frac{f(t + \Delta t) - f(t)}{\Delta t} = -f(t)$$

from which it follows that

$$f(t + \Delta t) = f(t) - \Delta t \times f(t). \qquad (8.2.3)$$

This is a simple algorithm from which the values of $f(t)$ can be determined in successive steps, starting from the initial value $f(0) = 1$. If $\Delta t$ is taken as 0.1, the series of values obtained up to $t = 1$ is

$f(0.0) = 1.0000$
$f(0.1) = 0.9000$
$f(0.2) = 0.8100$
$f(0.3) = 0.7290$
$f(0.4) = 0.6561$
$f(0.5) = 0.5905$
$f(0.6) = 0.5314$
$f(0.7) = 0.4783$
$f(0.8) = 0.4305$
$f(0.9) = 0.3874$
$f(1.0) = 0.3487$.

Actually, the exact solution of the differential equation (8.2.2) is $f(t) = \exp(-t)$ so that for $t = 1$ the exact result is $f(1) = 0.3679$. The approximation is not very good, but may be sufficient for practical purposes. The accuracy can be increased by taking smaller time steps. If $\Delta t = 0.01$ one obtains $f(1) = 0.3660$, and if $\Delta t = 0.001$, one obtains (after 1000 steps) $f(1) = 0.3677$.

This simple example demonstrates that the approximation of the differential quotient by a quotient of finite differences may lead to a very simple numerical algorithm. It also shows that the degree of accuracy depends strongly upon the magnitude of the steps taken, which seems natural.

It may be noted that in the definition (8.2.1) it was tacitly assumed that the limit does exist, and that the same limit is obtained for positive and negative values of the step $\Delta t$. When using finite differences, the limit is replaced by some value for a finite value of $\Delta t$ and the actual value may depend upon the magnitude of $\Delta t$, and also upon its sign. In the numerical process described by (8.2.3), the value of $\Delta t$ was assumed to be positive. This is not necessary, however, as one might have taken a negative value of $\Delta t$ in the approximation. Doing so would lead to the algorithm

$$f(t) = f(t - \Delta t)/(1 + \Delta t). \qquad (8.2.4)$$

Starting from the value $f(0) = 1$, and taking $\Delta t = 0.1$, one now obtains $f(1) = 0.3855$. This result is about as good (or bad) as in the previous case. Again, the accuracy is much better if smaller time steps are taken.

A third alternative for the algorithm can be obtained by noting that in the presentation above it was assumed, without much discussion, that the value of $t$ should be taken without any increment on the right-hand side of (8.2.2). It seems better balanced, however, when the value of the function $f$ on the right-hand side is taken as the average of the values $f(t)$ and $f(t + \Delta t)$. This leads to the algorithm

$$f(t + \Delta t) = f(t) \times (1 - \tfrac{1}{2}\Delta t)/(1 + \tfrac{1}{2}\Delta t). \qquad (8.2.5)$$

Application of this formula, with $\Delta t = 0.1$, gives $f(1) = 0.3676$, which is very close to the exact value. It thus appears that, with some attention for a careful and balanced approximation of the various terms, a very good approximation can be obtained.

The various possibilities for the approximation by finite differences are usually denoted as *forward*, *backward*, and *central* finite differences.

In Chapter 9 finite differences will be used to solve the differential equations of steady and unsteady groundwater flow in aquifers. The general principle is that the unknown variable, the head $\phi$, which is a function of $x$, $y$, and $t$, is represented, for every value of the time $t$, by a finite number of values $\phi(i, j)$, where $i$ and $j$ indicate counter variables in $x$- and $y$-direction, respectively. They represent mesh points in the region considered (see Figure 8.1). In this figure it has been assumed that all intervals are equal and constant. This is not necessary, but it may be a first assumption, and it usually leads to the simplest system of equations.

The form of the system of equations is that the values of the head in each point are related to the values in the surrounding points, in general, both the values in these points at the beginning and at the end of a time step. If the values at the beginning of the time step are known (and that is usually the case) the values at the end of the time step are the unknowns, and the resulting system of equations is a system of $N$ linear equations with $N$ unknowns. The number $N$ indicates the total number of mesh points. Thus, the mathematical problem to be solved is the solution of a linear system of equations. Many powerful and effective numerical

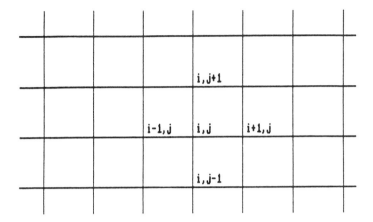

Fig. 8.1. Mesh points in plane.

methods are available for this task. It must be noted, however, that the system of equations may turn out to be rather large. For a system with 20 mesh points in $x$-direction and 20 in $y$-direction, the value of $N$ will be 400. This seems to indicate that a matrix of 400 by 400 elements will be needed to describe the system of equations, and this matrix contains 160 000 elements, which may require a large computer, even for a relatively small problem (see Figure 8.2).

In Chapter 9 we shall show that the actual solution of the system of equations does not require such a large matrix, because in each of the linear equations, only a few nonzero coefficients appear. This fact makes it possible to use solution procedures that are especially suited for such a system of equations. It will also enable us to solve large problems on a relatively small computer.

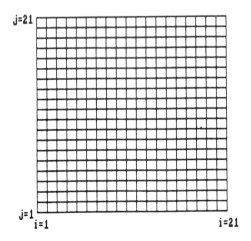

Fig. 8.2. Mesh of 400 points.

8.2.2. FINITE ELEMENTS

A second very powerful numerical method is the *finite element method* (see, e.g., Zienkiewicz, 1977). An elementary way of presenting this method is used in structural mechanics, where the elements are the actual parts of a structure, say the beams and columns in the framework of a building, or the grid of beams in the floor of a bridge. The deformation of each element is then expressed in terms of the forces acting upon it at the two ends. This enables us to express the displacement of each nodal point in terms of those of the neighboring nodes, and the deformation of the connecting elements. The final system of equations is obtained from the conditions of equilibrium at each node.

In groundwater flow problems, one could imagine that a region is subdivided into small elements, such that for each element the flow is described in terms of the head in the nodal points, and that then a system of equations is obtained from the condition that the flow must be continuous at each node.

The usual way of presenting the finite element method does not employ such a physical reasoning, however. Instead a mathematical argumentation is used, in which the system of equations is obtained by requiring that the differential equation be satisfied 'on the average', using certain weight functions.

The system of (linear) equations obtained in the finite element method has the same structure as in the finite difference method. Actually, the two methods are very similar and, for certain problems, it has been shown that they can be considered as two representations of the same model. However, the usual way of deriving and developing the equations introduces certain differences. For instance, the natural and simplest type of element is the triangle, which enables a flexible representation of the field (see Figure 8.3), whereas the simplest and most natural mesh in the finite difference method is a one of squares or rectangles, which is less flexible. Also, the standard formulation of the finite element method has the

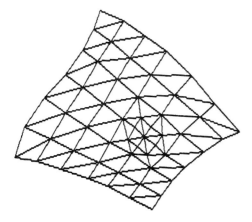

Fig. 8.3. Typical mesh of finite elements.

immediate property that each element can have its own values for such physical parameters as transmissivity and storativity.

The finite element method, in its standard form, is somewhat more flexible than the standard form of the finite difference method. This may explain its popularity. It must be noted, however, that variants exist for the finite difference method that are very similar to the finite element method (e.g., *integrated finite differences*, Narasimhan and Witherspoon, 1976). Thus, the two methods may be considered as more or less equivalent. The method to be used for any given problem should be determined by the characteristics of that problem, and also by such considerations as the convenience of a program, its availability and cost, and the level of experience of the user.

### 8.2.3. BOUNDARY ELEMENTS

A completely different numerical technique is the *boundary element method* (BEM), or *boundary integral equation method* (BIEM) (see, for example, Banerjee and Butterfield (1977), Liggett and Liu (1983)). The most common way of presenting this method is through integral theorems such as *Green's second identity*, or by starting from certain analytical solutions for typical singularities, such as a point source, or a dipole, in an infinite field. The system of equations is obtained by the addition of the contribution of singularity distributions along the entire boundary of the domain, and then requiring that the correct boundary conditions be satisfied along each part of the boundary, at least on the average. This leads to a system of $N$ equations with $N$ unknowns. These equations are usually linear. The number $N$ is here the number of points on the boundary and not, as in the finite difference and finite element methods, the number of points in the interior of the domain and its boundary. Thus, the number of unknowns, and the number of equations, is usually considerably lower. This fact is often considered as an important advantage of the method. It must be noted, however, that the matrix of the system of equations is completely filled with nonzero elements, whereas in the finite difference and finite element methods, the matrix is *sparse*, which means that only a small percentage of the elements of the matrix is actually different from zero. Thus, a direct and general comparison of memory requirements, and computation time requirements, is not well possible, and may depend upon the type of problem considered.

A disadvantage of the boundary element method is that it is basically restricted to regions with homogeneous soil and fluid properties, as it uses a fundamental solution for a singularity in a homogeneous field. With some additional effort, the method can be generalized to a body consisting of a number of smaller sub-regions, each of them having constant physical properties. However, it then seems more natural to use the finite element method. For this reason, the boundary element method will not be presented in this book. For more details, the reader is

INTRODUCTION TO NUMERICAL METHODS 223

referred, among others, to texts by Banerjee and Butterfield (1977) and Liggett and Liu (1983).

8.2.4. ANALYTICAL ELEMENTS

Another possible alternative to the finite difference and finite element methods is the *analytical element method* (Strack, 1987). Like the BIEM, this method also again applies primarily to homogeneous regions. It uses superposition to generate the analytical solution of a certain problem, expressed as a sum of basic solutions, each with a number of possibly unknown parameters. These parameters are then determined from the boundary conditions. Again, this leads to a system of algebraic equations. Because a large variety of basic solutions is available, or can be constructed, the method can be used to solve a rather wide class of problems. This includes problems involving infiltration pools, well systems, layered strata, free surfaces, etc. As the final expression obtained has an analytical form, it has all the advantages of such solutions, e.g., the possibility of obtaining detailed local information as well as global information, on quite a different scale. The method strongly relies on theoretical solutions and, therefore, requires a considerable amount of theoretical expertise. As it is outside the scope of this book, it will not be presented here. The reader is referred to a basic text on the subject by Strack (1987).

## 8.3. Computer Programming

Numerical methods exist only because of the availability of hardware (computers) and software. A numerical technique can be implemented on a computer by means of a *computer program*. The program instructs the computer to perform the necessary operations and, therefore, it must use certain standard statements only. Since a computer can perform only certain, very elementary, operations, special programming languages have been developed, which can be used as a link between the modeler and the computer. These languages enable the writing of a computer program in a form that conforms closely to the operations of the numerical model, as specified by the human modeler, and yet, the computer can interpret the commands included in the program in its own system language.

Well-known computer languages are FORTRAN, ALGOL, BASIC, PASCAL and C. Each of these has its advantages and disadvantages, and its own group of supporters. In this book, all programs are presented in BASIC. This is perhaps not the best language, in terms of flexibility, generality, and clarity. Nevertheless, for the purpose of this book, it possesses certain advantages compared to other languages. One determining factor is the wide-spread availability of BASIC interpreters and compilers on practically every computer down to inexpensive home computers. Thus, practically everybody can be assumed to have access to a computer that can run programs in BASIC. Another factor is that programs in BASIC tend to be short, much shorter than their counterparts in FORTRAN, and in other languages. Thus, much space is saved, and the student can easily copy a program from this book, simply by re-typing the code.

The use of BASIC in the present text does not necessarily mean that the authors advocate the use of BASIC for all computer programming in groundwater modeling. Actually, for professional programs, to be used intensively for large problems, the reader is encouraged to switch to be a more powerful language, such as FORTRAN, PASCAL or C. This will usually result in a much faster code, and the possibility of running the program with many more variables. Such programs may be developed on the basis of the BASIC programs listed in this book.

BASIC versions may slightly vary from computer to computer. The version used in this book is Microsoft's BASIC, as developed for the operating system MS–DOS. Care has been taken not to use machine-dependent coding statements, so that the programs should run with no or hardly any modifications on practically every computer. All programs can be compiled with Microsoft's compilers, which will make them run much faster.

All programs in this book have been tested extensively against analytic solutions of elementary problems, and their performance has been checked by running many sample problems. Nevertheless, errors cannot be completely excluded. It should also be noted that each program applies only to a certain restricted class of problems, and that the performance of a program may depend upon the combination of input data. In problems of nonsteady flow, for instance, time steps often cannot be taken too large, in order to avoid unstable behaviour of the solution. Furthermore, a very large contrast in soil properties (say a factor 1000 in permeabilities) may give rise to serious problems of accuracy. Such problems usually indicate that the model is not appropriate to the physical problem, and that a different schematization should be used. In a system of bodies of different permeability, for instance, sand and clay, it is often not realistic to try to determine the flow in both materials in the same model. It may be more realistic to consider the clay as completely impermeable, or the sand as a region of constant head.

With these general comments on numerical methods and the use of computers, we may now proceed to discuss in detail the two major methods, and employ them to present specific codes of practical interest.

CHAPTER NINE

# The Finite Difference Method

In this chapter the *finite difference method* is presented, for problems of steady and nonsteady groundwater flow. The presentation will be oriented towards the introduction of simple computer programs, written in BASIC, that can be run on personal computers.

The finite difference method was the first method to be used for the systematic numerical solution of partial differential equations. Although the fundamental ideas have been established and used by mathematicians of the 18th century, such as Taylor and Lagrange, the application of the finite difference method to the solution of engineering problems is usually considered to be an achievement of 20th century scientists (Southwell, 1940; Kantorovich and Krylov, 1964).

In general, the method consists of an approximation of partial derivatives by algebraic expressions involving the values of the dependent variable at a limited number of selected points. As a result of the approximation, the partial differential equation describing the problem is replaced by a finite number of algebraic equations, written in terms of the values of the dependent variable at the selected points. The equations are linear if the original partial differential equations are also linear. The values at the selected points become the unknowns, rather than the continuous spatial distribution of the dependent variable. The system of algebraic equations must then be solved. This may involve a large number of arithmetic operations. Originally all these calculations were performed manually, or by using mechanical devices. However, since the advent of the electronic computer, they are usually executed by means of a computer program.

In recent years, other powerful numerical methods have been developed, in particular the *finite element method*. This method has a greater flexibility and, therefore, is often preferred to the finite difference method. Because the basic theory of the finite difference method is much simpler, and it usually requires less computer memory and computation time, this method will be presented first here. The finite element method will be presented in Chapter 10.

## 9.1. Steady Flow

To demonstrate the method, let us consider the case of two-dimensional flow of a single fluid in a homogeneous isotropic confined aquifer, with no sources or sinks. For this case, the flow is described by the Laplace equation (4.2.25b)

$$\frac{\partial^2 \phi}{\partial x^2} + \frac{\partial^2 \phi}{\partial y^2} = 0. \tag{9.1.1}$$

This equation should be satisfied at all points within the considered aquifer domain $R$. On the boundary of $R$ the groundwater head, $\phi$, should satisfy certain boundary conditions. Throughout this section it will be assumed that the boundary conditions are, following the discussion in Section 4.3

$$\text{on } S_1: \quad \phi = f, \qquad (9.1.2)$$

$$\text{on } S_2: \quad Q_n = -T \frac{\partial \phi}{\partial n} = 0 \qquad (9.1.3)$$

where $S_1$ and $S_2$ are disjoint parts of the boundary, which together form the entire boundary of the region $R$. Equation (9.1.2) states that on the part $S_1$ of the boundary the head is prescribed. Equation (9.1.3) states that the remaining part of the boundary is impermeable. More general boundary conditions will be discussed later.

This basic idea of the finite difference method is that the region $R$ is covered with two families of straight parallel lines, in the $x$- and $y$-directions, respectively, which together form a mesh of rectangles. The simplest type of mesh is generated when all the intervals between the lines are equal. In that case the mesh consists of squares (Figure 9.1). The value of the variable $\phi$ in a nodal point of the mesh (or *node*) is now denoted by $\phi_{i,j}$, where $i$ indicates the position of the vertical mesh line (the *column*), and $j$ the horizontal mesh line (the *row*).

The set of values $\phi_{i,j}$ will be determined such they approximate the continuous function $\phi(x, y)$, at the points with coordinates $x = x_i$, $y = y_j$. This can be accomplished by an approximation of the partial derivatives appearing in (9.1.1).

In general, the approximation of the first derivative with respect to $x$ of a

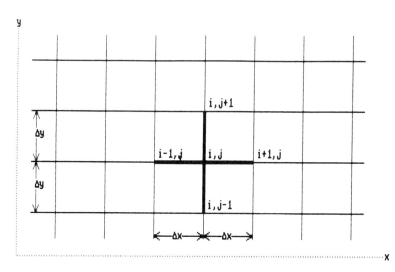

Fig. 9.1. A mesh of squares.

# THE FINITE DIFFERENCE METHOD

function $F(x, y)$, is given by

$$\frac{\partial F}{\partial x} \approx \frac{F(x + \Delta x, y) - F(x, y)}{\Delta x}. \tag{9.1.4}$$

Actually, this is the classical definition of the derivative, except that the condition that $\Delta x \to 0$ has been omitted. Equation (9.1.4) is said to be the *forward finite difference* approximation of the partial derivative $\partial F/\partial x$. If $\Delta x$ is replaced by $-\Delta x$, the *backward finite difference* is obtained, in the form

$$\frac{\partial F}{\partial x} \approx \frac{F(x, y) - F(x - \Delta x, y)}{\Delta x}. \tag{9.1.5}$$

When $\Delta x \to 0$, both expressions will be equal if the derivative exists. However, when $\Delta x \neq 0$ there will be a small difference between the two approximations. Because these approximations are biased with regard to either the region to the right or to the left of the point $x$, $y$, the *central finite difference* is often more accurate,

$$\frac{\partial F}{\partial x} \approx \frac{F(x + \tfrac{1}{2}\Delta x, y) - F(x - \tfrac{1}{2}\Delta x, y)}{\Delta x}. \tag{9.1.6}$$

The second derivative is the derivative of the first derivative. If a central finite difference approximation is used, we obtain

$$\begin{aligned}\frac{\partial^2 F}{\partial x^2} &\approx \frac{F(x + \Delta x, y) - 2F(x, y) + F(x - \Delta x, y)}{(\Delta x)^2} \\ &= \frac{F_{i+1,j} - 2F_{i,j} + F_{i-1,j}}{(\Delta x)^2}.\end{aligned} \tag{9.1.7}$$

This formula is illustrated in Figure 9.2. The function shown in this figure has a

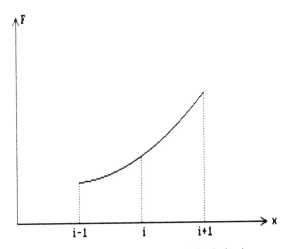

Fig. 9.2. Approximation of second the derivative.

positive second derivative, as is indicated by the slope increasing in the $x$-direction. The finite difference expression reflects this property, because the value at node $i$ is less than the average of the two values at the nodes $i-1$ and $i+1$. If the value in node $i$ would be increased, reaching the average of the two neighboring values, the finite difference approximation vanishes, and so does the second derivative, which vanishes if the values at the three nodes form a straight line.

Application of (9.1.7) to the partial derivatives in (9.1.1) now gives the following approximation of the Laplace operator, if, for reasons of simplicity, it is assumed that the intervals in the $x$- and $y$-directions are equal (i.e., $\Delta x = \Delta y = \Delta$).

$$\frac{\partial^2 \phi}{\partial x^2} + \frac{\partial^2 \phi}{\partial y^2} \approx \frac{\phi_{i,j-1} + \phi_{i,j+1} + \phi_{i-1,j} + \phi_{i+1,j} - 4\phi_{i,j}}{\Delta^2}. \qquad (9.1.8)$$

It now follows that the vanishing of the left-hand side, as required by the basic differential equation (9.1.1), can be approximated by requiring that

$$\phi_{i,j} = \tfrac{1}{4}(\phi_{i-1,j} + \phi_{i+1,j} + \phi_{i,j-1} + \phi_{i,j+1}). \qquad (9.1.9)$$

This is the basic equation of the finite difference method in its simplest form. It states that the value of the head, $\phi$, at a certain node must be equal to the average of the values at the four neighboring nodes.

Equation (9.1.9) should be satisfied at each node in the interior of the domain $R$. Thus, the method will lead to a system of linear algebraic equations, with the number of equations equal to the number of unknowns. At this point, this is not completely obvious, however, as (9.1.9) cannot be applied to nodes on the boundary. Nodes on the boundary require special attention, in order to accommodate the boundary conditions.

A possible boundary condition is the *Dirichlet condition* (9.1.2), which states that the groundwater head is specified along a certain part of the boundary (see Subsection 3.4.4 and Section 4.3). In this case, the value of the head at nodes on that part of the boundary are *a-priori* prescribed, rather than being unknown. When this is the case along the entire boundary, the system of Equations (9.1.9) constitutes a system of linear equations with just as many equations as unknowns, because this equation can be written for each interior node.

At a node on an impermeable boundary, along which the *Neumann boundary* condition (9.1.3) applies, the head is unknown, and the equation for that node should reflect the no-flow condition on the boundary (see Subsection 3.4.4 and Section 4.3). This can be done most simply by requiring that the central finite difference approximation of the normal derivative vanishes. In view of (9.1.6), this means that the value at the two nodes on either side of the boundary should be equal or, in other words, that the value in an (imaginary) node across the boundary should be equal to the value inside the boundary. For a node on a vertical boundary this can be expressed by the condition $\phi_{i-1,j} = \phi_{i+1,j}$. Substitution into

# THE FINITE DIFFERENCE METHOD

the general algorithm now gives

$$\phi_{i,j} = \tfrac{1}{4}(2\phi_{i+1,j} + \phi_{i,j-1} + \phi_{i,j+1}). \tag{9.1.10}$$

Equation (9.1.10) shows that for a node on an impermeable boundary, the general formula (9.1.9) can still be used, involving the four nodes surrounding it. For the value at the node outside the boundary, the value at the node inside the boundary should be substituted. As a result, this value in the interior of the domain counts twice in the algorithm.

A simple example, for a rectangular region, is shown in Figure 9.3. Along the upper boundary the head is given to be 100. In the lower left corner a zero head is specified. All other boundaries are impermeable. An initial estimate of all unknown values is required in order to solve the problem. For simplicity, these have all been assumed to be equal to the average of the boundary values, i.e., 50.

The first figure shows the initial data. By examining these values it must be concluded that they do not satisfy the conditions (9.1.9). However, they can now be corrected, successively, so that they do satisfy the basic equation. In the central part of the figure, the algorithm (9.1.9) has been applied once to all nodes, starting in the upper left corner, and rounding off to the nearest integer. Of course, subsequent corrections partly destroy the agreement reached in a previous correction, and the new values still do not satisfy (9.1.9). The results show a definite improvement on the initial values, however. After a number of *iterations*, in each of which all values are updated, the correct solution, represented in the right part of the figure, is obtained.

The method described above is known as the method of *relaxation* after Southwell (1940), who used the term to indicate that in each step one of the errors is relaxed. Experience shows that the method converges for all problems of potential flow. In mathematical terminology, the relaxation method is also known as the *Gauss–Seidel method* (see Appendix A).

Convergence of the relaxation method can be improved by starting from a better estimate, and by *over-relaxation*, which means that in each correction the local value is updated not to the value indicated by (9.1.9), but to a value obtained

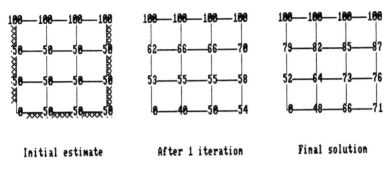

Fig. 9.3. Example of the finite difference method.

by correcting with a somewhat larger value than the error found. This over-relaxation factor should be less than 2 (and larger than 1).

Before presenting a computer program that will perform the finite difference calculations, the basic algorithm will be generalized to the case of a mesh of varying intervals. This can be done most conveniently by the application of Taylor's series expansion formula. If the value of a function $F(x)$ at the point $x_i$ is denoted by $F_i$, the values at the neighboring points $x_{i+1}$ and $x_{i-1}$ can be expressed in terms of the values of the function and its derivatives at point $x_i$, as follows

$$F_{i+1} = F_i + (x_{i+1} - x_i) \frac{dF}{dx} + \tfrac{1}{2}(x_{i+1} - x_i)^2 \frac{d^2F}{dx^2} + \cdots \qquad (9.1.11)$$

$$F_{i-1} = F_i + (x_i - x_{i-1}) \frac{dF}{dx} + \tfrac{1}{2}(x_i - x_{i-1})^2 \frac{d^2F}{dx^2} + \cdots \qquad (9.1.12)$$

Elimination of the first derivative $dF/dx$ from these equations gives

$$\frac{d^2F}{dx^2} = A_i F_{i+1} + B_i F_{i-1} - (A_i + B_i) F_i + \cdots \qquad (9.1.13)$$

where

$$A_i = 1/[(x_{i+1} - x_i)(x_{i+1} - x_{i-1})/2], \qquad (9.1.14)$$

$$B_i = 1/[(x_i - x_{i-1})(x_{i+1} - x_{i-1})/2]. \qquad (9.1.15)$$

Equation (9.1.13) is the generalization of (9.1.7) for variable distances between successive mesh points. If $x_{i+1} - x_i = x_i - x_{i-1} = \Delta x$, Equation (9.1.13) reduces to (9.1.7).

The approximation (9.1.13) can directly be used to approximate the partial derivative $\partial^2 F/\partial x^2$, and a similar formula can be derived for the partial derivative $\partial^2 F/\partial y^2$. In this way, the following generalization of (9.1.9) is obtained

$$(A_i + B_i + C_j + D_j)\phi_{i,j} = A_i \phi_{i+1,j} + B_i \phi_{i-1,j} + C_j \phi_{i,j+1} + D_j \phi_{i,j-1} \qquad (9.1.16)$$

where $A_i$ and $B_i$ are given by (9.1.14) and (9.1.15), and where

$$C_j = 1/[(y_{j+1} - y_j)(y_{j+1} - y_{j-1})/2], \qquad (9.1.17)$$

$$D_j = 1/[(y_j - y_{j-1})(y_{j+1} - y_{j-1})/2]. \qquad (9.1.18)$$

If all mesh sizes are equal the coefficients $A_i$, $B_i$, $C_j$ and $D_j$ are all equal and (9.1.16) reduces to (9.1.9).

A computer program, in BASIC, based on the algorithm (9.1.16) is reproduced as Program BV9-1.

For the convenience of operation the program uses interactive input. This means that all input data are generated by the response of the user to questions printed by the program on the screen of the monitor. Thus, no manual is needed, as the user is informed continuously by the program about the meaning of each input variable to be entered. As an example, the program can be used to solve the

# THE FINITE DIFFERENCE METHOD

```
100 DEFINT I-N:KEY OFF:GOSUB 540
110 PRINT"---   Bear & Verruijt - Groundwater Modeling"
120 PRINT"---   Plane steady groundwater flow"
130 PRINT"---   Finite differences":PRINT"---   Program 9.1"
140 PRINT"---   Rectangular area, with irregular mesh":PRINT:PRINT
150 DIM X(21),Y(21),F(21,21),IP(21,21),A(21),B(21),C(21),D(21)
160 INPUT"Number of mesh lines in x-direction (max. 20) ";NX
170 INPUT"Number of mesh lines in y-direction (max. 20) ";NY:PRINT
180 FOR I=1 TO NX:PRINT"x-coordinate of line ";I;:INPUT X(I):NEXT I
190 PRINT:FOR I=1 TO NY:PRINT"y-coordinate of line ";I;:INPUT Y(I)
200 NEXT I:X(0)=X(1)-(X(2)-X(1)):X(NX+1)=X(NX)+(X(NX)-X(NX-1))
210 Y(0)=Y(1)-(Y(2)-Y(1)):Y(NY+1)=Y(NY)+(Y(NY)-Y(NY-1)):GOSUB 540
220 A=0:K=0:PRINT"The head must be given in at least one point"
230 PRINT:PRINT"   i = ";:INPUT I:PRINT"   j = ";:INPUT J
240 PRINT"   F = ";:INPUT F(I,J):IP(I,J)=1:A=A+F(I,J):K=K+1
250 PRINT:PRINT"Repeat input of given head (Y/N) ? ";:GOSUB 510
260 IF A$="Y" THEN 230
270 A=A/K:FOR I=0 TO NX+1:FOR J=0 TO NY+1:IF IP(I,J)=0 THEN F(I,J)=A
280 NEXT J:NEXT I
290 FOR I=1 TO NX:A(I)=2/((X(I+1)-X(I))*(X(I+1)-X(I-1)))
300 B(I)=2/((X(I)-X(I-1))*(X(I+1)-X(I-1))):NEXT I
310 FOR J=1 TO NY:C(J)=2/((Y(J+1)-Y(J))*(Y(J+1)-Y(J-1)))
320 D(J)=2/((Y(J)-Y(J-1))*(Y(J+1)-Y(J-1))):NEXT J:NI=NX*NY:RX=1.4
330 PRINT:PRINT"Iteration";:FOR IT=1 TO NI:PRINT IT;
340 FOR I=1 TO NX:F(I,0)=F(I,2)
350 FOR J=1 TO NY:F(0,J)=F(2,J):IF IP(I,J)>0 THEN 380
360 A=A(I)*F(I+1,J)+B(I)*F(I-1,J)+C(J)*F(I,J+1)+D(J)*F(I,J-1)
370 A=A/(A(I)+B(I)+C(J)+D(J)):F(I,J)=F(I,J)+RX*(A-F(I,J))
380 F(NX+1,J)=F(NX-1,J):NEXT J:F(I,NY+1)=F(I,NY-1):NEXT I
390 FOR J=1 TO NY:F(NX+1,J)=F(NX-1,J)
400 FOR I=1 TO NX:F(I,NY+1)=F(I,NY-1):IF IP(I,J)>0 THEN 430
410 A=A(I)*F(I+1,J)+B(I)*F(I-1,J)+C(J)*F(I,J+1)+D(J)*F(I,J-1)
420 A=A/(A(I)+B(I)+C(J)+D(J)):F(I,J)=F(I,J)+RX*(A-F(I,J))
430 F(I,0)=F(I,2):NEXT I:F(0,J)=F(2,J):NEXT J
440 NEXT IT:GOSUB 540:PRINT"Output":PRINT:A$="#####.###"
450 FOR I=1 TO NX:FOR J=1 TO NY
460 PRINT" x = ";:PRINT USING A$;X(I);
470 PRINT"  -   y = ";:PRINT USING A$;Y(J);
480 PRINT"  -   F = ";:PRINT USING A$;F(I,J):NEXT J,I:PRINT
490 PRINT"Repeat iterations (Y/N) ? ";:GOSUB 510:IF A$="Y" THEN 330
500 PRINT:END
510 A$=INPUT$(1):IF A$="Y" OR A$="y" THEN A$="Y":PRINT"Yes":RETURN
520 IF A$="N" OR A$="n" THEN A$="N":PRINT"No":RETURN
530 GOTO 510
540 CLS:LOCATE 1,28,1:COLOR 0,7:PRINT" Finite Differences - 1 ";
550 COLOR 7,0:PRINT:PRINT:RETURN
```

Program BV9-1. Steady flow by finite differences.

problem illustrated in Figure 9.4. This is a problem of flow satisfying the Laplace equation (9.1.1) in a square region. Boundary conditions are a prescribed head, $\phi = 1$, along the upper and right boundary, and a given head, $\phi = 0$, in the lower left corner. The other boundaries are impermeable. As computer languages do not recognize Greek symbols, the head $\phi$ is represented by $F$ in the program.

The program starts by asking for the number of mesh lines in the $x$- and $y$-

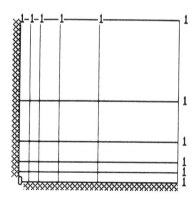

Fig. 9.4. Example solved by the finite difference method.

directions (lines 160—170). The maximum capacity has been defined, arbitrarily, as 20. If the program is to be used for larger networks, this value may be replaced by a larger number. The dimension statements in line 150 then should also be adjusted. In the case of Figure 9.4 there are six lines, including the boundaries, hence the number 6 must be entered. Next the program asks for the $x$-coordinates of the vertical lines. For the case shown in Figure 9.4, the user's response should be: 0, 1, 2, 4, 8, 16, successively. The response to the prompts for the $y$-coordinates should be the same sequence, if the figure is to be symmetric. Next, the boundary conditions should be given. The program assumes that all boundaries are impermeable, unless the head is specified in a certain node. In the case illustrated in Figure 9.4, there are 12 such nodes. For instance, for $i = 1$, $j = 1 : \phi = 0$, for $i = 6, j = 6 : \phi = 1$, etc.

In the program, the nodes with a specified head are distinguished from the others by a value 1 for the (integer) variable $IP$. When all input data have been entered by the user the program performs $N$ iterations, where $N$ is the number of nodes (36 in this case). The over-relaxation factor $RX$ is taken as 1.4 (see line 320). The actual calculations are performed in lines 330—430. Here the boundary conditions along an impermeable boundary are taken into account by the introduction of image nodes outside the boundary (for instance $I = 0$ or $I = NX + 1$). The boundary condition at nodes with specified head is taken into account by jumping over the relaxation algorithm, if $IP(I, J) = 1$, in lines 350 and 400. It should be noted that, actually, all values are updated twice in each iteration. The first time the order of running through the mesh is by verticals, the second time the updating procedure is performed along horizontal lines. In this way (performing a *double sweep*) the convergence of the process is less dependent upon the actual boundary conditions.

After completing the iterations the program prints the values in all mesh points on the screen. This is the simplest form of output presentation. It is suggested that the user replaces the output statements (lines 450—480) with some other output

# THE FINITE DIFFERENCE METHOD

procedure, for example, one that prints the output on a printer. The output data for the nodes on the impermeable boundaries are:

$\phi = 0.000$, 0.363, 0.530, 0.686, 0.839, 1.000.

Program BV9-1 may be used as the basis for a more general program, that can account for a variable transmissivity, distributed natural replenishment, localized pumpage, and artificial recharge through wells. Such generalizations can be introduced by starting from the basic equation (4.2.23), with $\partial \phi / \partial t = 0$. This may lead to a program in which the transmissivity may be different in each rectangle of the mesh, and water can be supplied to the system in each rectangle and at each node. These generalizations are more easily and more naturally introduced, however, in the finite element method, presented in Chapter 10.

## 9.2. Unsteady Flow

The simplicity and power of the finite difference method become even more evident when considering problems of non-steady flow. Two computer programs for this class of problems are presented in this section, again only for homogeneous isotropic aquifers. For reasons of simplicity, the considerations will be restricted to meshes of equidistant mesh lines, in both directions.

The basic partial differential equation for unsteady flow in an aquifer, with the possibility of water supplied to the aquifer by infiltration, is (4.2.25). It can be rewritten in the form

$$S \frac{\partial \phi}{\partial t} = T \left( \frac{\partial^2 \phi}{\partial x^2} + \frac{\partial^2 \phi}{\partial y^2} \right) + I \quad (9.2.1)$$

where $I$ is a given source function, representing the net infiltration, due to natural or artificial replenishment of the aquifer (negative for evapotranspiration), $T$ is the transmissivity and $S$ is the storativity. The complete formulation of the problem requires the specification of boundary conditions, and, in this case of unsteady flow, of initial conditions. As in the case of steady flow, it is assumed that along part of the boundary the head is given, and that the rest of the boundary is impermeable, i.e.,

on $S_1$:    $\phi = f$, $\quad$ (9.2.2)

on $S_2$:    $Q_n = -T \dfrac{\partial \phi}{\partial n} = 0$. $\quad$ (9.2.3)

The initial condition is assumed to be

$t = 0$:    $\phi = \phi^0$ $\quad$ (9.2.4)

where $\phi^0$ is a known function, specified throughout the entire domain.

The spatial derivatives in the right-hand side of (9.2.1) can be approximated by expressions of the form (9.1.7), as in the previous section. It seems most natural to approximate the time derivative by a forward finite difference, because the problem actually is to predict future values of the head from the initial values. Accordingly, we introduce the approximation

$$\frac{\partial \phi}{\partial t} \approx \frac{\phi'_{i,j} - \phi^0_{i,j}}{\Delta t} \tag{9.2.5}$$

where $\Delta t$ is the magnitude of the time step, and $\phi'$ is the value of the head at the end of the time step.

Equation (9.2.5) contains two values of the head, at node $i, j$ of the mesh. It is not immediately evident, however, which of these values should be used in the spatial approximation. In fact, there are several possibilities, some of which will be considered in some more detail below.

### 9.2.1. EXPLICIT METHOD

The simplest choice for the approximation of the spatial derivatives is to assume that in the spatial approximation, Equation (9.1.8), all values of the head are to be considered at the initial value of time. Then, after substitution of the various approximations in (9.2.1), we obtain

$$\phi'_{i,j} = \phi^0_{i,j} + I \, \Delta t/S + \alpha(\phi^0_{i-1,j} + \phi^0_{i+1,j} - 2\phi^0_{i,j}) +$$
$$+ \beta(\phi^0_{i,j-1} + \phi^0_{i,j+1} - 2\phi^0_{i,j}) \tag{9.2.6}$$

where $\alpha$ and $\beta$ are constants defined by

$$\alpha = T \, \Delta t/[S(\Delta x)^2], \tag{9.2.7}$$
$$\beta = T \, \Delta t/[S(\Delta y)^2]. \tag{9.2.8}$$

Here $\Delta x$ and $\Delta y$ are the (constant) distances between the mesh lines in the two directions.

Equation (9.2.6) expresses the new value of the head at the mesh point $i, j$ in terms of the initial values at that node and at its immediate neighbors. As all these values are known, the process is called *explicit*. For all interior nodes (9.2.6) specifies how to determine the new values of the head.

The values along the boundary can be determined in the same way as for the steady flow case. Nodes at which the head is prescribed present no difficulty, as their value is given. For nodes on an impermeable boundary, the no-flow condition again can be expressed most simply by the use of an image node across the boundary. At such a node, the head is then set equal to the value at the node located at the same distance in the interior. Alternatively this condition can be satisfied by using an expression of the form (9.1.10) for the approximation of the Laplacian at an impervious boundary node.

# THE FINITE DIFFERENCE METHOD

An elementary computer program that performs the finite difference calculations for the case of a rectangular aquifer, with a constant head boundary, and uniform infiltration, is reproduced below, as Program BV9-2.

Again, the program uses interactive input, with self-explanatory input prompts. The program first asks for some geometrical data, which define the mesh. Then it asks for the physical parameters: the initial head (before replenishment starts), the infiltration rate, the transmissivity, and the storativity. All these parameters are considered constant throughout the region. The program then asks for a value for the time step, after printing a suggestion for this time step (based on a criterion to be presented in Section 9.3), and for the desired number of time steps.

The actual calculations are performed in lines 270—310 of the program, in which the algorithm (9.2.6) is executed for all nodes in the interior of the region. Because the head has been assumed to be constant in time along the entire boundary of the rectangular region, the values along the boundary remain unchanged. After performing the calculations, the head in the center of the aquifer is printed on the screen. To avoid a flood of data on the screen, not all heads are printed. The user may, of course, introduce other forms of output presentation.

The program can be illustrated by considering the example of a square region of

```
100 DEFINT I-N:KEY OFF:GOSUB 370
110 PRINT"---   Bear & Verruijt - Groundwater Modeling"
120 PRINT"---   Non-steady Groundwater Flow"
130 PRINT"---   Explicit Finite Differences":PRINT"---   Program 9.2"
140 PRINT"---   Homogeneous infiltration in rectangular aquifer"
150 PRINT:PRINT:DIM F(50,50),FA(50,50):TT=0
160 INPUT"Dimension in x-direction : ";XT
170 INPUT"   Subdivisions ........ : ";NX:DX=XT/NX:A=1/(DX*DX)
180 INPUT"Dimension in y-direction : ";YT
190 INPUT"   Subdivisions ........ : ";NY:DY=YT/NY:B=1/(DY*DY)
200 INPUT"Initial head ........... : ";H
210 INPUT"Infiltration rate ...... : ";P
220 INPUT"Transmissivity ......... : ";T
230 INPUT"Storativity ............ : ";S:DT=S/(2*T*(A+B))
240 PRINT"Suggestion for time step : ";DT
250 INPUT"Time step .............. : ";DT:PP=P*DT/S
260 INPUT"Number of time steps ... : ";NS:GOSUB 370
270 FOR I=0 TO NX:FOR J=0 TO NY:F(I,J)=H:FA(I,J)=H:NEXT J,I
280 AA=T*DT/S:A=A*AA:B=B*AA:II=INT(NX/2+.1):JJ=INT(NY/2+.1)
290 FOR IS=1 TO NS:TT=TT+DT:FOR I=1 TO NX-1:FOR J=1 TO NY-1
300 D=F(I-1,J)-2*F(I,J)+F(I+1,J):E=F(I,J-1)-2*F(I,J)+F(I,J+1)
310 FA(I,J)=F(I,J)+PP+A*D+B*E:NEXT J,I
320 PRINT"Time : ";:PRINT USING "#####.###";TT;
330 PRINT"   ---   Head in the center : ";
340 PRINT USING "###.###";FA(II,JJ)
350 FOR I=1 TO NX-1:FOR J=1 TO NY-1:F(I,J)=FA(I,J):NEXT J,I,IS
360 PRINT:END
370 CLS:LOCATE 1,28,1:COLOR 0,7:PRINT" Finite Differences - 2 ";
380 COLOR 7,0:PRINT:PRINT:RETURN
```

Program BV9-2. Nonsteady flow by explicit finite differences.

dimensions 100 m by 100 m, having a transmissivity $T = 10$ m/d, a storativity $S = 0.4$, and an infiltration rate of 0.001 m/d. In this case, the program suggests a time step of 1 day.

A possible mesh for this problem is shown in the left part of Figure 9.5. The data to be entered in this case are, in the order in which they have to be given: 100, 10, 100, 10, 0, 0.001, 10, 0.4, 1, 100. Here the initial head has been assumed to be zero, and the number of time steps is 100. The results for the head in the center (in mm) as a function of time are shown in the right part of the figure.

In order to verify the results, one may consider the final steady state. The solution for steady infiltration on a circular aquifer can be shown to be $\phi = IR^2/4T$, where $R$ is the radius of the aquifer. The solution for a square aquifer must be larger than the solution for a circular aquifer having the inscribed circle as its perimeter, and smaller than the one for the circumscribed circle. These radii are 50 and $50\sqrt{2}$ m, respectively. For these cases, one obtains $\phi = 62.5$ mm, and $\phi = 125$ mm, respectively. The value of 73 mm obtained by running Program BV9-2 indeed is between these two limits. That it is closer to the solution for the inscribed circle seems to be physically reasonable.

Another verification is that in the beginning of the process the water level increases linearly with time, except near the boundary, according to the formula $\phi \approx It/S$. This is the solution in the absence of drainage to the boundaries. It can be expected that in the beginning, the head at and near the center will increase according to this formula, because drainage has not yet been developed in this region. The numerical solution indeed shows exactly this behaviour. The reader should verify this by running the program.

It may be noted that Program BV9-2 can be used to solve problems for a very large network, even on a small computer. The dimension statements in line 150 state that the network may consist of 50 × 50 elementary squares. In that case, the number of mesh points is 51 × 51 = 2601. Thus the program actually solves a system of 2601 equations with 2601 unknowns, in each time step. The reader may verify that many computers will accommodate even larger networks. The analysis

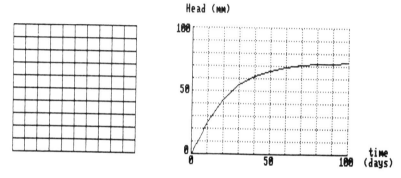

Fig. 9.5. Example of results obtained by Program BV9-2.

# THE FINITE DIFFERENCE METHOD

of such a system on a small computer may require a rather long computation time. On a system based on the 8088 processor, computation time for a network of 10 × 10 squares (Figure 9.5) is about 5 s per time step. As computation time depends approximately linearly on the number of mesh points, it follows that computation time may well be the limiting factor for the solution of large problems on a small computer. A considerable improvement can be obtained by using a compiler which transforms the BASIC program into faster code. The reduction in computation time can be as large as a factor 20. Alternatively, one may translate the program into a computer language that executes faster, such as FORTRAN or C, or use a computer based on a more powerful processor.

### 9.2.2. IMPLICIT METHOD

The time steps in Program BV9-2 are restricted by a stability criterion (see Section 9.3). This implies that even when, in a transient process, the behavior is very slow, for instance, for large values of time, when approaching the steady state, the magnitude of the time steps must remain small. This means that a large number of steps is required. This situation can be improved by using a more sophisticated procedure, as used in the *implicit method*.

In the analysis leading to Program BV9-2, the approximations of the spatial derivatives have been defined at the beginning of the time step (see Equation (9.2.6)). As an alternative approach, we may take them at the end of the time step, halfway through the time step, or, in general, at some intermediate point, such that

$$\phi_{i,j} = \varepsilon \phi^0_{i,j} + (1 - \varepsilon) \phi'_{i,j} \qquad (9.2.9)$$

where $\varepsilon$ is an *interpolation parameter*, having a value between 0 and 1. For $\varepsilon = 1$ the value of $\phi_{i,j}$ reduces to the initial value. This leads to (9.2.6). For $\varepsilon = 0$ the value of $\phi_{i,j}$ equals the value at the end of the time step, $\phi'_{i,j}$. In that case substituting the approximations into the basic equation (9.2.1) gives

$$\phi'_{i,j} = \phi^0_{i,j} + I\,\Delta t/S + \alpha(\phi'_{i-1,j} + \phi'_{i+1,j} - 2\phi'_{i,j}) +$$
$$+ \beta(\phi'_{i,j-1} + \phi'_{i,j+1} - 2\phi'_{i,j}). \qquad (9.2.10)$$

This is the basic equation for the *fully implicit* method. The new value at mesh point $i, j$ is expressed in terms of the initial value at that node and the new values at the four nodes surrounding that node. Because these values are also unknown, it follows that the process is no longer explicit, but rather all unknown values must be determined simultaneously from a system of linear equations of the form (9.2.10). The success of the iterative Gauss–Seidel (or relaxation) method used for the solution of steady-state problems in Section 9.1 suggests that (9.2.10) can be formulated as an equation for $\phi'_{i,j}$, in the form

$$\phi'_{i,j} = [\phi^0_{i,j} + I\,\Delta t/S + \alpha(\phi'_{i-1,j} + \phi'_{i+1,j}) +$$
$$+ \beta(\phi'_{i,j-1} + \phi'_{i,j+1})]/(1 + 2\alpha + 2\beta). \qquad (9.2.11)$$

The values of $\phi'_{i,j}$ can be determined iteratively, on the basis of an initial estimate. A convenient estimate is the value at the beginning of the time interval.

The iterative procedure necessary to solve the system of equations defined by the algorithm (9.2.11) means that the number of computations per time step is greatly increased, compared to the explicit method presented above. This is balanced, however, by the fact that the process is stable for all sizes of the time step. The process is said to be *unconditionally stable*. Thus, the time steps can be taken larger when the variations in the process become slower, for instance, when a steady state is approached.

Program BV9-3 performs the calculations for the fully implicit method. This computer program is applicable to the same class of problems as Program BV9-2, viz. homogeneous infiltration on the rectangular region, bounded by a boundary of

```
100 DEFINT I-N:KEY OFF:GOSUB 460
110 PRINT"---   Bear & Verruijt - Groundwater Modeling"
120 PRINT"---   Non-steady Groundwater Flow"
130 PRINT"---   Implicit Finite Differences":PRINT"---   Program 9.3"
140 PRINT"---   Homogeneous infiltration in rectangular aquifer"
150 PRINT:PRINT:DIM F(50,50),FA(50,50):TT=0
160 INPUT"Dimension in x-direction : ";XT
170 INPUT"    Subdivisions ........ : ";NX:DX=XT/NX:A=1/(DX*DX)
180 INPUT"Dimension in y-direction : ";YT
190 INPUT"    Subdivisions ........ : ";NY:DY=YT/NY:B=1/(DY*DY)
200 INPUT"Initial head ........... : ";H
210 INPUT"Infiltration rate ...... : ";P
220 INPUT"Transmissivity ......... : ";T
230 INPUT"Storativity ............ : ";S:DT=S/(2*T*(A+B))
240 PRINT"Suggestion for time step :  ";DT
250 INPUT"Time step .............. : ";DT
260 INPUT"Number of time steps ... : ";NS
270 INPUT"Number of iterations ... : ";NI
280 INPUT"Relaxation factor ...... : ";RX:GOSUB 460
290 FOR I=0 TO NX:FOR J=0 TO NY:F(I,J)=H:FA(I,J)=H:NEXT J,I
300 A=T*A:B=T*B:FF=2*A+2*B+S/DT:II=INT(NX/2+.1):JJ=INT(NY/2+.1)
310 FOR IS=1 TO NS:TT=TT+DT
320 FOR IT=1 TO NI:FOR I=1 TO NX-1:FOR J=1 TO NY-1
330 C=A*(FA(I-1,J)-2*FA(I,J)+FA(I+1,J))
340 D=B*(FA(I,J-1)-2*FA(I,J)+FA(I,J+1))
350 AA=P+C+D-S*(FA(I,J)-F(I,J))/DT:FA(I,J)=FA(I,J)+RX*AA/FF
360 NEXT J,I:FOR J=1 TO NY-1:FOR I=1 TO NX-1
370 C=A*(FA(I-1,J)-2*FA(I,J)+FA(I+1,J))
380 D=B*(FA(I,J-1)-2*FA(I,J)+FA(I,J+1))
390 AA=P+C+D-S*(FA(I,J)-F(I,J))/DT:FA(I,J)=FA(I,J)+RX*AA/FF
400 NEXT I,J,IT
410 FOR I=1 TO NX-1:FOR J=1 TO NY-1:F(I,J)=FA(I,J):NEXT J,I
420 PRINT"Time : ";:PRINT USING "#######.###";TT;
430 PRINT"    ---   Head in the center : ";
440 PRINT USING "###.###";F(II,JJ):IF IS>1 THEN DT=2*DT
450 NEXT IS:END
460 CLS:LOCATE 1,28,1:COLOR 0,7:PRINT" Finite Differences - 3 ";
470 COLOR 7,0:PRINT:PRINT:RETURN
```

Program BV9-3. Nonsteady flow by implicit finite differences.

constant head. The response to the input prompts us to solve the problem illustrated in Figure 9.5 should be: 100, 10, 100, 10, 0, 0.001, 10, 0.4, 1, 10, 20, 1.5, if the number of iterations in each time step is 20, and the relaxation factor is 1.5. In Program BV9-3, the length of the time steps is continuously doubled, so that after 10 time steps the value of the time parameter is 512. This is equivalent to 512 time steps in the explicit process. By running the program with these data, the same results are obtained as from Program BV9-2 (see Figure 9.5). Although the program can be run with larger time steps than those needed for stability in Program BV9-2, thus making the total number of time steps considerably smaller, the total computation time may be as large or larger than for that program, because of the iterations needed in each time step.

Instead of taking $\varepsilon = 0$ in (9.2.5), which leads to the fully implicit algorithm (9.2.11), one might take $\varepsilon = \frac{1}{2}$, which seems to be least biased towards either the initial value or the final value in a time step. This will result in a more accurate formulation, known as the *Crank–Nicholson scheme*. This is unconditionally stable, and very accurate (see Section 9.3). This alternative will not be elaborated here, as its advantages, compared to the fully implicit method, are not too large. The fully implicit method has the important advantage that it can also be used to study steady flow problems. This can be seen from the fact that the basic equation (9.2.10) reduces to the steady-state equivalent in the limit $S \to 0$. Thus, one can use a single program for both unsteady and steady flow problems.

## 9.3. Accuracy and Stability

An approximate numerical solution will never yield the exact solution of a problem and, in general, it is impossible to precisely evaluate its accuracy, if only for the reason that the exact solution is usually unknown. Nevertheless, some insight on the accuracy of a method can still be obtained, as will be demonstrated here. Special attention will be paid to the magnitude of the time steps in an unsteady process.

A numerical process must satisfy certain conditions in order to be reliable. These are usually denoted as the conditions of *consistency, convergence*, and *stability*. Consistency is the requirement that the numerical equations should reduce to the exact continuum equations when all finite intervals approach zero. All procedures considered in this chapter are indeed consistent, as can be seen from an inspection of the basic algebraic equations, by letting the finite differences approach zero.

It is much more difficult to verify the condition of convergence. This condition states that the solution of the numerical equations should approach that of the original partial differential equation if all finite intervals tend to zero. This can be shown in a general way only in some simple cases, such as one-dimensional problems, for which the numerical solution can be expressed in closed form. In many applications in engineering practice, it is impossible to prove convergence in a rigorous way. Therefore, it is usually considered sufficient if the numerical procedure has been verified against a variety of analytical solutions. For the procedures of the preceding sections, this has been done for a large number of cases, some of which have been included in the text. Thus, convergence of the procedures may be assumed, at least in principle.

### 9.3.1. STABILITY

A necessary condition for convergence is that errors, for instance those due to round-off, do not increase in magnitude with time. This is referred to as the *stability condition*. It is an extremely important condition that imposes certain restrictions on the size of the time step in an explicit process. A comprehensive discussion of stability of finite difference equations is beyond the scope of this book. Therefore, only some elementary cases will be considered, which apply to the most important problems of groundwater flow.

The first case to be considered is the explicit method for unsteady flow problems, described by (9.2.6),

$$\phi'_{i,j} = \phi^0_{i,j} + I\,\Delta t/S + \alpha(\phi^0_{i-1,j} + \phi^0_{i+1,j} - 2\phi^0_{i,j}) + \\ + \beta(\phi^0_{i,j-1} + \phi^0_{i,j+1} - 2\phi^0_{i,j}) \tag{9.3.1}$$

where $\alpha$ and $\beta$ are defined by

$$\alpha = T\,\Delta t/[S(\Delta x)^2], \tag{9.3.2}$$

$$\beta = T\,\Delta t/[S(\Delta y)^2]. \tag{9.3.3}$$

If this process is to be stable, any distribution of errors should gradually dissipate in time, and should certainly not grow in magnitude. As the system of equations is linear, it is sufficient to investigate the propagation of a certain error distribution, considered as a deviation from the particular solution $\phi = 0$ of the homogeneous equation (with $I = 0$). In order to maximize the effect of all terms in the right-hand side of (9.3.1), it is assumed that at a certain time the errors are

$$\phi^0_{i-1,j} = \phi^0_{i+1,j} = \phi^0_{i,j-1} = \phi^0_{i,j+1} = -\varepsilon,\ \phi^0_{i,j} - \varepsilon. \tag{9.3.4}$$

From (9.3.1), with $I = 0$, we now obtain

$$\phi'_{i,j} = (1 - 4\alpha - 4\beta)\varepsilon. \tag{9.3.4}$$

In order that the errors will not grow, this must be smaller than $\varepsilon$, and larger than $-\varepsilon$. Otherwise each error would be larger than the previous one, and they would grow without limit in time. With (9.3.2) and (9.3.3) this leads to the following condition for the time step $\Delta t$

$$0 < \Delta t < \frac{1}{2}\frac{S}{T}\frac{(\Delta x)^2(\Delta y)^2}{(\Delta x)^2 + (\Delta y)^2}. \tag{9.3.5}$$

# THE FINITE DIFFERENCE METHOD

The first condition, which states that $\Delta t$ must be positive, is trivial. The second condition, however, is very restrictive. It provides an upper limit for the magnitude of the time step in relation to the spatial intervals $\Delta x$ and $\Delta y$. It must be expected that when this condition is violated, the process will be unstable: errors (which can never be completely avoided, for instance, due to round-off) will gradually increase in magnitude, and after a number of time steps, the deviations from the true solution may become very large. Such a behavior is indeed observed if the time steps are taken too large.

Condition (9.3.5) is a necessary condition for stability. According to its present derivation, based on a particular distribution of errors, it is not certain that another combination of errors will not give a more severe condition. Nevertheless, more refined considerations (see, e.g., Forsythe and Wasow, 1960; Fox, 1962), which are usually based on the representation of the error distribution by a Fourier series, lead to the same condition.

The stability criterion (9.3.5) has been incorporated in Program BV9-2, where it is used to suggest a value for the time step to the user. In Program BV9-3, it is used to suggest a value for the first time step.

The implicit process presented in Section 9.2 is unconditionally stable, which means that for all (positive) values of the time step errors will dissipate in time. This can be demonstrated by showing that the amplitude of any component of a Fourier series will decrease in time. The simplest case is the one-dimensional equivalent of (9.2.10), in the absence of infiltration,

$$\phi'_{i,j} = \phi^0_{i,j} + \alpha(\phi'_{i-1,j} + \phi'_{i+1,j} - 2\phi'_{i,j}). \tag{9.3.6}$$

Now consider a component of the error which can be described by

$$\phi^0 = A \exp(i\omega x), \qquad \phi' = B \exp(i\omega x) \tag{9.3.7}$$

where $\omega$ is the frequency of this component of the Fourier series representation of the error, $A$ is its initial amplitude, and $B$ is its amplitude after the time step. Substituting (9.3.7) into (9.3.6) gives

$$\frac{A}{B} = 1 + 2\alpha[1 - \cos(\omega \Delta x)]. \tag{9.3.8}$$

This is always greater than 1, for all values of the frequency $\omega$ or of the dimensionless time step $\alpha$. Hence, we may conclude that the process is indeed always stable. For considerations of stability for more general processes, the reader is referred to the literature on the mathematical foundations of the finite difference method (e.g., Forsythe and Wasow, 1960; Fox, 1962).

### 9.3.2. ACCURACY

If a numerical process is consistent, convergent, and stable, it is still possible that the deviations from the true exact solution of the original partial differential equation are very large. Intuitively, this can be expected to be the case if a process with rapid variations is approximated by a coarse mesh, that is incapable of representing the local variations. Thus, the *accuracy* of the numerical solution deserves some attention.

In order to investigate the parameters influencing accuracy of a numerical solution, the one-dimensional finite difference representation of the steady flow problem will be considered. Let the numerical equivalent of the differential equation $d^2F/dx^2 = 0$ be determined by the use of a Taylor series expansion, see (9.1.11) and (9.1.12). If it is assumed that the intervals on either side of a nodal point are equal, and if the series are continued up to the fourth derivative, we obtain

$$F(x+h) = F(x) + hF'(x) + \frac{h^2}{2}F''(x) + \frac{h^3}{6}F'''(x) + \frac{h^4}{24}F''''(x) + \cdots$$

$$F(x-h) = F(x) - hF'(x) + \frac{h^2}{2}F''(x) - \frac{h^3}{6}F'''(x) + \frac{h^4}{24}F''''(x) + \cdots$$

where $h$ is the spatial interval, $h = \Delta x$. By adding the two equations, we obtain

$$F''(x) = \frac{F(x+h) - 2F(x) + F(x-h)}{h^2} - \frac{h^2}{12}F''''(x) + \cdots \quad (9.3.9)$$

This equation shows that the numerical expression for the second derivative that has generally been used in the programs entails an error which is proportional to the second power of the mesh size, and to the fourth-order derivative of the function itself. The value of the fourth derivative depends on the problem itself, and may be large at, or close to isolated singular points, e.g., corner points in the flow domain, or pumping wells. It cannot be influenced by the modeler. Nevertheless, the accuracy can be increased by taking smaller space steps, as seems natural. It is important to note that the error is proportional to the second power of the mesh size, and not to the mesh size itself. This is due to the fact that the coefficient of the third derivative in Equation (9.3.9) is zero.

The coefficient of the third derivative does not vanish when the mesh sizes on the two sides of a node are different. This shows that the gain in flexibility when using a variable mesh size is partly balanced by a loss of accuracy.

## 9.4. Generalizations

The finite difference method has been presented in this chapter for the simplest two-dimensional case only: a homogeneous isotropic aquifer, bounded by straight boundaries, parallel to the $x$- and $y$-axes. Without much difficulty, the method can be extended to more general types of problems. Some of these generalizations are briefly discussed in this section.

# THE FINITE DIFFERENCE METHOD

## 9.4.1. ANISOTROPY

The basic differential equation for steady flow without replenishment, in a homogeneous anisotropic aquifer, is rewritten here for convenience from (4.2.24), in the form

$$T_x \frac{\partial^2 \phi}{\partial x^2} + T_y \frac{\partial^2 \phi}{\partial y^2} = 0. \tag{9.4.1}$$

The finite difference approximation of this equation can easily be obtained by employing the approximations (9.1.14) and (9.1.15). This leads to an algorithm of the form (9.1.16), except that the coefficients $A_i$ and $B_i$ must be multiplied by $T_x$, and the coefficients $C_j$ and $D_j$ must be multiplied by $T_y$. This can easily be taken into account in a computer program.

## 9.4.2. NONHOMOGENEITY

The incorporation of nonhomogeneous soil properties requires some more effort. If, for reasons of simplicity, it is assumed that the aquifer is isotropic, the basic differential equation describing steady flow without replenishment is

$$\frac{\partial}{\partial x}\left(T \frac{\partial \phi}{\partial x}\right) + \frac{\partial}{\partial y}\left(T \frac{\partial \phi}{\partial y}\right) = 0. \tag{9.4.2}$$

In order to find a rational approximation of the terms in this equation by finite differences, we must first decide how to define the variable transmissivity, $T$, and the head, $\phi$. In the previous sections, the head was always defined at the nodes of a rectangular grid. It seems most natural to consider the transmissivity to be defined in these same points. The finite difference approximation of the first derivative in the $x$-direction to the right and to the left of node $i, j$ can then be written as follows, if the mesh size is constant,

$$T \frac{\partial \phi}{\partial x} \approx \tfrac{1}{2}(T_{i,j} + T_{i+1,j}) \frac{\phi_{i+1,j} - \phi_{i,j}}{\Delta x},$$

$$T \frac{\partial \phi}{\partial x} \approx \tfrac{1}{2}(T_{i-1,j} + T_{i,j}) \frac{\phi_{i,j} - \phi_{i-1,j}}{\Delta x}.$$

This leads to the following approximation of the second derivative

$$\frac{\partial}{\partial x}\left(T \frac{\partial \phi}{\partial x}\right) \approx A_i f_{i+1,j} - (A_i + B_i) f_{i,j} + B_i f_{i-1,j} \tag{9.4.3}$$

where

$$A_i = \tfrac{1}{2}(T_{i+1,j} + T_{i,j})/(\Delta x)^2, \qquad (9.4.4)$$

$$B_i = \tfrac{1}{2}(T_{i,j} + T_{i-1,j})/(\Delta x)^2. \qquad (9.4.5)$$

In the same way, the second derivative in the $y$-direction can be approximated by

$$\frac{\partial}{\partial y}\left(T\frac{\partial \phi}{\partial y}\right) \approx C_j f_{i,j+1} - (C_j + D_j)f_{i,j} + D_j f_{i,j-1} \qquad (9.4.6)$$

where

$$C_j = \tfrac{1}{2}(T_{i,j+1} + T_{i,j})/(\Delta y)^2, \qquad (9.4.7)$$

$$D_j = \tfrac{1}{2}(T_{i,j} + T_{i,j-1})/(\Delta y)^2. \qquad (9.4.8)$$

The final approximation of the partial differential Equation (9.4.2) is thus

$$(A_i + B_i + C_j + D_j)\phi_{i,j}$$
$$= A_i \phi_{i+1,j} + B_i \phi_{i-1,j} + C_j \phi_{i,j+1} + D_j \phi_{i,j-1}. \qquad (9.4.9)$$

where the coefficients are given by (9.4.4), (9.4.5), (9.4.7), and (9.4.8).

Equation (9.4.9) is of precisely the same form as (9.1.16), which was the basic finite difference equation for the homogeneous case. The only modification resulting from the nonhomogeneity of the aquifer is that the coefficients are different. The generalizations to an anisotropic nonhomogeneous aquifer, and to meshes with variable intervals between the mesh lines are left to the reader.

### 9.4.3. INTEGRATED FINITE DIFFERENCES

An interesting alternative to the method described above for nonhomogeneous aquifers is the method of *integrated finite differences* (Narasimhan and Witherspoon, 1976). In this method, the domain is subdivided into small subdomains or *elements* of arbitrary shape (Figure 9.6), and the basic set of algebraic equations is obtained by considering the mass balance for every subdomain. The groundwater head is most conveniently defined at the centroid of the element, and the fluxes are calculated by a finite difference approximation of the first derivative of the head. The transmissivity is also defined for each element, and in calculating the fluxes the harmonic mean transmissivity is used. An advantage of this method is that the approximation of the domain is more flexible, with smaller elements in certain regions and larger ones in others. Another advantage is that it is relatively easy to take into account supply or withdrawal of water over each element, as the basic equation is nothing but a water balance.

The method of integrated finite differences closely resembles the finite element method, which will be presented in great detail in Chapter 10. For that reason the

# THE FINITE DIFFERENCE METHOD

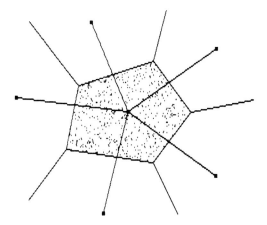

Fig. 9.6. Integrated finite differences.

method of integrated finite differences will not be presented in detail here. The main difference between the two methods is in the derivation of the equations.

### 9.4.4. CURVED BOUNDARIES

The finite difference method in its standard form uses a mesh of squares, or a mesh of rectangles. This means that curved boundaries cannot be represented accurately (Figure 9.7). Of course, when the mesh is sufficiently fine any curved boundary can be approximated rather well by straight line segments, but this may be a rather costly procedure, in terms of computer memory and computation time. For this reason, a local modification of the mesh is sometimes used, in which the nodes near the boundary are moved to that boundary, and a special numerical equation is derived for that particular node, to approximate the boundary condition. This can be done by using the Taylor series expansion method to

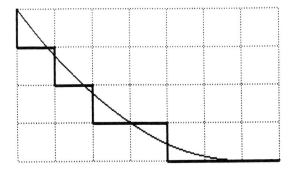

Fig. 9.7. Approximation of curved boundary.

approximate the partial derivatives in a mesh of irregular points (Forsythe and Wasow, 1960; Fox, 1962).

Curved boundaries can be incorporated without any difficulty in the finite element method. Therefore, problems in which the curvature of the boundaries is considered essential are usually solved by the finite element method.

CHAPTER TEN

# The Finite Element Method

In this chapter the principles of the *finite element method* will be presented through the application to problems of steady and nonsteady groundwater flow. The method was developed in the 1950's, first for problems of aeronautical engineering (the construction of airplanes), mechanical engineering (nuclear reactor vessels), and civil engineering (bridges). In later years, the method was generalized to practically all areas of engineering, including groundwater flow, where the solution to field equations is required.

The finite element method can be presented in several ways. In many of the original applications the *finite elements* were distinct parts of the structure, such as a beam in a frame. The mathematical problem to be solved could be formulated on the basis of such elementary principles as equilibrium of a single element. Later it was found that the basic concepts could conveniently be formulated by using a *variational formulation* of the original mathematical problem. This also made it possible to generalize the applicability of the method to a wide class of field problems in engineering science. A further generalization has been obtained by formulating the basic principles in terms of *weighted residuals*. This approach will be followed here, using the simplest type of element, a triangle. Some simple but powerful computer programs will also be given for various types of problems such as steady and nonsteady groundwater flow.

## 10.1. Steady Flow

Consider the steady flow of groundwater in the leaky aquifer of transmissivity $T$, shown in Figure 10.1. The aquifer is phreatic, with leakage taking place into it

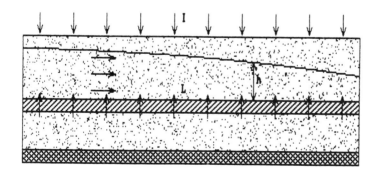

Fig. 10.1. Leaky phreatic aquifer.

from the lower aquifer, in which the head is assumed to be given. The basic differential equation is (4.2.5). For steady flow, this equation can be written as

$$\frac{\partial}{\partial x}\left(T\frac{\partial \phi}{\partial x}\right) + \frac{\partial}{\partial y}\left(T\frac{\partial \phi}{\partial y}\right) + I - \frac{\phi - \phi'}{c} = 0 \qquad (10.1.1)$$

where $I$ is the net infiltration or replenishment (assumed to be known), $\phi$ is the head in the aquifer, $\phi'$ is the (given) head in the lower aquifer, and $c$ is the resistance of the aquiclude separating the two aquifers. The last term on the left-hand side of (10.1.1) represents the leakage into the aquifer. The transmissivity in the upper aquifer is $T = kh$, where $h$ is the variable thickness of the layer of water which may be identified with the groundwater head. For the moment, we shall assume that the variations in groundwater level in the leaky phreatic aquifer are so small that the transmissivity $T$ may be considered as *a-priori* known, at least as a first approximation. It will appear later that a time-dependent value of the transmissivity of a phreatic aquifer can be incorporated iteratively into the calculations by updating the value of $T$. Thus, (10.1.1) may be considered to be valid, approximately, for a phreatic aquifer, as well as a confined one.

Equation (10.1.1) must be satisfied throughout a specified region $R$ in the $x$, $y$-plane. Boundary conditions must be specified on the boundary of that region. A general formulation of the boundary conditions is

on $S_1$: $\quad \phi = f,$ \hfill (10.1.2)

on $S_2$: $\quad T\dfrac{\partial \phi}{\partial n} = qh,$ \hfill (10.1.3)

where $S_1$ and $S_2$ are boundary segments, which together constitute the entire boundary of the region $R$. On $S_1$ the head is given, while on $S_2$ the groundwater flux normal to the boundary is prescribed.

In the finite element method, the region $R$, in which the flow takes place, is subdivided into a large number of small elements, in each of which the groundwater head is approximated by some simple function. The simplest way of subdividing the region is by using triangular elements (Figure 10.2). Such elements make it possible to closely follow natural (curved) boundaries. This kind of subdivision also facilitates using a dense mesh in subregions of great interest, and a coarse mesh in areas where the flow is of less interest. An altenative way is to use quadrangular elements of an arbitrary shape. Such elements will be introduced at a later stage, by a suitable combination of triangles.

The simplest way of approximating the variation of the head within a triangular element is by assuming that the head varies linearly within each element (Figure 10.3). The piezometric surface is thus approximated by a diamond-shaped surface,

# THE FINITE ELEMENT METHOD

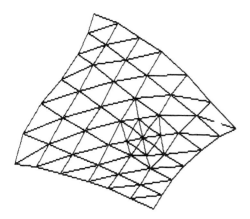

Fig. 10.2. Domain $R$ divided into triangular elements.

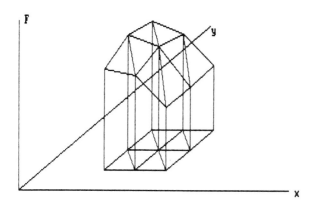

Fig. 10.3. Linear interpolation of head within elements.

such that in each element the head is represented by a planar surface. The surface generated by such small planar elements, defined by the values at the nodal points is a *continuous* surface; slopes are discontinuous across the element boundaries. The groundwater head at a point inside an element is defined by a linear interpolation between the values at the mesh points, the *nodes*.

Formally, the piezometric head $\phi$ throughout the entire region can be expressed by

$$\phi = \sum_{i=1}^{n} N_i(x, y)\phi_i, \tag{10.1.4}$$

where $\phi_i$ is the head at node $i$, and $N_i$ is a *shape function*, or *base function*, defined by

$$N_j = 1, \quad \text{if } j = i, \qquad N_j = 0, \quad \text{if } j \neq i \tag{10.1.5}$$

with linear interpolation within each element. A typical shape function is shown in Figure 10.4.

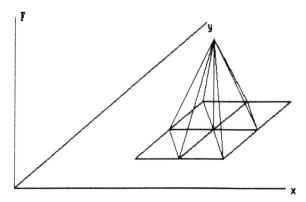

Fig. 10.4. A typical shape function.

In (10.1.4), values of $\phi_i$ are unknown when $i$ denotes an interior node, or when node $i$ is located on an $S_2$-type boundary segment. If node $i$ is located on an $S_1$-type boundary segment, the value of $\phi_i$ is known.

The interpolation function (10.1.4) can also be used for the known head in the lower aquifer, so that one may write

$$\phi' = \sum_{i=1}^{n} N_i(x, y)\phi'_i \qquad (10.1.6)$$

where $\phi'_i$ is the known value of the head in the lower aquifer, at node $i$.

In general, the approximation (10.1.4) will not exactly satisfy the partial differential equation (10.1.1). Therefore, this condition is relaxed by requiring that the differential equation be satisfied only *on the average*, using a number of *weight functions*, equal to the number of unknowns. This is called the *method of weighted residuals* (Zienkiewicz, 1977). It is most convenient to use the shape functions $N_i$ as weight functions. This leads to the conditions

$$\int_R \left\{ \left[ \frac{\partial}{\partial x}\left( T \frac{\partial \phi}{\partial x} \right) + \frac{\partial}{\partial y}\left( T \frac{\partial \phi}{\partial y} \right) + I - \frac{\phi - \phi'}{c} \right] N_i \right\} dx\, dy = 0$$

$(i \in C)$ \hfill (10.1.7)

to be satisfied for each value of $i$ for which $\phi_i$ is unknown. These values of $i$

# THE FINITE ELEMENT METHOD 251

constitute a certain class, denoted by $C$. Satisfaction of (10.1.7) for all $i \in C$ will lead to just as many equations as there are unknown values.

The integral in (10.1.7) can be separated into two parts by noting that

$$\left[\frac{\partial}{\partial x}\left(T\frac{\partial \phi}{\partial x}\right) + \frac{\partial}{\partial y}\left(T\frac{\partial \phi}{\partial y}\right)\right]N_i$$

$$= \frac{\partial}{\partial x}\left[N_i T \frac{\partial \phi}{\partial x}\right] + \frac{\partial}{\partial y}\left[N_i T \frac{\partial \phi}{\partial y}\right] -$$

$$- T\frac{\partial N_i}{\partial x}\frac{\partial \phi}{\partial x} - T\frac{\partial N_i}{\partial y}\frac{\partial \phi}{\partial y}. \tag{10.1.8}$$

Substituting (10.1.8) into (10.1.7), with (10.1.4) and (10.1.6), gives

$$J_1 + J_2 + J_3 = 0 \quad (i \in C) \tag{10.1.9}$$

where $J_1, J_2$ and $J_3$ are three integrals, defined, for all $i \in C$, by

$$J_1 = \int_R \left\{\frac{\partial}{\partial x}\left[N_i T \frac{\partial \phi}{\partial x}\right] + \frac{\partial}{\partial y}\left[N_i T \frac{\partial \phi}{\partial y}\right]\right\} dx\, dy \quad (i \in C), \tag{10.1.10}$$

$$J_2 = -\int_R \left\{T \sum_j \phi_j \left[\frac{\partial N_i}{\partial x}\frac{\partial N_j}{\partial x} + \frac{\partial N_i}{\partial y}\frac{\partial N_j}{\partial y}\right]\right\} dx\, dy \quad (i \in C), \tag{10.1.11}$$

$$J_3 = \int_R \left\{IN_i - \frac{1}{c}\sum_j N_i N_j (\phi_j - \phi'_j)\right\} dx\, dy \quad (i \in C). \tag{10.1.12}$$

The summation in the second and third integrals should be performed over all values of $j$ from $j = 1$ to $j = n$, where $n$ is the number of nodes. Equation (10.1.9) is the basic equation of the method of finite elements. Each of the three integrals will now be evaluated separately.

## 10.1.1. THE FIRST INTEGRAL, $J_1$

The first integral, $J_1$, as expressed by (10.1.10), can be transformed into a line integral along the boundary $S$ of the region $R$ by the so-called *divergence theorem* (or *Gauss's theorem*). This gives

$$J_1 = \int_R \left\{ \frac{\partial}{\partial x}\left[ N_i T \frac{\partial \phi}{\partial x} \right] + \frac{\partial}{\partial y}\left[ N_i T \frac{\partial \phi}{\partial y} \right] \right\} dx\, dy$$

$$= \int_S \left\{ N_i T \frac{\partial \phi}{\partial n} \right\} dS \quad (i \in C). \tag{10.1.13}$$

Because in (10.1.13) the values of $i$ are restricted to those belonging to the class $C$, i.e., the class of node numbers in which the value of the head is unknown, the values of $i$ for all points on the boundary segment $S_1$ are excluded, see (10.1.2). Thus, in the integral on the right-hand side of (10.1.13) the values of $i$ are restricted to points located on the boundary $S_2$. The corresponding shape functions $N_i$ are zero on $S_1$ and, therefore, only a contribution from the integral along $S_2$ remains. On that part of the boundary, the value of $T(\partial \phi/\partial n)$ is known, see (10.1.2). Let the value of the supply function along a typical boundary element be denoted by $q_k h$, and the length of that part of the boundary by $L_k$. Contributions to the integral in (10.1.13) can be expected only from those parts of the boundary on which $N_i$ is different from zero. This is the case only along the two boundary elements on the two sides of node $i$. Along these two line segments the average value of $N_i$ is $\frac{1}{2}$ and, thus, the integral will be the sum of two values $\frac{1}{2} q_k h L_k$, one to the left and one to the right of node $i$. This sum will be denoted by $Q_i$. Physically, this means that the total water supply along a boundary element is attributed evenly to the two nodes at its ends. Hence,

$$J_1 = \int_R \left\{ \frac{\partial}{\partial x}\left[ N_i T \frac{\partial \phi}{\partial x} \right] + \frac{\partial}{\partial y}\left[ N_i T \frac{\partial \phi}{\partial y} \right] \right\} dx\, dy = Q_i$$

$$(i \in C) \tag{10.1.14}$$

where $Q_i$ is the amount of water supplied to the system at node $i$. The water supply along any part of the boundary segment $S_2$, an element side identified by a number $k$, with length $L_k$, is divided evenly among the nodes on its two ends. This means that the corresponding discharge $Q_i$ in (10.1.14) can be expressed as

$$Q_i = \tfrac{1}{2} q_k h L_k. \tag{10.1.15}$$

It seems natural to assume that any concentrated water supply at a node on the boundary can be added to the value of $Q_i$. This enables us to introduce wells.

## 10.1.2. THE SECOND INTEGRAL, $J_2$

The second integral, $J_2$ defined by (10.1.11), can be written formally as

$$J_2 = -\int_R \left\{ T \sum_j \phi_j \left[ \frac{\partial N_i}{\partial x} \frac{\partial N_j}{\partial x} + \frac{\partial N_i}{\partial y} \frac{\partial N_j}{\partial y} \right] \right\} dx\, dy$$

$$= -\sum_j P_{ij} \phi_j, \tag{10.1.16}$$

where the summation in the right-hand side means summation over all elements $R_p$ included in the domain $R$, and where

$$P_{ij} = \int_{R_p} \left\{ T \left[ \frac{\partial N_i}{\partial x} \frac{\partial N_j}{\partial x} + \frac{\partial N_i}{\partial y} \frac{\partial N_j}{\partial y} \right] \right\} dx\, dy \quad (i \in C). \tag{10.1.17}$$

In order to evaluate this integral, we first note that contributions to this integral can be expected only if element $R_p$ contains both nodes $i$ and $j$. If either node $i$ or node $j$ do not belong to this element, one of the shape functions is zero and, thus, no contribution to the integral is made. Thus restriction can be made to elements containing both nodes $i$ and $j$.

Because the shape functions are linear, we may write

$$N_i(x, y) = p_i x + q_i y + r_i, \qquad N_j(x, y) = p_j x + q_j y + r_j. \tag{10.1.18}$$

where the coefficients $p_i$, $p_j$, etc., are constants. Let the three nodes of the element $R_p$ be denoted by $j$, $k$ and $l$ (where $i$ may be either $j$, $k$ or $l$). Then $N_j$ must be 1 in node $j$, and 0 in nodes $k$ and $l$, hence,

$$p_j x_j + q_j y_j + r_j = 1, \qquad p_j x_k + q_j y_k + r_j = 0, \qquad p_j x_l + q_j y_l + r_j = 0.$$

This is a system of three linear equations with three unknowns, $p_j$, $q_j$ and $r_j$. The solution of this system is

$$p_j = b_j/D, \qquad q_j = c_j/D, \qquad r_j = d_j/D \tag{10.1.19}$$

where

$$b_j = y_k - y_l, \qquad b_k = y_l - y_j, \qquad b_l = y_j - y_k,$$
$$c_j = x_l - x_k, \qquad c_k = x_j - x_l, \qquad c_l = x_k - x_j,$$
$$d_j = x_k y_l - x_l y_k, \qquad d_k = x_l y_j - x_j y_l, \qquad d_l = x_j y_k - x_k y_j,$$
$$D = x_j b_j + x_k b_k + x_l b_l. \tag{10.1.20}$$

The quantity $D$ represents the determinant of the system of equations. The other quantities are sub-determinants of the system.

For the evaluation of the integral (10.1.17), the quantities needed are

$$\frac{\partial N_i}{\partial x} = p_i, \qquad \frac{\partial N_j}{\partial x} = p_j, \qquad \frac{\partial N_i}{\partial y} = q_i, \qquad \frac{\partial N_j}{\partial x} = q_j.$$

It is now assumed that the transmissivity $T$ is constant, $T_p$, throughout the element $R_p$. This assumption can be made without much loss of generality, because the subdivision of the region into elements can always be done such that, in each element, the transmissivity is practically constant. Then we obtain

$$P_{ij} = T_p A_p \{b_i b_j + c_i c_j\}/D^2$$

where $A_p$ is the area of the element $R_p$. This area can be expressed into the coordinates of the nodal points of the element by several standard formulas. A suitable formula is $A_p = \frac{1}{2}|D|$. This formula is especially convenient because the quantity $D$ has already been introduced at an earlier stage, see (10.1.20) and, thus, no additional expression for the area of a triangle is needed. The formula can be derived by transforming the elementary surface integral into a line integral along the boundary by Gauss's integral formula.

The final expression for the coefficients $P_{ij}$ is

$$P_{ij} = \frac{T_p}{2|D|} \{b_i b_j + c_i c_j\} \tag{10.1.21}$$

It should be noted that these coefficients can easily be calculated if the coordinates of the vertices of an element, and the transmissivity in that element, are known. Substituting (10.1.21) into (10.1.16) finally gives, for the second integral,

$$J_2 = -\sum_j P_{ij} \phi_j \quad (i \in C), \tag{10.1.22}$$

where the coefficients $P_{ij}$ are defined by (10.1.21). The summation over all elements must still be performed. We may consider this is an *administrative* problem, which can be most conveniently solved by a summation in the computer program, as will be shown below.

### 10.1.3. THE THIRD INTEGRAL, $J_3$

The third integral, $J_3$, defined by (10.1.12), may be regarded as consisting of two parts. The first part is an integral of the infiltration function $I$,

$$J_{3-1} = \int_R \{I N_i\} \, dx \, dy \quad (i \in C). \tag{10.1.23}$$

For any particular value of $i$, the shape function $N_i$ is different from zero only in

# THE FINITE ELEMENT METHOD 255

the surrounding elements. If in all these elements the infiltration function $I$ is assumed to be constant (which is only a minor restriction if the elements are sufficiently small), the integral over an element $R_p$ actually expresses the average of the product $I_p N_i$ over that element, multiplied by the area of that element. Because $I_p$ is constant, the average value of $N_i$ is $\frac{1}{3}$, and the area of a triangle is $\frac{1}{2}|D|$, the first part of the third integral takes the form

$$J_{3-1} = I_p |D|/6 \quad (i \in R_p). \tag{10.1.24}$$

Here the condition $i \in R_p$ indicates that, for a particular value of $p$ (i.e., for the infiltration over element $R_p$), a contribution is obtained only if node $i$ belongs to that element. Physically it means that the total infiltration over an element, $\frac{1}{2} I_p |D|$, is distributed evenly over the three nodes of that triangular element, just as in the case of the first integral a (boundary) surface supply was distributed evenly over the two nodes of a surface element. Actually, one may also write

$$J_{3-1} = Q_i \quad (i \in C) \tag{10.1.25}$$

where $Q_i$ now represents that part of the infiltration that is attributed to node $i$. For all elements to which node $i$ belongs, we have

$$Q_i = I_p |D|/6 \quad (i \in R_p). \tag{10.1.26}$$

From (10.1.12), it follows that the second part of the third integral is

$$J_{3-2} = -\int_R \left\{ \frac{1}{c} \sum_j N_i N_j (\phi_j - \phi_j') \right\} dx\, dy \quad (i \in C). \tag{10.1.27}$$

Again, it is assumed that the physical parameter, in this case the resistance $c$, is constant in each element, say $c_p$ in element $R_p$. Because the values of $\phi_j$ and $\phi_j'$ are also constant in $R_p$, we may write

$$J_{3-2} = -\sum_j \frac{1}{c_p} (\phi_j - \phi_j') \int_{R_p} \{N_i N_j\}\, dx\, dy \quad (i \in C). \tag{10.1.28}$$

Using (10.1.18), the two shape functions $N_i$ and $N_j$ can be expressed as

$$N_i(x, y) = p_i x + q_i y + r_i, \qquad N_j(x, y) = p_j x + q_j y + r_j.$$

Employing (10.1.19) and (10.1.20), the coefficients $p_i \ldots r_j$ can be expressed in terms of the geometrical data of the nodes. As each term will contain a factor linear or quadratic in $x$ and $y$, the integrals in (10.1.28) give rise to integrals of first- and second-order powers over an element. These integrals take the simplest form if the origin of the coordinate system coincides with the centroid of the element. It can be shown that this can be assumed without loss of generality. In that case, the first-order areal moments vanish,

$$\int_{R_p} x \, dx \, dy = 0, \quad \int_{R_p} y \, dx \, dy = 0. \tag{10.1.29}$$

The second-order moments can now be expressed as

$$\int_{R_p} x^2 \, dx \, dy = \{x_j^2 + x_k^2 + x_l^2\}|D|/24 = \tfrac{1}{2}|D|Z_{xx},$$

$$\int_{R_p} y^2 \, dx \, dy = \{y_j^2 + y_k^2 + y_l^2\}|D|/24 = \tfrac{1}{2}|D|Z_{yy},$$

$$\int_{R_p} xy \, dx \, dy = \{x_j y_j + x_k y_k + x_l y_l\}|D|/24 = \tfrac{1}{2}|D|Z_{xy} \tag{10.1.30}$$

where $Z_{xx}$, $Z_{xy}$ and $Z_{yy}$ are coefficients defined by these expressions. They can be derived by elementary mathematical techniques. The simplest method is by transforming the surface integrals into line integrals along the boundary, using Gauss's integral theorem.

Using (10.1.19), (10.1.29), and (10.1.30), the integral (10.1.28) can be written as

$$J_{3-2} = -\sum_j R_{ij}(\phi_j - \phi_j') \tag{10.1.31}$$

where

$$R_{ij} = \{b_i b_j Z_{xx} + c_i c_j Z_{yy} + (b_i c_j + b_j c_i) Z_{xy} + d_i d_j\}/\{2|D|c_p\}. \tag{10.1.32}$$

This completes the evaluation of the third integral.

With (10.1.14), (10.1.22), (10.1.24), and (10.1.31), the basic formula (10.1.9) now becomes

$$\sum_j \{P_{ij}\phi_j + R_{ij}(\phi_j - \phi_j')\} = Q_i \quad (i \in C), \tag{10.1.33}$$

where the coefficients $P_{ij}$ and $R_{ij}$ are given by (10.1.21) and (10.1.32). All these coefficients can easily be calculated for an element if the values of the transmissivity $T$, the infiltration $I$ and the resistance $c$ are given, and if the geometrical data (i.e., the coordinates of the nodes) are known.

Equation (10.1.33) is the basic algebraic equation of the finite element method for plane steady groundwater flow. In view of (10.1.15) and (10.1.26), it may be regarded as a continuity equation for the flow in the neighborhood of node $i$, with the right-hand side representing all the possible supplies of water at this node, either due to infiltration or from a supply along the boundary.

Equation (10.1.33) constitutes a linear system of equations. Various effective methods of solution exist for such a system (see Appendix A). The actual solution will be discussed in the next section.

## 10.2. Steady Flow in a Confined Aquifer

### 10.2.1. A SIMPLE PROGRAM

Let us demonstrate the solution method, and a simple computer program, by considering the case of steady flow in a completely confined aquifer. For such a case, the basic differential equation (10.1.1) can be simplified by taking $I = 0$ and $c = \infty$. The resulting equation is

$$\frac{\partial}{\partial x}\left(T\frac{\partial \phi}{\partial x}\right) + \frac{\partial}{\partial y}\left(T\frac{\partial \phi}{\partial y}\right) = 0. \tag{10.2.1}$$

For this case, the system of equations (10.1.33) reduces to

$$\sum_j \{P_{ij}\phi_j\} = Q_i \quad (i \in C), \tag{10.2.2}$$

where, according to (10.1.21), the coefficients $P_{ij}$ are composed of a summation over all elements, each of which makes a contribution of the form

$$P_{ij} = \frac{T_p}{2|D|}\{b_i b_j + c_i c_j\}. \tag{10.2.3}$$

The coefficients $b_j$ and $c_j$ and the determinant $D$ are defined by (10.1.20).

Once the coefficients $P_{ij}$ have been calculated, the remaining step is to solve the system of linear equations (10.2.2). The simplest way to do this is by the Gauss–Seidel method, in which an initial estimate is gradually updated (see Appendix A). Let the initial estimate of the solution be $\phi_i = f_i$, for all values of $i \in C$. In general these values do not satisfy Equation (10.2.2). Substituting the initial estimate into (10.2.2) will give rise to *residuals* $R_i$,

$$R_i = Q_i - \sum_j \{P_{ij} f_j\}.$$

The residual in equation $i$ can be reduced to zero by updating the value of $f_i$ by an amount $\Delta f_i$ such that the $i$th equation is satisfied, i.e.,

$$Q_i - \sum_j \{P_{ij} f_j\} - P_{ii} \Delta f_i = 0.$$

From this equation it follows that the increment $\Delta f_i$ is given by

$$\Delta f_i = [Q_i - \sum_j \{P_{ij} f_j\}]/P_{ii}. \tag{10.2.3}$$

This is the basic algorithm of the Gauss–Seidel method. Starting from an arbitrary initial estimate for all the unknown values $\phi_i$, these values are continuously updated according to (10.2.3), until the residuals are sufficiently small. The process is of an iterative character, because in each step the residuals in the surrounding nodes are influenced, and if one of them has been reduced to zero, it will not remain so when other values are updated. However, for a certain class of matrices, namely positive definite matrices, it can be shown (Carnahan *et al.*, 1969) that the iterative process will always converge. The matrix $P$ in the present system indeed belongs to that class, provided that the transmissivity is not negative, which is a physical necessity. The actual proof will not be given here. For the present purpose, it may suffice to assume that the Gauss–Seidel process will converge, and then observe whether this is the case by performing the calculations. It will indeed be found convergent, as the reader may verify by running the programs.

The Gauss–Seidel method requires some initial estimate for the values of the unknown variables. If the calculations were to be performed manually, it would probably be cost-effective to try and use a good guess for these initial estimates. In a computer program, this is not really necessary, as the effort spent in finding a good initial guess may not be balanced by the gain in computer time. Similarly, the number of iterations is often not determined by checking the actual values of the residuals, because this in itself uses computer time. Instead, it is often set to a predetermined value, say 50 or 100, or a certain multiple of the number of nodes.

It is generally found that convergence of the Gauss–Seidel process can be improved if in each step the value of the unknown parameter is updated by an amount somewhat larger than $\Delta f_i$, say $R \times \Delta f_i$, where $R$ is a factor somewhat larger than 1. This procedure is called *over-relaxation*, and the factor $R$ is called the *over-relaxation factor*. It can be shown that this factor must be smaller than 2. The optimal value of the relaxation factor depends upon the eigenvalues of the matrix $P$. Usually, it is determined by trial and error. Values such as 1.5 or 1.6 have been reported to give good results.

An elementary program that performs the calculations described above is listed below, as Program BV10-1.

# THE FINITE ELEMENT METHOD

```
100 DEFINT I-N:KEY OFF:OPTION BASE 1:GOSUB 510
110 PRINT"---  Bear & Verruijt - Groundwater Modeling"
120 PRINT"---  Plane steady Groundwater Flow"
130 PRINT"---  Finite Elements":PRINT"---  Program 10.1"
140 PRINT"---  Triangular elements, interactive input":PRINT:PRINT
150 DIM X(50),Y(50),IP(50),F(50),Q(50),P(50,50),NP(75,3),T(75)
160 DIM XJ(3),YJ(3),B(3),C(3)
170 INPUT"Number of nodes ........ ";N
180 INPUT"Number of elements ..... ";M
190 INPUT"Number of iterations ... ";NI
200 INPUT"Relaxation factor ...... ";RX
210 FOR I=1 TO N:GOSUB 510:PRINT"Node ";I:PRINT
220 INPUT" x ................ ";X(I)
230 INPUT" y ................ ";Y(I)
240 PRINT" Head given (Y/N) .. ? ";:GOSUB 480
250 IF A$="Y" THEN IP(I)=1:INPUT" Head .............. ";F(I)
260 IF A$="N" THEN IP(I)=-1:INPUT" Supply ............ ";Q(I)
270 NEXT I:FOR J=1 TO M:GOSUB 510:PRINT"Element ";J:PRINT
280 INPUT"Node 1 ............. ";NP(J,1)
290 INPUT"Node 2 ............. ";NP(J,2)
300 INPUT"Node 3 ............. ";NP(J,3)
310 INPUT"Transmissivity ..... ";T(J):NEXT J:GOSUB 510
320 PRINT"Generation of system matrix":PRINT"   Element ........ ";
330 FOR J=1 TO M:PRINT J;:FOR I=1 TO 3:K=NP(J,I):XJ(I)=X(K):YJ(I)=Y(K)
340 NEXT I:B(1)=YJ(2)-YJ(3):B(2)=YJ(3)-YJ(1):B(3)=YJ(1)-YJ(2)
350 C(1)=XJ(3)-XJ(2):C(2)=XJ(1)-XJ(3):C(3)=XJ(2)-XJ(1)
360 D=XJ(1)*B(1)+XJ(2)*B(2)+XJ(3)*B(3):D=T(J)/(2*ABS(D))
370 FOR K=1 TO 3:KK=NP(J,K):FOR L=1 TO 3:LL=NP(J,L)
380 P(KK,LL)=P(KK,LL)+D*(B(K)*B(L)+C(K)*C(L)):NEXT L,K,J:PRINT:PRINT
390 PRINT"Solution of equations":PRINT"   Iteration ..... ";
400 FOR IT=1 TO NI:PRINT IT;:FOR I=1 TO N:IF IP(I)>0 THEN 420
410 A=Q(I):FOR J=1 TO N:A=A-P(I,J)*F(J):NEXT J:F(I)=F(I)+RX*A/P(I,I)
420 NEXT I,IT:FOR I=1 TO N:Q(I)=0:FOR J=1 TO N:Q(I)=Q(I)+P(I,J)*F(J)
430 NEXT J,I:D$="########.###":GOSUB 510
440 PRINT"   i          x          y          F          Q":PRINT
450 FOR I=1 TO N:PRINT USING "###";I;:PRINT USING D$;X(I);
460 PRINT USING D$;Y(I);:PRINT USING D$;F(I);:PRINT USING D$;Q(I)
470 NEXT I:PRINT:END
480 A$=INPUT$(1):IF A$="Y" OR A$="y" THEN A$="Y":PRINT "Yes":RETURN
490 IF A$="N" OR A$="n" THEN A$="N":PRINT"No":RETURN
500 GOTO 480
510 CLS:LOCATE 1,30,1:COLOR 0,7:PRINT" Finite Elements - 1 ";
520 COLOR 7,0:PRINT:PRINT:RETURN
```

Program BV10-1. Steady flow in confined aquifer — elementary program.

The program is written in Microsoft BASIC, for the MS-DOS operating system. For other systems (e.g., CP/M systems) some minor modifications may be necessary, for example for the screen control statements.

As this program will be used as a basis for more advanced versions, it may be worthwhile to study its operations in some detail.

*Input*

The input data are generated interactively, with the program first asking (in lines 170–200) for some general data: the number of nodes, the number of elements, the number of iterations, and the relaxation factor. Then the data of all nodes must be entered: the two coordinates, and either the head or the local rate of water supply, expressed as a discharge. Actually, the program will ask whether the head in a certain node is given. If this is the case (for a node located on the boundary part $S_1$), the value of a type-indicator $IP(I)$ is set equal to 1, and the program asks for the value of the head, which is then known. If the head is not given, the value of $IP(I)$ is set equal to $-1$, and the program asks for the rate of the local water supply. Such a node belongs to the class $C$ mentioned in the previous section, the class of nodes in which the head is unknown, and must be determined. Finally, the data of all elements must be given: the three node numbers of the nodes of an element, and the transmissivity in that element.

It may be noted that in the dimension statements, fixed dimensions have been used (50 nodes, and 75 elements). Actually, BASIC allows for dynamic array definitions, with dimensions statements of the form DIM $X(N)$ after the entry of a value for $N$. This facility is not used here, however, because compilers usually do not accept it, and it is often very useful to compile the programs. This makes them run much faster.

*Generation of System Matrix*

When all the input data have been entered, the system matrix $P$ can be generated. This is performed in lines 320–380. For a particular element, as indicated by the variable $J$, the coordinates of the three nodes are stored in the variables $XJ(I)$ and $YJ(I)$. Then the coefficients $B(I)$ and $C(I)$, and the determinant $D$ are calculated, using formulas (10.1.20). This then enables us to calculate the nine values of the matrix $P$ for element $J$, and to store them in the appropriate positions of the matrix, as defined by $KK$ and $LL$. It should be noted that the summation over all elements is performed by adding the contributions from each element consecutively to the matrix $P$, which is initially filled with zeros.

*Solution of Equations*

The system of equations is solved, by the Gauss–Seidel method, in lines 390–420. The residual is denoted by $A$. It should be noted that the restriction of the values of $I$ to the class $C$, indicating the unknown values, is taken into account by simply skipping the algorithm if $IP(I)$ is positive.

After the completion of $NI$ iterations, the values of $Q(I)$ are calculated. This enables to verify the convergence, as it must be equal to the prescribed value at all points where the head is unknown. The solution is finally presented as a list on the screen (lines 440–470).

It is suggested that the reader first use the program to solve a simple problem such as the one illustrated in Figure 10.5. This problem concerns the uniform flow from left to right. The horizontal and vertical dimensions of the elements are equal

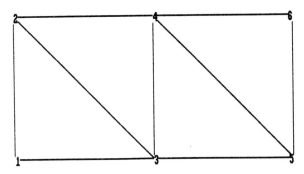

Fig. 10.5. Elementary problem.

to 1. The head on the left side is 10, while the head on the right side is 0. The transmissivity is 1 in all elements. If the number of iterations is 50, and the relaxation factor is chosen as 1.5, the input data are, in the order in which they have to be entered,

6, 4, 50, 1.5,
0, 0, Y, 10, 0, 1, Y, 10, 1, 0, N, 0, 1, 1, N, 0, 2, 0, Y, 0, 2, 1, Y, 0,
1, 2, 3, 1, 2, 3, 4, 1, 3, 4, 5, 1, 4, 5, 6, 1

The list gives the answers to the questions about input data prompted by the program. The user will note that after entering the data, the program will indicate its progress until the presentation of the list of output data. Because the flow in this case is uniform, the value of the head in nodes 3 and 4 should be 5. When the program is executed these values are indeed obtained.

The program also prints the discharges at the nodes. These represent the amount of water passing through the system. The total discharge is found to be 5, as it should be, because the hydraulic gradient is 5, the transmissivity is 1, and the width is 1.

### 10.2.2. AN IMPROVED PROGRAM

The Program BV10-1, presented above, may well serve as a first introduction to the finite element method. However, it has several serious shortcomings. Some of these will now be eliminated.

*Input and Output*

Some of the disadvantages of a simple program such as Program BV10-1 are related to the input and output procedures. Interactive input may be convenient for the user, because it eliminates the need for a manual of instructions, but it creates the need for a possibility to correct typing errors. After entering a large number of data, it is rather ineffective to have to start anew if a single error is made. This enhancement can be introduced by having the program present the input data after their entry, with the possibility of correction. This option is introduced in Program BV10-2. A further improvement, which is left to the reader, would be to write a separate program for the generation of input data, which then are stored in a datafile (e.g., on diskette). This enables us to modify the datafile and to use it as input for the main program.

An improved way of presenting the output data is to have them printed on a printer, together with the input data, in order to check the latter. This is a simple modification (use LPRINT instead of PRINT), which is left for implementation by the reader. Further improvements may involve a graphical presentation of output data, in the form of a fishnet diagram, or a contour plot. Examples of these are shown in Figures 10.12 and 10.13.

*Pointer Matrix*

The major shortcomings of Program BV10-1 are the large amount of memory needed to store the matrix $P$, especially for a large network, and the large amount of multiplications that have to be performed. For a system of 50 nodes, the matrix contains 2500 elements, and the number of multiplications in the main algorithm (in line 410) is $IT \times N \times N$, which is 125 000. As each real number requires about 4 bytes for storage, this means that about 10 kB of storage is needed for the storage of such a matrix. For a system of 100 nodes (which is not really a large network, as it may be obtained by considering a rectangle with 10 columns and 10 rows of nodes), 40 kB of memory is needed. In many systems, the maximum amount available is about 60 kB, thus leaving very little memory for other variables, or for the program itself.

Actually, computer memory is used very ineffectively if all coefficients of the matrix are really stored, because most of them are zeros. In fact, a coefficient in the position described by row $I$ and column $J$ is different from zero only if nodes $I$ and $J$ occur jointly in at least one element. As for each value of $I$ this will usually be the case for not more than six or perhaps seven values of $J$, it follows that in each row there are only six or seven nonzero coefficients. Thus, most of the memory is used to store zeros, and most of the calculations involve multiplications by zero.

In order to eliminate most of this waste of memory and computation time, a *pointer matrix* may be introduced. This is a matrix of integers describing the positions of the nonzeros in the matrix. These can then be stored in a matrix of limited width, say $N \times 7$. Such a pointer matrix, denoted by $KP$, can be generated by the following subroutine.

```
250 PRINT"Generation of pointer matrix":DIM KP(500,8):NZ=8
260 FOR I=1 TO N:KP(I,1)=I:KP(I,NZ)=1:NEXT I
270 PRINT"   Element ...... ";:FOR J=1 TO M:PRINT J;
280 FOR K=1 TO 3:KK=NP(J,K):FOR L=1 TO 3:LL=NP(J,L)
290 IA=0:FOR II=1 TO KP(KK,NZ):IF KP(KK,II)=LL THEN IA=1
300 NEXT II:IF IA=0 THEN KB=KP(KK,NZ)+1:KP(KK,NZ)=KB:KP(KK,KB)=LL
310 IF KB=NZ THEN PRINT"Pointer width (NZ) too small.":PRINT:END
320 NEXT L,K,J
```

The maximum dimensions of the pointer matrix are assumed to be 500 and 8. The first seven columns of row $I$ are used to store the node numbers of nodes that appear in an element that also contains node $I$. The last column is used to store the actual number of such nodes, which may not be greater than 7. In line 260 of the partial program listed above the numbers $I$ are stored in the first column of $KP$, while the last column is filled with 1's. Then all elements are considered consecutively, checking for each node of an element whether it is represented in the rows for the other two nodes of the element. If this is not the case ($IA = 0$), the node number is added to the matrix, and the value in the last column is increased by 1. If this value exceeds 7 an error statement is given. The user should then increase the value of $NZ$ in line 250.

For the network shown in Figure 10.5 the pointer matrix is

| | | | | | | | |
|---|---|---|---|---|---|---|---|
| 1 | 2 | 3 | 0 | 0 | 0 | 0 | 3 |
| 2 | 1 | 3 | 4 | 0 | 0 | 0 | 4 |
| 3 | 1 | 2 | 4 | 5 | 0 | 0 | 5 |
| 4 | 2 | 3 | 5 | 6 | 0 | 0 | 5 |
| 5 | 3 | 4 | 6 | 0 | 0 | 0 | 4 |
| 6 | 4 | 5 | 0 | 0 | 0 | 0 | 3 |

Once the pointer matrix has been generated, the nonzero coefficients of the matrix $P$ can be stored in a matrix of width 8, by storing the coefficient $P_{31}$ in the third row, second column, because $KP(3, 2) = 1$. This means that the memory requirement is greatly reduced, especially for large systems. There is a small price to be paid, however, and that is the need to store the pointer matrix, which consists of integers. If the width of the matrices is 8, and assuming that a real number uses 4 bytes and an integer number 2 bytes, the total memory requirement can be estimated to be $N \times 8 \times 6$ bytes. For a system of 500 nodes this is 24 kB. Thus, it can be expected that meshes with 500, or perhaps up to 1000 nodes, can be accommodated in a computer with 64 kB storage.

The part of the program in which the matrix coefficients are stored must be modified, because the storage locations are different. This can best be illustrated by inspection of Program BV10-2, presented below.

The multiplication algorithm in the Gauss—Seidel procedure also needs some adjustment. Its modified form is

```
390 PRINT"Solution of equations":PRINT"    Iteration ..... ";
400 FOR IT=1 TO NI:PRINT IT;:FOR I=1 TO N:IF IP(I)>0 THEN 425
410 A=Q(I):FOR J=1 TO KP(I,NZ):A=A-P(I,J)*F(KP(I,J))
420 NEXT J:F(I)=F(I)+RX*A/P(I,1)
425 NEXT I
```

Thus, the most inner loop now no longer runs from $J = 1$ to $J = N$, but rather from $J = 1$ to $J = KP(I, NZ)$, which on the average will be about 6. The number of multiplications is reduced by a factor $6/N$. For a large network, this may mean a great improvement.

*Quadrangular Elements*

The triangular elements used in Program BV10-1 lead to a certain asymmetry in most networks, because of the diagonal in a system of squares. This can be avoided by adding a second diagonal, and an additional node in the center. However, this leads to almost twice the number of nodes and elements in the mesh. It is more effective to compose a quadrangular element by considering it to be the sum of four triangular ones (Figure 10.6). The program now has actually

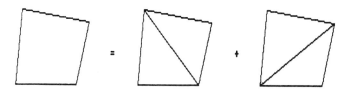

Fig. 10.6. Quadrangular element.

only half as many elements, each with four nodes instead of three. As a consequence, the matrix must be generated in four steps, one for each of the composing triangles. Again, this procedure requires some adjustment of the part of the program in which the matrix is generated, see Program BV10-2. Because a quadrangular element is considered to be the sum of two sets of two triangles, the effective transmissivity is doubled in the process. In order to account for this effect, all transmissivities should be reduced by a factor 2 in the program.

A network of quadrangular elements may appear to be less flexible than a system of triangles, because it seems to be less amenable to local refinements. However, a triangle may also be considered to be a quadrangle, if we allow the area of one of the composing triangles to be zero. Some elements of this type are shown in Figure 10.7. The incorporation of such elements in a program requires that each element be checked for the possibility of its area being zero (because the algorithm involves division by the determinant $D$, which is twice the

# THE FINITE ELEMENT METHOD

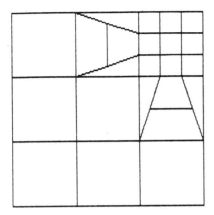

Fig. 10.7. Network of quadrangular elements.

area). If the area is found to be zero, the element can simply be skipped in the part of the program in which the system matrix is generated.

## Conjugate Gradient Method

The number of iterations necessary to obtain convergence of the Gauss–Seidel method is usually in the order of the number of nodes, and sometimes even more. This means that the computation time may be inconveniently large. Direct methods of solution, such as the *Cholesky decomposition method*, or *Gaussian elimination* with a *wave front technique* may be used as an alternative (Appendix A). However, these methods generally have the disadvantage that the width of the matrix is modified during the elimination process. It, thus, may grow, depending on the numbering sequence. In recent years, the *method of conjugate gradients* has been found to be effective both with regard to memory requirement and to computation time (Hestenes and Stiefel, 1952; Reid, 1971; Gambolati and Perdon, 1984). The method uses an iterative procedure, where in each step the variables are updated so that a measure of error is minimized. If there are no round-off errors, the procedure converges exactly in $N$ steps, where $N$ is the number of unknowns. In many cases, however, it is found that the number of iterations needed to obtain a sufficient accuracy is much less than the number of nodes. This may be especially true for time-dependent processes, in which the solution for the previous time step is a good estimate for the solution in the time step considered. The method of conjugate gradients allows for the use of a pointer matrix and, like in the case of the Gauss–Seidel method, it is very simple to account for the boundary conditions without the need to modify the system matrix. For these reasons, the method of conjugate gradients is used in the finite element programs to be presented in the sequel. Details of the method can be found in the literature (e.g., Reid, 1971), and in Appendix A.

## Computer Program

A computer program containing the improvements described above is listed below, as Program BV10-2.

```
100 DEFINT I-N:KEY OFF:OPTION BASE 1:GOSUB 980
110 PRINT"---   Bear & Verruijt - Groundwater Modeling"
120 PRINT"---   Plane steady Groundwater Flow"
130 PRINT"---   Finite Element Method":PRINT"---   Program 10.2"
140 PRINT"---   Quadrangular elements, interactive input":PRINT:PRINT
150 DIM XJ(4),YJ(4),B(3),C(3),KS(4,3)
160 DIM X(500),Y(500),IP(500),F(500),Q(500),U(500),V(500),W(500)
170 DIM P(500,11),KP(500,11),NP(400,4),T(400):NZ=11
180 INPUT"Number of nodes .......... ";N
190 INPUT"Number of elements ....... ";M
200 FOR I=1 TO N:GOSUB 660:NEXT I:FOR J=1 TO M:GOSUB 730:NEXT J
210 GOSUB 980:PRINT:PRINT"Present input data (Y/N) ........ ? ";
220 GOSUB 950:IF A$="N" THEN 240
230 FOR I=1 TO N:GOSUB 790:NEXT I:FOR J=1 TO M:GOSUB 860:NEXT J
240 FOR I=1 TO 4:FOR J=1 TO 3:K=I+J-1:IF K>4 THEN K=K-4
250 KS(I,J)=K:NEXT J,I:GOSUB 980:PRINT"Generation of pointer matrix"
260 FOR I=1 TO N:KP(I,1)=I:KP(I,NZ)=1:NEXT I
270 PRINT"   Element ...... ";:FOR J=1 TO M:PRINT J;
280 FOR K=1 TO 4:KK=NP(J,K):FOR L=1 TO 4:LL=NP(J,L)
290 IA=0:FOR II=1 TO KP(KK,NZ):IF KP(KK,II)=LL THEN IA=1
300 NEXT II:IF IA=0 THEN KB=KP(KK,NZ)+1:KP(KK,NZ)=KB:KP(KK,KB)=LL
310 IF KB=NZ THEN 650
320 NEXT L,K,J:PRINT:PRINT:EE=.00001
330 PRINT"Generation of system matrix":PRINT"   Element ...... ";
340 FOR J=1 TO M:PRINT J;:FOR KW=1 TO 4:FOR I=1 TO 3
350 K=NP(J,KS(KW,I)):XJ(I)=X(K):YJ(I)=Y(K):NEXT I
360 B(1)=YJ(2)-YJ(3):B(2)=YJ(3)-YJ(1):B(3)=YJ(1)-YJ(2)
370 C(1)=XJ(3)-XJ(2):C(2)=XJ(1)-XJ(3):C(3)=XJ(2)-XJ(1)
380 D=ABS(XJ(1)*B(1)+XJ(2)*B(2)+XJ(3)*B(3)):IF D<EE THEN 450
390 D=T(J)/(4*D)
400 FOR K=1 TO 3:KK=NP(J,KS(KW,K)):II=KP(KK,NZ):FOR LL=1 TO II:L=1
410 KV=KS(KW,L):IF NP(J,KV)=KP(KK,LL) THEN 430
420 L=L+1:IF L<4 THEN 410 ELSE 440
430 P(KK,LL)=P(KK,LL)+D*(B(K)*B(L)+C(K)*C(L))
440 NEXT LL,K
450 NEXT KW,J:PRINT:PRINT:EE=EE*EE
460 PRINT"Solution of equations":PRINT"   Iteration .... ";:IT=1
470 FOR I=1 TO N:IZ=KP(I,NZ):U(I)=0:IF IP(I)>0 THEN 490
480 U(I)=Q(I):FOR J=1 TO IZ:L=KP(I,J):U(I)=U(I)-P(I,J)*F(L):NEXT J
490 V(I)=U(I):NEXT I:UU=0:FOR I=1 TO N:UU=UU+U(I)*U(I):NEXT I
500 PRINT IT;:FOR I=1 TO N:W(I)=0:IZ=KP(I,NZ)
510 FOR J=1 TO IZ:W(I)=W(I)+P(I,J)*V(KP(I,J)):NEXT J,I
520 VW=0:FOR I=1 TO N:VW=VW+V(I)*W(I):NEXT I
530 AA=UU/VW:FOR I=1 TO N:IF IP(I)>0 THEN 550
540 F(I)=F(I)+AA*V(I):U(I)=U(I)-AA*W(I)
550 NEXT I:WW=0:FOR I=1 TO N:WW=WW+U(I)*U(I):NEXT I
560 BB=WW/UU:FOR I=1 TO N:V(I)=U(I)+BB*V(I):NEXT I:UU=WW
570 IT=IT+1:IF IT<=N AND UU>EE THEN 500
580 FOR I=1 TO N:Q(I)=0:FOR J=1 TO KP(I,NZ)
590 Q(I)=Q(I)+P(I,J)*F(KP(I,J)):NEXT J,I
600 D$="########.###":GOSUB 980
610 PRINT"     i          x           y            F           Q":PRINT
```

```
620 FOR I=1 TO N:PRINT USING "###";I;:PRINT USING D$;X(I);
630 PRINT USING D$;Y(I);:PRINT USING D$;F(I);
640 PRINT USING D$;Q(I):NEXT I:PRINT:END
650 GOSUB 980:PRINT"Pointer width (NZ) too small.":PRINT:END
660 GOSUB 980:PRINT"Data of node ";I:PRINT:PRINT
670 INPUT"   x ................ ";X(I)
680 INPUT"   y ................ ";Y(I)
690 PRINT"   Head given (Y/N) .. ? ";:GOSUB 950
700 IF A$="Y" THEN IP(I)=1:INPUT"   F ................ ";F(I)
710 IF A$="N" THEN IP(I)=-1:INPUT"   Q ................ ";Q(I)
720 RETURN
730 GOSUB 980:PRINT"Data of element ";J:PRINT:PRINT
740 INPUT"Node 1 ............. ";NP(J,1)
750 INPUT"Node 2 ............. ";NP(J,2)
760 INPUT"Node 3 ............. ";NP(J,3)
770 INPUT"Node 4 ............. ";NP(J,4)
780 INPUT"Transmissivity ..... ";T(J):RETURN
790 GOSUB 980:PRINT"Node ";I:PRINT:PRINT
800 PRINT"   x ................ ";X(I)
810 PRINT"   y ................ ";Y(I)
820 IF IP(I)>0 THEN PRINT"   F ................ ";F(I)
830 IF IP(I)<0 THEN PRINT"   Q ................ ";Q(I)
840 GOSUB 940:IF A$="N" THEN GOSUB 660:GOTO 790
850 RETURN
860 GOSUB 980:PRINT"Element ";J:PRINT:PRINT
870 PRINT"Node 1 ............. ";NP(J,1)
880 PRINT"Node 2 ............. ";NP(J,2)
890 PRINT"Node 3 ............. ";NP(J,3)
900 PRINT"Node 4 ............. ";NP(J,4)
910 PRINT"Transmissivity ..... ";T(J)
920 GOSUB 940:IF A$="N" THEN GOSUB 730:GOTO 860
930 RETURN
940 PRINT:PRINT:PRINT"Correct (Y/N) ......? ";
950 A$=INPUT$(1):IF A$="Y" OR A$="y" THEN PRINT"Yes":A$="Y":RETURN
960 IF A$="N" OR A$="n" THEN PRINT"No":A$="N":RETURN
970 GOTO 950
980 CLS:LOCATE 1,30,1:COLOR 0,7:PRINT" Finite Elements - 2 ";
990 COLOR 7,0:PRINT:PRINT:RETURN
```

Program BV10-2. Steady flow in confined aquifer — improved program.

Like Program BV10-1, this program also uses interactive input. However, it will present the input data on the screen before the calculations are performed, so that corrections can be made. The program also uses a pointer matrix to reduce memory requirement and computation time. It also uses the more effective conjugate gradient method to solve the system of equations. Output is still in the simplest form: a list on the screen. It is left to the user to introduce more permanent forms of output, such as a list of input and output data on a printer. The simplest extension is to replace PRINT by LPRINT in lines 610—640.

It may be noted that the use of the quadrangular elements described above, has lead to some modifications, the most important of which is that in the generation of the system matrix a summation over four triangles is performed (by the variable $KW$). The $4 \times 3$ matrix $KS$ describes the shape of the four triangles. It may also be noted that because of this particular composition of the elements, the transmissivity is to be reduced by a factor of 2. This is taken into account in line 390.

The width of the matrices $P$ and $KP$ is assumed to be 11, which means that each node may be in contact with 10 nodes, including itself. As mentioned before, the last column of the matrix $KP$ is used to store the number of nonzeros. Similarly, the last column of the matrix $P$ is used to store the right-hand side of the system of equations, which in this case consists only of the vector $Q$. The maximum capacity of the program is set to 500 nodes and 400 elements. Larger systems may be accommodated, depending upon the type of hardware or software used.

## 10.3. Steady Flow with Infiltration and Leakage

The two programs presented in the previous section apply only to the simplest type of problem: steady flow in a confined aquifer with supply of water only at the nodes (representing local sources or sinks) and along the boundary. Infiltration over the areal surface, or replenishment by leakage, from an overlying or underlying aquifer, were not included. The basic equations derived in Section 10.1 are more general, as they include both infiltration and leakage. In this section, these features will be included in a finite element program, but still only for steady flow.

The basic algebraic equation for steady flow with infiltration and leakage is Equation (10.1.33), which can be written as

$$\sum_j \{P_{ij}\phi_j + R_{ij}(\phi_j - \phi'_j)\} = Q_i \quad (i \in C) \tag{10.3.1}$$

where the coefficients $P_{ij}$ and $R_{ij}$ are given by (10.1.21) and (10.1.32)

$$P_{ij} = \{T_p/2|D|\}\{b_i b_j + c_i c_j\}, \tag{10.3.2}$$

$$R_{ij} = \frac{\{b_i b_j Z_{xx} + c_i c_j Z_{yy} + (b_i c_j + b_j c_i) Z_{xy} + d_i d_j\}}{\{2|D|c_p\}}. \tag{10.3.3}$$

These expressions give the contributions of a single element. A summation over all elements is implied. The coefficients $b_i$, $c_i$, $Z_{xx}$, etc. are given by (10.1.20) and (10.1.30).

In order to take into account the additional terms due to the matrix $R$, the part of the program in which the system matrix is generated must be modified. Of course, the solution of the system of equations should also be modified to account for the new system matrix $(P+R)$, and for the additional terms in the expressions for the supply function $Q$ on the right-hand side, such as the contribution due to the surface infiltration over the elements, see (10.1.26). All these modifications are elementary.

A computer program for this type of problem is listed in Program BV10-3.

# THE FINITE ELEMENT METHOD 269

```
100 DEFINT I-N:KEY OFF:OPTION BASE 1:GOSUB 720
110 PRINT"---  Bear & Verruijt - Groundwater Modeling"
120 PRINT"---  Plane steady Flow with Infiltration and Leakage"
130 PRINT"---  Finite Element Method":PRINT"---  Program 10.3"
140 PRINT"---  Quadrangular elements, input by DATA statements"
150 DIM XJ(4),YJ(4),B(3),C(3),D(3),KS(4,3)
160 DIM X(300),Y(300),IP(300),F(300),FA(300),Q(300)
170 DIM NP(250,4),T(250),S(250),PP(250),CC(250)
180 DIM U(300),V(300),W(300),P(300,11),R(300,11),KP(300,11):NZ=11
190 PRINT:PRINT:PRINT"Reading data":READ N,M
200 FOR I=1 TO N:READ X(I),Y(I),IP(I),FA(I),Q(I):F(I)=FA(I):NEXT I
210 FOR J=1 TO M:READ NP(J,1),NP(J,2),NP(J,3),NP(J,4),T(J),PP(J),CC(J)
220 NEXT J:FOR I=1 TO 4:FOR J=1 TO 3:K=I+J-1:IF K>4 THEN K=K-4
230 KS(I,J)=K:NEXT J,I:GOSUB 720:PRINT"Generation of pointer matrix"
240 FOR I=1 TO N:KP(I,1)=I:KP(I,NZ)=1:NEXT I
250 PRINT"   Element ...... ";:FOR J=1 TO M:PRINT J;
260 FOR K=1 TO 4:KK=NP(J,K):FOR L=1 TO 4:LL=NP(J,L)
270 IA=0:FOR II=1 TO KP(KK,NZ):IF KP(KK,II)=LL THEN IA=1
280 NEXT II:IF IA=0 THEN KB=KP(KK,NZ)+1:KP(KK,NZ)=KB:KP(KK,KB)=LL
290 IF KB=NZ THEN 710
300 NEXT L,K,J:PRINT:PRINT:EE=.00001
310 PRINT"Generation of system matrices":PRINT"   Element ...... ";
320 FOR J=1 TO M:PRINT J;:FOR KW=1 TO 4:ZX=0:ZY=0
330 FOR I=1 TO 3:K=NP(J,KS(KW,I)):XJ(I)=X(K):YJ(I)=Y(K)
340 ZX=ZX+X(K):ZY=ZY+Y(K):NEXT I:ZX=ZX/3:ZY=ZY/3
350 FOR I=1 TO 3:XJ(I)=XJ(I)-ZX:YJ(I)=YJ(I)-ZY:NEXT I
360 B(1)=YJ(2)-YJ(3):B(2)=YJ(3)-YJ(1):B(3)=YJ(1)-YJ(2)
370 C(1)=XJ(3)-XJ(2):C(2)=XJ(1)-XJ(3):C(3)=XJ(2)-XJ(1)
380 D(1)=XJ(2)*YJ(3)-XJ(3)*YJ(2):D(2)=XJ(3)*YJ(1)-XJ(1)*YJ(3)
390 D(3)=XJ(1)*YJ(2)-XJ(2)*YJ(1)
400 D=ABS(D(1)+D(2)+D(3)):IF D<EE THEN 530
410 DD=T(J)/(4*D):DE=1/(4*CC(J)*D)
420 XX=(XJ(1)*XJ(1)+XJ(2)*XJ(2)+XJ(3)*XJ(3))/12
430 XY=(XJ(1)*YJ(1)+XJ(2)*YJ(2)+XJ(3)*YJ(3))/12
440 YY=(YJ(1)*YJ(1)+YJ(2)*YJ(2)+YJ(3)*YJ(3))/12
450 FOR K=1 TO 3:KK=NP(J,KS(KW,K)):II=KP(KK,NZ):FOR LL=1 TO II:L=1
460 KV=KS(KW,L):IF NP(J,KV)=KP(KK,LL) THEN 480
470 L=L+1:IF L<4 THEN 460 ELSE 510
480 P(KK,LL)=P(KK,LL)+DD*(B(K)*B(L)+C(K)*C(L))
490 AA=XX*B(K)*B(L)+XY*(B(K)*C(L)+B(L)*C(K))+YY*C(K)*C(L)+D(K)*D(L)
500 R(KK,LL)=R(KK,LL)+DE*AA
510 NEXT LL,K
520 FOR I=1 TO 3:K=NP(J,KS(KW,I)):Q(K)=Q(K)+D*PP(J)/12:NEXT I
530 NEXT KW,J:PRINT:PRINT:EE=EE*EE
540 PRINT"Solution of equations":PRINT"   Iteration .... ";:IT=1
550 FOR I=1 TO N:IZ=KP(I,NZ):U(I)=0:IF IP(I)>0 THEN 580
560 U(I)=Q(I):FOR J=1 TO IZ:CC=R(I,J):L=KP(I,J)
570 U(I)=U(I)-(P(I,J)+CC)*F(L)+CC*FA(L):NEXT J
580 V(I)=U(I):NEXT I:UU=0:FOR I=1 TO N:UU=UU+U(I)*U(I):NEXT I
590 PRINT IT;:FOR I=1 TO N:W(I)=0:IZ=KP(I,NZ)
600 FOR J=1 TO IZ:W(I)=W(I)+(P(I,J)+R(I,J))*V(KP(I,J)):NEXT J,I
610 VW=0:FOR I=1 TO N:VW=VW+V(I)*W(I):NEXT I
620 AA=UU/VW:FOR I=1 TO N:IF IP(I)>0 THEN 640
630 F(I)=F(I)+AA*V(I):U(I)=U(I)-AA*W(I)
640 NEXT I:WW=0:FOR I=1 TO N:WW=WW+U(I)*U(I):NEXT I
650 BB=WW/UU:FOR I=1 TO N:V(I)=U(I)+BB*V(I):NEXT I:UU=WW
660 IT=IT+1:IF IT<=N AND UU>EE THEN 590
```

```
670 GOSUB 720:PRINT:A$="#####.###":FOR I=1 TO N
680 PRINT" x = ";:PRINT USING A$;X(I);:PRINT" -  y = ";
690 PRINT USING A$;Y(I);:PRINT" -  F = ";:PRINT USING A$;F(I):NEXT I
700 PRINT:END
710 GOSUB 720:PRINT"Pointer width (NZ) too small.":GOTO 700
720 CLS:LOCATE 1,30,1:COLOR 0,7:PRINT" Finite Elements - 3 ";:COLOR 7,0
730 COLOR 7,0:PRINT:PRINT:RETURN
740 DATA 22,10
750 DATA 0,0,-1,0,0,0,100,-1,0,0
760 DATA 100,0,-1,0,0,100,100,-1,0,0
770 DATA 200,0,-1,0,0,200,100,-1,0,0
780 DATA 300,0,-1,0,0,300,100,-1,0,0
790 DATA 400,0,-1,0,0,400,100,-1,0,0
791 DATA 500,0,-1,0,0,500,100,-1,0,0
792 DATA 600,0,-1,0,0,600,100,-1,0,0
793 DATA 700,0,-1,0,0,700,100,-1,0,0
794 DATA 800,0,-1,0,0,800,100,-1,0,0
795 DATA 900,0,-1,0,0,900,100,-1,0,0
800 DATA 1000,0,1,0,0,1000,100,1,0,0
810 DATA 1,2,3,4,100,0.001,10000
820 DATA 3,4,5,6,100,0.001,10000
830 DATA 5,6,7,8,100,0.001,10000
840 DATA 7,8,9,10,100,0.001,10000
850 DATA 9,10,11,12,100,0.001,10000
860 DATA 11,12,13,14,100,0.001,10000
870 DATA 13,14,15,16,100,0.001,10000
880 DATA 15,16,17,18,100,0.001,10000
890 DATA 17,18,19,20,100,0.001,10000
900 DATA 19,20,21,22,100,0.001,10000
```

Program BV10-3. Steady flow with infiltration and leakage.

The program uses input by READ and DATA statement, rather than interactive input, in order to reduce its length. Interactive input facilities may be added by the user, when desired. The meaning of the input variables is as follows:

| | |
|---|---|
| $N$ | Number of nodes, |
| $M$ | Number of elements, |
| $X(I)$ | X-coordinate of node $I$, |
| $Y(I)$ | Y-coordinate of node $I$, |
| $IP(I)$ | Type indicator of node $I$, |
| $FA(I)$ | Head in adjacent aquifer, in node $I$, |
| $Q(I)$ | Water supply in node $I$, |
| $NP(J, 1)$ | Node 1 of element $J$, |
| $NP(J, 2)$ | Node 2 of element $J$, |
| $NP(J, 3)$ | Node 3 of element $J$, |
| $NP(J, 4)$ | Node 4 of element $J$, |
| $T(J)$ | Transmissivity in element $J$, |
| $PP(J)$ | Infiltration in element $J$, |
| $CC(J)$ | Resistance of aquitard in element $J$. |

The type indicator $IP(I)$ must be greater than 0 (e.g., 1) if the head at node $I$ is given. For such nodes the value of $F(I)$ should be that given value. If the local head is unknown, $IP(I)$ must be less than 0 (e.g., −1). The input value of $F(I)$ is then used as the first estimate in the iteration procedure. It can be taken as zero. In all these nodes, the local water supply must be given. It must be zero for an interior node if there is no source or sink at that node. If the head is given at node $I$, the READ statement also asks for a value for the discharge $Q(I)$. Any arbitrary value may be assigned to $Q(I)$ for such nodes (e.g., 0), as the program ignores the value.

Program BV10-3 contains data for a strip of 1000 × 100 m, with a uniform infiltration of 0.001 m/d (Figure 10.8). The transmissivity of the aquifer is 100 m²/

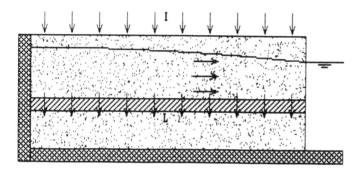

Fig. 10.8. Problem solved by Program BV10-3.

d, the resistance of the clay layer is 10 000 days, and the head in the lower aquifer ($FA$ in the program) is zero. The left side boundary ($x = 0$) is impermeable. At the right side boundary ($x = 1000$), the head is zero. The program uses 22 nodes and 10 elements.

The exact solution of this problem, which is actually one-dimensional, even though it has been solved by a two-dimensional program, can easily be found by integrating the differential equation. The solution is

$$\phi = Ic \left\{ \frac{1 - \cosh(x/\lambda)}{\cosh(L/\lambda)} \right\}$$

where $\lambda$ is the leakage factor, $\lambda = \sqrt{(Tc)}$.

The exact solution and the numerical one, obtained by running Program BV10-3, are compared in Table 10.1. The agreement appears to be excellent (errors smaller than 0.1%).

It may be mentioned that the computation time for a simple problem such as the one described above, with 22 nodes and 10 elements, is about 3 min on a common Personal Computer. This can be reduced by using a BASIC compiler.

Then, the computation time on the same system is only about 12 sec. Such a gain is very attractive for large systems.

Table 10.1. Solutions obtained by running Program 10.3.

| x | Exact solution | Numerical solution |
|---|---|---|
| 0.0 | 3.519 | 3.522 |
| 100.0 | 3.487 | 3.489 |
| 200.0 | 3.389 | 3.391 |
| 300.0 | 3.226 | 3.228 |
| 400.0 | 2.994 | 2.996 |
| 500.0 | 2.692 | 2.694 |
| 600.0 | 2.318 | 2.319 |
| 700.0 | 1.866 | 1.867 |
| 800.0 | 1.333 | 1.334 |
| 900.0 | 0.713 | 0.713 |
| 1000.0 | 0.000 | 0.000 |

## 10.4. Steady Flow through a Dam

An elegant application of the finite element method, in which the power of the method becomes clearly evident, is the flow of groundwater in a vertical plane, with a phreatic surface, such as the flow through a dam, shown in Figure 10.9. We shall limit the discussion to the case of steady flow, without accretion. The plane in which the flow takes place will be denoted as the $x$, $y$-plane, with the $y$-axis pointing vertically upward.

As explained in Subsection 3.4.6, the main complication in this type of problem is that the location of the free surface (or phreatic surface) is *a-priori* unknown. In fact, one of the main objectives of the solution is to determine the elevation of this surface. If capillary effects are disregarded, the pressure along the free surface is

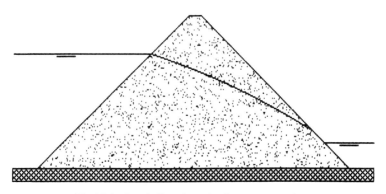

Fig. 10.9. Steady flow through a homogeneous dam.

zero, so that it may be considered as a boundary with a given head, namely $\phi = y$. On the other hand, the free surface is also a stream line, because the flow is steady and there is no accretion. All this means that the free surface is a boundary with two boundary conditions (both Dirichlet and Neumann), with an unknown location. This complication can be overcome by an iterative procedure, in which the location of the free surface is updated in each step. First, an initial estimate for the location of the free surface is introduced, for instance in the shape of a straight line, or according to Dupuit's approximate formula. The free surface is then regarded as a stream line (i.e., an impermeable boundary), with an unknown head distribution. The distribution of the piezometric head in the interior of the domain, and on its boundary, can be calculated by using the finite element method. This will lead to values for the head at the points on the free surface. These heads can then be compared with the elevation of these points. In general, these values will not be the same, especially in the beginning of the process, and a new position for the free surface can be assumed, for instance, by taking $y = \phi$. This procedure has been applied successfully in the graphical method of sketching flow nets, and it is also a proven technique when using an electrical analogue. Because the finite element method uses a network of flexible shapes, and curved boundaries can easily be followed, the shape of the free surface can always be represented by the upper boundary of the finite element network. Thus, the familiar iterative technique can be directly applied to the finite element method (Taylor and Brown, 1967).

It is, of course, most convenient to let the adjustments of the location of the nodal points be performed by the computer program itself. It is also attractive to let the program generate the initial mesh. An example of a program incorporating these procedures is Program BV10-4. This is a program that solves the flow

```
1000 DEFINT I-N:KEY OFF:OPTION BASE 1:GOSUB 1760
1010 PRINT"---   Bear & Verruijt - Groundwater Modeling"
1020 PRINT"---   Plane steady flow in triangular dam"
1030 PRINT"---   Finite Element Method":PRINT"---   Program 10.4"
1040 PRINT"---   Quadrangular elements, interactive input":PRINT
1050 DIM X(240),Y(240),IP(240),F(240),U(240),V(240),W(240)
1060 DIM KP(240,10),P(240,10),NP(200,4),T(200):NZ=10
1070 DIM XJ(3),YJ(3),PJ(3,3),B(3),C(3),KS(4,3)
1080 DIM XB(21),XT(21),YT(21),CT(21)
1090 INPUT"Width of dam at the bottom (m) ........ ";W
1100 INPUT"Inclination of slope left (degrees) ... ";AL
1110 INPUT"Inclination of slope right (degrees) .. ";AR
1120 INPUT"Water level left (m) .................. ";HL
1130 INPUT"Water level right (m) ................. ";HR
1140 INPUT"Number of columns of elements ......... ";NL
1150 IF NL>20 THEN NL=20:PRINT"    Maximum value : 20"
1160 INPUT"Number of rows of elements ............ ";NH
1170 IF NH>10 THEN NH=10:PRINT"    Maximum value : 10"
1180 PRINT:A=4*ATN(1)/180:AL=A*AL:AR=A*AR:E=HL/500
1190 N=(NH+1)*(NL+1):M=NH*NL
1200 CL=COS(AL)/SIN(AL):CR=COS(AR)/SIN(AR):IS=0
1210 FOR I=1 TO NL+1:A=(I-1)/NL:XB(I)=A*W:CT(I)=CL+A*(CR-CL)
```

```
1220 YT(I)=HL+.8*A*(HR-HL):XT(I)=XB(I)+CT(I)*YT(I):NEXT I
1230 FOR I=1 TO NL:K=(I-1)*(NH+1):L=(I-1)*NH:FOR J=1 TO NH:JA=L+J
1240 NP(JA,1)=K+J:NP(JA,2)=NP(JA,1)+NH+1:NP(JA,3)=NP(JA,1)+1
1250 NP(JA,4)=NP(JA,2)+1:T(JA)=1:NEXT J,I
1260 PRINT"Calculation of pointer matrix":PRINT"     Element ..... ";
1270 FOR I=1 TO N:KP(I,1)=I:KP(I,NZ)=1:NEXT I:FOR J=1 TO M:PRINT J;
1280 FOR K=1 TO 4:KK=NP(J,K):FOR L=1 TO 4:LL=NP(J,L):IA=0
1290 FOR II=1 TO KP(KK,NZ):IF KP(KK,II)=LL THEN IA=1
1300 NEXT II:IF IA=0 THEN KB=KP(KK,NZ)+1:KP(KK,NZ)=KB:KP(KK,KB)=LL
1310 NEXT L,K,J:PRINT:LL=1
1320 FOR I=1 TO 4:FOR J=1 TO 3:K=I+J-1:IF K>4 THEN K=K-4
1330 KS(I,J)=K:NEXT J,I
1340 FOR J=1 TO NH+1:F(J)=HL:IP(J)=2:IP(N+1-J)=2:F(N+1-J)=HR:NEXT J
1350 IP(N)=0:FOR I=1 TO NL+1:K=(I-1)*NH+I
1360 FOR J=1 TO NH+1:L=K+J-1:Y(L)=(J-1)*YT(I)/NH
1370 IF IS=0 THEN F(L)=YT(I)
1380 X(L)=XB(I)+Y(L)*CT(I):NEXT J,I
1390 GOSUB 1760:PRINT"Free surface after";IS;"cycle";
1400 IF IS<>1 THEN PRINT"s";
1410 PRINT:PRINT:FOR I=1 TO NL+1:K=I*(NH+1):A$="####.###"
1420 PRINT"   x = ";:PRINT USING A$;X(K);
1430 PRINT"   y = ";:PRINT USING A$;Y(K):NEXT I:IF LL<0 THEN 1620
1440 FOR I=N-NH TO N:F(I)=HR:IF Y(I)>HR THEN F(I)=Y(I)
1450 NEXT I:PRINT:PRINT"Generation of Matrix":PRINT"     Element ..... ";
1460 FOR I=1 TO N:FOR J=1 TO NZ:P(I,J)=0:NEXT J,I
1470 FOR J=1 TO M:PRINT J;:FOR KW=1 TO 4:FOR I=1 TO 3
1480 K=NP(J,KS(KW,I)):XJ(I)=X(K):YJ(I)=Y(K)
1490 NEXT I:B(1)=YJ(2)-YJ(3):B(2)=YJ(3)-YJ(1):B(3)=YJ(1)-YJ(2)
1500 C(1)=XJ(3)-XJ(2):C(2)=XJ(1)-XJ(3):C(3)=XJ(2)-XJ(1)
1510 D=ABS(XJ(1)*B(1)+XJ(2)*B(2)+XJ(3)*B(3)):IF D<EE THEN 1580
1520 D=T(J)/(4*D)
1530 FOR K=1 TO 3:KK=NP(J,KS(KW,K)):II=KP(KK,NZ):FOR LL=1 TO II:L=1
1540 KV=KS(KW,L):IF NP(J,KV)=KP(KK,LL) THEN 1560
1550 L=L+1:IF L<4 THEN 1540 ELSE 1570
1560 P(KK,LL)=P(KK,LL)+D*(B(K)*B(L)+C(K)*C(L))
1570 NEXT LL,K
1580 NEXT KW,J:PRINT:PRINT:GOSUB 1630
1590 LL=-1:FOR I=1 TO NL+1:K=I*(NH+1)
1600 A=F(K)-Y(K):YT(I)=YT(I)+A:IF ABS(A)>E THEN LL=1
1610 NEXT I:IS=IS+1:PRINT:GOTO 1350
1620 PRINT:END
1630 PRINT"Solution of equations":PRINT"   Iteration ... ";:IT=1
1640 EE=.00001:EE=EE*EE:FOR I=1 TO N:U(I)=0:IF IP(I)>0 THEN 1660
1650 FOR J=1 TO KP(I,NZ):U(I)=U(I)-P(I,J)*F(KP(I,J)):NEXT J
1660 V(I)=U(I):NEXT I:UU=0:FOR I=1 TO N:UU=UU+U(I)*U(I):NEXT I
1670 PRINT IT;:FOR I=1 TO N:W(I)=0
1680 FOR J=1 TO KP(I,NZ):W(I)=W(I)+P(I,J)*V(KP(I,J)):NEXT J,I
1690 VW=0:FOR I=1 TO N:VW=VW+V(I)*W(I):NEXT I
1700 AA=UU/VW:FOR I=1 TO N:IF IP(I)>0 THEN 1720
1710 F(I)=F(I)+AA*V(I):U(I)=U(I)-AA*W(I)
1720 NEXT I:WW=0:FOR I=1 TO N:WW=WW+U(I)*U(I):NEXT I
1730 BB=WW/UU:FOR I=1 TO N:V(I)=U(I)+BB*V(I):NEXT I:UU=WW
1740 IT=IT+1:IF IT<=N AND UU>EE THEN 1670
1750 RETURN
1760 CLS:LOCATE 1,30,1:COLOR 0,7:PRINT" Finite Elements - 4 ";
1770 COLOR 7,0:PRINT:PRINT:RETURN
```

Program BV10-4. Steady flow in triangular dam.

through a homogeneous dam of triangular shape, with constant water levels on both sides. The flow is assumed to be from left to right, with the water level on the left side being higher than that on the right side. It can then be expected that there will be a seepage surface on the right side slope. An initial estimate for the free surface and the length of the seepage surface is generated by assuming that the initial estimate for the free surface has the shape of a straight line, connecting the upper water level with a point on the downstream slope located at a height which is one-fifth of the total head difference above the downstream water level.

The iterative procedure described above is performed in the program until the difference between the new height of a point of the free surface and its previous value is less than 0.2% of the greatest water level (see line 1180).

An example of the result of the computations is shown in Figure 10.10, for a

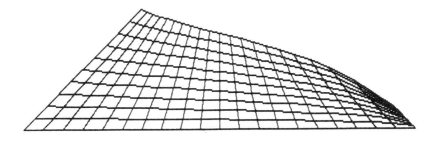

Fig. 10.10. Example of network generated by Program BV10-4.

dam with slopes of 45° on both sides. To generate this network, the input data to be given to the program, in response to the questions posed by it, are

100, 45, 135, 30, 5, 20, 10.

The first value is the width of the dam as its toe, the second and third values indicate that the slopes are 45 and 135°, respectively, with respect to an axis pointing to the right. The water level at the left side is 30 m, and the water level at the right side is 5 m, both above the impervious base used as a datum level. The last two values (20 and 10) indicate that the mesh to be generated should consist of 20 columns of elements and 10 rows. The number of elements is then 200, and the number of nodes is 231. On the basis of these data the computer program generates a mesh, by estimating a free surface, and subdividing the domain into 20 sections, using equidistant points along the lower boundary. Each section of the domain contains 10 elements in an approximately vertical direction, and each of these sections has a triangular shape, with its top angle in the intersection point of the upstream and downstream slope. During subsequent iterations, the nodes of the network move along the lines separating the segments. As the segments remain unchanged in the process, the program needs little modification to account for a variable permeability. As it is, the permeability is constant.

After performing the finite element calculations, the values of the head along the free surface are compared with the elevation y (in line 1600). If one of the differences is greater than the accuracy $E$, the value of the switch parameter $LL$ is set to 1, which indicates that another iteration will be made (see line 1430). The accuracy $E$ is set to $HL/500$ in line 1180. The user may, of course, replace this by any other acceptable value.

## 10.5. Unsteady Flow in an Aquifer

An important generalization of the models discussed in the previous sections is the introduction of storage and the unsteady behaviour that it entails. In this section, the unsteady effects will be considered for the case of horizontal flow in an aquifer with distributed infiltration, point sources and sinks, and elastic or phreatic storage.

The basic differential equation is, see (4.2.24), rewritten here, for an isotropic aquifer, in the form

$$S \frac{\partial \phi}{\partial t} = \frac{\partial}{\partial x}\left(T \frac{\partial \phi}{\partial x}\right) + \frac{\partial}{\partial y}\left(T \frac{\partial \phi}{\partial y}\right) + I \qquad (10.5.1)$$

where $S$ denotes the storativity, and $I$ denotes the net infiltration.

The boundary conditions are assumed to be of the first or second kind. Thus, everywhere along the boundary, either the head or the rate of water supply is given. Unlike the steady flow case, initial conditions are now also needed. This initial condition is written as

$$t = 0: \quad \phi = \phi^0(x, y) \qquad (10.5.2)$$

where $\phi^0(x, y)$ is a known function.

The problem is solved by a stepwise integration in time. A procedure will be developed in which the values at the end of a first time step can be obtained, starting from the initial values. A next time step can then be made by considering the values at the end of the first time step as the initial values for the second step, etc.

A simple way of deriving the basic algebraic equations of the numerical method is to integrate the differential equation (10.5.1) from $t = 0$ to $t = \Delta t$. Doing so, we obtain

$$S \frac{\phi' - \phi^0}{\Delta t} = \frac{\partial}{\partial x}\left(T \frac{\partial \phi}{\partial x}\right) + \frac{\partial}{\partial y}\left(T \frac{\partial \phi}{\partial y}\right) + I \qquad (10.5.3)$$

where $\phi'$ is the value of the head at the end of the time interval considered, while the values of $\phi$ and $I$ are averages over the time interval. They are obtained by integration over $\Delta t$, and dividing the result by $\Delta t$.

It is now assumed that the average value of the head, $\phi$, during the time interval can be expressed in terms of the values at its beginning and its end, in the form

$$\phi = \varepsilon\phi^0 + (1-\varepsilon)\phi' \tag{10.5.4}$$

where $\varepsilon$ is an interpolation constant, with $0 \leqslant \varepsilon \leqslant 1$. Formula (10.5.4) states that the average value is a linear combination of the initial value and the final one, possibly with different weights. For $\varepsilon = 0$ the average value is equal to the final value. This corresponds to a *backward finite difference*. For $\varepsilon = 1$ the average value is equal to the initial value, corresponding to a *forward finite difference* in time. The unbiased value $\varepsilon = \frac{1}{2}$ corresponds to a *central finite difference* (Subsection 8.2.1). In terms of finite differences, this would lead to the *Crank–Nicholson process* (Section 9.2). In many practical cases, the transient processes are of a slowly decaying nature. This suggests that the average value should be slightly biased towards the final value, with $\varepsilon$ being somewhat smaller than $\frac{1}{2}$. At this stage, we leave the value of $\varepsilon$ unspecified. It will be an input parameter of the computer program.

From (10.5.4) we obtain

$$\phi' - \phi^0 = \frac{\phi - \phi^0}{1 - \varepsilon} \tag{10.5.5}$$

so that (10.5.3) can be rewritten as

$$\frac{\partial}{\partial x}\left(T\frac{\partial\phi}{\partial x}\right) + \frac{\partial}{\partial y}\left(T\frac{\partial\phi}{\partial y}\right) + I - S\frac{\phi - \phi^0}{\Delta t(1-\varepsilon)} = 0. \tag{10.5.6}$$

This is the final form of the differential equation after integration over the time step. The time derivative has been eliminated by the process of integrating over the time step. As a result, an equation in terms of the average value $\phi$ has been obtained.

By comparing (10.5.6) with (10.1.1), it is found that both are formally identical. The only differences are that in the latter the coefficient $1/c$ is replaced by $S/\Delta t(1-\varepsilon)$, and that the given variable $\phi'$ is replaced by $\phi^0$. These are merely differences of notation. It follows that the techniques developed for the solution of (10.1.1) can be directly applied to the solution of (10.5.6).

Altogether, it may be concluded that a finite element program for the solution of nonsteady flow in an aquifer can be based on a slightly modified form of (10.3.1), i.e.,

$$\sum_j \{P_{ij}\phi_j + R_{ij}(\phi_j - \phi_j^0)\} = Q_i \quad (i \in C) \tag{10.5.6}$$

where now the coefficients $P_{ij}$ and $R_{ij}$ are given by

$$P_{ij} = \{T_p/2|D|\}\{b_i b_j + c_i c_j\} \tag{10.5.7}$$

$$R_{ij} = \frac{\{b_i b_j Z_{xx} + c_i c_j Z_{yy} + (b_i c_j + b_j c_i)Z_{xy} + d_i d_j\}S_p}{\{2|D|\Delta t(1-\varepsilon)\}} \tag{10.5.8}$$

A computer program for this system of equations can be derived from Program BV10-3 by an appropriate modification of the coefficients. The only complication is that, in general, it is not desirable to assume that the value of $\Delta t$ remains constant during the process. A variable value of $\Delta t$ seems to mean that the matrix $R$ has to be recalculated in each time step. However, as $\Delta t$ is the same for all elements, it can be considered as a variable multiplication factor of a constant matrix, thus saving much computation time.

The computer program for unsteady flow is listed in Program BV10-5.

```
1000 DEFINT I-N:KEY OFF:OPTION BASE 1:GOSUB 1660
1010 PRINT"---  Bear & Verruijt - Groundwater Modeling"
1020 PRINT"---  Plane non-steady Groundwater Flow"
1030 PRINT"---  Finite Element Method":PRINT"---  Program 10.5"
1040 PRINT"---  Quadrangular elements, input by DATA statements"
1050 PRINT"---  Solution by conjugate gradient method"
1060 DIM XJ(4),YJ(4),B(3),C(3),D(3),KS(4,3)
1070 DIM X(300),Y(300),IP(300),F(300),FA(300),Q(300)
1080 DIM NP(250,4),T(250),S(250),PP(250)
1090 DIM U(300),V(300),W(300),P(300,11),R(300,11),KP(300,11):NZ=11
1100 PRINT:PRINT:PRINT"Reading data":READ N,M,E,NS
1110 FOR I=1 TO N:READ X(I),Y(I),IP(I),FA(I),Q(I):F(I)=FA(I):NEXT I
1120 FOR J=1 TO M:READ NP(J,1),NP(J,2),NP(J,3),NP(J,4),T(J),S(J),PP(J)
1130 NEXT J:FOR I=1 TO 4:FOR J=1 TO 3:K=I+J-1:IF K>4 THEN K=K-4
1140 KS(I,J)=K:NEXT J,I:GOSUB 1660:PRINT"Generation of pointer matrix"
1150 FOR I=1 TO N:KP(I,1)=I:KP(I,NZ)=1:NEXT I
1160 PRINT"   Element ...... ";:FOR J=1 TO M:PRINT J;
1170 FOR K=1 TO 4:KK=NP(J,K):FOR L=1 TO 4:LL=NP(J,L)
1180 IA=0:FOR II=1 TO KP(KK,NZ):IF KP(KK,II)=LL THEN IA=1
1190 NEXT II:IF IA=0 THEN KB=KP(KK,NZ)+1:KP(KK,NZ)=KB:KP(KK,KB)=LL
1200 IF KB=NZ THEN 1650
1210 NEXT L,K,J:PRINT:PRINT:EE=.00001
1220 PRINT"Generation of system matrices":PRINT"   Element ...... ";
1230 FOR J=1 TO M:PRINT J;:FOR KW=1 TO 4:ZX=0:ZY=0
1240 FOR I=1 TO 3:K=NP(J,KS(KW,I)):XJ(I)=X(K):YJ(I)=Y(K)
1250 ZX=ZX+X(K):ZY=ZY+Y(K):NEXT I:ZX=ZX/3:ZY=ZY/3
1260 FOR I=1 TO 3:XJ(I)=XJ(I)-ZX:YJ(I)=YJ(I)-ZY:NEXT I
1270 B(1)=YJ(2)-YJ(3):B(2)=YJ(3)-YJ(1):B(3)=YJ(1)-YJ(2)
1280 C(1)=XJ(3)-XJ(2):C(2)=XJ(1)-XJ(3):C(3)=XJ(2)-XJ(1)
1290 D(1)=XJ(2)*YJ(3)-XJ(3)*YJ(2):D(2)=XJ(3)*YJ(1)-XJ(1)*YJ(3)
1300 D(3)=XJ(1)*YJ(2)-XJ(2)*YJ(1)
1310 D=ABS(D(1)+D(2)+D(3)):IF D<EE THEN 1440
1320 DD=T(J)/(4*D):DE=S(J)/(4*(1-E)*D)
1330 XX=(XJ(1)*XJ(1)+XJ(2)*XJ(2)+XJ(3)*XJ(3))/12
1340 XY=(XJ(1)*YJ(1)+XJ(2)*YJ(2)+XJ(3)*YJ(3))/12
1350 YY=(YJ(1)*YJ(1)+YJ(2)*YJ(2)+YJ(3)*YJ(3))/12
```

```
1360 FOR K=1 TO 3:KK=NP(J,KS(KW,K)):II=KP(KK,NZ):FOR LL=1 TO II:L=1
1370 KV=KS(KW,L):IF NP(J,KV)=KP(KK,LL) THEN 1390
1380 L=L+1:IF L<4 THEN 1370 ELSE 1420
1390 P(KK,LL)=P(KK,LL)+DD*(B(K)*B(L)+C(K)*C(L))
1400 AA=XX*B(K)*B(L)+XY*(B(K)*C(L)+B(L)*C(K))+YY*C(K)*C(L)+D(K)*D(L)
1410 R(KK,LL)=R(KK,LL)+DE*AA
1420 NEXT LL,K
1430 FOR I=1 TO 3:K=NP(J,KS(KW,I)):Q(K)=Q(K)+D*PP(J)/12:NEXT I
1440 NEXT KW,J:PRINT:PRINT:TN=0:EE=EE*EE
1450 FOR IS=1 TO NS:READ TT:DT=TT-TN:B=1/DT:TN=TT
1460 PRINT"Solution of equations":PRINT"    Iteration .... ";:IT=1
1470 FOR I=1 TO N:U(I)=0:IZ=KP(I,NZ):IF IP(I)>0 THEN 1500
1480 U(I)=Q(I):FOR J=1 TO IZ:CC=B*R(I,J):L=KP(I,J)
1490 U(I)=U(I)-(P(I,J)+CC)*F(L)+CC*FA(L):NEXT J
1500 V(I)=U(I):NEXT I:UU=0:FOR I=1 TO N:UU=UU+U(I)*U(I):NEXT I
1510 PRINT IT;:FOR I=1 TO N:W(I)=0:IZ=KP(I,NZ)
1520 FOR J=1 TO IZ:W(I)=W(I)+(P(I,J)+B*R(I,J))*V(KP(I,J)):NEXT J,I
1530 VW=0:FOR I=1 TO N:VW=VW+V(I)*W(I):NEXT I
1540 AA=UU/VW:FOR I=1 TO N:IF IP(I)>0 THEN 1560
1550 F(I)=F(I)+AA*V(I):U(I)=U(I)-AA*W(I)
1560 NEXT I:WW=0:FOR I=1 TO N:WW=WW+U(I)*U(I):NEXT I
1570 BB=WW/UU:FOR I=1 TO N:V(I)=U(I)+BB*V(I):NEXT I:UU=WW
1580 IT=IT+1:IF IT<=N AND UU>EE THEN 1510
1590 FOR I=1 TO N:FA(I)=FA(I)+(F(I)-FA(I))/(1-E):F(I)=FA(I):NEXT I
1600 GOSUB 1660:PRINT"Time =";TN:PRINT:A$="#####.###":FOR I=1 TO N
1610 PRINT" x = ";:PRINT USING A$;X(I);:PRINT"   -   y = ";
1620 PRINT USING A$;Y(I);:PRINT"   -   F = ";:PRINT USING A$;FA(I)
1630 NEXT I,IS
1640 PRINT:END
1650 GOSUB 1660:PRINT"Pointer width (NZ) too small.":GOTO 1640
1660 CLS:LOCATE 1,30,1:COLOR 0,7:PRINT" Finite Elements - 5 ";
1670 COLOR 7,0:PRINT:PRINT:RETURN
1680 DATA 22,10,0.5,13
1690 DATA 0,0,-1,0,0,0,100,-1,0,0
1700 DATA 100,0,-1,0,0,100,100,-1,0,0
1710 DATA 200,0,-1,0,0,200,100,-1,0,0
1720 DATA 300,0,-1,0,0,300,100,-1,0,0
1730 DATA 400,0,-1,0,0,400,100,-1,0,0
1740 DATA 500,0,-1,0,0,500,100,-1,0,0
1750 DATA 600,0,-1,0,0,600,100,-1,0,0
1760 DATA 700,0,-1,0,0,700,100,-1,0,0
1770 DATA 800,0,-1,0,0,800,100,-1,0,0
1780 DATA 900,0,-1,0,0,900,100,-1,0,0
1790 DATA 1000,0,1,0,0,1000,100,1,0,0
1800 DATA 1,2,3,4,1000,0.4,0.001
1810 DATA 3,4,5,6,1000,0.4,0.001
1820 DATA 5,6,7,8,1000,0.4,0.001
1830 DATA 7,8,9,10,1000,0.4,0.001
1840 DATA 9,10,11,12,1000,0.4,0.001
1850 DATA 11,12,13,14,1000,0.4,0.001
1860 DATA 13,14,15,16,1000,0.4,0.001
1870 DATA 15,16,17,18,1000,0.4,0.001
1880 DATA 17,18,19,20,1000,0.4,0.001
1890 DATA 19,20,21,22,1000,0.4,0.001
1900 DATA 1,2,3,5,10,20,30,50,100,200,300,500,1000
```

Program BV10-5. Unsteady flow with infiltration and leakage.

Program BV10-5 contains data for a problem similar to the one illustrated in Figure 10.8, except that there is no leakage, and that the flow is unsteady. It is assumed that at time $t = 0$, uniform infiltration starts on the entire area of the aquifer at a constant rate $I = 0.001$ m/d, and that the initial head is zero. The storativity is $S = 0.4$. The transmissivity is $T = 1000$ m$^2$/d.

As in all transient solutions, the magnitude of the time steps deserves some special attention. Because the finite element method is very similar to the finite difference method, it can be expected that the same general ideas (presented in Section 9.3) are valid. Thus, it can be expected that for a value of $\varepsilon$ in the range from $\varepsilon = \frac{1}{2}$ to $\varepsilon = 1$ the numerical process is unconditionally stable. This is indeed observed when the program is executed. Also, in order to obtain sufficient accuracy it is suggested, following (9.3.5), that the first time step be of the order of magnitude of

$$\Delta t \approx S(\Delta x)^2/(2T) \tag{10.5.9}$$

where $\Delta x$ denotes a characteristic measure for the mesh size. Because, in the present case, the smallest mesh size is 100 m, it follows that the first time step should be about 2 days. In the program the first time step is actually 1 day, and subsequent time steps are gradually increased, so that in 13 time steps a value of 1000 days is reached.

The solution obtained by running Program BV10-5 is compared with the analytical solution of this problem in Figure 10.11. This analytical solution, which can be obtained by the Laplace transform method, is

$$\phi = \frac{I(L^2 - x^2)}{2T} - \frac{16IL^2}{\pi^3 T} \Sigma \left\{ \frac{(-1)^k}{(2k+1)^3} \cos\left[(2k+1)\frac{\pi x}{2L}\right] \times \right.$$

$$\left. \times \exp\left[-(2k+1)^2 \frac{\pi^2 Tt}{4SL^2}\right] \right\} \tag{10.5.10}$$

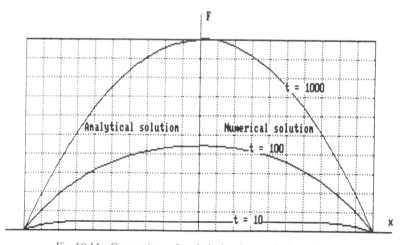

Fig. 10.11. Comparison of analytical and numerical solutions.

In this expression, the summation should be from $k = 0$ to $k = \infty$ or, in practice, from $k = 0$ to a value of $k$ such that the exponent is larger than 10. The analytical solution is represented in the left half of the figure and the numerical solution in the right half. As can be seen from the figure, the agreement is very good, with errors of less than 1%.

## 10.6. Generalizations

In this section some possible extensions and generalizations of the computer programs presented in Sections 10.1—10.5 are discussed. Some of these will be described in some detail and additional generalizations are included in the Problems.

### 10.6.1. LAYERED SYSTEM OF AQUIFERS

All programs presented above apply to the flow in a single aquifer. In practice, a layered system of two or three aquifers with variable heads, and with exchange of water through the intermediate aquitards, is often encountered. The extension to systems of this type is not too difficult. It is suggested that the same mesh be used for all aquifers, so that the major difference with respect to the programs presented in this chapter will be that several variables are defined for each node, one for the head in each aquifer. The simplest procedure is to generate a system of linear equations for each of these variables, similar to Program BV10-3, with the coupling terms due to leakage being accounted for through matrices of the form of matrix $R$ in that program. The solution for the heads in the various aquifers can then be obtained iteratively. As the solution procedure is iterative anyway, this modification can easily be incorporated. Of course, the memory requirement is significantly increased, as a system matrix has to be stored for each layer. Also, input and output data must be given for each layer. However, this is mainly an administrative problem.

Other relatively simple generalizations include: the introduction of an anisotropic transmissivity, the introduction of leakage into Program BV10-5, and the introduction of another mesh generator in Program BV10-4, for instance, to account for a nonhomogeneous permeability, or a dam having a nontriangular shape.

### 10.6.2. EXTENDED FACILITIES FOR INPUT AND OUTPUT

For professional applications of the finite element method it is often convenient to divide the program in at least three different parts: input, the actual calculations, and output.

The input program may be operating interactively, with the program asking for all required data to be entered. Such a program should also have the option of presenting the data on the screen, in order to enable modifications or corrections, when necessary. These data can then be stored in a datafile, on a disk. In this way, a dataset can be used as a basis for various alternative datasets, thus enabling the investigation of the sensitivity of the solution to the values of certain data. Editing of a datafile may be done by a special program, or by a commercial editor, for example, a word-processor, a data base program, or a spreadsheet program. Input facilities may include a mesh generator, which permits the generation of a large number of data for nodes and elements on the basis of a limited number of input data. Such a mesh may then be edited to account for local inhomogeneities.

If a separate input program is used to generate a dataset, the second part of the program, in which the actual calculations are executed, needs only to contain some statements for reading the dataset from the disk. Many arithmetic computations have to be performed in this part of the program. It may be worthwhile to use special facilities for this purpose, such as a special language, or a special processor. BASIC, as used in all programs presented above, allows for user-friendly input and sophisticated screen control but, in general, it is not the fastest language, even when compiled. Also, most versions of BASIC can only address 64 kB of memory on the majority of small computer systems. Other languages, e.g., FORTRAN, PASCAL, or C, may be better suited to perform these calculations. For the analysis of large systems, it may also be worthwhile to install a special arithmetic co-processor.

The results of the calculations can also be stored in a datafile, which can then be used as input for a special output program. Like the input program, this may be a user-friendly program, presenting the data on the screen, on a printer, or in graphical form on the screen or the printer.

In order to demonstrate the possibilities of a program with extended output facilities, two examples of applications using such a program are shown below.

EXAMPLE 1. The first example concerns steady flow in a rectangular aquifer of $1000 \times 600$ m. The transmissivity is $T = 20$ m$^2$/d. Two wells of discharge 250 m$^3$/d are located at the points $x = 300$ m, $y = 300$ m, and $x = 700$ m, $y = 300$ m. The head is constant along the entire boundary. The results are shown in Figure 10.12 in the form of a fishnet diagram. Such a diagram can be presented on the screen by projecting the point $x, y, F$ in three-dimensional space on the two-dimensional screen at the point with coordinates

$$x' = S_x x + S_y y \cos(\alpha), \qquad y' = S_y y \sin(\alpha) + S_f F$$

where $\alpha$ is the angle between the $x$- and $y$-axes in the horizontal plane, and $S_x$, $S_y$, and $S_f$ are scale-factors which can be chosen such that the figure fills a selected window on the screen. The fishnet diagram is then obtained by connecting the four points representing the values of the head in the four vertices of each element by straight lines.

The results shown in Figure 10.12 were obtained by an extended version of Program BV10-5, taking $S = 0$ to obtain the steady-state solution. The mesh, consisting of 1581 nodes and 1500 elements, was generated by a mesh generator

THE FINITE ELEMENT METHOD 283

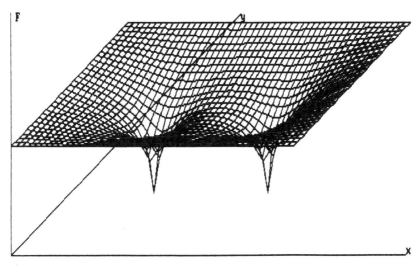

Fig. 10.12. Example 1, Fishnet diagram.

in a separate program. The actual calculations took a little more than one hour, on an 8088 computer, with 512 kB memory, using a PASCAL compiler.

EXAMPLE 2. The results of a second example are shown in Figure 10.13. The figure shows contours of the steady groundwater head for the case of a rectangular aquifer, of transmissivity 20 m$^2$/d, with a constant head boundary. A pond, in which the head remains equal to the head along the outer boundary, is located near the center. Over the entire region a uniform infiltration rate of 0.001 m/d is

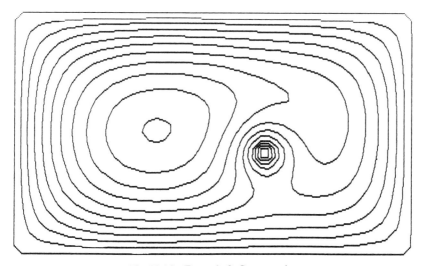

Fig. 10.13. Example 2, Contour plot.

supplied to the aquifer. It appears that most of the water flows towards the outer boundary. The pond drains only a small area.

Groundwater level contours are often considered as an attractive form of graphical presentation of the output data. Such a contour plot can be produced by taking a certain value, say $F_0$, for the head, and then checking for each of the four boundaries of each element whether the value $F_0$ is between the two values of the head at the two nodes at either end of that element boundary. If this is the case, the contour will intersect that element boundary, and the exact location of the intersection point can be determined by linear interpolation. As such intersection points always occur in pairs, for each element, a part of the contour line can then be drawn by connecting the two intersection points found on the boundary of the element considered. Repetition of this process for each element, and for a number of contour values, will produce a figure of the type shown in Figure 10.13.

In addition to a graphical presentation of the output data, a list of output data may be produced by a printer, to obtain detailed numerical information.

CHAPTER ELEVEN

# Transport by Advection

The basic theory of pollution transport by advection, dispersion, and diffusion, in groundwater, has been presented in Chapter 6. Numerical models for that general case are considered in Chapter 12. In this chapter, numerical models are presented for an approximate model (Section 6.6), in which the transport of a polluting component is taking place by advection only, neglecting transport by mechanical dispersion and molecular diffusion. In such a model the pollutant is transported at the average velocity of the water, which acts as its carrier.

Two numerical techniques are presented in this chapter: a semi-analytical method, based on an analytical solution of the groundwater flow problem, and a fully numerical method based on a finite element solution of the problem. In the latter case, it will turn out to be advantageous, from a viewpoint of accuracy, to use a formulation in terms of the stream function. All examples to be given in this chapter are basically two-dimensional, in the sense that one of the three components of the water velocity vector is so much smaller than either of the other two, that the velocity can be described by a vector in a plane.

## 11.1. Basic Equations

Before proceeding to the presentation of the numerical models, the basic equations for transport by advection only will be recalled from Chapter 6. As seen in that chapter, the basic transport phenomena of a tracer or a pollutant in groundwater consist of advection, which is transport at the average water velocity, and diffusion and dispersion, which constitute mechanisms for transport of the pollutant relative to the water. If the groundwater contains a single polluting component, the basic variable is its concentration in the water, which is denoted by $c$, the mass of tracer per unit volume of the pore water. The flux of this tracer, expressed as a mass flux through a unit area in the soil, is expressed by (6.2.19). Assuming that the soil is completely saturated with water, and neglecting dispersion and diffusion, this equation can be written as

$$\mathbf{q}_c = nc\mathbf{V} \tag{11.1.1}$$

where $n$ is the porosity of the soil, and $\mathbf{V}$ is the average velocity of the groundwater.

The second basic equation is the equation of conservation of mass of the tracer (6.3.1). In the absence of losses due to adsorption, ion exchange, decay, etc., this equation can be written as

$$\frac{\partial(nc)}{\partial t} = -\nabla \cdot \mathbf{q}_c. \qquad (11.1.2)$$

Substitution of (11.1.1) into (11.1.2) gives

$$n\frac{\partial c}{\partial t} = -n\mathbf{V} \cdot \nabla c + c \cdot \nabla(n\mathbf{V}). \qquad (11.1.3)$$

Assuming that the flow is steady, the second term in the right-hand member vanishes and, thus, the basic equation becomes (see (6.6.8)),

$$\frac{\partial c}{\partial t} = -\mathbf{V} \cdot \nabla c = -V_x \frac{\partial c}{\partial x} - V_y \frac{\partial c}{\partial y} - V_z \frac{\partial c}{\partial z}. \qquad (11.1.4)$$

In the two-dimensional case, this equation further reduces to (6.6.9), i.e.,

$$\frac{\partial c}{\partial t} = -V_x \frac{\partial c}{\partial x} - V_y \frac{\partial c}{\partial y}. \qquad (11.1.5)$$

This is the mathematical problem that is to be solved, with the concentration $c$ being the unknown quantity, and the velocity field assumed to be given. It should be noted that this simplified problem has been obtained on the basis of a large amount of restrictive assumptions, such as steady flow, no dispersion, no adsorption, and no coupling through feedback of the tracer transport on the fluid transport. Thus, various important phenomena, including the influence of density differences, have been excluded. Nevertheless, it is believed that in many practical cases the major transport phenomenon is advection, so that at least the major component of the transport process is taken into account.

It was shown in Section 6.6 that an equation of the form (11.1.5) can be solved by the method of *characteristics*. The concentration remains constant in the direction of the characteristics, which coincides with the direction of the flow. Thus, in an infinitesimal time step $dt$, the pollution is transported in spatial steps $dx$ and $dy$, such that

$$dx = V_x \, dt, \qquad dy = V_y \, dt. \qquad (11.1.6)$$

The main problem left to be solved now is to determine the streamlines. When these have been found, it still remains to trace the progress of the particles along them, in accordance with (11.1.6). For the determination of the streamlines, a semi-analytic method will be presented in the next section. A fully numerical solution will be presented in Section 11.6.

## 11.2. Semi-Analytic Solution

Analytic solutions have been obtained for a limited class of groundwater flow problems, mainly for cases with simple domain geometry and homogeneous permeabilty. Among those reported in the literature, we find analytic solutions for flow under the influence of an arbitrary number of wells in aquifers of infinite extent, in a sector of a plane, and in an infinite or semi-infinite strip. Solutions of this type can be found in the many existing textbooks and handbooks on the theory of flow through porous media, e.g., Muskat (1937), Kober (1957), Harr (1962), Polubarinova–Kochina (1962), Aravin and Numerov (1965), Bear (1979), and Verruijt (1982). These solutions are usually expressed in terms of an analytic function for the piezometric head, or in terms of a complex potential. The velocity fields can be determined by inserting the solution into Darcy's law. As can be expected, these expressions are usually so complicated that analytic integration of (11.1.6) is impossible and, therefore, they must be integrated numerically.

The simplest method of integrating (11.1.6) is by using *Euler's method*, which is equivalent to a forward finite difference. If the location of a particle at a certain instant of time is denoted by its coordinates $x_0$ and $y_0$, Euler's method gives the following expressions for the displacement components during a time step of length $\Delta t$

$$\Delta x = V_x(x_0, y_0)\, \Delta t, \qquad \Delta y = V_y(x_0, y_0)\, \Delta t. \tag{11.2.1}$$

It should be noted that in this simplest possible method, the velocity is taken as constant during the entire time interval, and equal to the initial velocity. It can be expected that this will be especially inaccurate in the vicinity of a source or a sink, where the velocity changes rapidly, and also in a region where the streamlines are strongly curved. In such regions, the approximation by straight lines, following the initial velocity, may not be able to follow the curvature. The effect of the deviations may be cumulative and, thus, the error may grow with time.

Despite its simplicity and possible inaccuracy, the Euler method has been used with considerable success, provided that the time steps are taken small enough (Javandel et al., 1984, who include the listing of a computer program, in FORTRAN). A more refined method, developed by Van den Akker (1982), for a variety of problems, is to use one of the Runge–Kutta methods (Abramowitz and Stegun, 1965; Korn and Korn, 1968). In these methods, the displacement components are calculated in several steps (two, three, or even more), with increasing accuracy.

In the computer program listed in this section, the two-point Runge–Kutta formula is used, which can be formulated for the present problem in the following form. First the velocity is determined at the initial particle location $x_0$, $y_0$, and a point $x_1$, $y_1$ is determined halfway along the path emanating from the initial point,

$$x_1 = x_0 + V_x(x_0, y_0)\, \Delta t/2, \qquad y_1 = y_0 + V_y(x_0, y_0)\, \Delta t/2. \tag{11.2.2}$$

Next, the velocity at the point $x_1$, $y_1$ is determined (using the analytical solution), and this velocity is used for a new estimate for the entire time step, starting again at the initial location,

$$x_2 = x_0 + V_x(x_1, y_1)\,\Delta t, \qquad y_2 = y_0 + V_y(x_1, y_1)\,\Delta t. \tag{11.2.3}$$

This can be expected to give more accurate results because the velocity used is more representative of the average velocity along the incremental displacement.

## 11.3. System of Wells in an Infinite Field

In many cases, pollution is transported within an aquifer of large areal extent, in which the flow regime is determined by the superposition of the natural flow in the aquifer and the flow produced by a number of wells, each having its own positive or negative rate of production. As an example, consider the case of steady uniform flow in a confined aquifer, in which $m$ pumping and recharging wells are operating. For this case, the analytical solution is, in terms of the piezometric head $\phi$,

$$\phi = -\frac{n}{K}(xV_{0x} + yV_{0y}) + \sum_{j=1}^{m} \frac{Q_j}{2\pi KB} \ln(r_j/R) \tag{11.3.1}$$

where $V_{0x}$ and $V_{0y}$ are the components of the uniform (natural) flow, $Q_j$ is the production of well number $j$, and $r_j$ is the distance from the point at which $\phi$ is being determined to that well,

$$r_j = \sqrt{\{(x - x_j)^2 + (y - y_j)^2\}}.$$

The components of the (average) velocity vector can be determined by using Darcy's law,

$$V_x = -\frac{K}{n}\frac{\partial \phi}{\partial x}, \qquad V_y = -\frac{K}{n}\frac{\partial \phi}{\partial y}. \tag{11.3.2}$$

This gives

$$V_x = V_{0x} - \sum \frac{Q_j}{2\pi nB}\frac{x - x_j}{r_j^2}, \qquad V_y = V_{0y} - \sum \frac{Q_j}{2\pi nB}\frac{y - y_j}{r_j^2}. \tag{11.3.3}$$

These equations give the components of the velocity vector at any point of the field. When the number of wells, $m$, is very large, the actual computation of the velocity components in a point may suitably be performed by a computer program.

A computer program for the determination of streamlines will be presented below. Again, the program is of an interactive character, with all data being generated by the response of the user to questions posed by the computer program. For the sake of convenience, the program is divided into two parts: one for the generation of input data, and one for the actual calculations. This enables the use of the same input data for various types of problems. The first part of the program is listed as Program BV11-1-1.

```
100 DEFINT I-N:KEY OFF:OPTION BASE 1:GOSUB 530
110 PRINT"---   Bear & Verruijt - Groundwater Modeling"
120 PRINT"---   Plane steady Groundwater Flow"
130 PRINT"---   Advection using analytic solution"
140 PRINT"---   Program 11.1 - Part 1"
145 PRINT"---   Generation of dataset":PRINT
150 DIM XX(10),YY(10),QQ(10),AX(100),AY(100),MK(100)
170 NWM=10:NSM=100:PRINT:PRINT
180 INPUT"Name of dataset ................... ";D$:PRINT
190 PRINT"Properties of screen :":PRINT
200 INPUT" Dots in x-direction (640 ?) ....... ";NLX
210 INPUT" Dots in y-direction (200 ?) ....... ";NLY
220 INPUT" Scale factor x/y (2 ?) ............ ";SXY
230 INPUT" Graphics screen parameter (2 ?) ... ";NSCR
240 GOSUB 530:PRINT"Dimensions of field (m) :":PRINT
250 INPUT" x = 0 - ........................... ";XT
260 INPUT" y = 0 - ........................... ";YT
270 PRINT:PRINT"Natural velocity (m/d) :":PRINT
280 INPUT" Vx ................................ ";VX
290 INPUT" Vy ................................ ";VY:PRINT
300 INPUT" Time step (d) ..................... ";DT
310 INPUT" Steps between markers ............. ";NT
320 INPUT" Porosity .......................... ";PP
330 INPUT" Thickness of aquifer (m) .......... ";HH
340 INPUT" Number of wells ................... ";NW
350 INPUT" Number of stream lines ............ ";NS
360 IF NW>NWM THEN NW=NWM
370 IF NS>NSM THEN NS=NSM
380 FOR I=1 TO NW:GOSUB 530:PRINT"Well";I;":":PRINT
390 INPUT" x = ............................... ";XX(I)
400 INPUT" y = ............................... ";YY(I)
410 INPUT" Q = ............................... ";QQ(I)
420 NEXT I:FOR I=1 TO NS:GOSUB 530
430 PRINT"First point of stream line";I;":":PRINT
440 INPUT" x = ............................... ";AX(I)
450 INPUT" y = ............................... ";AY(I)
460 PRINT" Place markers (Y/N) .............. ? ";:GOSUB 550
470 IF A$="Y" THEN MK(I)=1 ELSE MK(I)=0
480 NEXT I:GOSUB 530:PRINT"Writing dataset ";D$:PRINT
490 OPEN"O",1,D$:PRINT#1,D$:PRINT#1,NLX;NLY;SXY;NSCR
500 PRINT#1,XT;YT;VX;VY;DT;NT;NW;PP;HH;NS
510 FOR I=1 TO NW:PRINT#1,XX(I);YY(I);QQ(I):NEXT I
520 FOR I=1 TO NS:PRINT#1,AX(I);AY(I);MK(I):NEXT I:CLOSE 1:END
530 CLS:LOCATE 1,25,1:COLOR 0,7:PRINT" Semi-analytic advection - Part 1 ";
540 COLOR 7,0:PRINT:PRINT:RETURN
550 A$=INKEY$:IF A$="Y" OR A$="y" THEN A$="Y":PRINT"Yes":RETURN
560 IF A$="N" OR A$="n" THEN A$="N":PRINT"No":RETURN
570 GOTO 550
```

Program BV11-1-1. Semi-analytic advection: generation of input data.

This program generates a datafile, the name of which is to be supplied by the user in response to the first question. The program then asks for the characteristics of the monitor, for the presentation of graphics. A standard form of screen may consist of 640 × 200 dots, and graphics may be represented on such a screen after the BASIC command SCREEN 2. On such a screen lines may be drawn by the command LINE (X1, Y1)—(X2, Y2). In this graphics mode the horizontal and vertical scales are unequal, so that a square would be represented by a rectangle which is approximately twice as high as it is wide. In order to correct this representation, the coordinates of the physical field may be adjusted by a factor *SXY*, which is read in line 220. For the command SCREEN 2, the appropriate value of *SXY* is 2. It may be mentioned that on some computers graphics of higher resolution may be represented, for instance, 640 × 400 dots. The corresponding values for the parameters entered in lines 200—230 are 640, 400, 1, 3, indicating a command SCREEN 3 to generate graphics on a screen with 640 × 400 dots, on which the horizontal and vertical scales are practically equal.

The next data to be entered are the dimensions of the physical field that is to be represented on the screen, say a region of 1000 meters by 500 meters. The flow may occur in a field of larger dimensions, perhaps even a region of infinite extent, but only the part of the field specified by the user will be represented on the screen. The program itself should determine (in its second part, to be presented below) the scales in *x*- and *y*-direction, so that a region of at least the given width and height will fit on the screen. The values for *VX* and *VY* entered in lines 280—290 represent the flow at infinity, or the natural flow, if there were no wells operating. They should be considered as the components of the average velocity vector, not the specific discharge. The next data to be given are the time step (e.g. 1 day, if the unit of time is a day), the number of time steps after which a marker should be set on a streamline (e.g., 50 to indicate markers after every 50 time steps), the porosity of the soil (e.g., 0.4), the thickness of the aquifer (e.g., 10 meter), the number of wells and the number of streamlines to be drawn. The maximum number of wells is 10, and the maximum number of streamlines is 100, as set in line 170. For each well the user should give the coordinates of its location and its discharge, positive for a pumping well, negative for a recharge well. Finally, for each of the streamlines the initial point must be given, and the user must indicate whether markers are actually to be set along this stream line. This enables to place markers on certain streamlines, and not on others.

DATA11.1 is shown as an example of a datafile.

```
DATA11.1
640    200    2    2
1000   600    0    0    2    100   2   .4   10   12
250    300   -1920
750    300    1920
251.932   300.518   0
251.414   301.414   0
250.518   301.932   0
249.482   301.932   0
248.586   301.414   0
248.068   300.518   0
248.068   299.482   0
248.586   298.586   0
```

```
         249.482    298.068   0
         250.518    298.068   0
         251.414    298.586   0
         251.932    299.482   0
                 Datafile DATA11.1.
```

These data refer to a problem of a source and a sink, at the points (250,300) and (750,300). A field of 1000 meter by 600 meter should be shown on the screen, and 12 streamlines should be drawn, starting at points encircling the source, without placing markers.

The program that will actually draw the streamlines on the screen is reproduced as Program BV11-1-2.

```
100 DEFINT I-N:KEY OFF:OPTION BASE 1:GOSUB 530
110 PRINT"---   Bear & Verruijt - Groundwater Modeling"
120 PRINT"---   Plane steady Groundwater Flow"
130 PRINT"---   Advection using analytic solution"
140 PRINT"---   Program 11.1 - Part 2"
150 PRINT"---   Stream lines for wells in infinite field":PRINT
160 DIM XX(10),YY(10),QQ(10),AX(100),AY(100),MK(100)
170 DIM PL(100),TA(100,10),TB(100,10)
180 PRINT:INPUT"Name of dataset ............. ";D$:PRINT
190 PRINT"Reading dataset ";D$:PRINT
200 OPEN"I",1,D$:INPUT#1,A$:INPUT#1,NLX,NLY,SXY,NSCR
210 INPUT#1,XT,YT,VX,VY,DT,NT,NW,PP,HH,NS
220 FOR I=1 TO NW:INPUT#1,XX(I),YY(I),QQ(I):NEXT I
230 FOR I=1 TO NS:INPUT#1,AX(I),AY(I),MK(I):NEXT I:CLOSE 1
240 SX=NLX/XT:SY=SXY*NLY/YT:IF SY<SX THEN SX=SY
250 SY=SX/SXY:PI=4*ATN(1):PA=.5*PI:PB=1.5*PI:IT=1:GOSUB 530
260 PRINT"Stream lines will be drawn on the screen":PRINT:PRINT
270 COLOR 0,7:PRINT" B ";:COLOR 7,0:PRINT" .... Begin":PRINT
280 COLOR 0,7:PRINT" S ";:COLOR 7,0:PRINT" .... Stop":PRINT
290 GOSUB 550:CLS:SCREEN NSCR
300 LINE(0,0)-(NLX-1,0):LINE(NLX-1,0)-(NLX-1,NLY-1)
310 LINE(NLX-1,NLY-1)-(0,NLY-1):LINE(0,NLY-1)-(0,0)
320 B$=INKEY$:IF B$="s" OR B$="S" THEN 520
330 FOR K=1 TO NS:IF PL(K)<0 THEN 500
340 XA=AX(K):YA=AY(K):GOSUB 560
350 FOR J=1 TO NW:IF QQ(J)<0 THEN 440
360 A=XB-XX(J):B=YB-YY(J):T=ATN(B/A):IF A<0 THEN T=T+PI
370 IF T<0 THEN T=T+2*PI
380 A=SX*A:B=SY*B:R=SQR(A*A+B*B):IF IT=1 THEN TA(K,J)=T:GOTO 440
390 TB(K,J)=TA(K,J):TA(K,J)=T:TC=ABS(TA(K,J)-TB(K,J))
400 IF (TC>PA AND TC<PB) OR R<5 THEN 450
410 A=XD-XX(J):B=YD-YY(J):T=ATN(B/A):IF A<0 THEN T=T+PI
420 IF T<0 THEN T=T+2*PI
430 TC=ABS(T-TB(K,J)):IF TC>PA AND TC<PB THEN 450
440 NEXT J:GOTO 460
450 XB=XX(J):YB=YY(J):PL(K)=-1:MS=MS-1
460 LINE(SX*XA,SY*YA)-(SX*XB,SY*YB):AX(K)=XB:AY(K)=YB
470 IF IT<NT OR PL(K)<0 OR MK(K)=0 THEN 500
480 AA=SX*XB:BB=SY*YB:FOR II=-1 TO 1
490 LINE(AA-2,BB+II)-(AA+2,BB+II):NEXT II
500 NEXT K:IT=IT+1:IF IT>NT THEN IT=1
```

```
510 GOTO 320
520 SCREEN 0:END
530 CLS:LOCATE 1,25,1:COLOR 0,7:PRINT" Semi-analytic advection - Part 2 ";
540 COLOR 7,0:PRINT:PRINT:RETURN
550 A$=INKEY$:IF A$="B" OR A$="b" THEN RETURN ELSE 550
560 REM   Runge-Kutta
570 X=XA:Y=YA:GOSUB 590:XD=XA+U*DT/2:YD=YA+V*DT/2
580 X=XD:Y=YD:GOSUB 590:XB=XA+U*DT:YB=YA+V*DT:RETURN
590 REM   Calculation of velocities
600 U=VX:V=VY:FOR KK=1 TO NW:D=QQ(KK)/(2*PI*PP*HH)
610 A=X-XX(KK):B=Y-YY(KK)
620 RR=A*A+B*B:U=U-D*A/RR:V=V-D*B/RR:NEXT KK:RETURN
```

Program BV11-1-2. Semi-analytic advection: wells in an infinite field.

Before presenting the output data, it may be instructive to give a general outline of this second part of the program. First the datafile, generated by Part 1, is read, in lines 180—230. Then, the scales are calculated, and the program will start after the key 'B' has been touched. A number of streamlines will be drawn in subsequent steps, based on a Runge—Kutta approximation of the spatial steps (the subroutine in lines 560—580), which uses the velocity field, as determined by (11.3.3). These velocities are calculated in the subroutine starting at line 590. The actual drawing of an incremental section of a streamline, from (XA, YA) to (XB, YB), is performed in line 460, taking into account the appropriate scale factors. The program can be interrupted at any time by touching the key 'S', as this is checked before each time step.

It may be noted that the programs have been written such that they admit compilation. Thus, the dimensions of all arrays have been fixed, avoiding dynamic dimensioning. Because of the large amount of calculations to be performed computer time may be rather large, up to several hours. Running them in compiled form will greatly reduce computer time.

The streamlines drawn on the screen, using the datafile DATA11.1 as input, are shown in Figure 11.1.

In the absence of natural flow in the aquifer, the streamlines for a system of a source and a sink in an infinite field should be circles. Figure 11.1 shows that this is indeed the case.

A special feature of the second part of Program BV11-1 is that it contains a number of criteria for the termination of a certain streamline, in particular, in the vicinity of a well. As a particle approaches a pumping well, the velocity becomes very large and, without special measures, it may happen that a particle is transported far beyond the well. In the graph this would lead to a blurred picture, or to particles being shot all across the field. Two special features have been incorporated in the program to eliminate this effect. One is to let a point coincide with a well when the distance has become smaller than a certain small amount (5 dots, see line 400). The other check is that during the entire process the angle of the line from a particle to each pumping well is stored, in the two-dimensional array TA. After each step, these angles are re-calculated, and if the difference is

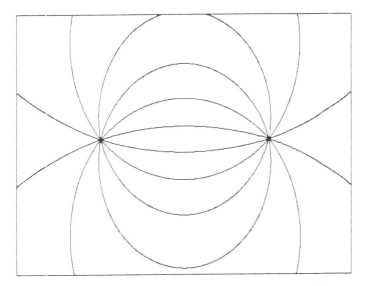

Fig. 11.1. Example DATA11.1. Streamlines for source and sink.

larger than $\pi/2$, or smaller than $3\pi/2$, this is interpreted to indicate the crossing of a well. The particle is then located at this well, and the drawing of this streamline is terminated. This is a time-consuming procedure, especially if the number of wells and streamlines is large. It may be omitted, at the risk of the picture being spoiled unexpectedly. The modifications of the program to omit these checks are left as exercises.

A second example is shown in Figure 11.2, for datafile DATA11.2.

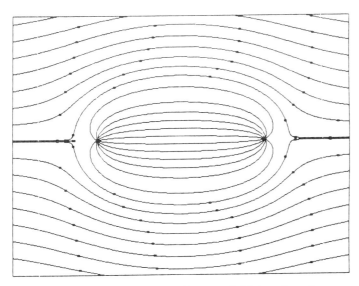

Fig. 11.2. Example DATA11.2. Source and sink in uniform flow.

```
DATA11.2
640   200   2  2
1000  600   1  0   2  100  2  .4  10  29
250   300   -1920
750   300    1920
251.932  300.518  0
251.414  301.414  0
250.518  301.932  0
249.482  301.932  0
248.586  301.414  0
248.068  300.518  0
248.068  299.482  0
248.586  298.586  0
249.482  298.068  0
250.518  298.068  0
251.414  298.586  0
251.932  299.482  0
0    20   1
0    60   1
0   100   1
0   140   1
0   180   1
0   220   1
0   260   1
0   299   1
0   300   1
0   301   1
0   340   1
0   380   1
0   420   1
0   460   1
0   500   1
0   540   1
0   580   1
```

Datafile DATA11.2.

This example concerns a source and a sink of equal strength in a uniform flow. Markers are placed on the streamlines coming from the left edge. As the two wells are located along the line $y = 300$, a streamline starting with that $y$-coordinate will pass through a stagnation point. For that reason, streamlines have been entered for $y = 299$, $y = 300$ and $y = 301$. It can be seen from the figure that the central streamline of these three is indeed trapped in the stagnation point, and stops moving. The two others flow around the water body circulating from the source to the sink, and they meet again at the right-hand side. These observations may serve as demonstrations of the accuracy of the procedure.

A final example is shown in Figure 11.3, describing the flow under the influence of two wells which intercept part of a flow field.

The datafile for the case illustrated in Figure 11.3 is DATA11.3, in which most of the initial points of the markers on the streamlines are omitted.

TRANSPORT BY ADVECTION                                                          295

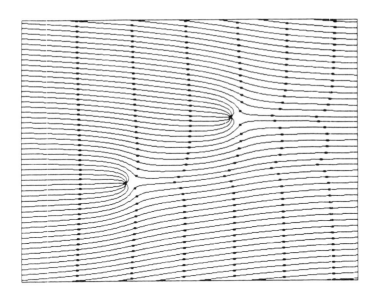

Fig. 11.3. Example DATA11.3. Two wells in an infinite field.

It should be pointed out that the programs can determine streamlines outside the region actually shown on the screen. This enables for streamlines to appear across the upper and lower boundaries. In Figure 11.3 this indeed occurs, because the wells intercept so much of the flow.

The data for example DATA11.3 are such that each well should intercept about

```
DATA11.3
640    200   2   2
1280   800   2   0   2   50   2   0.4   10   74
400    500   1280
800    300   1280
0   -184   1
0   -168   1
0   -152   1
. . . . . . .
. . . . . . .
. . . . . . .
0    952   1
0    968   1
0    984   1
```

Datafile DATA11.3.

10 streamlines, as the discharge of the wells is 1280 m$^3$/d, and the discharge between two streamlines at a distance of 16 meter at infinity is 128 m$^3$/d. This is indeed approximately observed in Figure 11.3, in which each well intercepts nine

streamlines. The difference may be due to the fact that the left side boundary is too close to the wells, and perhaps due to inaccuracies in the numerical procedures.

## 11.4. System of Wells in an Infinite Strip

The case of a system of wells in an infinite field, considered in the previous section, is only one of a series of problems that can be studied by the semi-analytic method. The applicability of this method depends on the availability of an analytical expression for the velocity field. Another class of problems for which such an analytical solution is available is the flow in an infinite strip between two parallel impermeable boundaries. The basic solution for this class is the solution for a single well in an arbitrary point between two impermeable boundaries (Figure 11.4).

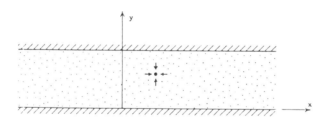

Fig. 11.4. Well between two parallel impermeable boundaries.

The solution of the flow problem in this case can be obtained by the use of a conformal transformation of the infinite strip on a half-plane, and then using the method of images to obtain the solution in the transformed plane. After inverse transformation, the solution is found to be as follows, in terms of the complex potential $\Omega = \Phi + i\Psi$, where $\Phi$ is the potential, defined as $\Phi = k\phi$, and $\Psi$ is the stream function.

$$\Omega = \frac{Q}{2\pi H} \{\ln[\exp(\pi z/a) - \exp(\pi z_0/a)] +$$

$$+ \ln[\exp(\pi z/a) - \exp(\overline{\pi z_0}/a)] - \ln[\exp(\pi z/a)]\} \qquad (11.4.1)$$

In this formula $a$ is the width of the strip, $z$ is the complex variable, $z = x + iy$, and $z_0$ is the location of the well, $z_0 = x_0 + iy_0$.

The components of the velocity vector can be obtained by differentiation, using

# TRANSPORT BY ADVECTION

Equations (11.3.2). After some elementary calculations we obtain

$$V_x = -\frac{Q}{4anB}\left\{\frac{\sinh[\pi(x-x_0)/a]}{\cosh[\pi(x-x_0)/a] - \cos[\pi(y-y_0)/a]} + \right.$$

$$\left. + \frac{\sinh[\pi(x-x_0)/a]}{\cosh[\pi(x-x_0)/a] - \cos[\pi(y+y_0)/a]}\right\} \qquad (11.4.2)$$

$$V_y = -\frac{Q}{4anB}\left\{\frac{\sin[\pi(y-y_0)/a]}{\cosh[\pi(x-x_0)/a] - \cos[\pi(y-y_0)/a]} + \right.$$

$$\left. + \frac{\sin[\pi(y+y_0)/a]}{\cosh[\pi(x-x_0)/a] - \cos[\pi(y+y_0)/a]}\right\} \qquad (11.4.3)$$

This is the basic solution for a single well between the two parallel boundaries $y = 0$ and $y = a$. It can easily be verified that the vertical velocity component is zero along these two boundaries, and that at the two ends of the strip at infinity ($x \to \pm\infty$) one half of the total discharge $Q$ is supplied.

By superposition of this elementary solution, and an eventual uniform flow through the strip, the velocity field for an arbitrary number of wells in an infinite strip can be obtained. This basic solution can then be used for the generation of streamlines by a computer program. The computer program can be practically the same as Program BV11-2-2, except for some unimportant modifications in the header. The only necessary modification is the subroutine in which the velocity vector is calculated. The last part of the program is then as follows.

```
   . . . . . . . . . .
   . . . . . . . . . .
   . . . . . . . . . .
530 CLS:LOCATE 1,25,1:COLOR 0,7:PRINT" Semi-analytic advection - Part 3 ";
540 COLOR 7,0:PRINT:PRINT:RETURN
550 A$=INKEY$:IF A$="B" OR A$="b" THEN RETURN ELSE 550
560 REM   Runge-Kutta
570 X=XA:Y=YA:GOSUB 590:XD=XA+U*DT/2:YD=YA+V*DT/2
580 X=XD:Y=YD:GOSUB 590:XB=XA+U*DT:YB=YA+V*DT:RETURN
590 REM   Calculation of velocities
600 U=VX:V=VY:FOR KK=1 TO NW:D=QQ(KK)/(4*YT*PP*HH)
610 A=PI*(X-XX(KK))/YT:B=PI*(Y-YY(KK))/YT:C=PI*(Y+YY(KK))/YT
620 EA=EXP(A):EB=1/EA:CH=(EA+EB)/2:SH=(EA-EB)/2
630 CB=COS(B):SB=SIN(B):CC=COS(C):SC=SIN(C)
640 U=U-D*(SH/(CH-CB)+SH/(CH-CC)):V=V-D*(SB/(CH-CB)+SC/(CH-CC))
650 NEXT KK:RETURN
```

Program BV11-1-3. Semi-analytic advection: wells in infinite strip.

Formulas (11.4.2) and (11.4.3) have been implemented in the subroutine starting at line 600, with the variable YT representing the width $a$ of the strip. As an example, the program may been run with the same datafile as used for Figure 11.3. The results are shown in Figure 11.5.

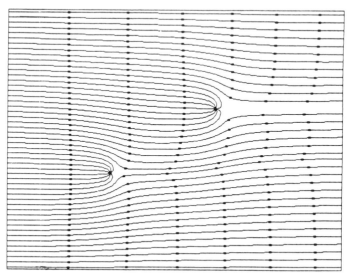

Fig. 11.5. Example DATA11.3. Two wells in infinite strip.

By comparing Figures 11.3 and 11.5, the effect of the impermeable boundaries at the top and bottom of the latter figure can be seen.

The programs and examples shown in this section and in the previous one, may serve to demonstrate the power of the semi-analytic method. With some effort, the method can be extended to such cases as flow in a semi-confined aquifer (the basic solution for a well then involves the Bessel function $K_0(x)$ and its derivatives), and confined aquifers of a shape that can be mapped conformally onto a half-plane, such as a circle, a rectangle, or a semi-infinite strip. In general, the method is restricted to homogeneous aquifers because general solutions for nonhomogeneous aquifers are scarce.

It may be noted that the separation of the program into a number of parts enables us to use various types of input of the datasets. Interactive input, which is a user-friendly form of input, especially for inexperienced users, is possible by running Part 1 of Program BV11-1. For large systems this may be inconvenient, if only for the fact that the program does not contain a facility for the correction of errors. A dataset can also be generated, however, once the user is familiar with the structure of such a dataset, by an editing program, for instance a word-processor. The separation of the program into various parts also enables us to replace the calculating part of the program by a compiled one, or even to replace it by a program in some other, faster, computer language.

## 11.5. Numerical Solution in Terms of the Piezometric Head

The semi-analytic method presented in the previous section fails in the case of problems for aquifers of complicated shape, and especially for nonhomogeneous aquifers. Therefore, a fully numerical method may be used for determining the piezometric head, and from it the velocity distribution. The procedure that suggests itself is to use the solution of a groundwater flow problem by either the finite difference method or the finite element method as the basis for the analysis of the advective transport.

The basic principles of a fully numerical method for the analysis of transport in a porous medium by advection only, are that first the velocity field is determined by solving the flow problem, either in terms of the piezometric head or in terms of the stream function. In a second stage, the transport in the direction of flow is considered, in the same way as in the previous sections, by tracing the path of particles.

If the groundwater flow problem is solved in terms of the piezometric head, by the finite difference method or by the finite element method, the velocity field must be determined by numerical differentiation. This may lead to an unacceptable loss of accuracy. In the simplest type of finite element model, as presented in Chapter 10, the head is approximated by linear functions, such that the velocity is constant within each element. This means that any curvature of a streamline must be generated by discontinuities across the boundaries of the elements. An example of the resulting flow field is shown in Figures 11.6 and 11.7 for the case of flow towards a well on the boundary. Figure 11.6 shows the network of elements, while Figure 11.7 shows the flow field. The streamlines in this figure are not satisfactory. It appears that some of the water seems to leave the aquifer along the impermeable boundary near the well. In general, the behaviour near the well is not

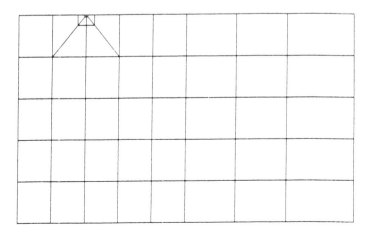

Fig. 11.6. Network of elements.

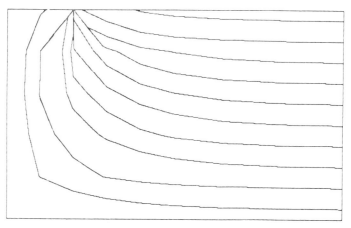

Fig. 11.7. Stream lines determined from the piezometric head.

realistic: streamlines may even intersect each other. This is due to the fact that the physical condition of the continuity of velocities does not have priority in the solution of the groundwater flow problem. Actually, the velocity is discontinuous across the boundaries of the elements, and this is physically unrealistic.

The approximation can be improved by using a finer mesh and by using an approximation of a higher order. A realistic flow field may be obtained if the numerical approximation is such that both the head and the velocity component normal to the element boundaries are continuous across these boundaries. This requirement is not easy to satisfy, however, and it will lead to a rather complex computer program. A simpler way to improve the accuracy of the flow field is to give priority to the condition of continuity rather than to Darcy's law, and to use a formulation in terms of the stream function. This will be presented in the next section. A further alternative would be to use a formulation in terms of the actual velocity components (Zijl, 1984).

## 11.6. Numerical Solution in Terms of the Stream Function

An alternative formulation for the problem of two-dimensional flow of groundwater is in terms of a (modified) stream function, as presented in Section 4.6. For the sake of convenience, some of the main results are recalled here.

The starting point is the equation of continuity, for the simplest case of steady two-dimensional flow,

$$\frac{\partial(Bq_x)}{\partial x} + \frac{\partial(Bq_y)}{\partial y} = 0 \tag{11.6.1}$$

where $B$ is the thickness of the aquifer. Equation (11.6.1) can be satisfied by

TRANSPORT BY ADVECTION

introducing a stream function $\Psi$ such that

$$Bq_x = -\frac{\partial \Psi}{\partial y}, \qquad Bq_y = +\frac{\partial \Psi}{\partial x}. \tag{11.6.2}$$

Because the flow is a potential one (Subsection 4.6.4), the components of the specific discharge vector should satisfy the condition

$$\frac{\partial(q_x/K)}{\partial y} - \frac{\partial(q_y/K)}{\partial x} = 0. \tag{11.6.3}$$

Substituting (11.6.2) into (11.6.3) leads to the differential equation (4.6.15) for the stream function

$$\frac{\partial}{\partial x}\left(\frac{1}{T}\frac{\partial \Psi}{\partial x}\right) + \frac{\partial}{\partial y}\left(\frac{1}{T}\frac{\partial \Psi}{\partial y}\right) = 0 \tag{11.6.4}$$

where $T = kB$, the transmissivity of the aquifer. It should be noted that the stream function used in these equations applies to the flow through the entire thickness of the aquifer. The difference $\Delta\Psi$ expresses the flow between two adjacent streamlines, through the thickness of the aquifer.

Equation (11.6.4) has precisely the same form as the partial differential equation in terms of the head $\phi$, see for instance Equation (10.1.1), except that the transmissivity $T$ has now been replaced by $1/T$. This implies that the numerical models and computer programs developed for the head can be used to solve problems for the stream function with only minor modifications. Of course, the appropriate boundary conditions, in terms of $\Psi$, must be supplied. The stream function is constant along an impermeable boundary, while along a boundary of constant head, the component of specific discharge in the direction of that boundary is zero and, hence, the normal derivative of the stream function is zero. Thus, the two most common types of boundary conditions — an impermeable boundary and a boundary of constant head — must be interchanged when passing from a formulation in terms of the head to one in terms of the stream function.

As an example, the problem solved in the previous section (Figures 11.6 and 11.7) has also been solved in terms of the stream function, using the same network of elements. The result, using a computer program to be presented later, is shown in Figure 11.8. Compared with Figure 11.7, the results now seem to be much more realistic, even close to the well. All the water leaves the aquifer in the well, and streamlines do not intersect.

The obvious difference in quality of the results shown in Figures 11.8 and 11.7 seems to justify the presentation of a complete computer program on the basis of the stream function. The program will again be divided into several parts, in order to retain the convenience of the possibility of interactive input, in conjunction with the possibility of editing a datafile by some other editor, such as a word-processing program. In this way, the same dataset can also be used for several different

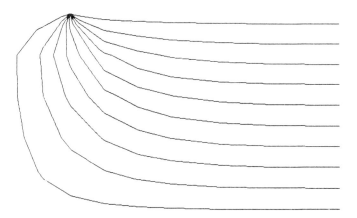

Fig. 11.8. Streamlines determined from the stream function.

purposes. Of course, the user may combine certain parts of the program to a single program, when desired.

The first part of the program is one that generates a dataset, in an interactive way. It is reproduced in Program BV11-2-1.

```
100 DEFINT I-N:KEY OFF:OPTION BASE 1:GOSUB 560
110 PRINT"---   Bear & Verruijt - Groundwater Modeling"
120 PRINT"---   Plane steady Groundwater Flow"
130 PRINT"---   Advection using numerical solution"
140 PRINT"---   Program 11.2 - Part 1"
150 PRINT"---   Generation of dataset":PRINT
160 DIM X(250),Y(250),IP(250),F(250),Q(250)
170 DIM NP(250,4),T(250),MP(100,2),AW(100),QW(100)
180 INPUT"Name of dataset ";N$:PRINT
190 PRINT"Properties of screen :":PRINT
200 INPUT" Dots in x-direction (640 ?) ........ ";NLX
210 INPUT" Dots in y-direction (    ?) ....... ";NLY
220 INPUT" Scale factor x/y (2 ?) ........... ";SXY
230 INPUT" Graphics screen parameter (2 ?) ... ";NSCR
240 PRINT:PRINT"General data :":PRINT
250 INPUT" Number of nodes ................. ";N
260 INPUT" Number of elements .............. ";M
270 INPUT" Points on discontinuity ......... ";MW
280 FOR I=1 TO N:GOSUB 560:PRINT"Data of node ";I:PRINT
290 INPUT" x ............................... ";X(I)
300 INPUT" y ............................... ";Y(I)
310 PRINT" Stream function given (Y/N) ....... ? ";:GOSUB 580
320 IF A$="Y" THEN IP(I)=2 ELSE IP(I)=0
330 IF A$="N" THEN 340
340 INPUT" Stream function ................. ";F(I)
350 Q(I)=0:NEXT I
360 FOR J=1 TO M:GOSUB 560:PRINT"Data of element ";J:PRINT
370 PRINT"Enter 4 nodal points counter-clockwise":PRINT
380 INPUT" Node 1 .......................... ";NP(J,1)
390 INPUT" Node 2 .......................... ";NP(J,2)
400 INPUT" Node 3 .......................... ";NP(J,3)
```

```
410 INPUT" Node 4 ........................... ";NP(J,4)
420 INPUT" Transmissivity .................. ";T(J)
430 NEXT J:IF MW=0 THEN 490
440 FOR J=1 TO MW:GOSUB 560:PRINT"Data of discontinuity ";J:PRINT
450 INPUT" Node 1 ........................... ";MP(J,1)
460 INPUT" Node 2 ........................... ";MP(J,2)
470 INPUT" Discontinuity .................. ";QW(J)
480 AW(J)=10000:NEXT J
490 GOSUB 560:PRINT"Writing datafile ";N$:PRINT
500 OPEN"O",1,N$:PRINT#1,NLX;NLY;SXY;NSCR;N;M;MW
510 FOR I=1 TO N:PRINT#1,X(I);Y(I);IP(I);F(I);Q(I):NEXT I
520 FOR J=1 TO M:PRINT#1,NP(J,1);NP(J,2);NP(J,3);NP(J,4);T(J):NEXT J
530 IF MW=0 THEN 550
540 FOR J=1 TO MW:PRINT#1,MP(J,1),MP(J,2),AW(J),QW(J):NEXT J
550 CLOSE 1:END
560 CLS:LOCATE 1,26,1:COLOR 0,7:PRINT" Advection using FEM - Part 1 ";
570 COLOR 7,0:PRINT:PRINT:RETURN
580 A$=INKEY$:IF A$="Y" OR A$="y" THEN A$="Y":PRINT"Yes":RETURN
590 IF A$="N" OR A$="n" THEN A$="n":PRINT"No":RETURN
600 GOTO 580
```

Program BV11-2-1. Advection by FEM: generation of input data.

Because of the interactive nature of the program, and its resemblance to Program BV11-1-1, it is not necessary to repeat the comments on its operation. An exception must be made for the variables involving the discontinuities, as entered in lines 270 and 440—480. These variables are needed to allow for the possibility of wells in the interior of the field. These variables are not needed for the production of Figure 11.8, i.e., the variable NW can be set equal to zero, so that statements 440—480 are skipped. It may be noted that the node numbers of the 4 nodes, that together constitute an element, must be entered in counterclockwise direction (see line 370). This is needed in order to simplify the later presentation of the network of elements on the screen.

As mentioned above, special care and some additional statements are needed to allow for the introduction of wells in the interior of a given region. The reason for this is that the stream function is double-valued in the case of an interior well, with a discontinuity that represents the discharge of the well. An example of such a problem is shown in Figure 11.9. This network may be used to solve the problem of flow towards a single well, from a boundary of constant head at the right side. In order to introduce the possibility of a discontinuity in the stream function when encircling the well, a branch cut is introduced, running from the well to the right side boundary (Van den Akker, 1982). On either side of the branch cut, separate nodes are located, and in the finite element mesh there is no connection whatever between the nodes on the two sides of the branch cut. In order to represent the jump in the value of the stream function across the branch cut, an additional equation is entered in the system of equations of the form

$$A \times (\Psi(I) - \Psi(J)) = A \times Q \tag{11.6.5}$$

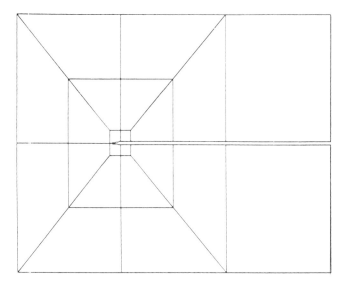

Fig. 11.9. Network of elements for a well in a rectangular region.

where $I$ and $J$ denote the node numbers of the nodes on opposite sides of a branch cut, $Q$ denotes the discontinuity (the discharge of the well), and $A$ is some arbitrary large number. By taking $A$ very large (say $A = 10000$) and adding (11.6.5) to the equations in the system for each of the nodes $I$ and $J$, this equation will overshadow all other conditions and, thus indeed, lead to the correct discontinuity. The dataset representing the data for the problem shown in Figure 11.9 is (in compact form) given as DATA11.9.

```
640 200 2 2 33 22 5
0 0 2 0 0 0 2 0 0 0 4 2 2 0
1 1 0 0 0 1 2 0 0 0 1 3 0 0 0
1.8 1.8 0 0 0 1.8 2 0 0 0 1.8 2.2 0 0 0
2 0 2 0 0 2 1 0 0 0 2 1.8 0 0 0 2 1.97 0 0 0
2 2.03 0 0 0 2 2.2 0 0 0 2 3 0 0 0 2 4 2 2 0
2.2 1.8 0 0 0 2.2 1.97 0 0 0 2.2 2.03 0 0 0 2.2 2.2 0 0 0
3 1 0 0 0 3 1.97 0 0 0 3 2.03 0 0 0 3 3 0 0 0
4 0 2 0 0 4 1.97 0 0 0 4 2.03 0 0 0 4 4 2 2 0
6 0 2 0 0 6 1.97 0 0 0 6 2.03 0 0 0 6 4 2 2 0
1 4 5 2 1 2 5 6 3 1 6 16 17 3 1 16 25 29 17 1
24 28 29 25 1 22 26 27 23 1 10 26 22 11 1 1 10 11 4 1
4 7 8 5 1 5 8 9 6 1 9 15 16 6 1 15 21 25 16 1
20 24 25 21 1 18 22 23 19 1 11 22 18 12 1 4 11 12 7 1
7 12 13 8 1 8 14 15 9 1 14 20 21 15 1 12 18 19 13 1
26 30 31 27 1 28 32 33 29 1
13 14 10000 1 19 20 10000 1 23 24 10000 1
27 28 10000 1 31 32 10000 1
```

Dataset DATA11.9.

The number of nodes is 33, the number of elements is 22 and the number of points of discontinuity is 5. The data for these discontinuities are represented in the last two lines. The vertical coordinates on the two sides of the branch cut are 1.97 and 2.03, respectively. Actually, these values may be taken as exactly equal. In the dataset shown above, a small difference has been introduced in order to show the branch cut in the figure.

Program BV11-2-2, reproduced below, will present the network on the screen.

```
100 DEFINT I-N:KEY OFF:OPTION BASE 1:GOSUB 380
110 PRINT"---  Bear & Verruijt - Groundwater Modeling"
120 PRINT"---  Plane steady Groundwater Flow"
130 PRINT"---  Advection using numerical solution"
140 PRINT"---  Program 11.2 - Part 2"
150 PRINT"---  Mesh on screen":PRINT
160 DIM X(250),Y(250),IP(250),F(250),Q(250)
170 DIM NP(250,4),T(250),MP(100,2),AW(100),QW(100)
180 INPUT"Name of dataset ";N$:OPEN "I",1,N$:PRINT
190 INPUT#1,NLX,NLY,SXY,NSCR,N,M,MW
200 FOR I=1 TO N:INPUT#1,X(I),Y(I),IP(I),F(I),Q(I):NEXT I
210 FOR J=1 TO M:INPUT#1,NP(J,1),NP(J,2),NP(J,3),NP(J,4),T(J):NEXT J
220 IF MW>0 THEN FOR J=1 TO MW:INPUT#1,MP(J,1),MP(J,2),AW(J),QW(J):NEXT J
230 CLOSE 1:NLX=NLX-1:NLY=NLY-1
240 XA=X(1):XB=XA:YA=Y(1):YB=YA:FOR I=1 TO N
250 IF X(I)<XA THEN XA=X(I)
260 IF X(I)>XB THEN XB=X(I)
270 IF Y(I)<YA THEN YA=Y(I)
280 IF Y(I)>YB THEN YB=Y(I)
290 NEXT I:SX=NLX/(XB-XA):SY=SXY*NLY/(YB-YA):IF SY<SX THEN SX=SY
300 SY=SX/SXY
310 SCREEN NSCR:CLS
320 FOR J=1 TO M:A1=NP(J,1):A2=NP(J,2):A3=NP(J,3):A4=NP(J,4)
330 X1=SX*(X(A1)-XA):Y1=SY*(Y(A1)-YA):X2=SX*(X(A2)-XA):Y2=SY*(Y(A2)-YA)
340 X3=SX*(X(A3)-XA):Y3=SY*(Y(A3)-YA):X4=SX*(X(A4)-XA):Y4=SY*(Y(A4)-YA)
350 LINE (X1,Y1)-(X2,Y2):LINE (X2,Y2)-(X3,Y3)
360 LINE (X3,Y3)-(X4,Y4):LINE (X4,Y4)-(X1,Y1):NEXT J
370 A$=INPUT$(1):SCREEN 0:END
380 CLS:LOCATE 1,26,1:COLOR 0,7:PRINT" Advection using FEM - Part 2 ";
390 COLOR 7,0:PRINT:PRINT:RETURN
```

Program BV11-2-2. Advection by FEM: mesh on screen.

The program first reads the dataset, in lines 180—230, and then, in lines 240—300, determines the scale factors so that the figure will fit on the screen. Finally, the figure is printed on the screen. Figure 11.9 will be drawn on the screen, if the dataset DATA11.9 is used as input, as the reader may verify by running the program with that dataset.

In order to present the streamlines on the screen, the values of the stream function $\Psi$ should first be calculated. This is performed by Program BV11-2-3, which is listed below.

```
100 DEFINT I-N:KEY OFF:OPTION BASE 1:GOSUB 830
110 PRINT"---   Bear & Verruijt - Groundwater Modeling"
120 PRINT"---   Plane steady Groundwater Flow"
130 PRINT"---   Advection using numerical solution"
140 PRINT"---   Program 11.2 - Part 3"
150 PRINT"---   Stream function by finite elements":PRINT
160 DIM XJ(3),YJ(3),B(3),C(3),KS(4,3)
170 DIM X(250),Y(250),IP(250),F(250),Q(250),U(250),V(250),W(250)
180 DIM P(250,12),KP(250,12):NZ=12
190 DIM NP(250,4),T(250),QX(250),QY(250),FC(250),MP(100,2),AW(100),QW(100)
200 INPUT"Name of dataset ";N$:OPEN "I",1,N$:PRINT
210 INPUT#1,NLX,NLY,SXY,NSCR,N,M,MW
220 FOR I=1 TO N:INPUT#1,X(I),Y(I),IP(I),F(I),Q(I)
230 Q(I)=0:IF IP(I)=0 THEN F(I)=0
240 NEXT I
250 FOR J=1 TO M:INPUT#1,NP(J,1),NP(J,2),NP(J,3),NP(J,4),T(J):NEXT J
260 IF MW>0 THEN FOR J=1 TO MW:INPUT#1,MP(J,1),MP(J,2),AW(J),QW(J):NEXT J
270 CLOSE 1
280 FOR I=1 TO 4:FOR J=1 TO 3:K=I+J-1:IF K>4 THEN K=K-4
290 KS(I,J)=K:NEXT J,I
300 PRINT"Generation of pointer matrix"
310 FOR I=1 TO N:KP(I,1)=I:KP(I,NZ)=1:NEXT I:IC=CSRLIN
320 FOR J=1 TO M:LOCATE IC,1,0:PRINT"       Element ......   ";J
330 FOR K=1 TO 4:KK=NP(J,K):FOR L=1 TO 4:LL=NP(J,L)
340 IA=0:FOR II=1 TO KP(KK,NZ):IF KP(KK,II)=LL THEN IA=1
350 NEXT II:IF IA=0 THEN KB=KP(KK,NZ)+1:KP(KK,NZ)=KB:KP(KK,KB)=LL
360 IF KB=NZ THEN 820
370 NEXT L,K,J
380 FOR I=1 TO N:FOR J=1 TO MW:FOR II=1 TO 2:IF MP(J,II)=I THEN 400
390 NEXT II:GOTO 430
400 FOR II=1 TO 2:IJ=MP(J,II):FOR L=1 TO KP(I,NZ):IF KP(I,L)=IJ THEN 420
410 NEXT L:K=KP(I,NZ)+1:KP(I,K)=IJ:KP(I,NZ)=K:IF K=NZ THEN 820
420 NEXT II
430 NEXT J,I
440 PRINT:PRINT"Generation of system matrix":EE=.000001:IC=CSRLIN
450 FOR J=1 TO M:LOCATE IC,1,0:PRINT"       Element ......   ";J:FOR KW=1 TO 4
460 FOR I=1 TO 3:K=NP(J,KS(KW,I)):XJ(I)=X(K):YJ(I)=Y(K):NEXT I
470 B(1)=YJ(2)-YJ(3):B(2)=YJ(3)-YJ(1):B(3)=YJ(1)-YJ(2)
480 C(1)=XJ(3)-XJ(2):C(2)=XJ(1)-XJ(3):C(3)=XJ(2)-XJ(1)
490 D=ABS(XJ(1)*B(1)+XJ(2)*B(2)+XJ(3)*B(3)):IF D<EE THEN 560
500 D=1/(4*D*T(J))
510 FOR K=1 TO 3:KK=NP(J,KS(KW,K)):II=KP(KK,NZ):FOR LL=1 TO II:L=1
520 KV=KS(KW,L):IF NP(J,KV)=KP(KK,LL) THEN 540
530 L=L+1:IF L<4 THEN 520 ELSE 550
540 P(KK,LL)=P(KK,LL)+D*(B(K)*B(L)+C(K)*C(L))
550 NEXT LL,K
560 NEXT KW,J
570 FOR J=1 TO MW:KK=MP(J,1):P(KK,1)=P(KK,1)+AW(J):Q(KK)=Q(KK)+AW(J)*QW(J)
580 II=KP(KK,NZ):FOR LL=2 TO II
590 IF MP(J,2)=KP(KK,LL) THEN P(KK,LL)=P(KK,LL)-AW(J)
600 NEXT LL:KK=MP(J,2):P(KK,1)=P(KK,1)+AW(J):Q(KK)=Q(KK)-AW(J)*QW(J)
610 II=KP(KK,NZ):FOR LL=2 TO II
620 IF MP(J,1)=KP(KK,LL) THEN P(KK,LL)=P(KK,LL)-AW(J)
630 NEXT LL
640 NEXT J:PRINT:PRINT"Solution of equations":EE=.000001:EE=EE*EE
650 IC=CSRLIN:IT=1:FOR I=1 TO N:U(I)=0:IF IP(I)>0 THEN 670
660 U(I)=Q(I):FOR J=1 TO KP(I,NZ):U(I)=U(I)-P(I,J)*F(KP(I,J)):NEXT J
```

```
670 V(I)=U(I):NEXT I:UU=0:FOR I=1 TO N:UU=UU+U(I)*U(I):NEXT I
680 LOCATE IC,1,0:PRINT"    Iteration .... ";IT:FOR I=1 TO N:W(I)=0
690 FOR J=1 TO KP(I,NZ):W(I)=W(I)+P(I,J)*V(KP(I,J)):NEXT J,I
700 VW=0:FOR I=1 TO N:VW=VW+V(I)*W(I):NEXT I
710 AA=UU/VW:FOR I=1 TO N:IF IP(I)>0 THEN 730
720 F(I)=F(I)+AA*V(I):U(I)=U(I)-AA*W(I)
730 NEXT I:WW=0:FOR I=1 TO N:WW=WW+U(I)*U(I):NEXT I
740 BB=WW/UU:FOR I=1 TO N:V(I)=U(I)+BB*V(I):NEXT I:UU=WW
750 IT=IT+1:IF IT< N AND UU>EE THEN 680
760 PRINT:PRINT"Rewriting datafile ";N$:PRINT
770 OPEN"O",1,N$:PRINT#1,NLX;NLY;SXY;NSCR;N;M;MW
780 FOR I=1 TO N:PRINT#1,X(I);Y(I);IP(I);F(I);Q(I):NEXT I
790 FOR J=1 TO M:PRINT#1,NP(J,1);NP(J,2);NP(J,3);NP(J,4);T(J):NEXT J
800 IF MW>0 THEN FOR J=1 TO MW:PRINT#1,MP(J,1);MP(J,2);AW(J);QW(J):NEXT J
810 CLOSE 1:END
820 PRINT:PRINT"Pointer width too small":PRINT:END
830 CLS:LOCATE 1,26,1:COLOR 0,7:PRINT" Advection using FEM - Part 3 ";
840 COLOR 7,0:PRINT:PRINT:RETURN
```

Program BV11-2-3. Advection by FEM: calculation of stream function.

This program is equivalent to the programs presented in Section 10.2. As input, it uses the dataset created by Program BV11-2-1 (or by some other editor). After execution of the finite element calculations, the data are rewritten in the same datafile. The variables $F(I)$ then represent the stream function $\Psi$. The data are then ready for further processing, for instance, the presentation of contour lines of the stream function (the streamlines) on the screen. Program BV11-2-4 will perform that task.

```
100 DEFINT I-N:KEY OFF:OPTION BASE 1:GOSUB 600
110 PRINT"---   Bear & Verruijt - Groundwater Modeling"
120 PRINT"---   Plane steady Groundwater Flow"
130 PRINT"---   Advection using numerical solution"
140 PRINT"---   Program 11.2 - Part 4"
150 PRINT"---   Stream lines on screen":PRINT
160 DIM X(250),Y(250),IP(250),F(250),Q(250),B(4),C(4)
170 DIM NP(250,4),T(250),MP(100,2),AW(100),QW(100)
180 INPUT"Name of dataset ";N$:OPEN "I",1,N$:PRINT
190 INPUT#1,NLX,NLY,SXY,NSCR,N,M,MW
200 FOR I=1 TO N:INPUT#1,X(I),Y(I),IP(I),F(I),Q(I):NEXT I
210 FOR J=1 TO M:INPUT#1,NP(J,1),NP(J,2),NP(J,3),NP(J,4),T(J):NEXT J
220 IF MW>0 THEN FOR J=1 TO MW:INPUT#1,MP(J,1),MP(J,2),AW(J),QW(J):NEXT J
230 CLOSE 1:NLX=NLX-1:NLY=NLY-1
240 XA=X(1):XB=XA:YA=Y(1):YB=YA:FOR I=1 TO N
250 IF X(I)<XA THEN XA=X(I)
260 IF X(I)>XB THEN XB=X(I)
270 IF Y(I)<YA THEN YA=Y(I)
280 IF Y(I)>YB THEN YB=Y(I)
290 NEXT I:SX=NLX/(XB-XA):SY=SXY*NLY/(YB-YA):IF SY<SX THEN SX=SY
300 SY=SX/SXY
310 FA=F(1):FB=FA:FOR I=1 TO N:IF F(I)<FA THEN FA=F(I)
320 IF F(I)>FB THEN FB=F(I)
330 NEXT I:GOSUB 600:PRINT"Maximum value of psi : ";FB
```

```
340 PRINT"Minimum value of psi : ";FA:PRINT
350 PRINT"Please enter minimum value of contour and interval":PRINT
360 INPUT"Minimum value ....... ";PA
370 INPUT"Interval ........... ";DF:NP=INT((FB-PA)/DF)+1
380 SCREEN NSCR:CLS:FOR JP=0 TO NP:FF=PA+JP*DF:IF FF<FA OR FF>FB THEN 580
390 FOR J=1 TO M:I1=NP(J,1):I2=NP(J,2):I3=NP(J,3):I4=NP(J,4)
400 F1=F(I1):F2=F(I2):F3=F(I3):F4=F(I4)
410 K=0:IF (F1<=FF AND F2<=FF) OR (F1>=FF AND F2>=FF) THEN 440
420 K=K+1:B(K)=X(I1)+(X(I2)-X(I1))*(FF-F1)/(F2-F1)-XA
430 C(K)=Y(I1)+(Y(I2)-Y(I1))*(FF-F1)/(F2-F1)-YA
440 IF (F2<=FF AND F3<=FF) OR (F2>=FF AND F3>=FF) THEN 480
450 K=K+1
460 B(K)=X(I2)+(X(I3)-X(I2))*(FF-F2)/(F3-F2)-XA
470 C(K)=Y(I2)+(Y(I3)-Y(I2))*(FF-F2)/(F3-F2)-YA
480 IF (F3<=FF AND F4<=FF) OR (F3>=FF AND F4>=FF) THEN 520
490 K=K+1:IF K>2 THEN 560
500 B(K)=X(I3)+(X(I4)-X(I3))*(FF-F3)/(F4-F3)-XA
510 C(K)=Y(I3)+(Y(I4)-Y(I3))*(FF-F3)/(F4-F3)-YA
520 IF (F4<=FF AND F1<=FF) OR (F4>=FF AND F1>=FF) THEN 560
530 K=K+1:IF K>2 THEN 560
540 B(K)=X(I4)+(X(I1)-X(I4))*(FF-F4)/(F1-F4)-XA
550 C(K)=Y(I4)+(Y(I1)-Y(I4))*(FF-F4)/(F1-F4)-YA
560 IF K>1 THEN LINE (SX*B(1),SY*C(1))-(SX*B(2),SY*C(2))
570 NEXT J
580 NEXT JP
590 A$=INPUT$(1):SCREEN 0:END
600 CLS:LOCATE 1,26,1:COLOR 0,7:PRINT" Advection using FEM    Part 4 ";
610 COLOR 7,0:PRINT:PRINT:RETURN
```

Program BV11-2-4. Advection by FEM: streamlines on screen.

For the datast DATA11.9 the final result is shown in Figure 11.10.

The streamlines are smooth, and one-third of them are intercepted by the well. This is in agreement with the fact that in the input data the stream function along

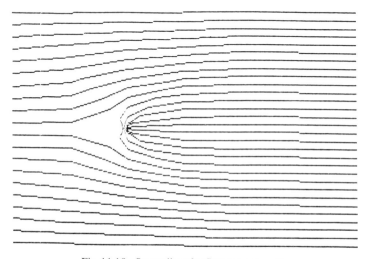

Fig. 11.10. Streamlines for flow towards well.

TRANSPORT BY ADVECTION

the upper and lower boundaries are defined as 0 and 2, respectively, and the value of the discontinuity along the branch cut is 1. Thus, twice as much water exits at the left side as is flowing to the well.

As a final example, the problem shown in Figure 11.5, which is solved in Section 11.4 by the semi-analytic method, is considered. The problem may be solved numerically, using the network of elements shown in Figure 11.11. The streamlines are shown in Figure 11.12. In this case, with 128 elements and 167 nodes, the dataset has not been created by Program BV11-2-1, listed above, but by a text-editor, which makes it much easier to apply corrections.

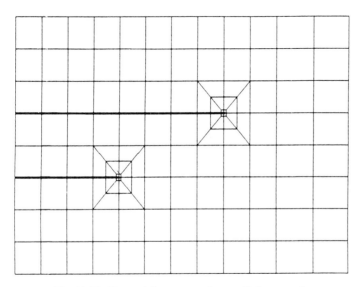

Fig. 11.11. Network for system of two wells in rectangle.

The branch cuts are shown in Figure 11.11 by a double line. In the dataset used for the actual calculations the thickness of the cut has been taken as zero, so that the streamlines appear as continuous lines across the branch cuts.

It appears from Figure 11.12 that the streamlines are indeed continuous and smooth, and their relative position seems to be physically realistic.

The main advantage of the numerical method presented in this section compared to the semi-analytical method of Section 11.2 is that it allows for a nonhomogeneous transmissivity. The only difference is that in the dataset in which the input data are stored, instead of a uniform value for the transmissivity, different values are entered for different elements. As an example, a lower transmissivity (by a factor 10) has been assigned to a part of the region of the problem shown in Figures 11.11 and 11.12. The modifications in the dataset can easily be entered by using a text editor. The resulting streamlines are shown in Figure 11.13. The effect of the zone of lower transmissivity can clearly be seen in

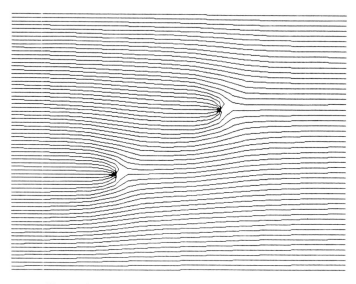

Fig. 11.12. Streamlines for system of two wells in rectangle.

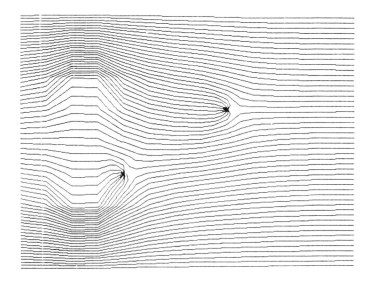

Fig. 11.13. Streamlines for system of two wells in a nonhomogeneous rectangle.

the figure. Streamlines try to avoid the less permeable zone, and they exhibit a refraction at the discontinuity.

The results obtained by the method presented in this section seem to be sufficiently accurate, and the method is reasonably flexible and powerful. However, the actual transport parameters are not yet fully described, because no information is obtained about travel times. The addition of time as a parameter is discussed in the next section.

## 11.7. Tracing Particles Along a Stream Line

In order to study the movement of particles along the streamlines, it is necessary to trace them from their source, as a function of time. This means that at a certain point the velocity should be determined, so that the displacement in a time step can be calculated. As the stream function is known at every point, the velocity components can be determined by (numerical) differentiation. A computer program that performs the necessary computations is reproduced as Program BV11-2-5.

```
100 DEFINT I-N:KEY OFF:OPTION BASE 1:GOSUB 960
110 PRINT"---   Bear & Verruijt - Groundwater Modeling"
120 PRINT"---   Plane steady Groundwater Flow"
130 PRINT"---   Advection using numerical solution"
140 PRINT"---   Program 11.3 - Part 5"
150 PRINT"---   Stream lines with markers":PRINT
160 DIM XJ(3),YJ(3),B(3),C(3),D(3),KS(4,3)
170 DIM X(250),Y(250),IP(250),F(250),Q(250),NP(250,4),T(250),DL(250)
180 DIM XN(100),YN(100),XP(100),YP(100)
190 DIM LP(100),ML(100),LQ(100),PT(100),IQ(100)
200 INPUT"Name of first dataset ..... ";N$:PRINT
210 OPEN"I",1,N$:INPUT#1,NLX,NLY,SXY,NSCR,N,M,MW
220 FOR I=1 TO N:INPUT#1,X(I),Y(I),IP(I),F(I),Q(I):NEXT I
230 FOR J=1 TO M:INPUT#1,NP(J,1),NP(J,2),NP(J,3),NP(J,4),T(J):NEXT J
240 CLOSE 1:INPUT"Name of second dataset .... ";M$:PRINT
250 OPEN "I",2,M$:INPUT#2,HH,PP,NP,DT,NT,NS:HP=HH*PP
260 IF NP>0 THEN FOR I=1 TO NP:INPUT#2,XN(I),YN(I)
270 LP(I)=INT(N/2):ML(I)=1:PT(I)=DT:NEXT I
280 IF NS>0 THEN FOR J=1 TO NS:INPUT#2,IQ(J):NEXT J
290 CLOSE 2:NLX=NLX-1:NLY=NLY-1
300 XA=X(1):XB=XA:YA=Y(1):YB=YA:FOR I=1 TO N
310 IF X(I)<XA THEN XA=X(I)
320 IF X(I)>XB THEN XB=X(I)
330 IF Y(I)<YA THEN YA=Y(I)
340 IF Y(I)>YB THEN YB=Y(I)
350 NEXT I:SX=NLX/(XB-XA):SY=SXY*NLY/(YB-YA):IF SY<SX THEN SX=SY
360 SY=SX/SXY
370 SCREEN NSCR:CLS:FOR J=1 TO M:XA=X(NP(J,1)):XB=XA:YA=Y(NP(J,1)):YB=YA
380 FOR K=2 TO 4:XX=X(NP(J,K)):YY=Y(NP(J,K)):IF XX<XA THEN XA=XX
390 IF XX>XB THEN XB=XX
400 IF YY<YA THEN YA=YY
410 IF YY>YB THEN YB=YY
420 NEXT K:DL(J)=XB-XA:BB=YB-YA:IF BB<DL(J) THEN DL(J)=BB
430 DL(J)=DL(J)/5:NEXT J
440 FOR I=1 TO 4:FOR J=1 TO 3:K=I+J-1:IF K>4 THEN K=K-4
450 KS(I,J)=K:NEXT J,I:JT=0
460 JT=JT+1:IF JT>NT THEN JT=1
```

```
470 FOR IP=1 TO NP:IF ML(IP)<0 THEN 630
480 JJ=LP(IP):XA=XN(IP):YA=YN(IP):GOSUB 660
490 IF L=0 THEN ML(IP)=-1:GOTO 630
500 FOR II=1 TO NS:IF J=IQ(II) THEN ML(IP)=-1
510 NEXT II:IF ML(IP)<0 THEN GOTO 630
520 LP(IP)=J:GOSUB 810:FM=FA:DW=DL(J)/10
530 XA=XN(IP)-DW:YA=YN(IP):GOSUB 810:FP=FA:UY=-(FP-FM)/(DW*HP)
540 XA=XN(IP):YA=YN(IP)+DW:GOSUB 810:FP=FA:UX=-(FP-FM)/(DW*HP):IW=0
550 XP(IP)=XN(IP)+UX*PT(IP):YP(IP)=YN(IP)+UY*PT(IP)
560 XA=XP(IP):YA=YP(IP):JJ=LP(IP):GOSUB 660:IF L>0 THEN 590
570 PT(IP)=.7*PT(IP):IW=IW+1:ML(IP)=-1
580 IF IW<9 THEN GOTO 550 ELSE GOTO 630
590 IX=SX*XN(IP):IY=SY*YN(IP):JX=SX*XP(IP):JY=SY*YP(IP)
600 LINE (IX,IY)-(JX,JY):IF JT<NT THEN 620
610 FOR II=-1 TO 1:LINE (JX-2,JY+II)-(JX+2,JY+II):NEXT II
620 XN(IP)=XP(IP):YN(IP)=YP(IP)
630 NEXT IP:KP=-1:FOR IP=1 TO NP:IF ML(IP)>0 THEN KP=1
640 NEXT IP:IF KP>0 THEN 460
650 A$=INPUT$(1):SCREEN 0:END
660 REM Locate point XA,YA in element J
670 II=0
680 J=JJ+II:IF (J<=M AND J>0) THEN GOSUB 720:IF L>0 THEN 710
690 J=JJ-II:IF (J<=M AND J>0) THEN GOSUB 720:IF L>0 THEN 710
700 II=II+1:IF II<=M THEN 680
710 RETURN
720 L=0:KW=0
730 KW=KW+1:FOR I=1 TO 3:K=NP(J,KS(KW,I)):XJ(I)=X(K):YJ(I)=Y(K):NEXT I
740 B(1)=YJ(2)-YJ(3):B(2)=YJ(3)-YJ(1):B(3)=YJ(1)-YJ(2)
750 C(1)=XJ(3)-XJ(2):C(2)=XJ(1)-XJ(3):C(3)=XJ(2)-XJ(1)
760 A1=(YA-YJ(1))*C(3)+(XA-XJ(1))*B(3)
770 A2=(YA-YJ(2))*C(1)+(XA-XJ(2))*B(1)
780 A3=(YA-YJ(3))*C(2)+(XA-XJ(3))*B(2)
790 IF A1*A2>=0 AND A1*A3>=0 AND A2*A3>=0 THEN L=J ELSE IF KW<4 THEN 730
800 RETURN
810 REM Isoparametric coordinates of (XA,YA) in element J
820 XZ=0:YZ=0:I1=NP(J,1):I2=NP(J,2):I3=NP(J,3):I4=NP(J,4):II=0:ZZ=.1
830 A1=(1-XZ)*(1-YZ)/4:A2=(1+XZ)*(1-YZ)/4
840 A3=(1+XZ)*(1+YZ)/4:A4=(1-XZ)*(1+YZ)/4
850 B1=(1-XZ-ZZ)*(1-YZ)/4:B2=(1+XZ+ZZ)*(1-YZ)/4
860 B3=(1+XZ+ZZ)*(1+YZ)/4:B4=(1-XZ-ZZ)*(1+YZ)/4
870 C1=(1-XZ)*(1-YZ-ZZ)/4:C2=(1+XZ)*(1-YZ-ZZ)/4
880 C3=(1+XZ)*(1+YZ+ZZ)/4:C4=(1-XZ)*(1+YZ+ZZ)/4
890 XB=A1*X(I1)+A2*X(I2)+A3*X(I3)+A4*X(I4)
900 XC=B1*X(I1)+B2*X(I2)+B3*X(I3)+B4*X(I4):XS=ZZ/(XC-XB)
910 YB=A1*Y(I1)+A2*Y(I2)+A3*Y(I3)+A4*Y(I4)
920 YC=C1*Y(I1)+C2*Y(I2)+C3*Y(I3)+C4*Y(I4):YS=ZZ/(YC-YB)
930 DX=XA-XB:DY=YA-YB:DD=SQR(DX*DX+DY*DY):IF DD<.002*DL(J) THEN 950
940 XZ=XZ+XS*DX:YZ=YZ+YS*DY:II=II+1:IF II<11 THEN 830
950 FA=A1*F(I1)+A2*F(I2)+A3*F(I3)+A4*F(I4):RETURN
960 CLS:LOCATE 1,26,1:COLOR 0,7:PRINT" Advection using FEM - Part 5 ";
970 COLOR 7,0:PRINT:PRINT:RETURN
```

Program BV11-2-5. Advection by FEM: particles along streamlines.

The program contains some special features that deserve further consideration. After reading of the first dataset ($N\$$), which is the same as in all previous parts of the program, and which describes the geometrical data and the values of the stream function at all the nodes, a second dataset is read, in which the initial points of the streamlines and the time step are read. The actual meaning of the input data of this second dataset ($M\$$) is as follows.

| | |
|---|---|
| HH | Thickness of the aquifer, |
| PP | Porosity of the soil, |
| NP | Number of points from which streamlines are originating, |
| DT | Magnitude of time step, |
| NT | Number of time steps between successive markers, |
| NS | Number of elements in which streamlines may end, |
| $XN(I)$ | $x$-coordinate of starting point $I$, |
| $YN(I)$ | $y$-coordinate of starting point $I$, |
| $IQ(J)$ | Element in which a streamline will end, |

These input data specify that particles will be marked along $NP$ streamlines, starting at the points ($XN(I)$, $YN(I)$), and that a marker will be set along each streamline after every $NT$ time steps. In order to improve the appearance of the figure, streamlines can be terminated as soon as they enter certain elements, identified by the parameters $IQ(J)$; for instance, the elements surrounding a well. Otherwise streamlines might cross the boundary of the considered region, or keep jumping across a well.

After reading the input data, the program determines the scales such that the figure will fit on the screen, initializes some variables, and for each element calculates a characteristic length $DL(J)$, which is used as a measure for the accuracy in locating points in an element. The actual tracing of the streamlines starts at line 470. First the element in which a certain initial point is located must be determined. This task is performed in the subroutine in lines 660—800. This subroutine determines whether a certain point is located on the same side of each element boundary, which is a criterion for the points in the interior of an element, provided that the nodes of the element have been numbered in counter-clockwise direction. The subroutine, which is frequently used in the program, uses an initial guess ($JJ$), and then searches in upward and downward direction. In this way it is possible to use the previous value as an initial guess, and to quickly identify a neighbouring element when a particle crosses an element boundary. Another subroutine that is frequently used is the interpolation subroutine, which starts at line 810. The velocity components are determined numerically, in lines 520—540, employing the relations (11.6.2), taking into account that the velocity components differ from the specific discharge by a factor $n$ (the porosity). The subroutine for the determination of the stream function at a certain point within an element uses an algorithm known as *isoparametric interpolation*, in which a certain quadrangular element is mapped onto a square in the $\xi$, $\eta$-plane, and any function (e.g. $x$, $y$ or

$\Psi$) is interpolated by the formula

$$F = [(1-\xi)(1-\eta)/4] \times F(1) + [(1+\xi)(1-\eta)/4] \times F(2) + \\ + [(1+\xi)(1+\eta)/4] \times F(3) + [(1-\xi)(1+\eta)/4] \times F(4) \quad (11.7.1)$$

where $F(1)$, $F(2)$, $F(3)$ and $F(4)$ are the values of the variable $F$ in the four corner points of the element (Figure 11.14), again numbered in counter-clockwise direction. It can easily be verified that in the corner points, for $\xi = \pm 1$, $\eta = \pm 1$, the original values are recovered. Even though the interpolation has not been used in the actual finite element calculations (Program 11.2.3), it can be used as a convenient and uniquely defined algorithm for the interpolation of the stream function in the elements.

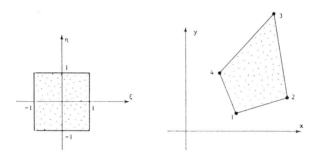

Fig. 11.14. Isoparametric interpolation.

As an example the streamlines for the two-well problem of Figure 11.11 are traced in Figure 11.15, with markers being set after every 100 time steps.

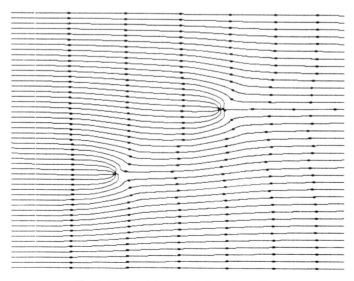

Fig. 11.15. Particles traced along streamlines.

It should be remarked that this part of the program uses a considerable amount of computer time, because of the large number of times that the program has to search for the element in which a point is located, and the large number of interpolations. Even when using the program in compiled form it may take more than one hour to complete a figure such as Figure 11.15. Of course, computation time can be reduced by taking larger time steps, but this reduces the accuracy. It may be worthwhile to modify the program so that a first program generates a datafile consisting of the location of all tracer points as a function of time, while a separate program is used to produce a graphical representation of these data. Such a graph can then be reproduced in a short time.

The agreement between Figures 11.5 and 11.15, which represent the semi-analytical and fully numerical solutions, respectively, of the same problem, is sufficient to consider both methods of about the same accuracy. The numerical method may deserve preference because of its flexibility with regard to the shape of the domain and the possibility to account for nonhomogeneous aquifer properties.

CHAPTER TWELVE

# Transport by Advection and Dispersion

In the previous chapter, numerical models for the main mechanism of transport in porous media — advection — were presented. In reality, the transport of particles of a component of a pollutant is also influenced by diffusion and dispersion, which cause a spreading of the pollutant over an ever-growing region. Thus, the groundwater may be polluted over a much larger region than in the case of pure advection, with a corresponding reduction in the maximum and average concentrations of the pollutant. The conceptual and mathematical modeling of these phenomena are discussed in Chapter 6. In this chapter, some numerical models for the quantitative analysis of transport by advection and dispersion are presented. These include a fully numerical two-dimensional method, and a semi-analytical method, using a random walk model. For didactic reasons, the simple one-dimensional case will be considered first. This case is discussed in order to show that the numerical analysis of dispersion may involve a rather disturbing influence of discretization, i.e., numerical dispersion which should be recognized and, if possible, reduced, or at least controlled.

## 12.1. Dispersion in One-Dimensional Flow

As shown in Chapter 6, the flux of a single pollutant, due to advection and dispersion, can be expressed by (6.2.19). When the soil is completely saturated this equation can be written for the one-dimensional case as

$$q_c = n \left( cV - D \frac{\partial c}{\partial x} \right) \tag{12.1.1}$$

where $V$ is the average velocity of the pore fluid, $n$ is the porosity of the soil, $D$ is the coefficient of dispersion, and $c$ is the concentration of the pollutant.

In the absence of chemical reactions and adsorption, the corresponding mass balance equation of the pollutant, Equation (6.3.2), with $\theta = n$, reduces to

$$\frac{\partial (nc)}{\partial t} = - \frac{\partial q_c}{\partial x}. \tag{12.1.2}$$

If the soil is assumed to be incompressible, the porosity $n$ is constant and, thus, $\partial n/\partial t = 0$. Substitution of (12.1.1) into (12.1.2) now gives, taking into account that the velocity $V$ must be constant in this case of uniform flow, and assuming

TRANSPORT BY ADVECTION AND DISPERSION

that the coefficient of dispersion $D$ is constant,

$$\frac{\partial c}{\partial t} = -V \frac{\partial c}{\partial x} + D \frac{\partial^2 c}{\partial x^2}. \tag{12.1.3}$$

The first term in the right-hand side is the change of the concentration due to advective transport, while the second term expresses the influence of dispersion on the concentration distribution. This equation should be solved, subject to certain initial and boundary conditions.

12.1.1. ANALYTICAL SOLUTION FOR A COLUMN OF INFINITE LENGTH

One of the simplest cases is that of flow in a homogeneous porous medium in a column of very large length, say from $x = 0$ to $x = \infty$, in which initially the concentration is zero, and which is polluted by a sudden increase of the concentration at the inflow end, $x = 0$, from time $t = 0$. The concentration distribution within the column is obtained by solving (12.1.3), subject to these initial and boundary conditions. The solution was obtained by Ogata and Banks (1961) in the form

$$\frac{c}{c_0} = \frac{1}{2} \left[ \text{erfc} \left\{ \frac{x - Vt}{\sqrt{4Dt}} \right\} + \exp\left(\frac{xV}{D}\right) \text{erfc} \left\{ \frac{x + Vt}{\sqrt{4Dt}} \right\} \right] \tag{12.1.4}$$

where $\text{erfc}(x)$ is the *complementary error function* (Abramowitz and Stegun, 1965). After a certain time, when the pollutant has moved sufficiently far into the column, the solution can be approximated by the first term only, i.e.,

$$\frac{c}{c_0} = \frac{1}{2} \text{erfc} \left\{ \frac{x - Vt}{\sqrt{4Dt}} \right\}. \tag{12.1.5}$$

This approximation satisfies the boundary condition at $x = 0$ ($c = c_0$) only if the value of the argument for $x = 0$ is sufficiently large. This occurs when $Vt$ is large compared to $D/V$. In Section 6.2, we have seen that in the one-dimensional case, the coefficient of dispersion can be expressed in terms of the longitudinal dispersivity $a_L$ by the relation (6.2.12), i.e.,

$$D = a_L V. \tag{12.1.6}$$

We recall that $a_L$ is a length that characterizes the inhomogeneity of the porous medium due to the presence of solids and void space. Thus, the requirement $Vt \gg D/V$ is equivalent to $Vt \gg a_L$. This means that (12.1.5) is applicable whenever the advective displacement is large compared to the dispersivity. For practical purposes this can be assumed to be the usual case.

In what follows, the concentration distribution, as expressed by (12.1.5), will be used for reference purposes. For this reason, a computer program that will show the concentration profile on the screen is reproduced as Program BV12-1.

```
100 DEFINT I-N:KEY OFF:GOSUB 340
110 PRINT"---   Bear & Verruijt - Groundwater Modeling"
120 PRINT"---   Dispersion in uniform groundwater flow"
130 PRINT"---   Program 12.1"
140 PRINT"---   Analytical solution":PRINT:PRINT
150 DIM C(641),CA(641):NN=640
160 INPUT"V*T (m) ....................................... ";VT
170 INPUT"Dispersivity (m) ............................. ";AL
180 C(0)=1:EE=.0001:EE=EE*EE:AA=SQR(4*AL*VT):IF AA<EE THEN AA=EE
190 FOR I=1 TO 639:X=ABS((I-VT)/AA):GOSUB 360
200 C(I)=.5*(1+ERF*SGN(VT-I)):NEXT I
210 SCREEN 2:LINE(0,20)-(639,20):LINE(639,20)-(639,120)
220 LINE(639,120)-(0,120):LINE(0,120)-(0,20)
230 FOR I=1 TO 639:J=INT(100*C(I)+.5):K=INT(100*C(I-1)+.5)
240 LINE(I-1,120-K)-(I,120-J):NEXT I
250 DX=1:X=VT:II=INT(X/DX+.5):PI=4*ATN(1):B$="####.####"
260 LOCATE 19,1,0:PRINT "VT = ";:PRINT USING B$;VT
270 LOCATE 20,1,0:PRINT "AL = ";:PRINT USING B$;AL
280 LOCATE 19,31,0:PRINT"Concentration at X=VT ......... ";
290 PRINT USING B$;C(II)
300 AA=(C(II+1)-C(II-1))/(2*DX):BL=1/(4*PI*X*AA*AA)
310 LOCATE 20,31,0:PRINT"Dispersivity from graph ....... ";
320 PRINT USING B$;BL
330 A$=INPUT$(1):SCREEN 0:END
340 CLS:LOCATE 1,32,1:COLOR 0,7:PRINT" Dispersion - 1 ";
350 COLOR 7,0:PRINT:PRINT:RETURN
360 REM Error Function
370 P=.47047:A1=.3480242:A2=-.0958798:A3=.7478556
380 X=ABS(X):IF X>2 THEN ERF=1:GOTO 400
390 TT=1/(1+P*X):ERF=1-(TT*(A1+TT*(A2+TT*A3)))*EXP(-X*X)
400 RETURN
```

Program BV12-1. One-dimensional dispersion — analytical solution.

The program will ask for values of the advective displacement, $VT$, and the dispersivity, $AL$, and then show the concentration profile over a length of 640 m from the left end of the column. The error function is calculated in the subroutine starting at line 360, using an approximation given by Abramowitz and Stegun (1965). As an example, such a concentration profile is shown in Figure 12.1, for the case that $VT = 320$ and $AL = 1$.

As could be expected, the relative concentration at $x = Vt$ (here $x = 320$), is just 0.5.

The solution (12.1.5) can be used to determine the dispersion coefficient from a laboratory experiment, in which the concentration at a fixed point is recorded, as a function of time. It can be shown that the first-order spatial derivative at the inflection point, where $c = \frac{1}{2} c_0$, is

$$\left(\frac{\partial c}{\partial x}\right)_{x=Vt} = \frac{c_0}{\sqrt{4\pi a_L x}}. \tag{12.1.7}$$

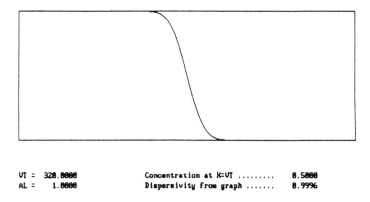

Fig. 12.1. Analytical solution for dispersion in column.

Thus, the dispersion length can be determined by measuring the slope of the concentration profile at the point $x$, where $c = \frac{1}{2}c_0$. In Program BV12-1, this relation is used to back-calculate the dispersivity from the values represented in the graph. The value of the concentration at $x = Vt$ and the dispersivity determined using (12.1.7) are also printed on the screen. As can be seen from Figure 12.1, these values are in close agreement with the input data.

### 12.1.2. NUMERICAL SOLUTION

As a first step towards the development of a numerical model for the general case of two- or three-dimensional flow, let us introduce a numerical model for the solution of the one-dimensional advection-dispersion equation (12.1.3). The two main numerical methods are the finite difference method and the finite element method. Although these methods have their own techniques, and their own supporters, they are basically very similar, especially in the one-dimensional case. Both use a spatial discretization and in one dimension this discretization is the same in the two methods. Because the finite difference method is somewhat simpler to derive and to understand, it will be used here to demonstrate the principles of the numerical solution of the dispersion equation, and the difficulties associated with it.

Various finite difference models for the solution of (12.1.3) can be developed, depending mostly upon the way in which the first-order derivatives are being approximated. The most natural, and the most accurate, approximation of the second-order derivative is a *central finite difference* (Section 9.1). This suggests the use of a central difference for the first-order spatial derivative, but it leaves open the choice of the difference approximation of the time derivative. The simplest numerical model is obtained when a forward finite difference approximation is used for this derivative. This can be called an *explicit-central approximation*. The complete finite difference equation is

$$\frac{c(x, t + \Delta t) - c(x, t)}{\Delta t}$$
$$= -V \frac{c(x + \Delta x, t) - c(x - \Delta x, t)}{2\Delta x} +$$
$$+ D \frac{c(x + \Delta x, t) - 2c(x, t) + c(x - \Delta x, t)}{(\Delta x)^2}. \qquad (12.1.8)$$

If all the values of the concentration at a certain instant of time are known, the values one time step later can be calculated explicitly from (12.1.8). Based on the considerations of Section 9.3.1, it can be expected that the process will be stable only if the time step satisfies the criterion

$$\Delta t < \tfrac{1}{2}(\Delta x)^2/D \qquad (12.1.9)$$

On the other hand, it seems physically inconsistent to attempt to let the process run faster than the velocity of the advective transport. Thus,

$$\Delta t \leq \Delta x/V. \qquad (12.1.10)$$

A program performing the numerical calculations described above, and incorporating conditions (12.1.9) and (12.1.10), is listed as Program BV12-2.

```
100 DEFINT I-N:DEFDBL A-H,O-Z:KEY OFF:GOSUB 440
110 PRINT"---   Bear & Verruijt - Groundwater Modeling"
120 PRINT"---   Dispersion in uniform groundwater flow"
130 PRINT"---   Program 12.2"
140 PRINT"---   Explicit central finite differences":PRINT:PRINT
150 DIM C(2001),CA(2001)
160 INPUT"Velocity (m/d) .............................. ";V
170 INPUT"Dispersivity (m) ........................... ";AL
180 INPUT"Space step (m) ............................. ";DX
190 D=V*AL:A=V/DX:B=2*D/(DX*DX):IF B>A THEN A=B
200 PRINT"Suggestion for time step (d) ............... ";1/A
210 INPUT"Time step (d) .............................. ";DT
220 INPUT"Number of time steps (1...2001) ............ ";N:PRINT
230 A=V*DT/(2*DX):B=D*DT/(DX*DX):C(0)=1:CA(0)=1:CM=2:IC=CSRLIN
240 FOR K=1 TO N:LOCATE IC,4,0:PRINT"Step ..... ";K:FOR I=1 TO K
250 CA(I)=C(I)-A*(C(I+1)-C(I-1))+B*(C(I+1)-2*C(I)+C(I-1)):NEXT I
260 FOR I=1 TO K:C(I)=CA(I):IF C(I)>CM THEN C(I)=CM
270 IF C(I)<0 THEN C(I)=0
280 NEXT I:NEXT K
290 SCREEN 2:LINE(0,20)-(639,20):LINE(639,20)-(639,120)
300 LINE(639,120)-(0,120):LINE(0,120)-(0,20)
310 FOR I=1 TO 639:J=INT(100*C(I)+.5):K=INT(100*C(I-1)+.5)
320 LINE(I-1,120-K)-(I,120-J):NEXT I
330 X=N*V*DT:II=INT(X/DX+.5):PI=4*ATN(1):B$="####.####"
340 LOCATE 19,1,0:PRINT "V  =  ";:PRINT USING B$;V
350 LOCATE 20,1,0:PRINT "D  =  ";:PRINT USING B$;D
```

```
360 LOCATE 21,1,0:PRINT "DX = ";:PRINT USING B$;DX
370 LOCATE 22,1,0:PRINT "DT = ";:PRINT USING B$;DT
380 LOCATE 19,31,0:PRINT"Concentration at X=N*V*DT ... ";
390 PRINT USING B$;C(II)
400 AA=(C(II+1)-C(II-1))/(2*DX):BL=1/(4*PI*X*AA*AA)
410 LOCATE 20,31,0:PRINT"Dispersivity from graph ..... ";
420 PRINT USING B$;BL
430 A$=INPUT$(1):SCREEN 0:END
440 CLS:LOCATE 1,32,1:COLOR 0,7:PRINT" Dispersion - 2 ";
450 COLOR 7,0:PRINT:PRINT:RETURN
```

Program BV12-2. One-dimensional dispersion — explicit central finite differences.

When the physical parameters that describe the problem have been entered, the program prints a suggestion for the magnitude of the time step, using the criteria (12.1.9) and (12.1.10). Then the program will execute the number of time steps prescribed by the user and represent the resulting concentration profile on the screen. The program contains two limits for the relative concentration: it cannot be larger than 2, and not smaller than 0 (see lines 260 and 270). This limitation is not physically necessary, but it is a useful mathematical limitation which restricts eventual errors. After presentation on the screen, the program also calculates the concentration at the point displaced at the average velocity, which should be 0.5, and it also uses a numerical equivalent of formula (12.1.7) to determine the apparent dispersivity. If the method were perfect, this value should be equal to the input value $AL$.

As an example, the output results for the case that $V = 1$, $AL = 1$, $DX = 1$, $DT = 0.5$ and $N = 640$ are shown in Figure 12.2. The data refer to the same problem as considered above, for which the analytical solution was shown in Figure 12.1. Although the results look rather good, they are not very accurate, as is illustrated by the value of the calculated dispersivity, which is found to be 0.7491 rather than the input value 1. A closer inspection of the two graphs will

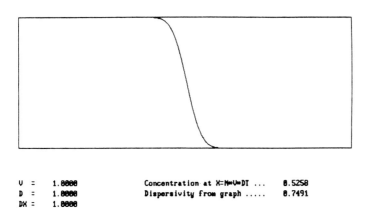

Fig. 12.2. Solution by explicit central finite differences.

also show that the front in Figure 12.2 is indeed steeper than that in Figure 12.1, indicating a smaller dispersivity. The difference of 25% is due to *numerical dispersion*, a phenomenon that will be explained in the next section. The program also records the value of the concentration at the center, as 0.5258 rather than 0.5000. This is not a serious error because the dispersivity is rather small compared to the spatial step. At the next point the concentration is 0.4984 so that this error seems acceptable.

The *stability criterion* (12.1.9) and the additional *accuracy condition* (12.1.10) do not guarantee that the solution will be satisfactory. For many combinations of parameters, the numerical solution will show unacceptable oscillations. The user may verify this statement by using, for instance, the values $V = 5$, $AL = 1$, $DX = 5$ and $DT = 1$. These oscillations are due to the strong discontinuity in the higher-order derivatives near the sharp front. For smooth initial distributions of concentration, the criteria are usually sufficient to ensure a smooth solution. In many cases, however, it may be necessary to use a time step which is up to 5 times smaller than that required by the theoretical stability criteria.

As an alternative approach, we may consider the use of a different type of finite difference approximation, for instance, a forward (*downstream*) or a backward (*upstream*) spatial finite difference. As an example, a program based on a backward finite difference scheme is presented below, as Program BV12-3.

```
100 DEFINT I-N:KEY OFF:GOSUB 440
110 PRINT"---  Bear & Verruijt - Groundwater Modeling"
120 PRINT"---  Dispersion in uniform groundwater flow"
130 PRINT"---  Program 12-3"
140 PRINT"---  Explicit backward finite differences":PRINT:PRINT
150 DIM C(2001),CA(2001)
160 INPUT"Velocity (m/d) ............................ ";V
170 INPUT"Dispersivity (m) .......................... ";AL
180 INPUT"Space step (m) ............................ ";DX
190 D=V*AL:DT=DX*DX/(2*D+V*DX)
200 PRINT"Suggestion for time step (d) .............. ";DT
210 INPUT"Time step (d) ............................. ";DT
220 INPUT"Number of time steps (1...2001) ........... ";N:PRINT
230 A=V*DT/DX:B=D*DT/(DX*DX):C(0)=1:CA(0)=1:CM=2:IC=CSRLIN
240 FOR K=1 TO N:LOCATE IC,1,0:PRINT"    Step ..... ";K:FOR I=1 TO K
250 CA(I)=C(I)-A*(C(I)-C(I-1))+B*(C(I+1)-2*C(I)+C(I-1)):NEXT I
260 FOR I=1 TO K:C(I)=CA(I):IF C(I)>CM THEN C(I)=CM
270 IF C(I)<0 THEN C(I)=0
280 NEXT I:NEXT K
290 SCREEN 2:LINE(0,20)-(639,20):LINE(639,20)-(639,120)
300 LINE(639,120)-(0,120):LINE(0,120)-(0,20)
310 FOR I=1 TO 639:J=INT(100*C(I)+.5):K=INT(100*C(I-1)+.5)
320 LINE(I-1,120-K)-(I,120-J):NEXT I
330 X=N*V*DT:II=INT(X/DX+.5):PI=4*ATN(1):B$="##.####"
340 LOCATE 19,1,0:PRINT "V  = ";:PRINT USING B$;V
350 LOCATE 20,1,0:PRINT "D  = ";:PRINT USING B$;D
360 LOCATE 21,1,0:PRINT "DX = ";:PRINT USING B$;DX
370 LOCATE 22,1,0:PRINT "DT = ";:PRINT USING B$;DT
380 LOCATE 19,31,0:PRINT"Concentration at X=N*V*DT ... ";
```

```
390 PRINT USING B$;C(II)
400 AA=(C(II+1)-C(II-1))/(2*DX):BL=1/(4*PI*X*AA*AA)
410 LOCATE 20,31,0:PRINT"Dispersivity from graph ..... ";
420 PRINT USING B$:BL
430 A$=INPUT$(1):SCREEN 0:GOSUB 440:END
440 CLS:LOCATE 1,32,1:COLOR 0,7:PRINT" Dispersion - 3 ";
450 COLOR 7,0:PRINT:PRINT:RETURN
```

Program BV12-3. One-dimensional dispersion — explicit backward finite differences.

It may be noted that there is very little difference between this program and the previous one. Actually, the only difference is in the basic algorithm in line 250, and in the suggestion for the time step in line 200. This time step is derived by the application of the usual stability criterion. Again, the program also prints the value of the concentration at the point travelling at the average velocity, and the dispersivity determined from the concentration profile by formula (12.1.7). The results for the case corresponding to that of Figures 12.1 and 12.2 are shown in Figure 12.3. This time the apparent dispersivity is larger than the actual one

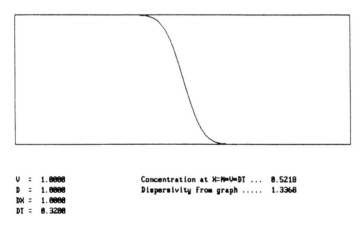

Fig. 12.3. Solution by explicit backward finite differences.

(1.3368 rather than 1.0000). Again, this is due to numerical dispersion. This phenomenon is discussed in the next section.

## 12.2. Numerical Dispersion

A characteristic and disturbing phenomenon appearing in most numerical procedures in which advective transport is combined with diffusive (or dispersive) transport, is that of *numerical dispersion*. This phenomenon is caused by the approximation of the first-order derivatives, which involves errors of the order of magnitude of the second-order derivative. The effect can be investigated by using Taylor series expansions to derive the approximations of the first-order derivatives. The forward spatial derivative, for instance, can be derived by writing

$$c(x+\Delta x) = c(x) + \Delta x \frac{\partial c}{\partial x} + \frac{1}{2}(\Delta x)^2 \frac{\partial^2 c}{\partial x^2} + \cdots \tag{12.2.1}$$

from which it follows that

$$\frac{\partial c}{\partial x} = \frac{c(x+\Delta x) - c(x)}{\Delta x} - \frac{1}{2}\Delta x \frac{\partial^2 c}{\partial x^2} + \cdots \tag{12.2.2}$$

This means that the term $-V(\partial c/\partial x)$ in the advection-dispersion equation (12.1.3) should be replaced by the expressions in the right-hand side of (12.2.2), multiplied by $-V$. This implies the addition of a dispersion-like term, having a coefficient $\frac{1}{2}V\Delta x$. Similarly, the derivation of the backward finite difference leads to

$$\frac{\partial c}{\partial x} = \frac{c(x) - c(x-\Delta x)}{\Delta x} + \frac{1}{2}\Delta x \frac{\partial^2 c}{\partial x^2} + \cdots \tag{12.2.3}$$

Thus, again, we note that an additional dispersion term will be generated, this time having a coefficient $-\frac{1}{2}V\Delta x$.

The approximation of the first-order derivative by a central finite difference does not involve such an error because the term involving the second-order derivative has a coefficient zero in the Taylor series expansion. Thus,

$$\frac{\partial c}{\partial x} = \frac{c(x+\Delta x) - c(x-\Delta x)}{2\Delta x} + O\left((\Delta x)^2, \frac{\partial^3 c}{\partial x^3}\right) \tag{12.2.4}$$

Another numerical dispersion term is generated, however, through the approximation of the time derivative. In general, one may derive the forward finite difference in time by writing

$$\frac{\partial c}{\partial t} = \frac{c(t+\Delta t) - c(t)}{\Delta t} - \frac{1}{2}\Delta t \frac{\partial^2 c}{\partial t^2} + \cdots \tag{12.2.5}$$

The second-order time derivative can be expressed in terms of a spatial derivative by using the original differential equation, with its main term, advection, only. The result is

$$\frac{\partial c}{\partial t} = \frac{c(t+\Delta t) - c(t)}{\Delta t} - \frac{1}{2}V^2\Delta t \frac{\partial^2 c}{\partial x^2} + \cdots \tag{12.2.6}$$

Equation (12.2.6) shows the appearance of an additional (numerical) dispersion, expressed by a coefficient of magnitude $-\frac{1}{2}V^2\Delta t$.

Summarizing the above results, the total dispersion, expressed by an apparent dispersion coefficient $D^*$, consists of the original physical dispersion, $D$, and numerical dispersion, which is different for the various approximation schemes.

We find for the various types of finite difference approximations

Explicit forward:
$$D^* = D - \tfrac{1}{2} V \Delta x - \tfrac{1}{2} V^2 \Delta t, \tag{12.2.7}$$

Explicit central:
$$D^* = D - \tfrac{1}{2} V^2 \Delta t, \tag{12.2.8}$$

Explicit backward:
$$D^* = D + \tfrac{1}{2} V \Delta x - \tfrac{1}{2} V^2 \Delta t. \tag{12.2.9}$$

Actually, these results are confirmed by the examples shown above. In the case of Figure 12.2, the data indicate an apparent dispersion of 0.75, which is very close to the calculated value of 0.7491, and in the case of Figure 12.3 the data indicate an apparent dispersion of 1.34, which is very close to the calculated value of 1.3368.

There are several ways to avoid, or at least reduce, numerical dispersion. Mathematically, the obvious solution is to use spatial steps and time steps that are sufficiently small, so that the numerical dispersion is small compared to the physical one. This may not always be feasible from a practical point of view, however, because the physical dispersion itself may be rather small. In many practical cases, the dispersion can be written, at least as a first approximation, as $D = a_L V$, where $a_L$ is a characteristic length of the inhomogeneity of the porous medium. In a homogeneous porous medium consisting of a single component, say sand, this dispersion length is of the order of the mean particle size, and it is not realistic, from an engineering point of view, to use a mesh size which is small compared to the size of the particles. Of course, if the physical dispersion is large, possibly because of large-scale inhomogeneities in the soil (*macro-dispersion*), numerical dispersion may well be suppressed by using a mesh that is fine compared to the size of the inhomogeneities.

The fact that numerical dispersion can so well be estimated from equations of the type (12.2.7)–(12.2.9) suggests that numerical dispersion can possibly be suppressed by modifying the original dispersion coefficient. Thus, if for instance $V = 1$ m/d, $\Delta t = 0.4$ d, and $D = 1.2$ m²/d we obtain from (12.2.8) that $D^* = 1$ m²/d. This means that correct results for a problem with dispersion 1 m²/d may be obtained by using a value of 1.2 m²/d as input to the computer program. The result of such a computation is shown in Figure 12.4. In this case, back calculation of the dispersion coefficient, using (12.1.7), gives $D = 0.9993$, which is indeed very close to the desired value of 1. Of course, this technique is applicable only if the magnitude of the numerical dispersion is fully known. This is the case only in such elementary problems as one-dimensional dispersion, for which an analytical solution exists. For the more general case of two- or three-dimensional transport, the magnitude of numerical dispersion is usually not fully known beforehand, and

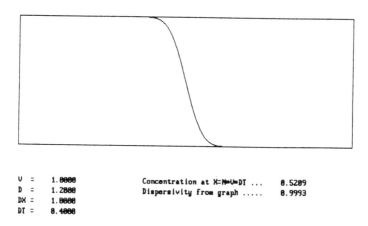

Fig. 12.4. Effect of modified dispersion coefficient.

may vary over the considered region. In such cases, the only way to suppress numerical dispersion is to use a network with mesh sizes small compared to the dispersivity.

## 12.3. A Finite Element Model for Two-Dimensional Problems

In this section a finite element model for the two-dimensional advection-dispersion problem is presented. The program may suffer from the negative effects of numerical dispersion if the mesh size is too large, but this has to be accepted when using this type of numerical method.

The basic differential equation to be solved is (6.3.27), which may be written, for the case of two-dimensional transport in an incompressible porous medium, and transport in the absence of sources, decay, etc., as

$$\frac{\partial c}{\partial t} = -V_i \frac{\partial c}{\partial x_i} + \frac{\partial}{\partial x_i}\left(D_{ij} \frac{\partial c}{\partial x_j}\right) \qquad (12.3.1)$$

where the summation convention has been used to indicate summation over a (dummy) repetitive index. It may be recalled that the first term in the right-hand side of (12.3.1) represents the advective transport component and the second term, the dispersive transport. If the porous medium is assumed to be isotropic, the components of the dispersion tensor **D** may be expressed by (6.2.10), i.e.,

$$D_{ij} = a_T V \delta_{ij} + (a_L - a_T) V_i V_j / V \qquad (12.3.2)$$

where $a_L$ and $a_T$ are the longitudinal and transverse dispersivities, respectively, and $V$ is the magnitude of the velocity vector.

As a simple example, let us consider the case of uniform flow in the $x$-direction, in a two-dimensional domain $R$ in the $x$, $y$-plane. In that case the only nonzero

TRANSPORT BY ADVECTION AND DISPERSION

coefficients of the dispersion tensor **D** are

$$D_{xx} = a_L V, \qquad D_{yy} = a_T V. \tag{12.3.3}$$

The basic differential equation for this case is

$$\frac{\partial c}{\partial t} = -V \frac{\partial c}{\partial x} + \frac{\partial}{\partial x}\left(D_{xx} \frac{\partial c}{\partial x}\right) + \frac{\partial}{\partial y}\left(D_{yy} \frac{\partial c}{\partial y}\right) \tag{12.3.4}$$

As in Chapter 10, the time derivative is now approximated by denoting the initial value, at the beginning of a time step of magnitude $\Delta t$, by $c_0$, and the value at the end of the interval by $c'$. The average value is assumed to be given by the interpolation formula

$$c = \varepsilon c_0 + (1 - \varepsilon) c' \tag{12.3.5}$$

where $\varepsilon$ is an interpolation parameter, with $\varepsilon = \tfrac{1}{2}$ corresponding to linear interpolation, $\varepsilon = 0$ to backward interpolation, and $\varepsilon = 1$ to forward interpolation.

With (12.3.5), Equation (12.3.4) becomes

$$\frac{\partial}{\partial x}\left(D_{xx} \frac{\partial c}{\partial x}\right) + \frac{\partial}{\partial y}\left(D_{yy} \frac{\partial c}{\partial y}\right) - V \frac{\partial c}{\partial x} - \frac{c - c_0}{(1 - \varepsilon)\Delta t} = 0. \tag{12.3.6}$$

This is the differential equation for which a finite element approximation is presented below.

### 12.3.1. FINITE ELEMENTS

As in Chapter 10, the simplest type of finite element approximation, using linear interpolation in triangular elements, is presented as the basis for a numerical model. The basic idea is that the physical variables (the piezometric head $\phi$ and the pollutant concentration $c$) are defined at the nodes of a network of triangles, and that the linear interpolation in the elements is performed by the definition of shape functions $N_i(x, y)$, such that

$$N_i(x, y) = 1, \quad \text{if } x = x_i, y = y_i,$$
$$N_i(x, y) = 0, \quad \text{if } x = x_j, y = y_j, j \neq i. \tag{12.3.7}$$

The groundwater head $\phi$ and the concentration $c$ can now be expressed as

$$\phi(x, y) = \sum N_i(x, y) \phi_i, \tag{12.3.8}$$
$$c(x, y) = \sum N_i(x, y) c_i. \tag{12.3.9}$$

This means that the velocity, which is defined by the derivative of the head, is constant throughout each element.

Because density and viscosity are not affected by changes in the concentration (a basic assumption of the model considered here, see Section 6.5.2), the velocity distribution can be solved as a separate problem, independent of the solution for the concentration. We say that the two problems, of determining $\phi(x, y, t)$ and $c(x, y, t)$, are *uncoupled*. Accordingly, we now assume that the groundwater flow problem has been solved, and that the velocity components in all elements can be considered as known.

The finite element equations can be derived by the Galerkin approach presented in Chapter 10. This means that the partial differential equation, (12.3.6), is multiplied by each of the shape functions, and that the result is integrated over the domain $R$. Each of the surface integrals can be evaluated by a summation of integrals over the triangular elements. By an appropriate rotation of the coordinate system in an element, the $x$-axis can always be made to coincide with the direction of flow, so that for the evaluation of the integral over a single triangle, the basic equation can always be written in the form (12.3.4) or, after approximation of the time derivative, in the form (12.3.6). Thus, the integral to be evaluated is

$$J_p = \int_{R_p} \left\{ \left[ \frac{\partial}{\partial x}\left(D_{xx}\frac{\partial c}{\partial x}\right) + \frac{\partial}{\partial y}\left(D_{yy}\frac{\partial c}{\partial y}\right) \right. \right.$$

$$\left. \left. - V\frac{\partial c}{\partial x} - \frac{c-c_0}{(1-\varepsilon)\Delta t} \right] N_i \right\} dx\, dy \qquad (12.3.10)$$

where $R_p$ is the area of element number $p$. The summation of the integrals $J_p$ over all elements should be zero

$$\sum_{p=1}^{m} J_p = 0. \qquad (12.3.11)$$

The first two terms in the integrand of (12.3.11) can be separated into two parts by noting that

$$\left[ \frac{\partial}{\partial x}\left(D_{xx}\frac{\partial c}{\partial x}\right) + \frac{\partial}{\partial y}\left(D_{yy}\frac{\partial c}{\partial y}\right) \right] N_i$$

$$= \frac{\partial}{\partial x}\left(N_i D_{xx}\frac{\partial c}{\partial x}\right) + \frac{\partial}{\partial y}\left(N_i D_{yy}\frac{\partial c}{\partial y}\right)$$

$$- D_{xx}\frac{\partial N_i}{\partial x}\frac{\partial c}{\partial x} - D_{yy}\frac{\partial N_i}{\partial y}\frac{\partial c}{\partial y}. \qquad (12.3.12)$$

# TRANSPORT BY ADVECTION AND DISPERSION

After substitution of (12.3.12) into (12.3.10), the resulting integral can be decomposed into four separate integrals,

$$J_p = J_1 + J_2 + J_3 + J_4 \tag{12.3.13}$$

where

$$J_1 = \int_{R_p} \left\{ \frac{\partial}{\partial x}\left(N_i D_{xx} \frac{\partial c}{\partial x}\right) + \frac{\partial}{\partial y}\left(N_i D_{yy} \frac{\partial c}{\partial y}\right) \right\} dx\, dy, \tag{12.3.14}$$

$$J_2 = -\int_{R_p} \left\{ D_{xx} \Sigma \left[ \frac{\partial N_i}{\partial x} \frac{\partial N_j}{\partial x} c_j \right] + \right.$$

$$\left. + D_{yy} \Sigma \left[ \frac{\partial N_i}{\partial y} \frac{\partial N_j}{\partial y} c_j \right] \right\} dx\, dy, \tag{12.3.15}$$

$$J_3 = \int_{R_p} \left\{ -VN_i \Sigma \left[ \frac{\partial N_j}{\partial x} c_j \right] \right\} dx\, dy, \tag{12.3.16}$$

$$J_4 = \int_{R_p} \left\{ \frac{1}{(1-\varepsilon)\Delta t} N_i \Sigma [N_j(c_j - c_j^0)] \right\} dx\, dy. \tag{12.3.17}$$

In all these integrals the summation $\Sigma$ is over all $j$, from $j = 0$ to $j = n$, where $n$ is the number of nodes. In the integrals $J_2$, $J_3$ and $J_4$ the concentration $c$ has been represented by its approximation (12.3.9). Each of the four integrals will be elaborated below.

## 12.3.2. THE FIRST INTEGRAL, $J_1$

The first integral can be transformed into a line integral along the boundary by the divergence theorem, as explained in Section 10.1, i.e.,

$$J_1 = \int \{-w_n N_i\}\, dS \tag{12.3.18}$$

where $w_n$ is the normal component of the dispersive flux vector. As in all the integrals, the value of $i$ should be restricted to the class of values representing the points in which the concentration is unknown. Hence, the contribution of a boundary segment with given concentration to the integral is zero, because $N_i = 0$

in these points. On the remaining part of the boundary the flux of the pollutant can be considered to be given. With very little loss of generality, the dispersive part of this flux can be assumed to be zero along these boundary segments, leaving only the advective transport. The first integral then vanishes,

$$J_1 = 0 \tag{12.3.19}$$

### 12.3.3. THE SECOND INTEGRAL, $J_2$

The second integral, defined by (12.3.15) can be elaborated in the same way as was done for the groundwater flow problem in Section 10.1 (see Equation (10.1.16)). The only difference is that in this case the terms in the $x$- and $y$-direction have different coefficients. This can easily be taken into account, and the final result is

$$J_2 = -\sum P_{ij} c_j \tag{12.3.20}$$

where

$$P_{ij} = \frac{1}{2|D|} (D_{xx} b_i b_j + D_{yy} c_i c_j). \tag{12.3.21}$$

The meaning of the coefficients $b_i$, $c_i$ and $D$ is the same as in Section 10.1 (see (10.1.20)).

It should be noted that for this expression to be valid, the direction of flow must be the positive $x$-direction. It may be necessary to first perform a coordinate transformation in order to obtain that property.

### 12.3.4. THE THIRD INTEGRAL, $J_3$

For the elaboration of the third integral it is recalled, from (10.1.18), that the shape function can be written as

$$N_i = p_i x + q_i y + r_i$$

so that

$$\partial N_i / \partial x = p_i.$$

Substitution of this result into (12.3.16) gives

$$J_3 = -V \sum [p_j c_j] \int N_i \, dx \, dy.$$

The integration of the shape function over an element is $\frac{1}{3}$ times the area of the triangle, which is just $\frac{1}{2}|D|$. Hence, for each of the three nodes in the element

TRANSPORT BY ADVECTION AND DISPERSION

considered, a contribution is obtained of the form

$$J_3 = - \sum F_{ij} c_j \tag{12.3.22}$$

where

$$F_{ij} = \tfrac{1}{6} |D| V p_j \tag{12.3.23}$$

Again, for this expression to be valid, the $x$-direction should be the direction of flow.

### 12.3.5. THE FOURTH INTEGRAL, $J_4$

The fourth integral, defined by (12.3.17), is of the same form as the integral representing leakage in a steady-flow problem, see (10.1.12), or the integral representing storage in Section 10.5. Thus, by analogy with the expression (10.1.31), we may write

$$J_4 = - \sum R_{ij}(c_j - c_j^0) \tag{12.3.24}$$

where

$$R_{ij} = \{ b_i b_j Z_{xx} + c_i c_j Z_{yy} + \\ + (b_i c_j + b_j c_i) Z_{xy} + d_i d_j \} / \{ 2|D| \Delta t (1 - \varepsilon) \} \tag{12.3.25}$$

The coefficients $Z_{xx}$, $Z_{yy}$ and $Z_{xy}$ are defined by (10.1.30).

### 12.3.6. THE COMPLETE SYSTEM OF EQUATIONS

The expressions for the four integrals can now be substituted into (12.3.12), using (12.3.14). After summation over all elements the following system of equations is obtained

$$\sum \{ (P_{ij} + F_{ij}) c_j + R_{ij}(c_j - c_j^0) \} = 0. \tag{12.3.26}$$

Here each of the matrices should be calculated by a summation over all elements, as is usual in the finite element method. The structure of the matrices **P**, **F** and **R** is given by Equations (12.3.21), (12.3.23), and (12.3.25). The system of equations can be solved by any standard method of solution.

### 12.3.7. COMPUTER PROGRAM

A computer program that solves the two-dimensional dispersion problem by the finite element method is listed in Program BV12-4.

```
100 DEFINT I-N:KEY OFF:OPTION BASE 1:GOSUB 900
110 PRINT"---   Bear & Verruijt - Groundwater Modeling"
120 PRINT"---   Transport by Dispersion"
130 PRINT"---   Finite Element Method":PRINT"---   Program 12.4"
140 PRINT"---   Quadrangular elements, input by DATA statements"
150 PRINT"---   Solution by Gauss-Seidel iteration"
160 DIM XJ(4),YJ(4),B(3),C(3),D(3),KS(4,3),X(350),Y(350),IP(350)
170 DIM CC(350),CA(350),Q(350),P(350,10),R(350,10),KP(350,10):NZ=10
180 DIM NP(300,4),VX(300),VY(300),D1(300),D2(300)
190 PRINT:PRINT:INPUT"Name of dataset ....... ";N$:PRINT:OPEN"I",#1,N$
200 INPUT#1,NLX,NLY,SXY,NSCR,N,M,DA,DB,DT,NS,EPS,NT,RX
210 FOR I=1 TO N:INPUT#1,X(I),Y(I),CA(I),IP(I):CC(I)=CA(I):Q(I)=0:NEXT I
220 FOR J=1 TO M:INPUT#1,NP(J,1),NP(J,2),NP(J,3),NP(J,4),VX(J),VY(J):NEXT J
230 CLOSE 1
240 XA=X(1):YA=Y(1):XB=XA:YB=YA:FOR I=1 TO N:IF X(I)<XA THEN XA=X(I)
250 IF X(I)>XB THEN XB=X(I)
260 IF Y(I)<YA THEN YA=Y(I)
270 IF Y(I)>YB THEN YB=Y(I)
280 NEXT I:SX=NLX/(XB-XA):SY=SXY*NLY/(YB-YA):IF SY<SX THEN SX=SY
290 SY=SX/SXY:GOSUB 920:FOR J=1 TO M:VV=SQR(VX(J)*VX(J)+VY(J)*VY(J))
300 D1(J)=DA*VV:D2(J)=DB*VV:NEXT J
310 FOR I=1 TO 4:FOR J=1 TO 3:K=I+J-1:IF K>4 THEN K=K-4
320 KS(I,J)=K:NEXT J,I:LOCATE 23,1,0:PRINT"Generation of pointer matrix"
330 FOR I=1 TO N:KP(I,1)=I:KP(I,NZ)=1:NEXT I:E=.000001
340 FOR J=1 TO M:LOCATE 24,4,0:PRINT"Element ...... ";J;
350 FOR K=1 TO 4:KK=NP(J,K):FOR L=1 TO 4:LL=NP(J,L)
360 IA=0:FOR II=1 TO KP(KK,NZ):IF KP(KK,II)=LL THEN IA=1
370 NEXT II:IF IA=0 THEN KB=KP(KK,NZ)+1:KP(KK,NZ)=KB:KP(KK,KB)=LL
380 IF KB=NZ THEN 890
390 NEXT L,K,J
400 LOCATE 23,1,0:PRINT"Generation of system matrices":FOR J=1 TO M
410 LOCATE 24,4,0:PRINT"Element ...... ";J;" ";:FOR KW=1 TO 4:ZX=0:ZY=0
420 FOR I=1 TO 3:K=NP(J,KS(KW,I)):XJ(I)=X(K):YJ(I)=Y(K)
430 ZX=ZX+X(K):ZY=ZY+Y(K):NEXT I:ZX=ZX/3:ZY=ZY/3
440 FOR I=1 TO 3:XJ(I)=XJ(I)-ZX:YJ(I)=YJ(I)-ZY:NEXT I
450 VV=SQR(VX(J)*VX(J)+VY(J)*VY(J)):C=VX(J)/VV:S=VY(J)/VV
460 FOR I=1 TO 3:XI=XJ(I):YI=YJ(I):XJ(I)=XI*C+YI*S:YJ(I)=YI*C-XI*S:NEXT I
470 B(1)=YJ(2)-YJ(3):C(1)=XJ(3)-XJ(2):D(1)=XJ(2)*YJ(3)-XJ(3)*YJ(2)
480 B(2)=YJ(3)-YJ(1):C(2)=XJ(1)-XJ(3):D(2)=XJ(3)*YJ(1)-XJ(1)*YJ(3)
490 B(3)=YJ(1)-YJ(2):C(3)=XJ(2)-XJ(1):D(3)=XJ(1)*YJ(2)-XJ(2)*YJ(1)
500 D=D(1)+D(2)+D(3):DD=ABS(D):IF DD<EE THEN 620
510 DA=D1(J)/(4*DD):DB=D2(J)/(4*DD):DC=DD/(12*D):DE=1/(4*(1-E)*DD)
520 XX=(XJ(1)*XJ(1)+XJ(2)*XJ(2)+XJ(3)*XJ(3))/12
530 XY=(XJ(1)*YJ(1)+XJ(2)*YJ(2)+XJ(3)*YJ(3))/12
540 YY=(YJ(1)*YJ(1)+YJ(2)*YJ(2)+YJ(3)*YJ(3))/12
550 FOR K=1 TO 3:KK=NP(J,KS(KW,K)):II=KP(KK,NZ):FOR LL=1 TO II:L=1
560 KV=KS(KW,L):IF NP(J,KV)=KP(KK,LL) THEN 580
570 L=L+1:IF L<4 THEN 560 ELSE 610
580 P(KK,LL)=P(KK,LL)+DA*B(K)*B(L)+DB*C(K)*C(L)+DC*B(L)*VV
590 AA=XX*B(K)*B(L)+XY*(B(K)*C(L)+B(L)*C(K))+YY*C(K)*C(L)+D(K)*D(L)
600 R(KK,LL)=R(KK,LL)+DE*AA
610 NEXT LL,K
620 NEXT KW,J:TN=0:FOR IS=1 TO NS:B=1/DT:TN=TN+DT:FOR IT=1 TO NT
630 FOR I=1 TO N:AA=0:IZ=KP(I,NZ):IF IP(I)>0 THEN 680
640 AA=Q(I):FOR J=1 TO IZ:CA=B*R(I,J):L=KP(I,J)
650 AA=AA-(P(I,J)+CA)*CC(L)+CA*CA(L):NEXT J
660 CC(I)=CC(I)+RX*AA/(P(I,1)+B*R(I,1)):AA=EP*CA(I):BB=AA+1-EP
```

# TRANSPORT BY ADVECTION AND DISPERSION

```
670 IF CC(I)<AA THEN CC(I)=AA ELSE IF CC(I)>BB THEN CC(I)=BB
680 NEXT I,IT:FOR I=1 TO N:CA(I)=CA(I)+(CC(I)-CA(I))/(1-E):CC(I)=CA(I)
690 NEXT I:GOSUB 920:AA=CA(1):FOR I=1 TO N:IF CA(I)>AA THEN AA=CA(I)
700 NEXT I:DC=AA/5:CA=DC/2:NC=4
710 LOCATE 24,4,0:PRINT"Time ........ ";:PRINT USING"###.###";TN;
720 FOR JC=0 TO NC:CF=CA+JC*DC
730 FOR J=1 TO M:I1=NP(J,1):I2=NP(J,2):I3=NP(J,3):I4=NP(J,4)
740 C1=CC(I1):C2=CC(I2):C3=CC(I3):C4=CC(I4)
750 K=0:IF (C1<=CF AND C2<=CF) OR (C1>=CF AND C2>=CF) THEN 780
760 K=K+1:B(K)=X(I1)+(X(I2)-X(I1))*(CF-C1)/(C2-C1)-XA
770 C(K)=Y(I1)+(Y(I2)-Y(I1))*(CF-C1)/(C2-C1)-YA
780 IF (C2<=CF AND C3<=CF) OR (C2>=CF AND C3>=CF) THEN 810
790 K=K+1:B(K)=X(I2)+(X(I3)-X(I2))*(CF-C2)/(C3-C2)-XA
800 C(K)=Y(I2)+(Y(I3)-Y(I2))*(CF-C2)/(C3-C2)-YA
810 IF (C3<=CF AND C4<=CF) OR (C3>=CF AND C4>=CF) THEN 840
820 K=K+1:B(K)=X(I3)+(X(I4)-X(I3))*(CF-C3)/(C4-C3)-XA
830 C(K)=Y(I3)+(Y(I4)-Y(I3))*(CF-C3)/(C4-C3)-YA
840 IF (C4<=CF AND C1<=CF) OR (C4>=CF AND C1>=CF) THEN 870
850 K=K+1:B(K)=X(I4)+(X(I1)-X(I4))*(CF-C4)/(C1-C4)-XA
860 C(K)=Y(I4)+(Y(I1)-Y(I4))*(CF-C4)/(C1-C4)-YA
870 IF K>1 THEN LINE (SX*B(1),SY*C(1))-(SX*B(2),SY*C(2))
880 NEXT J,JC,IS:A$=INPUT$(1):PRINT:END
890 GOSUB 900:PRINT"Pointer width (NZ) too small.":PRINT:END
900 CLS:LOCATE 1,25,1:COLOR 0,7:PRINT" Dispersion by Finite Elements ";
910 COLOR 7,0:PRINT:PRINT:RETURN
920 SCREEN NSCR:CLS:FOR J=1 TO M:I1=NP(J,1):I2=NP(J,2):I3=NP(J,3):I4=NP(J,4)
930 X1=SX*(X(I1)-XA):Y1=SY*(Y(I1)-YA):X2=SX*(X(I2)-XA):Y2=SY*(Y(I2)-YA)
940 X3=SX*(X(I3)-XA):Y3=SY*(Y(I3)-YA):X4=SX*(X(I4)-XA):Y4=SY*(Y(I4)-YA)
950 LINE (X1,Y1)-(X2,Y2):LINE (X2,Y2)-(X3,Y3)
960 LINE (X3,Y3)-(X4,Y4):LINE (X4,Y4)-(X1,Y1):NEXT J:RETURN
```

Program BV12-4. Two-dimensional dispersion by the finite element method.

The program uses input from a dataset, the name of which should be entered by the user. A dataset must be generated separately by the user, for instance using a word-processing program. The meaning of the input data, assembled in a dataset, is as follows

| | |
|---|---|
| *NLX* | Number of screen dots in horizontal direction (e.g., 640), |
| *NLY* | Number of screen dots in vertical driection (e.g., 200), |
| *SXY* | Ratio of horizontal and vertical scales (e.g., 2), |
| *NSCR* | Graphics screen parameter (e.g., 2), |
| *N* | Number of nodes, |
| *M* | Number of elements, |
| *DA* | Longitudinal dispersivity, |
| *DB* | Transverse dispersivity, |
| *DT* | Time step, |
| *NS* | Number of time steps, |
| *EPS* | Interpolation parameter, |
| *NT* | Number of Gauss—Seidel iterations, |
| *RX* | Relaxation factor, |

$X(I)$     $X$-coordinate of node $I$,
$Y(I)$     $Y$-coordinate of node $I$,
$CA(I)$    Initial concentration in node $I$,
$IP(I)$    Type indicator for node $I$,
$NP(J, 1)$ Node 1 of element $J$,
$NP(J, 2)$ Node 2 of element $J$,
$NP(J, 3)$ Node 3 of element $J$,
$NP(J, 4)$ Node 4 of element $J$,
$VX(J)$    $X$-component of velocity in element $J$,
$VY(J)$    $Y$-component of velocity in element $J$.

The dataset contains all geometrical data of the network, as defined by the coordinates of the nodes and the structure of the elements. The velocity field in the network of elements is supposed to be given, and must be entered by specification of the two components of the velocity in each element. The initial values of the concentration of the pollutant in all nodes must also be given. The type indicator $IP(I)$ indicates whether this concentration remains fixed in time (if $IP(I) = 1$), indicating a continuous source of pollution, or that no further pollutant is supplied (if $IP(I) = 0$). This enables us to follow the progress in time of a given initial distribution of the concentration of a pollutant. It also permits us to trace the path of a continuous pollution from nodes of constant concentration. Along an inflow boundary where fresh water is being supplied, the concentration can be fixed at the value 0.

The characteristics of the program are that it uses quadrangular elements, each composed of four triangles (see Chapter 10); that the system matrices are calculated only once, using a pointer matrix of limited width; and that the system of equations is solved by the simple method of Gauss–Seidel iteration. The program follows the procedures developed previously in this section, with some necessary changes in notation ($EPS$ for $\varepsilon$, $CA$ for $c_0$, etc.). Output of the computer program consists of contours of the concentration after each time step, for concentrations of 0.1, 0.3, 0.5, 0.7 and 0.9 times the maximum concentration in the field.

An elementary (small) dataset is shown below, as an example. This dataset refers to a network consisting of 45 nodes and 32 elements (Figure 12.5). The network consists of squares with a size of 10 m. The velocity of the fluid in the field is constant ($VX = 1$ m/d, $VY = 0$). The dispersivities in longitudinal and transverse directions are zero, which means that we neglect the effect of dispersion and leave advection only. The concentration at the node on the axis of the network, at a distance of 20 m from the left, is kept constant at a value 1. The figure shows the contours represented on the screen after the last (20th) time step, that is, after 50 days. The time step of 2.5 days expresses the fact that in each time step one quarter of an element is travelled by the carrying fluid. This seems to be a reasonable value for the time step.

# TRANSPORT BY ADVECTION AND DISPERSION

```
640 200  2  2  45  32  0  0  2.5  20   0.5  25   1.5
  0   0  0  1   0  10  0  1   0   20   0   1    0   30  0  1    0  40  0  1
 10   0  0  0  10  10  0  0  10   20   0   0   10   30  0  0   10  40  0  0
 20   0  0  0  20  10  0  0  20   20   1   1   20   30  0  0   20  40  0  0
 30   0  0  0  30  10  0  0  30   20   0   0   30   30  0  0   30  40  0  0
 40   0  0  0  40  10  0  0  40   20   0   0   40   30  0  0   40  40  0  0
 50   0  0  0  50  10  0  0  50   20   0   0   50   30  0  0   50  40  0  0
 60   0  0  0  60  10  0  0  60   20   0   0   60   30  0  0   60  40  0  0
 70   0  0  0  70  10  0  0  70   20   0   0   70   30  0  0   70  40  0  0
 80   0  0  0  80  10  0  0  80   20   0   0   80   30  0  0   80  40  0  0
  1   2  7  6   1   0  2  3   8    7   1   0
  3   4  9  8   1   0  4  5  10    9   1   0
  6   7 12 11   1   0  7  8  13   12   1   0
  8   9 14 13   1   0  9 10  15   14   1   0
 11  12 17 16   1   0 12 13  18   17   1   0
 13  14 19 18   1   0 14 15  20   19   1   0
 16  17 22 21   1   0 17 18  23   22   1   0
 18  19 24 23   1   0 19 20  25   24   1   0
 21  22 27 26   1   0 22 23  28   27   1   0
 23  24 29 28   1   0 24 25  30   29   1   0
 26  27 32 31   1   0 27 28  33   32   1   0
 28  29 34 33   1   0 29 30  35   34   1   0
 31  32 37 36   1   0 32 33  38   37   1   0
 33  34 39 38   1   0 34 35  40   39   1   0
 36  37 42 41   1   0 37 38  43   42   1   0
 38  39 44 43   1   0 39 40  45   44   1   0
```

Dataset DATA12.5, to be used with program BV12-4.

It can be seen from Figure 12.5 that, even though dispersion has been neglected, there is a certain spreading, which must be due to numerical dispersion. It should be noted that the lateral spreading, over a distance equal to the dimension of a single element, is a logical consequence of the numerical approximation. As the concentration is defined only at the nodal points, with linear

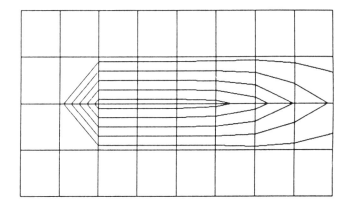

Time ......... 50.000

Fig. 12.5. Finite element solution for dataset DATA12.5.

interpolation in the elements, the fact that there is a source of pollution located at a single node automatically leads to a representation such that the concentration contours are distributed over the width of an element. The practical absence of numerical dispersion in a lateral direction is due to the fact that the velocity is strictly in the $x$-direction.

A second example is shown in Figure 12.6, for a larger network, consisting of 300 square elements of $10 \times 10$ m each. Again, the velocity field has been assumed to be uniform, at a rate of $V_x = 1$ m/d. At time $t = 0$ a local pollution concentration $c = 1$ has been assumed at the node at a distance of 30 m from the left side boundary. In all other nodes, the initial concentration is zero. The longitudinal and transverse dispersivities have been assumed to be 10 and 2 m, respectively. Figures 12.6 shows 5 contours of the concentration distribution calculated after 50 days, that is, after 20 time steps. As can be seen from the figure, there now is a considerable lateral spreading, due to dispersion.

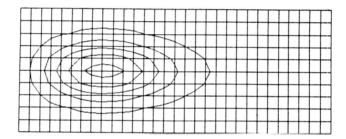

Fig. 12.6. Finite element solution, example 2.

It may be noted that, in general, the analysis of dispersion by the finite element method will be accurate only if numerical dispersion is small compared to the physical dispersion. As remarked above, this means that the network must be such that the dimensions of the elements must be at least of the order of magnitude of the dispersivity. For a large region, this may mean that a very large number of relativity small elements must be used, which may require a large computer and large costs for computation time. Actually, it can be argued that in such cases it may be more realistic to disregard dispersion together, and to consider advective transport only, at least as a first approximation. At the scale of the problem, this may give sufficiently accurate results.

## 12.4. Random Walk Model

Various alternative techniques have been proposed in order to avoid the effects of numerical dispersion. Many of these methods have the common feature that particles are being followed along their paths. These methods are, therefore, often referred to as *particle-tracking* methods. In contrast with the fixed networks

TRANSPORT BY ADVECTION AND DISPERSION

generally used in most standard numerical methods, these techniques involve a continuous adjustment of the coordinates of discrete particles. A simple method to analyze combined convection and dispersion is by using a *random walk model* to simulate the dispersive transport (Prickett *et al.*, 1981), superimposed upon a numerical or analytical solution for the convective transport. Actually, the use of a random element in the description of dispersive transport may be considered to be very well representative of the physical background of dispersion, which is due to the random character of the structure of the porous medium (Chapter 6). A model based upon these principles is presented in this section.

Consider a process in which a particle is travelling in discrete steps in the $x$-direction. The average length of each step is $A$, where $A$ is some constant. In each step there is a deviation from the average of maximum magnitude $B$, in both senses. Each step may be considered to consist of a deterministic part of magnitude $A$, and a random part of maximum magnitude $B$. If it is assumed that the distribution of the random part is homogeneous, the step can be described by a distribution function $p(x)$ of the following form

$$\begin{aligned} p(x) &= 0, & &\text{if } x < (A-B), \\ p(x) &= 1/2B, & &\text{if } (A-B) < x < (A+B), \\ p(x) &= 0, & &\text{if } x > (A+B). \end{aligned} \tag{12.4.1}$$

The statistical properties of the process can be described by the first and second moments of the distribution function. In this case these moments are

$$m = E(x) = \frac{1}{2b} \int [x p(x)] \, dx = A, \tag{12.4.2}$$

$$\sigma^2 = E(x^2 - m^2) = \frac{1}{2B} \int [(x^2 - m^2) p(x)] \, dx = B^2/3. \tag{12.4.3}$$

This means that the average distance covered in each step, $m$, is equal to $A$, as could be expected, and that the standard deviation, $\sigma$, which is a measure for the spreading, is $B/\sqrt{3}$. It is known from the theory of stochastic processes (Feller, 1966), that, according to the central limit theorem, the distribution function for the probability of travelling a certain distance after a great number of independent steps is *Gaussian*. This Gaussian or *normal distribution* can be written as

$$P(x) = \frac{1}{\sqrt{(2\pi S^2)}} \exp\left[-\frac{(x-M)^2}{2S^2}\right] \tag{12.4.4}$$

where

$$M = Nm = NA, \tag{12.4.5}$$

$$S^2 = N\sigma^2 = NB^2/3 \tag{12.4.6}$$

and $N$ is the number of steps. We note that the average distance travelled after $N$ steps is $N$ times the average distance travelled by the particle in each step, while the standard deviation increases with the square root of the number of steps.

This representation of a random walk process can be compared with the one-dimensional convection-dispersion problem, as described by the differential equation

$$\frac{\partial c}{\partial t} = -V \frac{\partial c}{\partial x} + D \frac{\partial^2 c}{\partial x^2}. \tag{12.4.7}$$

The solution for a unit of mass, injected at the point $x = 0$ at time $t = 0$, can be shown to be

$$\frac{c}{c_0} = \frac{1}{\sqrt{(4\pi Dt)}} \exp\left[ -\frac{(x - Vt)^2}{4Dt} \right]. \tag{12.4.8}$$

Comparison of (12.4.8) with (12.4.4) shows that both solutions are of the same form, and that they become formally identical if

$$M = Vt, \tag{12.4.9}$$

$$S^2 = 2Dt. \tag{12.4.10}$$

The random walk process can be considered to be a representation of transport by both convection and dispersion. From (12.4.5) and (12.4.9), it follows that $NA = Vt$. It then follows from (12.4.6) and (12.4.10) that

$$D = VB^2/(6A). \tag{12.4.11}$$

It can be concluded that the convection-dispersion process can be simulated by a random walk process, such that

$$B = \sqrt{6DA/V} \tag{12.4.12}$$

As the dispersion coefficient $D$ is related to the velocity $V$, expressed by the relation $D = a_L V$, where $a_L$ is the dispersivity, we obtain

$$B = \sqrt{6 a_L A} \tag{12.4.13}$$

The simulation procedure consists of deterministic steps of length $A$ (this represents the convective transport) and random steps of maximum length $B$, representing dispersion. As most computer languages support a simple random function, with a homogeneous distribution (usually a function of the form RND($X$), which varies between 0 and 1), such a simulation can easily be implemented in a computer program. The basis of such a numerical model can be a model for convective transport, as presented in Chapter 11. The transport process in such a model consists of discrete steps, which are calculated from a known velocity field, either by analytic formulas, or on the basis of a numerical approximation. The magnitude of a particular step can now be identified with the deterministic step $A$ in the random walk model, and a random step can be superimposed on each step in agreement with Equation (12.4.12) or (12.4.13). It should be noted that the magnitude of the random step depends upon the magnitude $A$ of the deterministic step. The result of this simulation is the realization of the random walk process for a single particle. To simulate dispersion, the process has to be repeated for a large number of particles.

Although the process has been explained here only for the one-dimensional case, it may be clear that a similar procedure can be used for the more general case of a two-dimensional flow field. The basic formula remains (12.4.12), except that in the longitudinal direction the dispersivity is usually larger than in the lateral one. This can easily be taken into account by considering the random walk to consist of two parts: one in the direction of flow, and one perpendicular to it.

### 12.4.1. COMPUTER PROGRAM

A computer program executing the deterministic and random steps representing the convection-dispersion process is presented below. The program consists of two parts, with the first part (Program BV12-5-1) creating a dataset in an interactive way, and the second part performing the actual calculations. The flow field refers to a system of wells and recharge wells, in a homogeneous aquifer of infinite extent, with a uniform flow at infinity.

```
100 DEFINT I-N:KEY OFF:OPTION BASE 1:GOSUB 570
110 PRINT"---  Bear & Verruijt - Groundwater Modeling"
120 PRINT"---  Plane steady Groundwater Flow"
130 PRINT"---  Dispersion by random walk model"
140 PRINT"---  Program 12.5 - Part 1"
150 PRINT"---  Generation of dataset":PRINT
160 DIM XX(10),YY(10),QQ(10),AX(100),AY(100)
170 NWM=10:NSM=100:NPM=100:PRINT:PRINT
180 INPUT"Name of dataset ................... ";D$:PRINT
190 PRINT"Properties of screen  :":PRINT
200 INPUT" Dots in x-direction (640 ?) ....... ";NLX
210 INPUT" Dots in y-direction (200 ?) ....... ";NLY
220 INPUT" Scale factor x/y (2 ?) ............ ";SXY
230 INPUT" Graphics screen parameter (2 ?) ... ";NSCR
240 GOSUB 570:PRINT"Dimensions of field (m) :":PRINT
250 INPUT" x = 0 - ........................... ";XT
260 INPUT" y = 0 - ........................... ";YT
270 PRINT:PRINT"Natural velocity (m/d) :":PRINT
280 INPUT" Vx ................................ ";VX
290 INPUT" Vy ................................ ";VY:PRINT
300 INPUT" Time step (d) ..................... ";DT
310 INPUT" Porosity .......................... ";PP
320 INPUT" Thickness of aquifer (m) .......... ";HH
330 INPUT" Number of wells ................... ";NW
340 INPUT" Number of stream lines ............ ";NS
```

```
350 GOSUB 570:PRINT"Location of tracer source :":PRINT
360 INPUT" x ................................. ";XE
370 INPUT" y ................................. ";YE
380 INPUT" Dispersivity ...................... ";DD
390 INPUT" Dispersion ratio ................. ";DR
400 INPUT" Number of particles .............. ";NP
410 IF NW>NWM THEN NW=NWM
420 IF NS>NSM THEN NS=NSM
430 IF NP>NPM THEN NP=NPM
440 FOR I=1 TO NW:GOSUB 570:PRINT"Well";I;":":PRINT
450 INPUT" x = ........................... ";XX(I)
460 INPUT" y = ........................... ";YY(I)
470 INPUT" Q = ........................... ";QQ(I)
480 NEXT I:FOR I=1 TO NS:GOSUB 570
490 PRINT"First point of stream line";I;":":PRINT
500 INPUT" x = ........................... ";AX(I)
510 INPUT" y = ........................... ";AY(I)
520 NEXT I:GOSUB 570:PRINT"Writing dataset ";D$:PRINT
530 OPEN"O",1,D$:PRINT#1,D$:PRINT#1,NLX;NLY;SXY;NSCR
540 PRINT#1,XT;YT;VX;VY;DT;NW;PP;HH;NS:PRINT#1,XE;YE;NP;DD;DR
550 FOR I=1 TO NW:PRINT#1,XX(I);YY(I);QQ(I):NEXT I
560 FOR I=1 TO NS:PRINT#1,AX(I);AY(I):NEXT I:CLOSE 1:END
570 CLS:LOCATE 1,26,1:COLOR 0,7:PRINT" Dispersion by random walk - 1 ";
580 COLOR 7,0:PRINT:PRINT:RETURN
```

Program BV12-5-1. Dispersion by random walk — generation of dataset.

Program BV12-5-1 will create a dataset, in an interactive way. First some data characterizing the properties of the screen must be given; then the program asks for the data describing the flow field. Finally, the location of the pollution source and the dispersivities must be given. As an example, a simple dataset created by this program is shown below. This dataset refers to a region of 640 × 400 m, with a uniform flow of 1 m/d in the $x$-direction, and a single well in the point $x = 300$ m, $y = 250$ m. The dataset describes that 31 streamlines will be traced (if they fit on the screen), and that initially at the point $x = 1$ m, $y = 310$ m a source of pollution is located, represented by 100 particles. The longitudinal dispersivity is 1 m, and the ratio of lateral to longitudinal dispersivities is 0.25.

```
DATA12.7
 640   200   2   2
 640   400   2   0   2   1   .4   10   31
   1   310  100   1   .25
 300   250 1600
   1  -100   1  -80   1  -60   1  -40   1  -20
   1     0   1   20   1   40   1   60   1   80   1  100
   1   120   1  140   1  160   1  180   1  200
   1   220   1  240   1  260   1  280   1  300
   1   320   1  340   1  360   1  380   1  400
   1   420   1  440   1  460   1  480   1  500
```

Dataset DATA12.7.

The second part of the program, in which the actual calculations are performed, and which traces the particles on the screen, is reproduced as Program BV12-5-2.

```
100 DEFINT I-N:KEY OFF:OPTION BASE 1:GOSUB 650
110 PRINT"---   Bear & Verruijt - Groundwater Modeling"
120 PRINT"---   Plane steady Groundwater Flow"
130 PRINT"---   Dispersion by random walk model"
140 PRINT"---   Program 12.5 - Part 2"
150 PRINT"---   Stream lines and particles on screen":PRINT
160 DIM XX(10),YY(10),QQ(10),AX(100),AY(100),XP(100),YP(100),PP(100)
170 DIM PL(100),TA(100,10),TB(100,10),SA(100,10),SB(100,10)
180 PRINT:INPUT"Name of dataset ............. ";D$:PRINT
190 PRINT"Reading dataset ";D$:PRINT
200 OPEN"I",1,D$:INPUT#1,A$:INPUT#1,NLX,NLY,SXY,NSCR
210 INPUT#1,XT,YT,VX,VY,DT,NW,PP,HH,NS:INPUT#1,XE,YE,NP,DD,DR
220 FOR I=1 TO NW:INPUT#1,XX(I),YY(I),QQ(I):NEXT I
230 FOR I=1 TO NS:INPUT#1,AX(I),AY(I):NEXT I:CLOSE 1
240 FOR I=1 TO NP:XP(I)=XE:YP(I)=YE:NEXT I
250 SX=NLX/XT:SY=SXY*NLY/YT:IF SY<SX THEN SX=SY
260 SY=SX/SXY:PI=4*ATN(1):PA=.5*PI:PB=1.5*PI:IT=1:GOSUB 650
270 PRINT"Stream lines and particles on screen":PRINT:PRINT
280 COLOR 0,7:PRINT" B ";:COLOR 7,0:PRINT" .... Begin":PRINT
290 COLOR 0,7:PRINT" S ";:COLOR 7,0:PRINT" .... Stop":PRINT
300 GOSUB 670:CLS:SCREEN NSCR
310 LINE(0,0)-(NLX-1,0):LINE(NLX-1,0)-(NLX-1,NLY-1)
320 LINE(NLX-1,NLY-1)-(0,NLY-1):LINE(0,NLY-1)-(0,0)
330 B$=INKEY$:IF B$="s" OR B$="S" THEN SCREEN 0:END
340 FOR K=1 TO NS:IF PL(K)<0 THEN 470
350 XA=AX(K):YA=AY(K):GOSUB 680:FOR J=1 TO NW:IF QQ(J)<0 THEN 440
360 A=XB-XX(J):B=YB-YY(J):T=ATN(B/A):IF A<0 THEN T=T+PI
370 IF T<0 THEN T=T+2*PI
380 A=SX*A:B=SY*B:R=SQR(A*A+B*B):IF IT=1 THEN TA(K,J)=T:GOTO 440
390 TB(K,J)=TA(K,J):TA(K,J)=T:TC=ABS(TA(K,J)-TB(K,J))
400 IF (TC>PA AND TC<PB) OR R<5 THEN 450
410 A=XD-XX(J):B=YD-YY(J):T=ATN(B/A):IF A<0 THEN T=T+PI
420 IF T<0 THEN T=T+2*PI
430 TC=ABS(T-TB(K,J)):IF TC>PA AND TC<PB THEN 450
440 NEXT J:GOTO 460
450 XB=XX(J):YB=YY(J):PL(K)=-1
460 LINE(SX*XA,SY*YA)-(SX*XB,SY*YB):AX(K)=XB:AY(K)=YB
470 NEXT K
480 FOR K=1 TO NP:IF PP(K)<0 THEN 640
490 XA=XP(K):YA=YP(K):GOSUB 680:A=XB-XA:B=YB-YA:R=SQR(A*A+B*B)
500 DA=2*SQR(6*R*DD)*(RND(1)-.5):DB=2*SQR(6*R*DR*DD)*(RND(1)-.5)
510 XB=XB+DA*A/R-DB*B/R:YB=YB+DA*B/R+DB*A/R
520 FOR J=1 TO NW:IF QQ(J)<0 THEN 610
530 A=XB-XX(J):B=YB-YY(J):T=ATN(B/A):IF A<0 THEN T=T+PI
540 IF T<0 THEN T=T+2*PI
550 A=SX*A:B=SY*B:R=SQR(A*A+B*B):IF IT=1 THEN SA(K,J)=T:GOTO 610
560 SB(K,J)=SA(K,J):SA(K,J)=T:TC=ABS(SA(K,J)-SB(K,J))
570 IF (TC>PA AND TC<PB) OR R<5 THEN 620
580 A=XD-XX(J):B=YD-YY(J):T=ATN(B/A):IF A<0 THEN T=T+PI
590 IF T<0 THEN T=T+2*PI
600 TC=ABS(T-SB(K,J)):IF TC>PA AND TC<PB THEN 620
610 NEXT J:GOTO 630
620 XB=XX(J):YB=YY(J):PP(K)=-1
```

```
630 PSET(SX*XB,SY*YB):XP(K)=XB:YP(K)=YB
640 NEXT K:IT=IT+1:GOTO 330
650 CLS:LOCATE 1,26,1:COLOR 0,7:PRINT" Dispersion by random walk - 2 ";
660 COLOR 7,0:PRINT:PRINT:RETURN
670 A$=INKEY$:IF A$="B" OR A$="b" THEN RETURN ELSE 670
680 REM   Runge-Kutta
690 X=XA:Y=YA:GOSUB 710:XD=XA+U*DT/2:YD=YA+V*DT/2
700 X=XD:Y=YD:GOSUB 710:XB=XA+U*DT:YB=YA+V*DT:RETURN
710 REM   Calculation of velocities
720 U=VX:V=VY:FOR KK=1 TO NW:D=QQ(KK)/(2*PI*PP*HH)
730 A=X-XX(KK):B=Y-YY(KK)
740 RR=A*A+B*B:U=U-D*A/RR:V=V-D*B/RR:NEXT KK:RETURN
```

Program BV12-5-2. Dispersion by random walk — tracing of particles.

The program applies to the case of an infinite field with a uniform velocity at infinity, and a number of wells at given points. Actually, the program is an extension (with a random walk procedure) of Program BV11-1, which applies to the convective transport only.

The results obtained with the dataset given above are shown in Figure 12.7. The

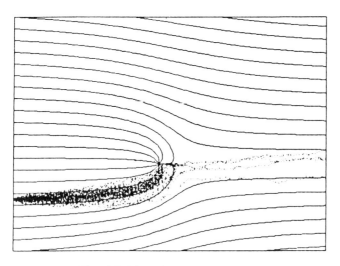

Fig. 12.7 Dispersion by random walk.

figure shows that a large part of the pollution is captured by the well, but not all of it. 4 out of the 100 particles (i.e., about 4%) escape to the right, even though the point at which the pollution source is located is clearly inside the region of influence (above the water divide) of the well.

The program can easily be modified to be applicable to different flow fields, for instance the flow between two parallel boundaries. All that is needed is to redefine the subroutine in which the velocities are calculated.

In principle, the same method can be used to superimpose a random walk process onto a numerical model for convective transport, such as the finite element model presented in Section 11.6. It may be mentioned again that in models of this nature, based on a particle-tracking technique, there is no numerical dispersion. This is an important advantage of these methods, compared to finite difference or finite element models, using a fixed network.

CHAPTER THIRTEEN

# Numerical Modeling of Seawater Intrusion

The conceptual and mathematical models for seawater intrusion are presented in Chapter 7. In this chapter, two numerical models are presented for the analysis of groundwater flow in an aquifer saturated with two fluids (e.g., fresh and salt water), separated by a sharp interface. The first model applies to the flow in a vertical plane, in which the location of the interface is time-dependent, because of boundary conditions, pumping wells, etc. In this model, the flow domain (in the vertical plane) is subdivided into small elements, such that the interface is represented by a series of elements, connecting certain nodes in the mesh. As the interface moves, the location of the nodes upon it changes, so that the entire mesh is changing as a function of time.

The second model is based on the vertically averaged equations developed in Section 7.3.1. This model consists of a system of two coupled equations. The equations are nonlinear because the coefficients depend upon one of the dependent variables — the depth of the interface. The equations are solved by the finite element method, formulated in such a way that the special case of no interface, which occurs when the interface intersects the bottom or the top of the aquifer, can be dealt with automatically. An elementary computer program, in BASIC, is included, and several examples are presented. These include problems with a moving toe or tip of the interface.

## 13.1. Model for Flow in a Vertical Plane

The general aspects of modeling seawater intrusion are presented in Chapter 7. Of special interest are problems in which the interface moves, thus allowing for a change in the storage of the fluids above and below it. The situation may be further complicated by the presence of impermeable or semi-permeable layers which may intersect off the interface. Much attention has been paid to the description of the problem, and its solution, mainly by numerical methods, in a vertical plane (Bear and Dagan, 1964a; Pinder and Cooper, 1970; Shamir and Dagan, 1971).

In a numerical model for flow in a vertical plane, the domain may be subdivided into a large number of small elements (Figure 13.1). Usually, the interface is assumed to consist of a series of element boundaries, such that the elements above it contain fresh water, and the ones below it contain salt water. In this way, the transient behaviour of the interface can be simulated by letting the nodes on the interface move, so that the actual configuration of the network is a function of time. This is basically the same procedure as used in Chapter 10 to model a phreatic surface.

# NUMERICAL MODELING OF SEAWATER INTRUSION

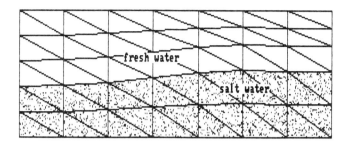

Fig. 13.1. Mesh of elements for an interface problem.

In order to concentrate on the main feature of the problem — the movement of the interface — other possible complications, such as elastic storage or storage due to a moving phreatic surface, will not be considered in this section. The physical justification for this assumption is that, when storage due to a moveable interface is possible, it largely overshadows the effects of both phreatic and elastic storage.

The basic equations, Darcy's law and the continuity equation, can be recalled from Section 7.2. In the present case of a variable fluid density, the appropriate form of Darcy's law is, assuming isotropy of the porous medium,

$$q_x = -\frac{k}{\mu}\left(\frac{\partial p}{\partial x}\right), \quad q_y = -\frac{k}{\mu}\left(\frac{\partial p}{\partial y} + \gamma\right) \qquad (13.1.1)$$

where $k$ is the permeability of the porous medium, $\mu$ is the viscosity of the fluid, and $\gamma$ its volumetric weight ($\gamma = \rho g$). The $y$-axis has been assumed to be pointing in an upward direction, that is, opposite to the action of gravity. Darcy's law has been expressed in terms of the fluid pressure, rather than in terms of fresh and salt water heads, because the pressure is continuous across the interface between fresh and salt water, and therefore it is a single-valued function in the entire domain, which makes it a convenient variable for a numerical approximation.

The conservation equation is, in the absence of distributed supply functions and elastic storage, and assuming that the fluid is incompressible,

$$\frac{\partial q_x}{\partial x} + \frac{\partial q_y}{\partial y} = 0. \qquad (13.1.2)$$

Substitution of (13.1.1) into (13.1.2) leads to the following basic differential equation, valid throughout the domain,

$$\frac{\partial}{\partial x}\left(\frac{k}{\mu}\frac{\partial p}{\partial x}\right) + \frac{\partial}{\partial y}\left(\frac{k}{\mu}\frac{\partial p}{\partial y}\right) = -\frac{\partial}{\partial y}\left(\frac{k}{\mu}\gamma\right). \qquad (13.1.3)$$

This equation is identical to (3.3.31) and (7.2.26).

To fully describe the problem, a set of boundary conditions should be added to this differential equation. For the sake of simplicity, it is assumed that all along the boundary either the pressure $p$ is described, or the flux of water is zero (impermeable boundary). It will further be assumed that all boundaries are either horizontal or vertical. Finally, the permeability, $k$, and the viscosity, $\mu$, are considered constant throughout the domain.

Compared to the problem considered in Section 10.1, the major, and only, complication of the problem defined by (13.1.3) and the boundary conditions, is the appearance of the term containing the volumetric weight $\gamma$. In the case of a sharp interface between fresh and salt water, the volumetric weight is constant everywhere, so that its derivative vanishes, except on the interface, where $\gamma$ is discontinuous, so that, strictly speaking, the derivative of $\gamma$ across the interface does not exist. However, a sharp interface may be considered to be the limiting case of a transition zone of thickness $d$, in which the volumetric weight changes from $\gamma_s$ to $\gamma_f$. Thus, within that transition zone

$$\frac{\partial}{\partial x}\left(\frac{k}{\mu}\frac{\partial p}{\partial x}\right) + \frac{\partial}{\partial y}\left(\frac{k}{\mu}\frac{\partial p}{\partial y}\right) = \frac{k}{\mu}\frac{\Delta\gamma}{d} \qquad (13.1.4)$$

where $\Delta\gamma = \gamma_s - \gamma_f$. The same differential equation applies everywhere else, except that the right-hand side is zero.

The right-hand side of (13.1.4) represents a distributed source, with a strength inversely proportional to the thickness of the transition zone. Such a distributed source can easily be introduced in a finite element model (see Section 10.1). If the transition zone is supposed to be represented by a series of line segments, the contribution of the sources along the line segment from node $i$ to node $j$ leads to source terms in node $i$ and node $j$ of a total strength equal to the strength of the distributed source, multiplied by the area, that is the thickness $d$ times the horizontal distance $L_{ij}$ of the two nodes. Thus, in each of these two nodes, a source term must be introduced of strength

$$Q = -\frac{1}{2}\frac{k}{\mu}\Delta\gamma L_{ij}. \qquad (13.1.5)$$

It is interesting to note that this is independent of the thickness of the transition zone which, therefore, can be considered to be zero.

A computer program for the solution of the interface problem described above is listed as Program BV13-1.

```
1000 DEFINT I-N:KEY OFF:OPTION BASE 1:GOSUB 2090
1010 PRINT"---   Bear & Verruijt - Groundwater Modeling"
1020 PRINT"---   Non-steady groundwater flow in vertical plane"
1030 PRINT"---   Fresh and salt water with sharp interface"
1040 PRINT"---   Program 13.1"
1050 PRINT"---   Upconing due to a single well":PRINT
1060 DIM X(360),Y(360),NT(360),NM(361),F(360),Q(360),KP(3000),P(3000)
1070 DIM T(320),NP(320,4),PX(320),PY(320),U(360),V(360),W(360)
```

# NUMERICAL MODELING OF SEAWATER INTRUSION 347

```
1080 DIM B(3),C(3),XJ(3),YJ(3),E(3,3),KS(4,3)
1090 DIM FS(21),NA(20),NB(20),DX(20),XN(21),YN(21),QX(21),QY(21),NK(21,7)
1100 PRINT:INPUT"Use default input data (Y/N) ";I$:I$=LEFT$(I$,1):PRINT
1110 IF I$="n" OR I$="N" THEN I$="N":GOTO 1140
1120 I$="Y":WW=400:HT=37.5:H1=10:H2=17.5:H3=10:NV=20:NW=21:N1=5:N2=8:N3=4
1130 GF=10:GS=10.25:DG=GS-GF:PM=1:PR=.35:QQ=5:GOTO 1280
1140 GOSUB 2090:INPUT"Length of aquifer .............. ";WW
1150 INPUT"Thickness of aquifer .......... ";HT
1160 INPUT"Interface level ............... ";H1
1170 INPUT"Well level .................... ";H2:H3=HT-H2:H2=H2-H1
1180 INPUT"Horizontal elements (3...20) ... ";NV:NW=NV+1
1190 INPUT"Vertical elements (4...16) ..... ";N6:N5=N6+1
1200 N1=INT(N5*H1/HT+1.1):IF N1<2 THEN N1=2
1210 N3=INT(N5*H3/HT+.1):IF N3<1 THEN N3=1
1220 N2=N5-N1-N3:IF N2<1 THEN N2=1
1230 PRINT"Density of fresh water ........      10.00":GF=10
1240 INPUT"Density of salt water ......... ";GS:DG=GS-GF
1250 INPUT"Permeability/viscosity ........ ";PM
1260 INPUT"Storativity ................... ";PR
1270 INPUT"Discharge of well ............. ";QQ
1280 GOSUB 2090:N5=N1+N2+N3:N6=N5-1:N=N5*NW:M=N6*NV
1290 FOR I=1 TO N5:NT(I*NW)=2:NEXT I
1300 FOR I=1 TO N6:L=(I-1)*NV:K=(I-1)*NW:FOR J=1 TO NV:LL=L+J:KK=K+J
1310 NP(LL,1)=KK:NP(LL,2)=KK+NW:NP(LL,3)=KK+NW+1:NP(LL,4)=KK+1:NEXT J,I
1320 FOR I=1 TO 4:FOR J=1 TO 3:K=I+J-1:IF K>4 THEN K=K-4
1330 KS(I,J)=K:NEXT J,I:PRINT"Generation of pointer vector":W=1
1340 CL=CSRLIN:FOR I=1 TO N:LOCATE CL,3,0:PRINT "Node";I:KP(W)=I:NM(I)=W
1350 K=0:FOR J=1 TO M:FOR H=1 TO 4:IF NP(J,H)=I GOTO 1370
1360 NEXT H:GOTO 1400
1370 FOR H=1 TO 4:G=NP(J,H):FOR L=0 TO K:IF KP(W+L)=G GOTO 1390
1380 NEXT L:K=K+1:KP(W+K)=G
1390 NEXT H
1400 NEXT J:W=W+K+1:NEXT I:PRINT:PRINT"Pointer length =";W:NM(N+1)=W
1410 FOR I=1 TO NW:NK(I,5)=4:NEXT I:NK(1,5)=2:NK(NW,5)=2
1420 L=(N1-2)*NV:FOR I=1 TO NV:NK(I,1)=L+I:NK(I,2)=L+I+NV:NEXT I
1430 L=(N1-2)*NV-1:FOR I=2 TO NV:NK(I,3)=L+I:NK(I,4)=L+I+NV:NEXT I
1440 L=(N1-1)*NV:NK(NW,1)=L:NK(NW,2)=L+NV:NN=W-1:DX(1)=1:DX(2)=1
1450 AA=2:FOR I=3 TO NV:DX(I)=1.4*DX(I-2):AA=AA+DX(I):NEXT I
1460 FOR I=1 TO NV:DX(I)=WW*DX(I)/AA:NEXT I:L=(N1-1)*NW:FOR I=1 TO NV
1470 NA(I)=L+I:NB(I)=L+I+1:NEXT I:FOR I=1 TO N5:L=(I-1)*NW+1:X(L)=0
1480 FOR J=1 TO NV:X(L+J)=X(L+J-1)+DX(J):NEXT J,I:FOR I=1 TO NW
1490 FS(I)=H1:XN(I)=1:YN(I)=0:NEXT I:FOR I=1 TO M:T(I)=PM:NEXT I
1500 SP=DX(1)*PR/(PM*GF):IF I$="Y" THEN SP=5:NS=8:GOTO 1540
1510 PRINT:PRINT"Suggestion for time step :";SP:PRINT
1520 INPUT"Time step ............. ";SP
1530 INPUT"Number of time steps ... ";NS
1540 FOR IS=0 TO NS:IF IS=0 THEN GOSUB 2050 ELSE GOTO 1570
1550 AA=(H2+H3)*GF:L=N1*NW:FOR I=1 TO L:F(I)=AA+(H1-Y(I))*GS:NEXT I
1560 L=L+1:FOR I=L TO N:F(I)=(Y(N)-Y(I))*GF:NEXT I:GOTO 1820
1570 FOR I=1 TO NN:P(I)=0:NEXT I:FOR I=1 TO M:FOR KW=1 TO 4:FOR J=1 TO 3
1580 K=NP(I,KS(KW,J)):XJ(J)=X(K):YJ(J)=Y(K):NEXT J
1590 B(1)=YJ(2)-YJ(3):B(2)=YJ(3)-YJ(1):B(3)=YJ(1)-YJ(2)
1600 C(1)=XJ(3)-XJ(2):C(2)=XJ(1)-XJ(3):C(3)=XJ(2)-XJ(1)
1610 D=T(I)/(4*ABS(XJ(1)*B(1)+XJ(2)*B(2)+XJ(3)*B(3)))
1620 FOR J=1 TO 3:FOR K=1 TO 3:E(J,K)=D*(B(J)*B(K)+C(J)*C(K)):NEXT K,J
1630 FOR J=1 TO 3:G=NP(I,KS(KW,J)):W=NM(G):H=NM(G+1)-W-1:FOR K=0 TO H
1640 FOR L=1 TO 3:IF NP(I,KS(KW,L))=KP(W+K) THEN P(W+K)=P(W+K)+E(J,L)
```

```
1650 NEXT L,K,J,KW,I:IT=1:EE=.000001:FOR I=1 TO N:Q(I)=0:NEXT I
1660 L=(N1+N2-1)*NW+1:Q(L)=-QQ:FOR K=1 TO NV:I=NA(K):J=NB(K)
1670 AA=.5*PM*DG*DX(K):Q(I)=Q(I)-AA:Q(J)=Q(J)-AA
1680 AA=.5*PM*GF*DX(K):Q(N-NW+K)=Q(N-NW+K)-AA:Q(N-NV+K)=Q(N-NV+K)-AA
1690 AA=.5*PM*GS*DX(K):Q(K)=Q(K)+AA:Q(K+1)=Q(K+1)+AA:NEXT K
1700 FOR I=1 TO N:H=NM(I):G=NM(I+1)-H-1:U(I)=0:IF NT(I)>1 THEN 1720
1710 U(I)=Q(I):FOR J=0 TO G:U(I)=U(I)-P(H+J)*F(KP(H+J)):NEXT J
1720 V(I)=U(I):NEXT I:UU=0:FOR I=1 TO N:UU=UU+U(I)*U(I):NEXT I
1730 LOCATE 5,54,0:PRINT"Iteration ";IT:LOCATE 6,54,0:PRINT"Error ";UU
1740 FOR I=1 TO N:W(I)=0:H=NM(I):G=NM(I+1)-H-1
1750 FOR J=0 TO G:W(I)=W(I)+P(H+J)*V(KP(H+J)):NEXT J,I
1760 VW=0:FOR I=1 TO N:VW=VW+V(I)*W(I):NEXT I
1770 AA=UU/VW:FOR I=1 TO N:IF NT(I)>1 THEN 1790
1780 F(I)=F(I)+AA*V(I):U(I)=U(I)-AA*W(I)
1790 NEXT I:WT=0:FOR I=1 TO N:WT=WT+U(I)*U(I):NEXT I
1800 BB=WT/UU:FOR I=1 TO N:V(I)=U(I)+BB*V(I):NEXT I:UU=WT
1810 IT=IT+1:IF IT<=N AND UU>EE THEN 1730
1820 FOR J=1 TO M:PX(J)=0:PY(J)=GS:IF J>(N1-1)*NV THEN PY(J)=GF
1830 FOR KW=1 TO 4:FOR K=1 TO 3:L=NP(J,KS(KW,K)):XJ(K)=X(L):YJ(K)=Y(L)
1840 NEXT K:B(1)=YJ(2)-YJ(3):B(2)=YJ(3)-YJ(1):B(3)=YJ(1)-YJ(2)
1850 C(1)=XJ(3)-XJ(2):C(2)=XJ(1)-XJ(3):C(3)=XJ(2)-XJ(1)
1860 AA=4*(XJ(1)*B(1)+XJ(2)*B(2)+XJ(3)*B(3)):FOR K=1 TO 3:L=NP(J,KS(KW,K))
1870 PX(J)=PX(J)+B(K)*F(L)/AA:PY(J)=PY(J)+C(K)*F(L)/AA:NEXT K,KW
1880 PX(J)=-PM*PX(J):PY(J)=-PM*PY(J):NEXT J:FOR I=1 TO NW:QX(I)=0:QY(I)=0
1890 FOR K=1 TO NK(I,5):II=NK(I,K):LL=NK(I,5):QX(I)=QX(I)+PX(II)/LL
1900 QY(I)=QY(I)+PY(II)/LL:NEXT K,I:AA=SP/PR:FOR I=1 TO NV:QX(I)=AA*QX(I)
1910 QY(I)=AA*QY(I):FS(I)=FS(I)+QY(I)-QX(I)*YN(I)/XN(I):NEXT I:GOSUB 2050
1920 FOR I=2 TO NV:AA=DX(I-1)+DX(I):AB=FS(I+1)-FS(I-1)
1930 A1=SQR(AA*AA+AB*AB):XN(I)=AA/A1:YN(I)=AB/A1:NEXT I
1940 SCREEN 2:CLS:SX=639/WW:SY=199/HT
1950 FOR I=1 TO M:FOR J=1 TO 4:JJ=NP(I,J):K=J+1:IF K>4 THEN K=1
1960 KK=NP(I,K):IA=SX*X(JJ):IB=SX*X(KK):JA=199-SY*Y(JJ):JB=199-SY*Y(KK)
1970 LINE(IA,JA)-(IB,JB):NEXT J,I:FOR I=1 TO NV:IA=SX*X(I)+1:IB=SX*X(I+1)
1980 JA=SY*FS(I):JB=SY*FS(I+1):IF IA>=IB THEN 2020
1990 FOR K=IA TO IB:AA=(K-IA)/(IB-IA):BB=JA+AA*(JB-JA)
2000 FOR L=1 TO 20:IC=199*RND(1):IF IC<=BB THEN PSET(K,199-IC)
2010 NEXT L,K
2020 NEXT I:LOCATE 4,54,0:PRINT"Time = ";:PRINT USING"####.###";T;
2030 T=T+SP:B$=INKEY$:IF B$="S" OR B$="s" THEN SCREEN 0:CLS:END
2040 NEXT IS:B$=INPUT$(1):SCREEN 0:CLS:END
2050 FOR I=1 TO NW:A1=FS(I)/(N1-1):A2=(H1+H2-FS(I))/N2:A3=H3/N3
2060 FOR J=1 TO N1:K=I+(J-1)*NW:Y(K)=(J-1)*A1:NEXT J:L=N1+1:LL=N1+N2
2070 FOR J=L TO LL:K=I+(J-1)*NW:Y(K)=FS(I)+(J-L+1)*A2:NEXT J:L=LL+1
2080 FOR J=L TO N5:K=I+(J-1)*NW:Y(K)=H1+H2+(J-L+1)*A3:NEXT J,I:RETURN
2090 CLS:LOCATE 1,26,1:COLOR 0,7:PRINT" Interface in vertical plane ";
2100 COLOR 7,0:PRINT:PRINT:RETURN
```

Program BV13-1. Nonsteady interface problem in a vertical plane.

Program BV13-1 is applicable to the problem of upconing due to a line sink, extracting fresh water in a point above an originally horizontal interface in a homogeneous aquifer of a constant thickness (Figure 13.2).

The program operates interactively, with the possibility of entering a set of default values for all the data. The program has been developed from the

# NUMERICAL MODELING OF SEAWATER INTRUSION

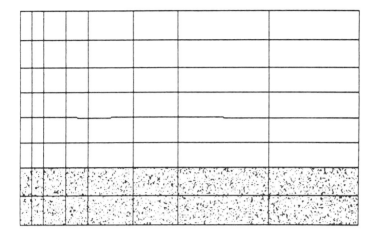

Fig. 13.2. Problem solved by Program BV13-1.

elementary programs presented in Chapter 10. Some special features are the following.

- The program uses the conjugated gradient method (Appendix A) to solve the system of equations, with a pointer vector to indicate the nonzero coefficients of the system matrix. All these nonzero coefficients are stored in a one-dimensional array $P$. The pointer vector is set up in the program itself, so that the user will not notice the arithmetic involved in this memory-saving procedure.
- The program generates a mesh of elements, based upon a small number of input data, such as the dimensions of the aquifer, the original position of the interface, the location of the sink, the number of elements in the two directions, and some physical data for the porous medium and the fluid. The program automatically generates a mesh such that the interface and the horizontal line through the well point are element boundaries. The height of the elements below the well may change as a function of time, depending on the movement of the interface. The dimensions of the elements in horizontal direction are small in the vicinity of the well, and increase gradually in magnitude towards the right side boundary of the aquifer. The maximum capacity of the program as listed here is a network of 320 elements.
- The boundary conditions along the upper and lower (impermeable) boundaries are that the normal derivative of the pressure is prescribed ($\gamma_f$ or $\gamma_s$). This is analogous to a given flux in the standard finite element formulation in terms of groundwater head. These boundary conditions can thus be incorporated by adding a term on the right-hand side of the system of equations (see lines 1680 and 1690 in the program).
- The interface condition (13.1.5) can be taken into account in a similar way, as has been explained above (lines 1660—1670).

— In order to determine the motion of the interface during a time step, the fluxes in the elements immediately below and above the interface are transformed into average velocities at the nodes along the interface. In the program, this is done by a simple arithmetic average, using the matrix NK to indicate the numbers of the elements surrounding a node. A more sophisticated method might be to use certain weight functions. As it is not obvious, however, what form of the weight functions would be less biased than simple averaging, the simplest form of averaging has been used.

— The motion of the nodes on the interface has been restricted to the vertical direction, to avoid too much distortion of the mesh. This means that the $x$- and $y$-components of the velocity in a node must be transformed into a vertical motion of the interface. The vertical displacement of the interface depends upon the slope of the interface (Figure 13.3). It follows from the figure that

$$\Delta y/\Delta t = v_y - v_x \times y_n/x_n \qquad (13.1.6)$$

where $x_n$ and $y_n$ are the components of a vector parallel to the interface. This expression can also be derived from the general interface condition (7.2.6). The relation (13.1.6) is used in the program in line 1910.

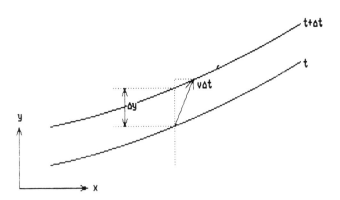

Fig. 13.3. Movement of the interface in a time step.

— Output of the program consists of a figure of the mesh on the screen. The user may replace this output (in lines 1940—2020) by output of numerical data on the screen or a printer.

The output of the program, using the default data of the program, is shown in Figure 13.4. It can be seen that a considerable, and very local, upconing is reached after 40 days. This is a common characteristic of the behaviour near wells above an impermeable boundary.

A finite element schematization in a vertical plane, as presented in this section, usually involves rather small elements, because the thickness of the plane is

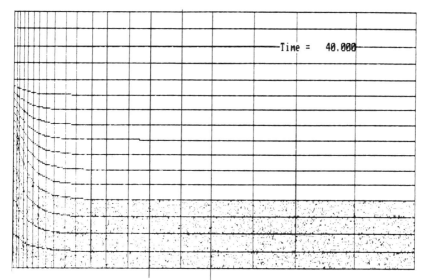

Fig. 13.4. Output of Program BV13-1.

generally of the order of magnitude of 10 m. This also means that the time steps must be rather small. This may acceptable for a problem of local upconing. However, such an approach is unattractive for the analysis of the long-term behaviour of a large extended aquifer. A fully three-dimensional model, for a region of arbitrary shape in the horizontal plane, is also unattractive because it requires a very large number of nodal points and elements. For this reason, the *hydraulic approach* (see Section 7.3) may be used, in which it is assumed that the vertical component of the flow rate is much smaller than the horizontal components. Because the horizontal dimensions of aquifers are, in general, much larger than the vertical dimensions, such a schematization may be justified in many practical problems. A model of this type is presented in the next section.

## 13.2. Basic Equations for a Regional Model of Seawater Intrusion

For the flow in aquifers of large horizontal extent, compared to their thickness, it may be assumed that the variations of the groundwater head in a vertical direction are so small that they can be neglected, which leads to averaged equations and a two-dimensional model in the horizontal plane. Models of this type have been used by various authors (Pinder and Page, 1977, Wilson and Sa da Costa, 1982). It sometimes appears that special difficulties arise in the case that the interface intersects an impermeable layer, and this may require fairly complicated toe-tracking techniques, especially for nonsteady problems.

In this section, a finite element model is presented for nonsteady problems, with storage due to the variable position of the interface. The problem is formulated in terms of two basic variables; namely, the head in one of the fluids (fresh water) and the depth of the interface. By using a combined balance equation for the conservation of total mass, a system of equations is obtained that can also be used in regions where there is no interface at all. In this way, a relatively simple general computer program can be developed.

Consider the flow of fresh and salt groundwater, separated by a sharp interface, in a confined aquifer, bounded by two, approximately horizontal, impermeable layers (Figure 13.5). The lower and upper surfaces of the aquifer are described by the surfaces $z = -H_1$ and $z = -H_2$, respectively, so that the thickness of the aquifer is $H = H_1 - H_2$. The permeability of the material in the aquifer is denoted by $k$, and the densities of the two fluids are denoted by $\rho_f$ and $\rho_s$, respectively. The two fluids are supposed to have the same dynamic viscosity $\mu$, and the location of the interface between the two fluids is described by the function $z = -h(x, y)$.

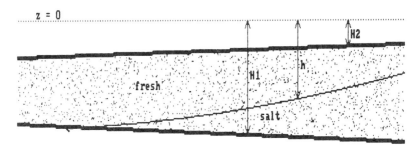

Fig. 13.5. Aquifer with fresh and salt groundwater.

In order to facilitate the numerical solution, we will use the approach of Section 7.3 to describe the problem, with some modifications with regard to the notation of the variables. In the hydraulic approach, use is made of the Dupuit assumption which states that the pressure distribution in vertical direction is hydrostatic

$$\partial p/\partial z = -\rho g \tag{13.2.1}$$

where $\rho = \rho_f$ in the fresh water, and $\rho = \rho_s$ in the salt water, and where $g$ is the strength of the gravity field. Equation (13.2.1) expresses that everywhere in the fluid the pressure in the fluid is in equilibrium with the local weight of the fluid. Integrating (13.2.1) in the vertical direction leads to expressions for the pressure in the two fluids

$$\text{fresh water:} \quad -h < z < -H_2 \text{:} \ p = \rho_f g \phi - \rho_f gz,$$
$$\text{salt water:} \quad -H_1 < z < -h \text{:} \ p = \rho_f g \phi - \rho_s gz - (\rho_s - \rho_f)gh. \tag{13.2.2}$$

It can easily be seen that these expressions satisfy (13.2.1), and that the pressure is continuous at the interface $z = -h$, as it should be. In (13.2.2) the pressure in the two fluids is actually expressed in terms of two variables, $h$ and $\phi$, which are both functions of the coordinates $x$ and $y$ in the horizontal plane, and of the time $t$. The variable $\phi$ is the fresh water head in the fresh groundwater at the level $z = 0$. All other heads can be expressed, when desired, in terms of $\phi$ and $h$. The variable $\phi$ is introduced because it has the same physical dimension as the interface depth $h$ (both are expressed in meters). It is emphasized that the basic formulation of the flow is in terms of the pressure $p$, which is a continuous function. The variables $h$ and $\phi$ are merely convenient reference quantities. An alternative set of basic variables might be the interface depth $h$ and the groundwater head at a certain level in the salt water. The use of a head in the fresh water seems to be preferable, however, because in many problems of practical interest, the main flow is in the fresh water.

The flow of the fluids in the $x, y$-plane is governed by Darcy's law

$$\mathbf{q} = -(k/\mu) \nabla p \qquad (13.2.3)$$

where $k$ is the permeability of the porous medium, and $\mu$ is the dynamic viscosity. Because viscosity differences usually have little influence on the shape of the interface, compared to the influence of density differences (Verruijt, 1980), it can be assumed without much loss of generality that the two fluids have the same viscosity. It now follows from (13.2.2) and (13.2.3) that

$$\text{fresh water:} \quad -h < z < -H_2 : \mathbf{q} = -K \nabla \phi,$$
$$\text{salt water:} \quad -H_1 < z < -h : \mathbf{q} = -K \nabla \phi + \alpha K \nabla h \qquad (13.2.4)$$

where $K$ is the hydraulic conductivity, in terms of the fresh water,

$$K = \frac{k}{\mu} \rho_f g \qquad (13.2.5)$$

and where $\alpha$ is the relative density difference,

$$\alpha = \frac{\rho_s - \rho_f}{\rho_f}. \qquad (13.2.6)$$

It can be seen from (13.2.4) that the flow rate $q$ is discontinuous at the interface, the discontinuity being proportional to the slope of the interface. Actually, more exact considerations avoiding the Dupuit assumption, demonstrate that this shear flow should be proportinal to the sine of the slope of the interface (Josselin de Jong, 1981). This means that the present approximation, in which the shear flow is proportional to the slope of the interface itself, is applicable only for relatively small slopes (a logical consequence of the Dupuit assumption).

The equations of continuity of the two fluids are

$$\text{fresh:} \quad S \frac{\partial h}{\partial t} = I_f - \nabla \cdot \{(h - H_2)\mathbf{q}\},$$

$$\text{salt:} \quad -S \frac{\partial h}{\partial t} = I_s - \nabla \cdot \{(H_1 - h)\mathbf{q}\} \qquad (13.2.7)$$

where $S$ is the storativity, and where $I_s$ and $I_f$ are supply functions, representing a distributed surface supply of salt and fresh water into the aquifer. In (13.2.7), the only storage term taken into account is due to the movement of the interface. Other forms of storage (elastic or phreatic) are disregarded here. The justification for this is that the main possibility for storage is the movement of the interface. Elastic storage is usually much smaller than phreatic storage (see Chapter 4) and phreatic storage is, in general, much smaller than interface storage, because the ratio of the vertical motion of the phreatic surface and an interface is of the order of magnitude of the relative density difference $\alpha$, which is small in the normal case of fresh and saline water (seawater).

Substitution of (13.2.4) into (13.2.7) leads to the following system of coupled differential equations

$$S \frac{\partial h}{\partial t} = I_f + \nabla \cdot \{K(h - H_2) \nabla \phi\} \tag{13.2.8}$$

$$-S \frac{\partial h}{\partial t} = I_s + \nabla \cdot \{K(H_1 - h) \nabla \phi\} - \nabla \cdot \{\alpha K(H_1 - h) \nabla h\} \tag{13.2.9}$$

where all spatial derivatives are in the $x$, $y$-plane only. The system of two coupled differential equations, in the two variables $h$ and $\phi$, should be solved subject to an appropriate set of boundary and initial conditions.

In order to obtain a more convenient system of equations, both in a physical as well as in a mathematical sense, the two equations are added. This leads to an equation without time derivatives,

$$-\nabla \cdot (T \nabla \phi) = I \tag{13.2.10}$$

where $T$ is the transmissivity of the aquifer as a whole,

$$T = K(H_1 - H_2) \tag{13.2.11}$$

and where $I = I(x, y, t)$ has the character of a source function,

$$I = I_f + I_s - \nabla \cdot \{\alpha K(H_1 - h) \nabla h\}. \tag{13.2.12}$$

Equation (13.2.10) can be regarded as a global continuity equation for the flow in the entire aquifer. It is of the same form as the usual differential equation for steady flow in a confined aquifer, for which many efficient numerical models have been developed.

The most effective way of solving the system of equations is that, first, the head $\phi$ is determined from Equation (13.2.10), using a given shape of the interface to estimate the value of the right-hand side $I$. This provides an estimate of the distribution of the head $\phi$, taking into account all external supplies of water, and the boundary conditions for the head. In a second step, an adjusted position of the interface is determined from either Equations (13.2.8) or (13.2.9). This procedure has been found to lead to a stable and rapidly converging procedure. The physical justification of the preference for the global continuity equation (13.2.10) in favor of one of the other equations is the notion that, in most cases, the head will be most affected by the total external supply, in combination with the boundary conditions for the head.

In the next section the two sets of equations will be solved simultaneously, with the possibility of an iterative adjustment of the coefficients. The need for such an iterative procedure is a consequence of the dependence of the coefficients on the variable $h$.

## 13.3. Finite Element Model for Regional Interface Problems

The solution of Equation (13.2.10) by the finite element method is a standard problem from finite element theory, see Chapter 10. In this section the simplest type of element will be used, namely triangular elements with linear interpolation of the variables $h$ and $\phi$ in each element. The standard procedures of the finite element method now lead to a system of linear equations which can be written as follows, see (10.4.9),

$$\sum_{j=1}^{n} P(i,j)\phi(j) = Q(i) \quad (i = 1, 2, \ldots, n) \tag{13.3.1}$$

where $n$ is the number of nodes. The matrix $P$ consists of a summation over all elements (numbered $k = 1, 2, \ldots, m$) of submatrices of the form

$$P_k(i,j) = \frac{T_k}{2|\Delta|} \{b(i)b(j) + c(i)c(j)\} \quad (i, j = 1, 2, 3) \tag{13.3.2}$$

where $b(1) = y(2) - y(3)$, $c(1) = x(3) - x(2)$, etc., and $\Delta = x(1)b(1) + x(2)b(2) + x(3)b(3)$. The vector $Q$ consists of contributions from each element of the form

$$Q_k(i) = I|\Delta|/6 \quad (i = 1, 2, 3). \tag{13.3.3}$$

Because $|\Delta|/2$ is the area of a triangular element, it follows that the right-hand side of (13.3.3) is one-third of the total supply to the element. It appears that this supply is distributed equally over the three nodes of the element, each node receiving one-third of the total supply. Thus, (13.3.1) can be considered as an expression of continuity of flow at node $i$. The expression $\Sigma P(i,j)\phi(j)$ is the numerical equivalent of the analytical expression $-\nabla \cdot (T\nabla\phi)$.

As mentioned above, it is most effective to use the total balance equation (13.2.10) in conjunction with one of the two other equations, (13.2.8) or (13.2.9). If the salt water equation is used as the second equation, the system of equations is

$$-\nabla \cdot (T\nabla\phi) = I_f + I_s - \nabla \cdot (\alpha T_a \nabla h), \tag{13.3.4}$$

$$S(\partial h/\partial t) - \nabla \cdot (\alpha T_a \nabla h) + \nabla \cdot (T_a \nabla \phi) = -I_s \tag{13.3.5}$$

where $T_a = T(H_1 - h)/(H_1 - H_2)$ represents a reduced transmissivity.

The system of Equations (13.3.4) and (13.3.5) is now slightly modified by introducing a new variable $f$, defined as

$$f = \phi/\alpha. \tag{13.3.6}$$

This variable differs from the head only by a scaling factor, chosen such that the variations in $f$ and $h$ are about the same. This will appear to be numerically advantageous. The system of equations now is

$$-\nabla \cdot (\alpha T \nabla f) + \nabla \cdot (\alpha T_a \nabla h) = I_f + I_s, \tag{13.3.7}$$

$$S(\partial h/\partial t) - \nabla \cdot (\alpha T_a \nabla h) + \nabla \cdot (\alpha T_a \nabla f) = -I_s. \tag{13.3.8}$$

One advantage of the change of variables is immediately obvious from these equations: the coupling terms now have the same coefficients or, in other words, the system is symmetric.

The numerical approximation of the differential equations (13.3.7) and (13.3.8) can be performed by the standard techniques of the finite element method, fully described in Chapter 10. Details of the derivations are omitted here. The resulting system of equations is

$$\sum_{j=1}^{n} P_a(i,j) f(j) - \sum_{j=1}^{n} P_b(i,j) h(j) = Q_f(i) + Q_s(i), \tag{13.3.9}$$

$$-\sum_{j=1}^{n} P_b(i,j) f(j) + \sum_{j=1}^{n} \{P_b(i,j) + P_c(i,j)\} h(j)$$

$$= -Q_s(i) + \sum_{j=1}^{n} P_c(i,j) h_a(j) \tag{13.3.10}$$

In these equations, the vectors $Q$ and the matrices $P$ are the numerical equivalents of the various terms in the differential equations (13.3.7) and (13.3.8). Their coefficients can be calculated in a similar way as explained in detail in Chapter 10.

An elementary computer program, in BASIC, that will perform the calculations outlined above, is listed in Program BV13-2.

```
1000 DEFINT I-N:KEY OFF:OPTION BASE 1:GOSUB 2000
1010 PRINT"---   Bear & Verruijt - Groundwater Modeling"
1020 PRINT"---   Regional groundwater flow in coastal aquifer"
1030 PRINT"---   Fresh and salt water with sharp interface"
1040 PRINT"---   Program 13.2"
1050 PRINT"---   Solution by finite elements":PRINT
1060 DIM X(200),Y(200),IP(200),H(200),F(200),HA(200),QF(200),QS(200)
1070 DIM PF(200),PS(200),PA(200,10),PB(200,10),PD(200,10),KP(200,10)
1080 DIM U(200),V(200),W(200),GF(200),GS(200):NS=200:MS=150:NZ=10
```

```
1090 DIM B(3),C(3),XJ(3),YJ(3),QA(3,3),QB(3,3),KS(4,3)
1100 DIM NP(150,4),TR(150),FF(150),FS(150),TT(100),RF(100)
1110 IS=1:A$="Reading input data":GOSUB 2040:READ N,M,NS,NC,GF,GS,HT,SS
1120 DG=(GS-GF)/GF:FOR I=1 TO N:READ X(I),Y(I),F(I),H(I),QF(I),QS(I),IP(I)
1130 F(I)=F(I)/DG:HA(I)=H(I):NEXT I
1140 FOR J=1 TO M:READ NP(J,1),NP(J,2),NP(J,3),NP(J,4),TR(J),FF(J),FS(J)
1150 NEXT J:FOR K=1 TO NS:READ TT(K),RF(K):NEXT K
1160 KS(1,1)=1:KS(1,2)=2:KS(1,3)=3:KS(2,1)=2:KS(2,2)=3:KS(2,3)=4
1170 KS(3,1)=3:KS(3,2)=4:KS(3,3)=1:KS(4,1)=4:KS(4,2)=1:KS(4,3)=2
1180 PRINT"Generation of pointer matrix":CL=CSRLIN:FOR I=1 TO N:KP(I,1)=I
1190 KP(I,NZ)=1:NEXT I:FOR J=1 TO M:LOCATE CL,4,0:PRINT"Element ....... ";
1200 PRINT USING "####";J;:FOR K=1 TO 4:KK=NP(J,K):FOR L=1 TO 4:LL=NP(J,L)
1210 IA=0:FOR II=1 TO KP(KK,NZ):IF KP(KK,II)=LL THEN IA=1
1220 NEXT II:IF IA=0 THEN KB=KP(KK,NZ)+1:KP(KK,NZ)=KB:KP(KK,KB)=LL
1230 NEXT L,K,J:DT=TT(IS):GOSUB 2000
1240 IC=1:A$="Step"+STR$(IS):GOSUB 2040:EE=.0000001
1250 FOR I=1 TO N:FOR J=1 TO NZ:PA(I,J)=0:PB(I,J)=0:PD(I,J)=0:NEXT J
1260 PF(I)=RF(IS)*(QF(I)+QS(I)):PS(I)=-RF(IS)*QS(I):NEXT I
1270 PRINT"Generation of system matrix":CL=CSRLIN:FOR J=1 TO M
1280 LOCATE CL,4,0:PRINT"Element ....... ";:PRINT USING "####";J;
1290 FOR KW=1 TO 4:HM=0:FOR I=1 TO 3:K=NP(J,KS(KW,I))
1300 XJ(I)=X(K):YJ(I)=Y(K):HM=HM+H(K)+HA(K):NEXT I:HM=HM/6
1310 B(1)=YJ(2)-YJ(3):B(2)=YJ(3)-YJ(1):B(3)=YJ(1)-YJ(2)
1320 C(1)=XJ(3)-XJ(2):C(2)=XJ(1)-XJ(3):C(3)=XJ(2)-XJ(1)
1330 D=ABS(XJ(1)*B(1)+XJ(2)*B(2)+XJ(3)*B(3)):IF D<EE THEN 1460
1340 FA=DG*TR(J)/(4*D):FB=FA*(HT-HM)/HT
1350 FOR K=1 TO 3:FOR L=1 TO 3:QA(K,L)=B(K)*B(L)+C(K)*C(L)
1360 QB(K,L)=FB*QA(K,L):QA(K,L)=FA*QA(K,L):NEXT L,K
1370 FOR K=1 TO 3:IU=NP(J,KS(KW,K)):IH=KP(IU,NZ):FOR L=1 TO IH:LV=1
1380 KV=KS(KW,LV):IF NP(J,KV)=KP(IU,L) THEN 1400
1390 LV=LV+1:IF LV<4 THEN 1380 ELSE 1420
1400 PA(IU,L)=PA(IU,L)+QA(K,LV):PB(IU,L)=PB(IU,L)+QB(K,LV)
1410 PD(IU,L)=PD(IU,L)+QB(K,LV)
1420 NEXT L,K:FOR I=1 TO 3:K=NP(J,KS(KW,I))
1430 PF(K)=PF(K)+RF(IS)*(FF(J)+FS(J))*D/12:PS(K)=PS(K)-RF(IS)*FS(J)*D/12
1440 PD(K,1)=PD(K,1)+SS*D/(12*DT):PD(K,NZ)=PD(K,NZ)+SS*D*HA(K)/(12*DT)
1450 NEXT I
1460 NEXT KW,J:PRINT:PRINT:EE=.00001:EE=EE*EE
1470 FOR I=1 TO N:PA(I,NZ)=PA(I,NZ)+PF(I):PD(I,NZ)=PD(I,NZ)+PS(I):NEXT I
1480 PRINT"Solution of equations for F":CL=CSRLIN:IT=1
1490 FOR I=1 TO N:U(I)=0:IF IP(I)>1 THEN 1520
1500 U(I)=PA(I,NZ):FOR J=1 TO KP(I,NZ):K=KP(I,J)
1510 U(I)=U(I)-PA(I,J)*F(K)+PB(I,J)*H(K):NEXT J
1520 V(I)=U(I):NEXT I:UU=0:FOR I=1 TO N:UU=UU+U(I)*U(I):NEXT I
1530 LOCATE CL,4,0:PRINT"Iteration ..... ";:PRINT USING "####";IT
1540 FOR I=1 TO N:W(I)=0:FOR J=1 TO KP(I,NZ)
1550 K=KP(I,J):W(I)=W(I)+PA(I,J)*V(K):NEXT J,I
1560 VW=0:FOR I=1 TO N:VW=VW+V(I)*W(I):NEXT I
1570 AA=UU/VW:FOR I=1 TO N:IF IP(I)>1 THEN 1590
1580 F(I)=F(I)+AA*V(I):U(I)=U(I)-AA*W(I)
1590 NEXT I:WW=0:FOR I=1 TO N:WW=WW+U(I)*U(I):NEXT I
1600 BB=WW/UU:FOR I=1 TO N:V(I)=U(I)+BB*V(I)
1610 NEXT I:UU=WW:IT=IT+1:IF UU>0 THEN E=LOG(UU)/LOG(10) ELSE E=-10
1620 PRINT"   Log(error) .... ";:PRINT USING "####.#";E
1630 IF (UU>EE AND IT<=2*N) THEN 1530
1640 FOR I=1 TO N:GF(I)=-PA(I,NZ):FOR J=1 TO KP(I,NZ)
1650 K=KP(I,J):GF(I)=GF(I)+PA(I,J)*F(K)-PB(I,J)*H(K):NEXT J,I
1660 PRINT:PRINT"Solution of equations for h":CL=CSRLIN:IT=1
```

```
1670 FOR I=1 TO N:U(I)=0:IF IP(I)>1 THEN 1700
1680 U(I)=PD(I,NZ):FOR J=1 TO KP(I,NZ):K=KP(I,J)
1690 U(I)=U(I)+PB(I,J)*F(K)-PD(I,J)*H(K):NEXT J
1700 V(I)=U(I):NEXT I:UU=0:FOR I=1 TO N:UU=UU+U(I)*U(I):NEXT I
1710 LOCATE CL,4,0:PRINT"Iteration ..... ";:PRINT USING "####";IT
1720 FOR I=1 TO N:W(I)=0:FOR J=1 TO KP(I,NZ)
1730 K=KP(I,J):W(I)=W(I)+PD(I,J)*V(K):NEXT J,I
1740 VW=0:FOR I=1 TO N:VW=VW+V(I)*W(I):NEXT I
1750 AA=UU/VW:FOR I=1 TO N:IF IP(I)>1 THEN 1770
1760 H(I)=H(I)+AA*V(I):U(I)=U(I)-AA*W(I)
1770 NEXT I:WW=0:FOR I=1 TO N:WW=WW+U(I)*U(I):NEXT I
1780 BB=WW/UU:FOR I=1 TO N:V(I)=U(I)+BB*V(I)
1790 NEXT I:UU=WW:IT=IT+1:IF UU>0 THEN E=LOG(UU)/LOG(10) ELSE E=-10
1800 PRINT"   Log(error) .... ";:PRINT USING "####.#";E
1810 IF (UU>EE AND IT<=2*N) THEN 1710
1820 FOR I=1 TO N:GS(I)=PD(I,NZ):FOR J=1 TO KP(I,NZ)
1830 K=KP(I,J):GS(I)=GS(I)+PB(I,J)*F(K)-PD(I,J)*H(K):NEXT J
1840 GF(I)=GF(I)-GS(I):NEXT I
1850 FOR I=1 TO N:IF H(I)<0 THEN H(I)=0 ELSE IF H(I)>HT THEN H(I)=HT
1860 NEXT I:IC=IC+1:A$="Step"+STR$(IS):IF IC>NC THEN 1880
1870 GOSUB 2000:A$=A$+", cycle"+STR$(IC):GOSUB 2040:GOTO 1250
1880 IS=IS+1:FOR I=1 TO N:HN=H(I)+H(I)-HA(I):HA(I)=H(I):H(I)=HN
1890 IF H(I)<0 THEN H(I)=0 ELSE IF H(I)>HT THEN H(I)=HT
1900 NEXT I:GOSUB 2000:A$="Output":GOSUB 2040:B$="###.###"
1910 FOR I=1 TO N:PRINT "i = ";:PRINT USING"###";I;
1920 PRINT"       f = ";:PRINT USING B$;DG*F(I);
1930 PRINT"       h =";:PRINT USING B$;HA(I);
1940 PRINT"       Qf =";:PRINT USING B$;GF(I);
1950 PRINT"       Qs =";:PRINT USING B$;GS(I)
1960 NEXT I:PRINT"Time = ";TT(IS-1);:GOSUB 2020:GOSUB 2000
1970 IF IS>NS THEN 1990
1980 DT=TT(IS)-TT(IS-1):GOTO 1240
1990 GOSUB 2000:A$="END":GOSUB 2040:END
2000 CLS:LOCATE 1,26,1:COLOR 0,7:PRINT" Regional interface model ";
2010 COLOR 7,0:PRINT:PRINT:RETURN
2020 LOCATE 25,27,0:COLOR 0,7:PRINT"  Touch any key to continue  ";
2030 COLOR 7,0:LOCATE 25,79,1:A$=INPUT$(1):RETURN
2040 COLOR 0,7:PRINT " ";A$;" ";:COLOR 7,0:PRINT:PRINT:RETURN
2050 DATA 22,10,17,3,1000,1025,20,0.4
2060 DATA 0,0,0,5,0,0,2,0,100,0,5,0,0,2
2070 DATA 100,0,0,5,0,0,0,100,100,0,5,0,0,0
2080 DATA 200,0,0,5,0,0,0,200,100,0,5,0,0,0
2090 DATA 300,0,0,5,0,0,0,300,100,0,5,0,0,0
2100 DATA 400,0,0,5,0,0,0,400,100,0,5,0,0,0
2110 DATA 500,0,0,5,0,0,0,500,100,0,5,0,0,0
2120 DATA 600,0,0,5,0,0,0,600,100,0,5,0,0,0
2130 DATA 700,0,0,5,0,0,0,700,100,0,5,0,0,0
2140 DATA 800,0,0,5,0,0,0,800,100,0,5,0,0,0
2150 DATA 900,0,0,5,0,0,0,900,100,0,5,0,0,0
2160 DATA 1000,0,0,5,0,0,0,1000,100,0,5,0,0,0
2170 DATA 1,2,3,4,2000,0.001,0,3,4,5,6,2000,0.001,0
2180 DATA 5,6,7,8,2000,0.001,0,7,8,9,10,2000,0.001,0
2190 DATA 9,10,11,12,2000,0.001,0,11,12,13,14,2000,0.001,0
2200 DATA 13,14,15,16,2000,0.001,0,15,16,17,18,2000,0.001,0
2210 DATA 17,18,19,20,2000,0.001,0,19,20,21,22,2000,0.001,0
2220 DATA 100,1,200,1,300,1,500,1,1000,1,2000,1,3000,1,5000,1
2230 DATA 10000,1,20000,1,30000,1,50000,1,100000,1
2240 DATA 200000,1,300000,1,500000,1,1000000,1
```

Program BV13-2. Regional interface problem.

The program has been kept as simple and short as possible, without sacrificing flexibility. The meaning of the input parameters is as follows.

| | |
|---|---|
| $N$ | Number of nodes, |
| $M$ | Number of elements, |
| $NS$ | Number of time steps, |
| $NC$ | Number of iterative cycles per time step, |
| $GF$ | Density of fresh water, |
| $GS$ | Density of salt water, |
| $HT$ | Thickness of aquifer, |
| $SS$ | Storativity, |
| $X(I)$ | $X$-coordinate of node $I$, |
| $Y(I)$ | $Y$-coordinate of node $I$, |
| $F(I)$ | Initial head, |
| $H(I)$ | Initial depth of interface, |
| $QF(I)$ | Local supply of fresh water, |
| $QS(I)$ | Local supply of salt water, |
| $IP(I)$ | Type of node $I$, |
| | $IP(I) = 2$ if head and interface depth are prescribed, |
| | $IP(I) = 0$ if the fresh and salt supplies are prescribed, |
| $NP(J, 1)$ | Node 1 of element $J$, |
| $NP(J, 2)$ | Node 2 of element $J$, |
| $NP(J, 3)$ | Node 3 of element $J$, |
| $NP(J, 4)$ | Node 4 of element $J$, |
| $TR(J)$ | Transmissivity in element $J$, |
| $FF(J)$ | Infiltration of fresh water in element $J$, |
| $FS(J)$ | Infiltration of salt water in element $J$, |
| $TT(K)$ | Value of time after time step $K$, |
| $RF(K)$ | Reduction factor during time step $K$. |

Some features of the program deserve some special attention. These are the following.

— The two variables, $f$ and $h$, are denoted in the program by $F$ and $H$. In the program the solution is obtained in an iterative way. First, the system is solved for the head $F$, and then for the depth $H$. In each cycle, the coefficients, which depend upon $H$, are updated. This order of solving the two systems of equations is chosen because the head $F$ is influenced by all inflow functions (fresh and salt), and it can be expected that these will significantly affect the values of the head $F$.
— In each time step the values of the variable $H$ are updated $NC$ times. This number of cycles may be taken as 3 or 4, to obtain sufficient accuracy.
— The program is based upon a fully implicit formulation for the variations in time (see Section 9.2). This has the advantage that the steady-state solution may be determined by the program in a single step, by taking the storativity as zero.

As a first example, the case of uniform infiltration in a homogeneous aquifer of 20 m thickness and 2000 m length is considered (Figure 13.6). Initially, at time $t = 0$, the interface is horizontal, at a depth of 5 m. At the two boundaries the interface is maintained at a constant depth. The transmissivity is 2000 m²/d, the

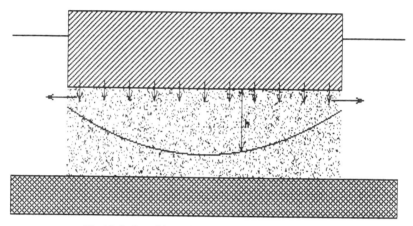

Fig. 13.6. Interface problem with uniform infiltration.

storativity (the effective porosity in this case) is 0.4 and the infiltration rate is 0.001 m/d. The input data of Program BV13-2 apply to this problem, using 22 nodes and 10 elements. Only one half of the aquifer is considered in the program, because of the symmetry of the problem. The number of time steps is 17, with the magnitude of the time steps gradually increasing from 100 d to 500 000 d.

Figure 13.7 shows the steady-state solution obtained by the program after 17

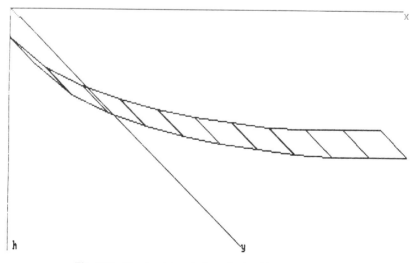

Fig. 13.7. Steady-state solution, obtained by Program 13.2.

time steps. This solution can also be obtained analytically, using the Ghyben–Dupuit approximation. It is then found that the interface depth is given by the formula

$$h^2 = h_0^2 + (Ix/\alpha K)(2L - x). \tag{13.3.11}$$

where $2L$ is the length of the aquifer. Equation (13.3.11) applies only in the zone where $h < H$. When the infiltration rate is sufficiently large, a zone in the middle of the aquifer will exist, in which the interface coincides with the bottom of the aquifer ($h = H$).

A comparison of the exact results (exact within the framework of the Ghyben–Dupuit approximation) and the numerical results using two subdivisions with 10 elements and 500 elements, respectively, is shown in Table 13.1.

Table 13.1. Comparison of exact and numerical results

| $x$ | Exact | FEM $M = 10$ | FEM $M = 500$ |
|---|---|---|---|
| 0 | 5.000 | 5.000 | 5.000 |
| 100 | 10.050 | 10.040 | 10.049 |
| 200 | 13.000 | 12.978 | 12.997 |
| 300 | 15.133 | 15.099 | 15.127 |
| 400 | 16.763 | 16.717 | 16.755 |
| 500 | 18.028 | 17.970 | 18.016 |
| 600 | 19.000 | 18.933 | 18.985 |
| 700 | 19.723 | 19.648 | 19.705 |
| 800 | 20.000 | 20.000 | 20.000 |
| 900 | 20.000 | 20.000 | 20.000 |
| 1000 | 20.000 | 20.000 | 20.000 |

The shape of the interface as determined by an advanced version of Program 13.2 and using a mesh of $50 \times 10 = 500$ elements, is shown in Figure 13.8. This version of the program uses a preprocessor and a postprocessor in Pascal, and a FORTRAN subprogram to perform the actual calculations. In this way the speed of FORTRAN can be combined with the graphic capabilities of Pascal.

From the results summarized in Table 13.1, it follows that the numerical solution is very accurate, at least in the limiting steady state. For the coarse mesh, the maximum error is 0.35%, and for the fine mesh it is less than 0.1%. It can be seen from the figures, and from the table, that the horizontal part of the interface, near the center where it intersects the lower boundary, is correctly obtained by the program.

As a further validation the accuracy of the nonsteady solution can be investigated by verifying whether the behaviour for small values of the time is correct. Uniform infiltration should result in a practically uniform lowering of the interface,

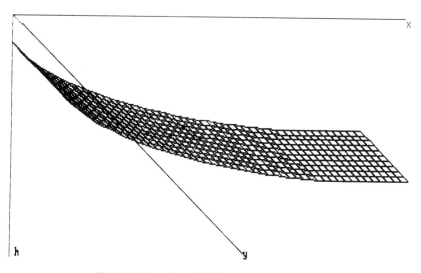

Fig. 13.8. Interface calculated with 500 elements.

except near the outflow boundaries. Such a behavior is indeed obtained, as the reader may verify by running Program BV13-2.

As a second example, let us consider the problem of the rotation of an initially vertical interface. Actually, this is not a problem for which the present method seems particularly suitable, as in this method the vertical flow rates are disregarded, while there will certainly be large vertical velocities in this case. Nevertheless, this problem may well serve to demonstrate the performance of Program BV13-2.

The problem is defined by the following data. In an aquifer of 10 m thickness, having a permeability of 10 m/d, and a storativity of 0.4, a vertical interface exists at time $t = 0$, between two fluids of densities 1000 and 1025 kg/m$^3$. The interface then starts to turn as a consequence of the density difference, and will ultimately reach a horizontal position. The shape of the interface, as calculated by Program BV13-2, after 30, 100, 300, and 1000 days, is shown in Figure 13.9. The total length of the aquifer has been assumed to be 2000 m, so that the center point (around which the interface rotates) is located at $x = 1000$ m, $y = 5$ m. The left and right side boundaries have been assumed to be impermeable. The results shown in Figure 13.9 have been determined by taking 40 elements of 50 m length (and 50 m width), using time steps in the sequence 10, 20, 30, 50, 70, 100, 150, 200, 300, 500, 700, and 1000, thus 12 steps in total.

Figure 13.9 also shows an approximate analytical solution of the problem (Wilson and Sa da Costa, 1982), which consists of a straight line rotating around the central point, such that the distance from the toe to the center is proportional to the square root of time. It appears that the agreement is reasonably good. The smearing zones near the toe and the tip of the interface are probably realistic,

# NUMERICAL MODELING OF SEAWATER INTRUSION

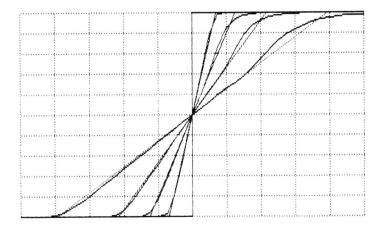

Fig. 13.9. Rotation of initially vertical interface.

because the exact theoretical solution for the initial velocities of the vertical interface also indicates such a behavior.

It may be noted that Program 13.2 does not contain any special facilities for the tracking of an eventual toe or tip of the interface. The tip and the toe of the interface are determined by simply limiting the values for the depth of the interface. Even with this simple procedure, the results obtained by the program are satisfactory.

APPENDIX

# Solution of Linear Equations

## A.1. Introduction

Numerical methods, such as the finite difference method or the finite element method, often lead to a set of linear equations. In order to solve such a set of equations on a small computer it seems worthwhile to consider methods of solution that use computer memory as efficient as possible. Because in most cases the matrices involved are sparse (i.e., contain a large amount of zeros) it can be expected that methods that account for that property, and, perhaps, also use the often encountered symmetry of the matrix, will be useful for the application of numerical methods on microcomputers.

In this appendix several algorithms will be developed, in Microsoft BASIC. These include the *Gauss–Seidel* method, the method of *LDU-decomposition*, the method of *Gaussian elimination* by a wave front technique, and the method of *conjugate gradients*.

## A.2. Reference Problem

Many of the methods to be presented make use of the physical background of the set of linear equations. The sparseness of the matrix is due, for instance, to the circumstance that in a network of nodes each node is connected to a very limited number of other nodes only. In many methods, this property is used in special data structures (pointers). In order to present these methods in full detail, it is convenient to consider a standard problem as an example. For this purpose the simple problem of a network of electrical resistors (Figure A.1) is selected.

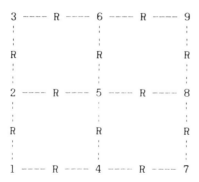

Fig. A.1. Reference problem.

# SOLUTION OF LINEAR EQUATIONS

The network consists of 9 nodes, connected by 12 resistors of equal resistance, $R$. The electric potential in node $I$ will be denoted by $F(I)$. In order to determine the potential in all points of the network, a set of linear equations can be derived, on the basis of Kirchhoff's first law. This law, which is nothing but a continuity equation, states that the algebraic sum of all currents flowing toward a junction is zero. It is assumed that the boundary conditions are that the potential in node 1 is fixed at a value $F(1) = 6$, and that the potential in node 9 is $F(9) = 0$. The solution of the problem then is

$$F(2) = F(4) = 4$$
$$F(3) = F(5) = F(7) = 3$$
$$F(6) = F(8) = 2$$

It can easily be verified that this set of values satisfies Kirchhoff's law at all nodes of the network, using Ohm's law to determine the current in each resistor.

The reference problem is solved by various numerical methods in this Appendix. Each of these methods should lead to the solution given above.

## A.3. Basic Equations

The generation of the system of equations that describe a physical problem usually comprises two distinct phases. In the first phase, a numerical equation is formulated for each node, employing fundamental physical principles, such as conservation of energy, and a constitutive equation. The boundary conditions are disregarded in this phase. In general, this phase results in a square matrix, which is often symmetric. This is true, for instance, in the example considered here. In the second phase, this matrix is modified in order to account for the boundary conditions. This may destroy the symmetry of the matrix. In many cases, however, symmetry can be restored by some simple operations.

In the case of the reference problem, the first phase leads to a system of equations described by the following matrix

$$\begin{matrix} 2 & -1 & 0 & -1 & 0 & 0 & 0 & 0 & 0 \\ -1 & 3 & -1 & 0 & -1 & 0 & 0 & 0 & 0 \\ 0 & -1 & 2 & 0 & 0 & -1 & 0 & 0 & 0 \\ -1 & 0 & 0 & 3 & -1 & 0 & -1 & 0 & 0 \\ 0 & -1 & 0 & -1 & 4 & -1 & 0 & -1 & 0 \\ 0 & 0 & -1 & 0 & -1 & 3 & 0 & 0 & -1 \\ 0 & 0 & 0 & -1 & 0 & 0 & 2 & -1 & 0 \\ 0 & 0 & 0 & 0 & -1 & 0 & -1 & 3 & -1 \\ 0 & 0 & 0 & 0 & 0 & -1 & 0 & -1 & 2 \end{matrix}$$

This is indeed a symmetric matrix. Its main diagonal is said to be *dominant*, because the coefficient on the main diagonal is equal to (or larger than) the sum of the absolute value of all other coefficients in that row. In some methods, this property can be used to great advantage. It is also important to note that many coefficients are zero. This is the case when the two nodes corresponding to the column and row numbers are not connected by a resistor. Furthermore, a banded structure can be discerned, with the nonzero coefficients concentrated in a zone around the main diagonal. This last property depends not only on the type of problem, but also on the way of numbering the nodes. For several methods of solution, it will be found advantageous to number the nodes such that the band width is as small as possible. For other methods, for instance the Gauss–Seidel method, and the method of conjugate gradients, the algorithms can be made independent of the numbering sequence. This may be an advantage for a general purpose program, as it eliminates the need of special numbering sequences.

In the example, the boundary conditions are $F(1) = 6$ and $F(9) = 0$. These conditions can be incorporated in various ways. One method is to modify the matrix such that the equations expressing continuity for nodes 1 and 9 are replaced by a statement of the boundary conditions. This means that in the system of equations

$$\sum \{P(I,J) \times F(J)\} = G(I),$$

the values of $G(1)$ and $G(9)$ are set equal to 6 and 0, respectively, and that the matrix is modified to the following form.

| 1  | 0  | 0  | 0  | 0  | 0  | 0  | 0  | 0  |
|----|----|----|----|----|----|----|----|----|
| −1 | 3  | −1 | 0  | −1 | 0  | 0  | 0  | 0  |
| 0  | −1 | 2  | 0  | 0  | −1 | 0  | 0  | 0  |
| −1 | 0  | 0  | 3  | −1 | 0  | −1 | 0  | 0  |
| 0  | −1 | 0  | −1 | 4  | −1 | 0  | −1 | 0  |
| 0  | 0  | −1 | 0  | −1 | 3  | 0  | 0  | −1 |
| 0  | 0  | 0  | −1 | 0  | 0  | 2  | −1 | 0  |
| 0  | 0  | 0  | 0  | −1 | 0  | −1 | 3  | −1 |
| 0  | 0  | 0  | 0  | 0  | 0  | 0  | 0  | 1  |

It can be seen that the first and the last equation of the system now state the boundary conditions. Solution of the modified system of equations indeed leads to the correct values of the potential $F$, as can be verified by performing the solution, or by substituting the assumed solution.

For some methods of solution it is a great disadvantage that the symmetry of the matrix has been lost. Therefore this symmetry is sometimes restored by elimination of all coefficients involving the known variables $F(1)$ and $F(9)$. This means that the right-hand side of the system of equations has to be adjusted. In the example it means that in the second equation the right-hand side should be increased by $-P(2,1) \times F(1) - P(2,9) \times F(9) = 6$. The matrix of the system of equations then becomes

$$\begin{pmatrix} 1 & 0 & 0 & 0 & 0 & 0 & 0 & 0 & 0 \\ 0 & 3 & -1 & 0 & -1 & 0 & 0 & 0 & 0 \\ 0 & -1 & 2 & 0 & 0 & -1 & 0 & 0 & 0 \\ 0 & 0 & 0 & 3 & -1 & 0 & -1 & 0 & 0 \\ 0 & -1 & 0 & -1 & 4 & -1 & 0 & -1 & 0 \\ 0 & 0 & -1 & 0 & -1 & 3 & 0 & 0 & 0 \\ 0 & 0 & 0 & -1 & 0 & 0 & 2 & -1 & 0 \\ 0 & 0 & 0 & 0 & -1 & 0 & -1 & 3 & 0 \\ 0 & 0 & 0 & 0 & 0 & 0 & 0 & 0 & 1 \end{pmatrix}$$

Now the matrix is again symmetric. It may be noted that the dominance of the main diagonal has been strengthened. Initially, the coefficient on the main diagonal was equal to the sum of the absolute values of all other coefficients in its row. Now in some rows, 1 and 9, the coefficient on the main diagonal is larger than the sum of the absolute values of all other coefficients. For some methods of solution this improves the accuracy, and the rate of convergence.

Some methods of solving the system of equations will be discussed in the following paragraphs. Because of its simplicity, the iterative Gauss–Seidel method will be presented first.

## A.4. Gauss–Seidel Iteration

In this section, the iterative Gauss–Seidel method is presented, emphasizing execution on a small computer. The theory of the method can be found in textbooks on numerical analysis (e.g. Carnahan *et al.*, 1969).

The system of linear equations is written as

$$\sum \{P(I,J) \times F(J)\} = G(I) \quad (I = 1 \ldots N) \tag{A.1}$$

where the summation must be performed from $J = 1$ to $J = N$.

In the Gauss–Seidel method an initial estimated solution is continuously updated by correcting the $I$th equation by modifying variable $I$. The initial estimate may be arbitrary, for example $F(I) = 0$ for all unknown values of $F(I)$. Alternatively, one may make some reasonable guess for the unknown values. In general, Equations (A.1) will not be satisfied by the estimated solution. The variable $F(I)$ is now corrected, by an amount $DF(I)$, such that equation $I$ is satisfied. If the estimated solution is denoted by $FA(I)$, this means that

$$\sum \{P(I,J) \times FA(J)\} + P(I,I) \times DF(I) = G(I)$$

The unknown correction $DF(I)$ can be determined from this equation

$$DF(I) = [G(I) - \sum \{P(I,J) \times FA(J)\}]/P(I,I) \tag{A.2}$$

The Gauss—Seidel algorithm now consists of a repeated execution of (A.2) for all values of $I$. That the process must be repeated follows from the fact that, by application of (A.2), the $I$th equation is satisfied, but this is disturbed later by updating other values. It can be shown that the Gauss—Seidel procedure converges, provided that the matrix is *positive-definite*. Convergence is reasonably fast if the main diagonal of the matrix is dominant.

The Gauss—Seidel algorithm can be executed by the following BASIC statements

```
1000 FOR K=1 TO NI
1010 FOR I=1 TO N
1020 A=G(I):FOR J=1 TO N:A=A-P(I,J)*F(J):NEXT J
1030 F(I)=F(I)+A/P(I,I):NEXT I:NEXT K
```

Here $NI$ is the number of iterations, which often is taken of the order of magnitude of the number of equations $N$. Sometimes the number of iterations needed may have to be considerably larger, especially if the main diagonal is not dominant. Then a different method may be more efficient.

A variant of the Gauss—Seidel algorithm is the Jacobi method, in which all corrections are calculated first, and then all values $F(I)$ are updated in one step. This can be executed as follows.

```
1000 FOR K=1 TO N
1010 FOR I=1 TO N
1020 A(I)=G(I):FOR J=1 TO N:A(I)=A(I)-P(I,J)*F(J):NEXT J,I
1030 FOR I=1 TO N:F(I)=F(I)+A(I)/P(I,I):NEXT I:NEXT K
```

This process requires a somewhat larger computation time, and uses more memory (because the vector of increments has to be stored). Convergence is usually somewhat slower. Therefore, the Gauss—Seidel method is usually preferred, except for special purposes.

Practical experience with the Gauss—Seidel method has shown that convergence can be improved by multiplying the correction in each updating step by a factor somewhat greater than 1. This means that in each step the error is not made equal to zero, but, by an extra kick, it is made to change sign, in anticipation of future corrections. The so-called *over-relaxation factor* should be smaller than 2, and at least equal to 1. The algorithm now is

```
1000 FOR K=1 TO NI
1010 FOR I=1 TO N
1020 A=G(I):FOR J=1 TO N:A=A-P(I,J)*F(J):NEXT J
1030 F(I)=F(I)+R*A/P(I,I):NEXT I:NEXT K
```

## A.5. Pointer Matrix

In the Gauss–Seidel algorithm as presented above, the central operation is the summation in statement 1020, in which all coefficients $P(I, J)$ from the $I$th row have to be multiplied by the values $F(J)$. In many physical applications, a large proportion of these coefficients is zero, which means that a considerable amount of computer memory and computation time is wasted by storing zeros, and by multiplications by zero. In fact, it can usually be expected that only a few coefficients (say 5, 6, or perhaps 10) are different from zero. This property can be used to save computer time and memory by separate storage of the precise location of the nonzero coefficients in a *pointer matrix*. This is a rectangular matrix, with length $N$ and width 10 (for example), in which each row contains the node numbers of the nonzero coefficients for that particular row. In the case of the example considered in this Appendix a suitable form of the pointer matrix is

| | | | | | | | |
|---|---|---|---|---|---|---|---|
| 1 | 2 | 4 | 0 | 0 | 0 | 0 | 3 |
| 2 | 1 | 3 | 5 | 0 | 0 | 0 | 4 |
| 3 | 2 | 6 | 0 | 0 | 0 | 0 | 3 |
| 4 | 1 | 5 | 7 | 0 | 0 | 0 | 4 |
| 5 | 2 | 4 | 6 | 8 | 0 | 0 | 5 |
| 6 | 3 | 5 | 9 | 0 | 0 | 0 | 4 |
| 7 | 4 | 8 | 0 | 0 | 0 | 0 | 3 |
| 8 | 7 | 5 | 9 | 0 | 0 | 0 | 4 |
| 9 | 6 | 8 | 0 | 0 | 0 | 0 | 3 |

Here the last column of the pointer matrix has been used to indicate the actual number of nonzero coefficients in each row. The numbers 1, 2, 4 in the first row indicate that of the coefficients $P(1, J)$ only those for which $J = 1, 2, 4$ are unequal to zero, which can immediately be seen from the structure of the problem (Figure A.1).

Using a pointer matrix, the algorithm for solving the system of equations is

```
1000 FOR K=1 TO NI
1010 FOR I=1 TO N
1020 A=P(I,Z):FOR J=1 TO K%(I,Z):A=A-P(I,J)*F(K%(I,J):NEXT J
1030 F(I)=F(I)+R*A/P(I,1):NEXT I,K
```

Here the pointer matrix is denoted by $K\%$. It should be noted that in statement 1030, the division is by the first element $P(I, 1)$, which contains the coefficient that in the original matrix was stored on the main diagonal, $P(I, I)$.

The application of this algorithm requires some preparation for the pointer matrix, and also for the system matrix $P$. If the coefficient $P(I, K\%(I, J))$ of the original matrix is stored in the position $P(I, J)$, all nonzero coefficients are stored in the left part of each row. The final matrix $P$ can have a width $Z$, where $Z$ is the width of the pointer matrix. It should also be noted that the right-hand side of the system of equations, $G(I)$, is assumed to be stored in the last column of the matrix $P$. The dimension statement in the beginning of the program must be

```
100 DIM P(N,Z),K%(N,Z)
```

where the order of magnitude of $Z$ is 10, or some other relatively small number, depending upon the complexity of the network used. The coefficients of the pointer matrix can usually be calculated by a simple subroutine in the computer program itself.

A complete program for the computations is presented as Program A.1. The meaning of the input variables should be clear from the interactive operation of the input statements.

```
10 CLS:PRINT"Gauss-Seidel with pointer matrix":PRINT
20 INPUT"Number of nodes .......... ";N:PRINT
30 INPUT"Number of elements ........ ";M:PRINT
40 INPUT"Estimated pointer width ... ";Z:PRINT
50 INPUT"Number of iterations ...... ";NI:PRINT
60 INPUT"Relaxation factor ......... ";R:PRINT
70 DIM T%(N),F(N),K%(N,Z),P(N,Z),N%(M,2),R(M)
80 FOR I=1 TO N:CLS:PRINT"Node ";I:PRINT
90 INPUT"x ...................... ";X(I):PRINT
100 INPUT"y ...................... ";Y(I):PRINT
110 INPUT"Potential given (y/n) ..... ";A$:PRINT
120 IF A$="y" OR A$="Y" THEN T%(I)=2:GOTO 150
130 IF A$="n" OR A$="N" THEN T%(I)=0:F(I)=0:GOTO 160
140 GOTO 110
150 INPUT"Potential ................ ";F(I):PRINT
160 NEXT I
170 FOR J=1 TO M:CLS:PRINT"Element ";J:PRINT
180 INPUT"Node 1 ................... ";N%(J,1):PRINT
190 INPUT"Node 2 ................... ";N%(J,2):PRINT
200 INPUT"Resistance ............... ";R(J):PRINT
210 NEXT J
220 CLS:PRINT"Pointer matrix":PRINT
230 FOR I=1 TO N:PRINT I;:K%(I,1)=I:K%(I,Z)=1:K=1
240 FOR J=1 TO M:FOR H=1 TO 2:IF N%(J,H)=I THEN 260
250 NEXT H:GOTO 290
260 FOR H=1 TO 2:U=N%(J,H):FOR L=1 TO K:IF K%(I,L)=U THEN 280
270 NEXT L:K=K+1:K%(I,K)=U:K%(I,Z)=K
280 NEXT H
290 NEXT J,I:PRINT:PRINT
300 PRINT"System matrix":PRINT:FOR J=1 TO M:PRINT J;
310 A=1/R(J):FOR I=1 TO 2:K=N%(J,I):L=N%(J,3-I)
320 P(K,1)=P(K,1)+A:FOR H=2 TO K%(K,Z):IF K%(K,H)=L THEN 340
330 NEXT H
340 P(K,H)=P(K,H)-A:NEXT I,J:PRINT:PRINT
350 PRINT"Iteration":PRINT:FOR K=1 TO NI:PRINT K;
360 FOR I=1 TO N:IF T%(I)>1 THEN 390
370 A=0:FOR J=1 TO K%(I,Z):A=A-P(I,J)*F(K%(I,J)):NEXT J
380 F(I)=F(I)+R*A/P(I,1)
390 NEXT I,K
400 CLS:PRINT"Solution":PRINT
410 FOR I=1 TO N:PRINT"I = ";I;" - F = ";F(I):NEXT I
420 END
```

Program A.1. Gauss–Seidel method.

It may be noted that a vector $T\%$ is used to denote whether the potential in a particular node is given (then $T\%(I) = 2$), or unknown (then $T\%(I) = 0$). This enables to incorporate the boundary conditions in a very simple way, without modifying the matrix. The Gauss—Seidel algorithm in statements 370 and 380 is simply skipped if $T\%(I) = 2$.

The program has been kept as simple as possible. It is left to the user to add such useful facilities as the possibility of correcting typing errors during input, to include default values for the relaxation factor, the refusal of unrealistic input (such as negative node numbers or resistances), and the printing of output on a line printer.

The advantage of using a pointer matrix is especially evident for large systems, say a network with 500 nodes. Then instead of a matrix of dimensions $500 \times 500$ one only needs two matrices of dimensions $500 \times 10$, one of which consists of integers. This results in a considerable saving in memory, and in computation time, because all multiplications by zero are avoided.

Actually, the pointer matrix does not completely eliminate the storage of zeros, because its width should be large enough to accommodate the widest row. The efficiency can be further improved by replacing the pointer matrix by a one-dimensional pointer vector, in which only the nonzero coefficients are collected. This will not be elaborated here.

## A.6. Gauss Elimination

A well known direct method for the solution of the system of linear equations

$$\sum \{P(I,J) \times F(J)\} = G(I)$$

is the method of *Gauss elimination* (see, e.g., Wilkinson, 1965). In this method the first equation is used to express the first variable in terms of all other variables, and then the first variable is eliminated from all subsequent equations. Then a system of $N - 1$ equations with $N - 1$ variables remains, and the process can be repeated until only one equation with one variable remains. This equation can easily be solved, and then all previous equations can be solved in a series of backward substitutions.

An elementary computer program that performs the process is listed below.

```
10 CLS:PRINT"Gauss elimination":PRINT
20 READ N:DIM P(N,N),F(N),G(N)
30 FOR I=1 TO N:FOR J=1 TO N:READ P(I,J):NEXT J
40 READ G(I):NEXT I
50 PRINT"Elimination":PRINT:FOR I=1 TO N:PRINT I;
60 A=1/P(I,I):G(I)=A*G(I)
70 FOR J=I TO N:P(I,J)=A*P(I,J):NEXT J:IF I=N THEN 100
80 FOR J=I+1 TO N:G(J)=G(J)-P(J,I)*G(I)
90 FOR K=I+1 TO N:P(J,K)=P(J,K)-P(J,I)*P(I,K):NEXT K,J,I
100 PRINT:PRINT:PRINT"Back substitution":PRINT
110 F(N)=G(N):FOR I=1 TO N-1:J=N-I:PRINT J;:F(J)=G(J)
120 FOR K=J+1 TO N:F(J)=F(J)-P(J,K)*F(K):NEXT K,I
```

```
130 CLS:PRINT"Solution":PRINT
140 FOR I=1 TO N:PRINT"I = ";I;" - F = ";F(I):NEXT I
150 END
160 DATA 9
170 DATA 1,0,0,0,0,0,0,0,0,6
180 DATA -1,3,-1,0,-1,0,0,0,0,0
190 DATA 0,-1,2,0,0,-1,0,0,0,0
200 DATA -1,0,0,3,-1,0,-1,0,0,0
210 DATA 0,-1,0,-1,4,-1,0,-1,0,0
220 DATA 0,0,-1,0,-1,3,0,0,-1,0
230 DATA 0,0,0,-1,0,0,2,-1,0,0
240 DATA 0,0,0,0,-1,0,-1,3,-1,0
250 DATA 0,0,0,0,0,0,0,0,1,0
```

Program A.2. Gauss elimination.

In this program input data are entered by reading the system matrix, modified to account for the boundary conditions. The program consists of two distinct parts, the elimination part and the back substitution part. It should be noted that the process only works when the coefficients on the main diagonal remain different from to zero. Otherwise, the division in line 60 would result in an error statement. Fortunately, in most physical processes described by linear equations, the coefficient on the main diagonal indeed is different from zero. If not, the system of equations has to be reordered (pivoting).

## A.7. Wave Front Method

Program A.2 takes no account of the property that the matrix that describes a physical problem is usually sparse. It, therefore, is very uneconomical, especially for large problems. An elegant and efficient way of using the sparseness of the system matrix is to use a *wave front* technique to solve the system of equations (Irons, 1970). In this technique, the nonzero coefficients are stored only, using a pointer matrix or a pointer vector. The elimination is performed in a special way, determined by the structure of the system. In fact, it is imagined that nodes are eliminated successively from the network, in such a way that a wave travels through the mesh. By taking care that the number of nodes on the wave front remains as small as possible, the number of multiplications remains limited.

A program that uses a wave front solution technique is reproduced below, as Program A.3.

```
10  CLS:PRINT"Gauss elimination with wave front":PRINT
20  INPUT"Number of nodes .......... ";N:PRINT
30  INPUT"Number of elements ....... ";M:PRINT
40  INPUT"Estimated pointer width ... ";Z:PRINT
50  DIM T%(N),F(N),K%(N,Z),P(N,Z),N%(M,2),R(M)
60  FOR I=1 TO N:CLS:PRINT"Node ";I:PRINT
70  INPUT"x ....................... ";X(I):PRINT
80  INPUT"y ....................... ";Y(I):PRINT
90  INPUT"Potential given (y/n) ..... ";A$:PRINT
100 IF A$="y" OR A$="Y" THEN T%(I)=2:GOTO 130
110 IF A$="n" OR A$="N" THEN T%(I)=0:F(I)=0:GOTO 140
```

# SOLUTION OF LINEAR EQUATIONS

```
120 GOTO 90
130 INPUT"Potential ................ ";F(I):PRINT
140 NEXT I
150 FOR J=1 TO M:CLS:PRINT"Element ";J:PRINT
160 INPUT"Node 1 .................. ";N%(J,1):PRINT
170 INPUT"Node 2 .................. ";N%(J,2):PRINT
180 INPUT"Resistance .............. ";R(J):PRINT
190 NEXT J
200 CLS:PRINT"Pointer matrix":PRINT
210 FOR I=1 TO N:PRINT I;:K%(I,1)=I:K%(I,Z)=1:K=1
220 FOR J=1 TO M:FOR H=1 TO 2:IF N%(J,H)=I THEN 240
230 NEXT H:GOTO 270
240 FOR H=1 TO 2:U=N%(J,H):FOR L=1 TO K:IF K%(I,L)=U THEN 260
250 NEXT L:K=K+1:K%(I,K)=U:K%(I,Z)=K
260 NEXT H
270 NEXT J,I:PRINT:PRINT
280 PRINT"System matrix":PRINT:FOR J=1 TO M:PRINT J;
290 A=1/R(J):FOR I=1 TO 2:K=N%(J,I):L=N%(J,3-I)
300 P(K,1)=P(K,1)+A:FOR H=2 TO K%(K,Z):IF K%(K,H)=L THEN 320
310 NEXT H
320 P(K,H)=P(K,H)-A:NEXT I,J:PRINT:PRINT
330 PRINT"Boundary conditions":PRINT:FOR I=1 TO N
340 PRINT I;:IF T%(I)<1 THEN 440
350 P(I,1)=1:P(I,Z)=F(I):L=K%(I,Z):IF L=1 THEN 380
360 FOR K=2 TO L:J=K%(I,K):P(J,Z)=P(J,Z)-P(I,K)*P(I,Z)
370 K%(I,K)=0:P(I,K)=0:NEXT K:K%(I,Z)=1
380 FOR J=1 TO N:L=K%(J,Z):IF L=1 THEN 430
390 FOR K=2 TO L:IF K%(J,K)=I THEN 410
400 NEXT K:GOTO 430
410 K%(J,K)=K%(J,L):K%(J,L)=0:P(J,K)=P(J,L):P(J,L)=0
420 K%(J,Z)=K%(J,Z)-1
430 NEXT J
440 NEXT I:PRINT:PRINT
450 PRINT"Elimination":PRINT:FOR I=1 TO N:PRINT I;
460 KC=K%(I,Z):IF KC=1 THEN 600
470 FOR J=2 TO KC:C=P(I,J)/P(I,1):JJ=K%(I,J)
480 P(JJ,Z)=P(JJ,Z)-C*P(I,Z):L=K%(JJ,Z)
490 FOR JK=2 TO L:IF K%(JJ,JK)=I THEN 510
500 NEXT JK
510 K%(JJ,JK)=K%(JJ,L):K%(JJ,L)=0:P(JJ,JK)=P(JJ,L)
520 P(JJ,L)=0:L=L-1:K%(JJ,Z)=L
530 FOR II=2 TO KC:FOR IJ=1 TO L
540 IF K%(JJ,IJ)=K%(I,II) THEN 570
550 NEXT IJ:L=L+1:IJ=L:IF L=Z THEN 690
560 K%(JJ,Z)=L:K%(JJ,IJ)=K%(I,II)
570 P(JJ,IJ)=P(JJ,IJ)-C*P(I,II)
580 NEXT II
590 NEXT J
600 C=1/P(I,1):FOR J=1 TO KC:P(I,J)=C*P(I,J):NEXT J
610 P(I,Z)=C*P(I,Z):NEXT I:PRINT:PRINT
620 PRINT"Back substitution":PRINT:FOR I=1 TO N:J=N-I+1
630 PRINT J;:L=K%(J,Z):IF L=1 THEN 650
640 FOR K=2 TO L:JJ=K%(J,K):P(J,Z)=P(J,Z)-P(J,K)*P(JJ,Z):NEXT K
650 NEXT I
660 CLS:PRINT"Solution":PRINT:FOR I=1 TO N:F(I)=P(I,Z)
670 PRINT"I = ";I;" -  F = ";F(I):NEXT I
680 END
690 CLS:PRINT"Front width too small":GOTO 680
```

Program A.3. Wave front method.

In this program, the pointer matrix introduced in Section A.5 is used to indicate the nonzero coefficients. The first part of the program is practically identical to Program A.1. In the main part of the program, the elimination procedure, a complication arises, in that during elimination, which is performed in the order of numbering of the nodes, a coefficient that was initially zero, may obtain a certain value. This means that the pointer matrix has to be adjusted. Its width may grow or decrease. For that reason an error statement has been added to warn for an eventual insufficient front width.

In the wave front method, the precise amount of memory needed depends on the numbering sequence. In general, numbering should be done following the shortest sides of the network.

Program A.3 does not use the symmetry of the matrix, and therefore can also be used for nonsymmetric matrices. It is left to the user to modify the program so that the symmetry of the matrix is used to further reduce memory requirements.

## A.8. LDU-Decomposition

A direct method of solution that is often used because of its efficiency is the *LDU-decomposition method* (see, e.g., Wilkinson, 1965). This method is based upon the circumstance that any square matrix $P$ can be decomposed as follows

$$P = L \times D \times U \tag{A.3}$$

in which $D$ is diagonal matrix, $L$ is a lower-diagonal matrix, and $U$ is an upper-diagonal matrix, i.e.,

$$L(I,J) = 0, \quad \text{if } J > 1,$$
$$L(I,J) = 1, \quad \text{if } J = I,$$
$$D(I,J) = 0, \quad \text{if } J \neq I,$$
$$U(I,J) = 0, \quad \text{if } J < I,$$
$$U(I,J) = 1, \quad \text{if } J = I.$$

The matrix $L$ contains coefficients different from zero only below the main diagonal. In $U$, these occur only above the main diagonal. Such a decomposition appears to be very advantageous in solving the system of equations.

It can be shown that if the matrix $P$ is symmetric, as is assumed here, the matrices $L$ and $U$ are each others transpose,

$$L(I,J) = U(J,I)$$

The decomposition algorithm will first be developed. Therefore, the decomposition (A.3) is written as

$$P = L \times A, \qquad A = D \times U$$

where $A$ is an auxiliary matrix. The matrix multiplications can be written in component form in the following way

```
FOR K=1 TO N:FOR J=1 TO N:A(K,J)=0:FOR M=1 TO N:
A(K,J)=A(K,J)+D(K,M)*U(M,J):NEXT M,J,K
```
(A.4)

```
FOR I=1 TO N:FOR J=1 TO N:P(I,J)=0:FOR K=1 TO N:
P(I,J)=P(I,J)+L(I,K)*A(K,J):NEXT K,J,I
```
(A.5)

In (A.4) the values $D(K, M)$ are unequal to zero only if $M = K$, so that this equation can be reduced to

```
FOR K=1 TO N:FOR J=1 TO N:A(K,J)=D(K,K)*U(K,J):NEXT J,K
```
(A.6)

Because $U(K, J) = 0$ if $J < K$, it follows that $A(K, J) = 0$ if $J < K$. Furthermore, in (A.5) $L(I, K)$ can be replaced by $U(K, I)$. The multiplication is now $U(K, I) \times A(K, J)$, with $K$ running from 1 to $N$. However, because $U(K, I) = 0$ if $K > I$, and $A(K, J) = 0$ if $K > J$, it follows that all terms for which $K$ is larger than the minimum of $I$ and $J$ are zero. Hence, the algorithm can be written as

```
FOR I=1 TO N:FOR J=1 TO N:P(I,J)=0:L=I:IF J<I THEN L=J:
FOR K=1 TO L:P(I,J)=P(I,J)+U(K,I)*D(K,K)*U(K,J):NEXT K,J,I
```
(A.7)

Elaboration of this algorithm now leads to the following system of equations, recalling that $U(K, I) = 1$ if $I = K$,

```
P(1,1)=D(1,1)
P(1,2)=D(1,1)*U(1,2)
.....
P(1,N)=D(1,1)*U(1,N)

P(2,2)=D(1,1)*U(1,2)*U(1,2)+D(2,2)
P(2,3)=D(1,1)*U(1,2)*U(1,3)+D(2,2)*U(2,3)
.....
P(2,N)=D(1,1)*U(1,2)*U(1,N)+D(2,2)*U(2,N)

.....
.....
.....

P(M,M)=D(1,1)*U(1,M)*U(1,M)+D(2,2)*U(2,M)*U(2,M)+...+D(M,M)
P(M,M+1)=D(1,1)*U(1,M)*U(1,M+1)+D(2,2)*U(2,M)*U(2,M+)+...+D(M,M)*U(M,M+1)
P(M,N)=D(1,1)*U(1,M)*U(1,N)+D(2,2)*U(2,M)*U(2,N)+...+D(M,M)*U(M,N)
```

Here only those coefficients of $P$ have been given for which $J \geq I$, the coefficients below the main diagonal follow from the symmetry property.

From the system of equations given above the coefficients of the matrices $D$ and $U$ can be calculated successively. In BASIC, the algorithm is as follows.

```
1000 FOR I=1 TO N:FOR J=1 TO N:A=P(I,J):P(I,J)=1
1010 II=I-1:IF I=1 THEN 1030
1020 FOR K=1 TO II:A=A-P(K,K)*P(K,I)*P(K,J):NEXT K
1030 P(I,J)=A/P(I,I):NEXT J,I
```

In order to save memory the coefficients $D(K, K)$ are stored as $P(K, K)$, and the coefficients $U(I, J)$ are stored as $P(I, J)$, with $J > I$.

In the algorithm $J \geq I$ and $K < I$, hence certainly $K < J$. Therefore, the algorithm involves only coefficients $P(I, J)$ to the right of the main diagonal, or on it. If now the original matrix $P$ has a banded structure, i.e.,

$$P(I, J) = 0 \quad \text{if } ABS(J - I) > BW$$

then the matrix $U$ will also have that property. For the algorithm this means that the number of multiplications can be reduced as follows. In line 1000 the value of $J$ need not go beyond $I + BW$, and not beyond $N$. Furthermore, we have $J \geq I$ in the algorithm and, hence, $J - K \geq I - K$. The value of $K$ in line 1020 runs from 1 to $I - 1$, but this range can now be reduced by noting that $P(K, J)$ is unequal to zero only if $J - K$ (which is always positive, because $K < I$ and $J >= I$) is smaller than, or equal to $BW$. This means that the smallest possible value of $K$ is the maximum of 1 and $J - BW$.

The algorithm now becomes

```
1000 FOR I=1 TO N:JL=I+BW:IF JL>N THEN JL=N
1010 FOR J=1 TO JL:A=P(I,J):P(I,J)=1:II=I-1:IF I=1 THEN 1040
1020 KA=J-BW:IF KA<1 THEN KA=1
1030 FOR K=KA TO II:A=A-P(K,K)*P(K,I)*P(K,J):NEXT K
1040 P(I,J)=A/P(I,I):NEXT J:NEXT I
```

The number of multiplications and, hence, computation time, has now been reduced by eliminating all multiplications by zero. The memory requirement can also be reduced by noting that the algorithm involves only coefficients $P(I, J)$ for which $J \geq I$ and $J \leq I + BW$. The band to the right of the main diagonal can now be stored in a narrow rectangular matrix by storing the coefficient $P(I, J)$ as $P(I, J - I + 1)$. The dimensions of the rectangular matrix needed are then $N \times (BW + 1)$. The algorithm now is

```
1000 FOR I=1 TO N:JL=I+BW:IF JL>N THEN JL=N
1010 FOR J=1 TO JL:A=P(I,J-I+1):P(I,J-I+1)=1:IF I=1 THEN 1040
1020 II=I-1:KA=J-BW:IF KA<1 THEN KA=1
1030 FOR K=KA TO II:A=A-P(K,1)*P(K,I-K+1)*P(K,J-K+1):NEXT K
1040 P(I,J-I+1)=A/P(I,1):NEXT J,I
```

This algorithm can be further streamlined by replacing $J - I + 1$ by $JJ$, and then again writing $J$ for $JJ$. Some of the expressions then are somewhat simpler. The result is

```
1000 FOR I=1 TO N:JL=BW+1:JN=N-I+1:IF JL>JN THEN JL=JN
1010 FOR J=1 TO JL:A=P(I,J):P(I,J)=1:IF I=1 THEN 1040
1020 II=I-1:KA=I+J-BW-1:IF KA<1 THEN KA=1
1030 FOR K=KA TO II:A=A-P(K,1)*P(K,I-K+1)*P(K,I-K+J):NEXT K
1040 P(I,J)=A/P(I,1):NEXT J,I
```

After execution of this algorithm the matrix $P$ contains the diagonal matrix $D$ in

SOLUTION OF LINEAR EQUATIONS

its first column, and the elements of $U$ above the main diagonal in the columns 2 to $BW + 1$. The system of equations can now easily be solved in consecutive steps.

The system of equations is

$$P \times F = G$$

in which $G$ is a given vector and $F$ is an unknown vector. This can also be written as

$$L \times D \times U \times F = G.$$

This system can be solved in three steps,

$$L \times V = G \rightarrow V,$$
$$D \times W = V \rightarrow W,$$
$$U \times F = W \rightarrow F.$$

Because of the special structure of the matrices $L$, $D$ and $U$, these systems of equations can easily be solved, in the first case from the top down, and in the last case, from the bottom upwards. This process is called *back substitution*. An algorithm in BASIC in which all coefficients are stored in the same matrix $P$, is

```
2000 FOR I=1 TO N:A=G(I):II=I-1:IF I=1 THEN 2020
2010 FOR K=1 TO II:A=A-P(K,I)*G(K):NEXT K
2020 G(I)=A:NEXT I:FOR I=1 TO N:G(I)=G(I)/P(I,I):NEXT I
2030 FOR I=1 TO N:J=N+1-I:A=G(J):JJ=J+1:IF J=N THEN 2050
2040 FOR K=JJ TO N:A=A-P(J,K)*G(K):NEXT K
2050 G(J)=A:NEXT I
```

The vector $G$ is used to store the known vector $G$ initially, and, later, to store the solution vector $F$.

Recalling that the matrix $P$ has a banded structure, and that the band is stored in a rectangular matrix, the back substitution can be executed by

```
2000 FOR I=1 TO N:A=G(I);II=I-1:IF I=1 THEN 2030
2010 KA=I-BW:IF KA<1 THEN KA=1
2020 FOR K=KA TO II:A=A-P(K,I-K+1)*G(K):NEXT K
2030 G(I)=A:NEXT I:FOR I=1 TO N:G(I)=G(I)/P(I,1):NEXT I
2040 FOR I=1 TO N:J=N+1-I:A=G(J):JJ=J+1:IF J=N THEN 2070
2050 KL=I+BW:IF KL>N THEN KL=N
2060 FOR K=JJ TO KL:A=A-P(J,K-J+1)*G(K):NEXT K
2070 G(J)=A:NEXT I
```

A computer program, in BASIC, that executes the various calculations described above, is reproduced as Program A.4. This program uses the symmetry

```
10 CLS:PRINT"LDU":PRINT
20 INPUT"Number of nodes ........... ";N:PRINT
30 INPUT"Number of elements ........ ";M:PRINT
40 INPUT"Band width ............... ";BW:PRINT
50 DIM T%(N),G(N),P(N,BW+1),N%(M,2),R(M)
60 FOR I=1 TO N:CLS:PRINT"Node ";I:PRINT
```

```
 70 INPUT"x ........................ ";X(I):PRINT
 80 INPUT"y ........................ ";Y(I):PRINT
 90 INPUT"Potential given (y/n) ..... ";A$:PRINT
100 IF A$="y" OR A$="Y" THEN T%(I)=2:GOTO 130
110 IF A$="n" OR A$="N" THEN T%(I)=0:G(I)=0:GOTO 140
120 GOTO 90
130 INPUT"Potential ................ ";G(I):PRINT
140 NEXT I
150 FOR J=1 TO M:CLS:PRINT"Element ";J:PRINT
160 INPUT"Node 1 ................... ";N%(J,1):PRINT
170 INPUT"Node 2 ................... ";N%(J,2):PRINT
180 INPUT"Resistance ............... ";R(J):PRINT
190 NEXT J
200 CLS:PRINT"System matrix":PRINT:FOR J=1 TO M:PRINT J;
210 A=1/R(J):FOR I=1 TO 2:K=N%(J,I):L=N%(J,3-I)
220 P(K,1)=P(K,1)+A:H=L-K+1:IF H>1 THEN P(K,H)=P(K,H)-A
230 NEXT I,J:PRINT:PRINT
240 PRINT"Boundary conditions":PRINT:FOR I=1 TO N:PRINT I;
250 IF T%(I)<1 THEN 310
260 P(I,1)=1:FOR J=1 TO BW:K=I+J:IF K>N THEN 280
270 G(K)=G(K)-P(I,J+1)*G(I):P(I,J+1)=0
280 NEXT J:II=I-1:IF I=1 THEN 310
290 FOR J=1 TO BW:K=I-J
300 G(K)=G(K)-P(K,J+1)*G(I):P(K,J+1)=0:NEXT J
310 NEXT I:PRINT:PRINT
320 PRINT"Decomposition":PRINT:FOR I=1 TO N:PRINT I;
330 JL=BW+1:JN=N-I+1:IF JL>JN THEN JL=JN
340 FOR J=1 TO JL:A=P(I,J):P(I,J)=1:II=I-1:IF I=1 THEN 380
350 KA=I+J-BW:IF KA<1 THEN KA=1
360 IF KA>II THEN 380
370 FOR K=KA TO II:A=A-P(K,1)*P(K,I-K+1)*P(K,I-K+J):NEXT K
380 P(I,J)=A/P(I,1):NEXT J,I:PRINT:PRINT
390 PRINT"Back substitution":PRINT:FOR I=1 TO N:PRINT I;
400 A=G(I):II=I-1:IF I=1 THEN 430
410 KA=I-BW:IF KA<1 THEN KA=1
420 FOR K=KA TO II:A=A-P(K,I-K+1)*G(K):NEXT K
430 G(I)=A:NEXT I:FOR I=1 TO N:G(I)=G(I)/P(I,1):NEXT I
440 PRINT:FOR I=1 TO N:J=N+1-I:PRINT J;
450 A=G(J):JJ=J+1:IF J=N THEN 480
460 KL=J+BW:IF KL>N THEN KL=N
470 FOR K=JJ TO KL:A=A-P(J,K-J+1)*G(K):NEXT K
480 G(J)=A:NEXT I
490 CLS:PRINT"Solution":PRINT
500 FOR I=1 TO N:PRINT"I = ";I;" -  F = ";G(I):NEXT I
510 END
```

Program A.4. LDU-decomposition.

of the matrix, and its banded structure. In the program, the bandwidth has to be specified by the user. It can be calculated from the largest difference of node numbers in any element (in the example this value is 3). If desired, this can also be calculated in the program itself.

## A.9. Conjugate Gradient Method

In recent years, the method of *conjugated gradients* has gained much popularity.

# SOLUTION OF LINEAR EQUATIONS

The method is of an iterative character, with all unknown values updated at the same time, until a certain error measure is reduced to zero. It can be shown that the method converges to the exact solution after a number of iterations equal to the number of equations, at least if there were no errors due to round-off. In practice, often much fewer iterations are needed to obtain a satisfactory accuracy. The theory of the method will not be presented here (see, e.g., Hestenes and Stiefel, 1952). In this section, an algorithm published by Reid (1971) will be presented, together with a program in BASIC.

It should be mentioned that the method, in the form presented here, is suitable only for systems having a positive definite symmetric matrix.

The system of equations to be solved is

$$\sum \{P(I,J) \times F(J)\} = G(I).$$

It is now assumed that an initial estimate for the unknown values $F(I)$ is represented by the vector $F_0(I)$. This estimate, in general, does not satisfy the system of equations, but gives rise to a residual $U(I)$, expressed by

$$U(I) = G(I) - \sum \{P(I,J) \times F_0(J)\}$$

A second vector $V(I)$ is now introduced, which is initially equal to $U(I)$,

$$V(I) = U(I)$$

Next an iterative process is executed, with the following operations,

$$UU = \sum \{U(I) \times U(I)\}$$
$$W(I) = \sum \{P(I,J) \times V(J)\}$$
$$VW = \sum \{V(I) \times W(I)\}$$
$$AA = UU/VW$$
$$F(I) = F(I) + AA \times V(I)$$
$$U(I) = U(I) - AA \times W(I)$$
$$WW = \sum \{U(I) \times U(I)\}$$
$$BB = WW/UU$$
$$V(I) = U(I) + BB \times V(I).$$

This process is repeated until the value of $UU$ (a measure for the absolute value of the error) is smaller than a given value.

A program that executes the computations presented above is reproduced as Program A.5. The program uses a pointer matrix, in order to restrict the memory requirements. During the entire process the system matrix is not modified, and its

```
10 CLS:PRINT"Method of conjugated gradients":PRINT
20 INPUT"Number of nodes .......... ";N:PRINT
30 INPUT"Number of elements ........ ";M:PRINT
40 INPUT"Estimated pointer width ... ";Z:PRINT
50 DIM T%(N),F(N),U(N),V(N),W(N),K%(N,Z),P(N,Z),N%(M,2),R(M)
60 FOR I=1 TO N:CLS:PRINT"Node ";I:PRINT
70 INPUT"x ....................... ";X(I):PRINT
80 INPUT"y ....................... ";Y(I):PRINT
90 INPUT"Potential given (y/n) ..... ";A$:PRINT
100 IF A$="y" OR A$="Y" THEN T%(I)=2:GOTO 130
110 IF A$="n" OR A$="N" THEN T%(I)=0:F(I)=0:GOTO 140
120 GOTO 90
130 INPUT"Potential ................ ";F(I):PRINT
140 NEXT I
150 FOR J=1 TO M:CLS:PRINT"Element ";J:PRINT
160 INPUT"Node 1 ................... ";N%(J,1):PRINT
170 INPUT"Node 2 ................... ";N%(J,2):PRINT
180 INPUT"Resistance ............... ";R(J):PRINT
190 NEXT J
200 CLS:PRINT"Pointer matrix":PRINT
210 FOR I=1 TO N:PRINT I;:K%(I,1)=I:K%(I,Z)=1:K=1
220 FOR J=1 TO M:FOR H=1 TO 2:IF N%(J,H)=I THEN 240
230 NEXT H:GOTO 270
240 FOR H=1 TO 2:U=N%(J,H):FOR L=1 TO K:IF K%(I,L)=U THEN 260
250 NEXT L:K=K+1:K%(I,K)=U:K%(I,Z)=K
260 NEXT H
270 NEXT J,I:PRINT:PRINT
280 PRINT"System matrix":PRINT:FOR J=1 TO M:PRINT J;
290 A=1/R(J):FOR I=1 TO 2:K=N%(J,I):L=N%(J,3-I)
300 P(K,1)=P(K,1)+A:FOR H=2 TO K%(K,Z):IF K%(K,H)=L THEN 320
310 NEXT H
320 P(K,H)=P(K,H) A:NEXT I,J:PRINT:PRINT
330 EPS=.000001:PRINT"Iteration":PRINT
340 K=1:FOR I=1 TO N:U(I)=0:IF T%(I)>1 THEN 360
350 U(I)=G(I):FOR J=1 TO K%(I,Z):U(I)=U(I)-P(I,J)*F(K%(I,J)):NEXT J
360 V(I)=U(I):NEXT I:UU=0:FOR I=1 TO N:UU=UU+U(I)*U(I):NEXT I
370 PRINT K;:IF UU<EPS THEN 450
380 FOR I=1 TO N:W(I)=0:FOR J=1 TO K%(I,Z):W(I)=W(I)+P(I,J)*V(K%(I,J))
390 NEXT J,I:VW=0:FOR I=1 TO N:VW=VW+V(I)*W(I):NEXT I
400 AA=UU/VW:FOR I=1 TO N:IF T%(I)>1 THEN 420
410 F(I)=F(I)+AA*V(I):U(I)=U(I)-AA*W(I)
420 NEXT I:WW=0:FOR I=1 TO N:WW=WW+U(I)*U(I):NEXT I
430 BB=WW/UU:FOR I=1 TO N:V(I)=U(I)+BB*V(I):NEXT I:UU=WW
440 K=K+1:IF K<=N AND UU>EPS THEN 370
450 CLS:PRINT"Solution":PRINT
460 FOR I=1 TO N:PRINT"I = ";I;" - F = ";F(I):NEXT I:PRINT
470 END
```

Program A.5. Conjugate gradient method.

original structure is maintained, with a very small bandwidth, independent of the order of numbering.

The number of iterations needed to obtain a satisfactory accuracy is often much smaller than the number of unknowns, so that the method is very efficient. This may occur especially in the case of time dependent problems, in which a similar system of equations has to be solved many times, and the previous solution can be used as a good estimate for the next one.

# References

Abramowitz, M. and Stegun, I. A., 1965, *Handbook of Mathematical Functions*, Dover, New York.
Adamson, A. W., 1967, *Physical Chemistry of Surfaces*, 2nd edn., Interscience, New York.
Akker, C. Van den, 1982, Numerical analysis of the stream function in plane groundwater flow, PhD Thesis, University of Technology, Delft.
Aravin, V. I. and Numerov, S. N., 1965, *Theory of Fluid Flow in Undeformable Porous Media*, Daniel Davey, New York.
Aris, R., 1962, *Vectors, Tensors and the Basic Equations of Fluid Mechanics*, Prentice-Hall, Englewood Cliffs, N.J.
Bachmat, Y. and Bear, J., 1986, Macroscopic modelling of transport phenomena in porous media. 1: The continuum approach, *Transport in Porous Media* **1**, 213—240.
Banerjee, P. K. and Butterfield, R., 1977, Boundary element methods in geomechanics, in G. Gudehus (ed.), *Finite Elements in Geomechanics*, Wiley, London, 529—570.
Barak, A. Z. and Bear, J., 1981, Flow at high Reynolds numbers through anisotropic porous media, *Adv. Water Resour.* **4**, 54—66.
Bear, J., 1960, The transition zone between fresh and salt waters in coastal aquifers, PhD Thesis, Univ. of California, Berkeley, California.
Bear, J., 1961, On the tensor form of dispersion, *J. Geophys. Res.* **66**(4), 1185—1197.
Bear, J., 1972, *Dynamics of Fluids in Porous Media*, American Elsevier, New York.
Bear, J., 1979, *Hydraulics of Groundwater*, McGraw-Hill, New York.
Bear, J. and Bachmat, Y., 1967, A generalized theory on hydrodynamic dispersion in porous media, *I.A.S.H. Symp. Artificial Recharge and Management of Aquifers*, Haifa, Israel, IASH **72**, pp. 7—16.
Bear, J. and Bachmat, Y., 1983, On the equivalence of areal and volumetric averages in transport phenomena, *Adv. Water Resour.* **6**(1), 59—62.
Bear, J. and Bachmat, Y., 1984, Transport phenomena in porous media — Basic equations, in J. Bear and M. Y. Corapcioglu (eds.), *Fundamentals of Transport Phenomena in Porous Media*, Martinus Nijhoff, Dordrecht, pp. 3—61.
Bear, J. and Bachmat, Y., 1986, Macroscopic modelling of transport phenomena in porous media, 2. Applications to mass, momentum and energy transport, *Transport in Porous Media* **1**, 241—269.
Bear, J. and Bachmat, Y., 1987, *Introduction to Transport Phenomena in Porous Media*, D. Reidel, Dordrecht (in press).
Bear, J., Braster, C., and Menier, P. C., 1987, Effective and relative permeabilities in anisotropic porous media, *Transport in Porous Media* **2**, 301—316.
Bear, J. and Corapcioglu, M. Y., 1981a, Mathematical model for regional land subsidence due to pumping, 1, Integrated aquifer subsidence equations based on vertical displacement only, *Water Resour. Res.* **17**, 937—946.
Bear, J. and Corapcioglu, M. Y., 1981b, Mathematical model for regional land subsidence due to pumping, 2, Integrated aquifer subsidence equations for vertical and horizontal displacements, *Water Resour. Res.* **17**, 947—958.
Bear, J. and Dagan, G., 1963, The transition zone between fresh water and salt water in a coastal aquifer, Prog. Rep. 2: A steady, flow to an array of wells above the interface approximate.
Bear, J. and Dagan, G., 1964a, Some exact solutions of interface problems by means of the hodograph method, *J. Geophys. Res.* **69**, 1563—1572.
Bear, J. and Dagan, G., 1964b, Moving interface in coastal aquifers, *Proc. ASCE* **99** (HY4), 193—215.
Bear, J. and Dagan, G., 1964c, The transition zone between fresh water and salt water in a coastal aquifer, Prog. Rep. 3: The unsteady interface below a coastal collector, 122 pp.

Bear, J. and Dagan, G., 1966, The transition zone between fresh water and salt water in a coastal aquifer, *Prog. Rep.* 4: Increasing the yield of a coastal collector by means of special operation techniques, 81 pp.

Bear, J. and Jacobs, M., 1965, On the movement of water bodies injected into aquifers, *J. Hydrology* **3**(1), 37—57.

Bear, J. and Shapiro, A. M., 1984, On the shape of the non-steady interface intersecting discontinuities in permeability, *Adv. Water Resour.* **7**, 106—112.

Bear, J. and Shapiro, A. M., 1986, Boundary conditions in transport phenomena in porous media, Internal Report, Technion, Haifa, Israel.

Bear, J., Zaslavsky, D., and Irmay, S., 1968, *Physical Principles of Water Percolation and Seepage*, UNESCO, Paris.

Biot, M. A., 1941, General theory of three-dimensional consolidation, *J. Appl. Phys.* **12**, 155—164.

Bishop, A. W., Alpan, I., Blight, G. E., and Donald, J. B., 1960, Factors controlling the strength of partially saturated cohesive soils, *Proc. Res. Conf. Shear Strength Cohesive Soils, A.S.C.E.*, pp. 503—532.

Bredehoeft, J. D. and Pinder, G. F., 1970, Digital analysis of areal flow in multiaquifer groundwater systems: a quasi three-dimensional model, *Water Resour. Res.* **6**(3), 883—888.

Brooks, R. H. and Corey, A. T., 1964, Hydraulic properties of porous media, *Colorado State Univ., Hydrology Papers* **3**, 1—27.

Burdine, N. T., 1953, Relative permeability calculations from pore size distribution data, *Trans. AIME* **198**, 71—77.

Carman, P. C., 1937, Fluid flow through a granular bed, *Trans. Inst. Chem. Engnrs.* **15**, 150—156.

Carnahan, B., Luther, H. A., and Wilkes, J. O., 1969, *Applied Numerical Methods*, Wiley, New York.

Carslaw, H. S. and Jaeger, J. C., 1959, *Conduction of Heat in Solids*, 2nd edn., Oxford Univ. Press, London, 510 pp.

Childs, E. C., 1969, *Introduction to the Physical Basis of Soil Water Phenomena*, Wiley, New York.

Childs, E. C. and Collis-George, N., 1950, The permeability of porous materials, *Proc. Roy. Soc.* **A201**, 392—405.

Conkling, H. *et al.*, 1934, Ventura County investigations, *California Div. Water Resour. Bull.* **6**, 244 pp.

Cooper, H. H. Jr., 1959, A hypothesis concerning the dynamic balance of fresh water and salt water in a coastal aquifer, *J. Geophys. Res.* **64**, 461—467.

Corapcioglu, M. Y. and Bear, J., 1983, A mathematical model for regional land subsidence due to pumping, 3, Integrated equations for a phreatic aquifer, *Water Resour. Res.* **19**, 895—908.

Corey, A. T., 1957, Measurement of water and air permeability in unsaturated soils, *Proc. Soil Sci. Soc. Amer.* **21**, 7—10.

Darcy, H., 1856, *Les Fontaines Publiques de la Ville de Dijon*, Dalmont, Paris.

Davis, S. N. and de Wiest, R. J. M., 1966, *Hydrogeology*, Wiley, New York.

Drabbe, J. and Badon Ghyben, W., 1889, Nota in verband met de voorgenomen putboring nabij Amsterdam (in Dutch), *Tijdschrift KIVI* **8**, 22.

Dupuit, J., 1863, *Etudes théoriques et pratiques sur le mouvement des eaux dans les canaux découverts et à travers les terrains perméables*, Dunod, Paris.

Ergun, S., 1952, Fluid flow through packed columns, *Chem. Engng. Prog.* **48**, 89—94.

Fair, G. M. and Hatch, L. P., 1933, Fundamental factors governing the streamline flow of water through sand, *J. Am. Water Works Assoc.* **25**, 1551—1565.

Feller, W., 1966, *An Introduction to Probability Theory and its Applications*, 2, Wiley, New York.

Forchheimer, P., 1901, Wasserbewegung durch Boden, *Z. Ver. Deutsche Ing.* **45**, 1782—1788.

Forsythe, G. E. and Wasow, W. R., 1960, *Finite Difference Methods for Partial Differential Equations*, Wiley, New York.

Fox, L., 1962, *Numerical Solution of Ordinary and Partial Differential Equations*, Pergamon Press, New York.

Freundlich, C. G. L., 1926, *Colloid and Capillary Chemistry*, Methuen, London.

Gambolati, G. and Perdon, A., 1984, The conjugate gradients in subsurface flow and land subsidence modelling, in J. Bear and M. Y. Corapcioglu (eds.), *Fundamentals of Transport Phenomena in Porous Media*, Martinus Nijhoff, Dordrecht, pp. 953—984.

Gardner, W. R., 1958, Some steady state solutions of the unsaturated moisture flow equations with application to evaporation from a water table, *Soil Sci.* **85**(4), 228—232.

Gardner, W. R. and Mayhugh, M. S., 1958, Solutions and tests on the diffusion equation for the movement of water in soil, *Proc. Soil Sci. Soc. Amer.* **22**, 197—201.

Gelhar, L. W., 1976, Stochastic analysis of flow in aquifers, *AWRA Symp. on Advances in Groundwater Hydrology*, Chicago, Ill.

Glover, R. E., 1959, The pattern of fresh water flow in coastal aquifer, *J. Geophys. Res.* **64**, 439—475.

Hantush, M. S., 1949, Plane potential flow of ground water with linear leakage, PhD Thesis, Univ. of Utah.

Hantush, M. S., 1964, Hydraulics of wells, *Adv. Hydrosci.* **1**, 281—442.

Harr, M. E., 1962, *Groundwater and Seepage*, McGraw-Hill, New York.

Hassanizadeh, M. and Gray, W. G., 1979a, General conservation equations for multiphase—systems, 1. Averaging procedure, *Adv. Water Resour.* **2**(3), 131—144.

Hassanizadeh, M. and Gray, W. G., 1979b, General conservation equations for multiphase—systems, 1. Mass, momentum and energy equations, *Adv. Water Resour.* **2**(4), 191—203.

Hendricks, D. W., 1972, Sorption in flow through porous media, in J. Bear (ed.), *Fundamentals of Transport Phenomena in Porous Media*, Elsevier, Amsterdam, pp. 384—392.

Henry, H. R., 1959, Salt intrusion into freshwater aquifers, *J. Geophys. Res.* **64**, 1911—1919.

Herrera, I. and Rodarte, L., 1973, Integrodifferential equations for systems of leaky aquifers and applications, 1, The nature of approximate theories, *Water Resour. Res.* **9**, 995—1005.

Herzberg, A., 1901, Die Wasserversorgung einiger Nordseebaden, *Z. Gasbeleuchtung und Wasserversorgung* **44**, 815—819.

Hestenes, M. R. and Stiefel, E., 1952, Methods of conjugate gradients for solving linear systems, *NBS J. Res.* **49**, 409—436.

Hubbert, M. K., 1940, The theory of ground-water motion, *J. Geol.* **4**, 785—944.

Irmay, S., 1954, On the hydraulic conductivity of unsaturated soils, *Trans. Amer. Geophys. Union* **35**, 463—468.

Irons, B. M., 1970, A frontal solution program for finite element analysis, *Int. J. Num. Meth. Eng.* **2**, 5—32.

Jacob, C. E., 1940, On the flow of water in an elastic artesian aquifer, *Trans. Amer. Geophys. Un.* **2**, 574—586.

Jacobs, M. and Schmorak, S., 1960, *Seawater Intrusion and Interface Determination Along the Coastal Plane of Israel*, State of Israel, Hydrological Service, Hydrological Paper No. 6, 12 pp.

Javandel, I., Doughty, C., and Tsang, C. F., 1984, *Groundwater Transport: Handbook of Mathematical Models*, American Geophysical Union, Washington.

Josselin de Jong, G. de, 1963, Consolidatie in drie dimensions (in Dutch), *L.G.M. Mededelingen* **7**, 57—73.

Josselin de Jong, G. de, 1969, Generating functions in the theory of flow through porous media, in R. J. M. de Wiest (ed.), *Flow through Porous Media*, Academic Press, New York, pp. 377—400.

Josselin de Jong, G. de, 1981, The simultaneous flow of fresh and salt water in aquifers of large horizontal extension determined by shear flow and vortex theory, in A. Verruijt and F. B. J. Barends (eds.), *Flow and Transport in Porous Media*, Balkema, Rotterdam, pp. 75—82.

Kantorovich, L. V. and Krylov, V. I., 1964, *Approximate Methods of Higher Analysis*, translated by C. D. Benster, Interscience Publishers, New York.

Kober, H., 1957, *Dictionary of Conformal Representations*, Dover, New York.

Korn, G. A. and Korn, T. M., 1968, *Mathematical Handbook for Scientists and Engineers*, 2nd edn., McGraw-Hill, New York.

Krumbein, W. C. and Monk, G. D., 1943, Permeability as a function of the size parameters of unconsolidated sands, *Trans. Amer. Inst. Min. Met. Engrs.* **151**, 153—163.

Langmuir, I., 1915, Chemical reactions at low temperatures, *J. Amer. Chem. Soc.* **37**, 1139.

Langmuir, I., 1918, The adsorption of gases on plane surfaces of glass, mica and platinum, *J. Amer. Chem. Soc.* **40**, 1361—1403.

Lapidus, L. and Amundson, N. R., 1952, Mathematics of absorption in beds VI. The effect of longitudinal diffusion in ion exchange and chromatographic columns, *J. Phys. Chem.* **56**, 984—988.

Liggett, J. A., and Liu, P. L. F., 1983, *The Boundary Integral Equation Method for Porous Media Flow*, Allen & Unwin, London.

Lindstrom, F. T., Boersma, L., and Stockard, D., 1971, A theory on the mass transport of previously distributed chemicals in a water saturated sorbing porous medium, *Soil Sci.* **112**, 291—300.

Mavis, F. T., and Tsui, T. P., 1939, Percolation and capillary movement of water through sand prisms, *Bull. 18, Univ. of Iowa, Studies in Eng.*, Iowa City.

Meinzer, O. E., 1942, *Hydrology*, Dover, New York.

Morel-Seytoux, H. J., 1973, Two-phase flows in porous media, *Adv. Hydroscience* **9**, 119—202.

Morse, P. M., and Feshbach, H., 1953, *Methods of Theoretical Physics*, McGraw-Hill, New York.

Mualem, Y., 1976, A new model for predicting the hydraulic conductivity of unsaturated porous media, *Water Resour. Res.* **12**, 513—522.

Mualem, Y., and Bear, J., 1974, The shape of the interface in steady flow in a stratified aquifer, *Water Resour. Res.* **10**, 1207—1215.

Muskat, M., 1937, *The Flow of Homogeneous Fluids through Porous Media*, McGraw-Hill, New York.

Narasimhan, T. N., and Witherspoon, P. A., 1976, An integrated finite difference method for analyzing fluid flow in porous media, *Water Resour. Res.* **12**, 57—64.

Neuman, S. P. and Witherspoon, P. A., 1969, Applicability of current theories of flow in leaky aquifers, *Water Resour. Res.* **5**, 817—829.

Neuman, S. P., and Witherspoon, P. A., 1971, Analysis of nonsteady flow with a free surface using the finite element method, *Water Resour. Res.* **7**, 611—623.

Nikolaevskii, V. N., 1959, Convective diffusion in porous media, *J. Appl. Math. Mech.* (*PMM*) **23**(6), 1042—1050.

Noblanc, A. and Morel-Seytoux, H. J., 1972, Perturbation analysis of two-phase infiltration, *Proc. Amer. Soc. Civil Engrs.* **98** (HY9), 1527—1541.

Ogata, A. and Banks, R. B., 1961, A solution of the differential equation of longitudinal dispersion in porous media, *USGS Prof. Paper* 411—A.

Pfannkuch, H. O., 1963, Contribution à l'étude des déplacements de fluides miscibles dans un milieu poreux, *Rev. Inst. Fr. Petrol.* **18**(2), 215—270.

Phuc Le Van and Morel-Seytoux, H. J., 1972, Effect of soil air movement and compressibility on infiltration rates, *Proc. Soil. Sci. Soc. Amer.* **36**, 237—241.

Pinder, G. F., and Cooper, H. H. Jr., 1970, A numerical technique for calculating the transient position of the saltwater front, *Water Resour. Res.* **6**, 875—882.

Pinder, G. F., and Page, R. H., 1977, Finite element simulation of salt water intrusion on the South Fork of Long Island, in W. G. Gray, G. F. Pinder and C. A. Brebbia (eds.), *Finite Elements in Water Resources*, Pentech Press, London, 2.51—2.69.

Polubarinova-Kochina, P. Ya., 1952, *Theory of Groundwater Movement* (in Russian), Gostekhizdat, Moscow. English trans. by Roger J. M. de Wiest, Princeton Univ. Press, Princeton, N.J., 1962.

Prickett, T. A., Naymik, T. G., and Lonnquist, C. G., 1981, A random-walk solute transport model for selected groundwater quality evaluations, *Bull. Illinois State Water Survey* **65**, Champaign.

Reid, J. K., 1971, On the method of conjugate gradients for the solution of large sparse systems of linear equations, in J. K. Reid (ed.), *Large Sparse Sets of Linear Equations*, Academic Press, London, pp. 231—253.

Richards, L. A., and Gardner, W., 1936, Tensiometers for measuring the capillary tension and soil water, *J. Agric. Res.* **69**, 215—235.

Saffman, P. G., 1960, Dispersion due to molecular diffusion and macroscoping mixing in flow through a network of capillaries, *J. Fluid Mech.* **7**(2), 194—208.

Scheidegger, A. E., 1960, *The Physics of Flow through Porous Media*, Univ. of Toronto Press, Toronto.

Scheidegger, A. E., 1961, General theory of dispersion in porous media, *J. Geophys. Res.* **66**, 3273—3278.

Schmorak, S., 1967, Salt water encroachment in the coastal plain of Israel, *IASH Symp. Artificial Recharge and Management of Aquifers, Haifa, Israel*, IASH **72**, 305—318.

Scriven, L. E., 1960, Dynamics of a fluid interface, *Chem. Engng. Sci.* **12**, 98—108.

Shamir, U., and Dagan, G., 1971, Motion of the seawater interface in coastal aquifers: a numerical solution, *Water Resour. Res.* **7**, 644—657.

Silin-Bekchurin, 1958, *Dynamics of Ground Water* (in Russian), Moscow Izdat., Moscow.
Slattery, J. C., 1967, Fundamentals, *Ind. Eng. Chem.* **6**, 108.
Slichter, C. S., 1905, Field measurement of the rate of movement of underground waters, *USGS Water Supply Paper*, 140.
Southwell, R. V., 1940, *Relaxation Methods in Engineering Science*, Oxford Univesity Press, London.
Spiegel, M. R., 1959, *Theory and Problems of Vector Analysis*, Schaum Outline Series, McGraw-Hill, New York.
Strack, O. D. L., 1972, Some cases of interface flow towards drains, *J. Engng. Math.* **6**, 175—191.
Strack, O. D. L., 1973, Many-valuedness encountered in groundwater flow, PhD Thesis, University of Technology, Delft.
Strack, O. D. L., 1987, *Groundwater Mechanics*, Prentice-Hall, Englewood Cliffs, N.J.
Taylor, R. L., and C. B. Brown, 1967, Darcy flow solutions with a free surface, *Proc. Amer. Soc. Civil Engs.* **93** (HY2), 25—33.
Terzaghi, K., 1925, *Erdbaumechanik auf Bodenphysikalische Grundlage*, Deuticke, Vienna.
Theis, C. V., 1935, The relation between the lowering of the piezometric surface and the rate and duration of discharge of a well using groundwater storage, *Trans. Amer. Geophys. Union* **16**, 519—524.
Todd, D. K., 1959, *Ground Water Hydrology*, Wiley, New York.
Vachaud, G., Gaudet, J. P., and Kuraz, V., 1974, Air and water flow during ponded infiltration in a vertical bounded column, *J. Hydrol.* **22**, 89—108.
Van Genuchten, M. Th., 1974, Mass transfer studies of sorbing porous media, PhD Thesis, New Mexico State Univ., La Cruz, NM.
Verruijt, A., 1969, Elastic storage of aquifers, in R. J. M. de Wiest (ed.), *Flow through Porous Media*, Academic Press, New York, pp. 331—376.
Verruijt, A., 1980, The rotation of a vertical interface in a porous medium, *Water Resour. Res.* **16**, 239—240.
Verruijt, A., 1982, *Theory of Groundwater Flow*, 2nd edn., Macmillan, London.
Verruijt, A., 1984, The theory of consolidation, in J. Bear and M. Y. Corapcioglu (eds.), *Fundamentals of Transport Phenomena in Porous Media*, Martinus Nijhoff, Dordrecht, pp. 351—368.
Ward, J. C., 1964, Turbulent flow in porous media, *Proc. Amer. Soc. Civil Engs.* **90** (HY5), 1—12.
Webster, A. G., 1955, *Partial Differential Equations of Mathematical Physics*, 2nd edn., Dover, New York.
Wilkinson, J. H., 1965, *The Algebraic Eigenvalue Problem*, Oxford University Press, London.
Wilson, J. L., and Sa da Costa, A., 1982, Finite element simulation of a saltwater—freshwater interface with interface toe tracking, *Water Resour. Res.* **18**, 1078—1080.
Wyckoff, R. D., and Botset, H. G., 1936, The flow of gas—liquid mixture through unconsolidated sands, *Physics* **7**, 325—345.
Zienkiewicz, O. C., 1977, *The Finite Element Method*, 3rd edn., McGraw-Hill, London.
Zijl, W., 1984, Finite element methods based on a transport velocity representation for groundwater motion, *Water Resour. Res.* **20**, 137—145.

# Problems

## Chapter 2

2.1. Water at 20°C flows through a vertical sand column of length 200 cm and cross-sectional area 200 cm$^2$. The depth of water in a reservoir above the sand column is 20 cm. The column drains into a reservoir in which the water table is located at a height $H$ above the column's bottom. The sand's grain diameter is 0.5 mm. The porosity of the sand is $n = 0.30$. The hydraulic conductivity is $K = 10$ m/day.

*Required*: (a) Is Darcy's law applicable? (b) What is the specific discharge through the column when the height of water in the lower reservoir, $H$, is equal to 10 cm? (c) What is the hydraulic gradient along the column? (d) How will the answers to (b) change when the water temperature is raised to 30°C.

*Ans.*: (a) yes, (b) $1.215 \cdot 10^{-4}$ m/s, (c) 1.05, (d) 30% higher.

2.2. In the column of Problem 2.1, draw the piezometric line along the column and determine the piezometric head and the pressure at a point 40 cm above the column's bottom, (a) for the data of Problem 2.1, (b) when the depth of the water in the upper reservoir is increased to 40 cm, (c) when the depth of the water in the lower reservoir is lowered from 10 cm to 5 cm above the bottom. (d) Draw a conclusion about the relation between direction of flow and pressure gradient.

*Ans.*: (a) 0.52 m, (b) 0.56 m, (c) 0.48 m.

2.3. The column of Prob. 2.1 in filled with sand of three kinds: bottom 30 cm with sand of $K = 15$ m/d $n = 0.35$, then 60 cm with sand of $K = 5$ m/d $n = 0.25$, and finally 110 cm with sand of $K = 10$ m/d, $n = 0.30$.

*Required*: (a) Specific discharge through column. (b) How long will it take for a labeled particle moving at average velocity to travel through the column. (c) Piezometric head and pressure at a point 40 cm above the column's bottom.

*Ans.*: (a) 8.4 m/d, (b) 1.67 h, (c) 27 cm.

2.4. Given a horizontal confined aquifer of constant thickness, $B = 50$ m, and porosity, $n = 0.25$. Flow in the aquifer is everywhere from East to West. Two observation wells are located a distance of 2000 m apart along the direction of the flow. The piezometric heads are +105 m at the eastern well and +100 m at the western one.

*Required*: (a) The discharge through the aquifer when the hydraulic conductivity is $K = 40$ m/d. (b) The same when $K$ varies between the two wells according to $K = 10 + 0.02x$, where $x$ (in m) $= 0$ at the eastern well. (c) The same as (a), but the aquifer's thickness varies from 20 m at the eastern well to 60 m at the western one. (d) The travel time for a labeled particle between the two wells in case (b).

*Ans.*: (a) 5 m$^3$/md, (b) 3.1 m$^3$/md, (c) 2.5 m$^3$/md, (d) 8064 days.

2.5. Using a tracer, the average flow velocity at a point in an aquifer was found to be 0.5 m/d. The aquifer's porosity is 0.25. The slope of the piezometric surface at this point was 0.002.

*Required*: Determine the hydraulic conductivity.

*Ans.*: 62.5 m/day.

2.6. A well of diameter 500 mm pumps 300 m³/h from a confined aquifer of thickness $B = 75$ m. No change in the piezometric head, +150 m, is observed at 200 m from the well. The aquifer's porosity is $n = 0.2$; its hydraulic conductivity is $K = 10$ m/d.
Required: (a) Determine the piezometric head at a point located at $r = 100$ m from the well. (b) Determine the travel time from $r = 100$ m to the pumping well. Hint: write Darcy's law for radial flow and integrate.
Ans.: (a) 149 m, (b) 65.4 days.

2.7. Given the piezometric heads in three observation wells located in a homogeneous confined aquifer of constant transmissivity $T = 5000$ m²/d.
Well A: $x = 0, y = 0, = +10.0$ m.
Well B: $x = 0, y = 300$ m, $= +8.4$ m.
Well C: $x = 200, y = 0, = +12.5$ m.
Required: (a) Draw the contours of the piezometric surface. (b) Determine the discharge through the aquifer (magnitude and direction) per unit width.
Ans.: (b) 68 m³/md, $-23°$ from x axis.

2.8. The equation of state for a slightly compressible fluid is (a) $\rho = \rho_0 \exp(\beta(p - p_0))$, or, (b) $\rho = \rho_0(1 + \beta(p - p_0))$, where $\rho_0$ is the density that corresponds to $p_0$ and $\beta$ is the coefficient of compressibility.
Required: (a) Hubbert's potential $\phi^* = \phi^*(z, p)$ and expressions for: (b) $\nabla \phi^*$ and (c) $\partial \phi^*/\partial t$.
Ans.:

(a) $\phi^* = \dfrac{1}{g\beta}\left(\dfrac{1}{\rho_0} - \dfrac{1}{\rho(p)}\right) + z,$  $\phi^* = \dfrac{\rho_0}{\beta} \ln(\rho(p)/\rho_0) + z,$

(b) $\nabla \phi^* = \dfrac{1}{\rho g} \nabla p + \nabla z,$

(c) $\dfrac{\partial \phi^*}{\partial t} = \dfrac{1}{\rho g} \dfrac{\partial p}{\partial t}.$

2.9. Repeat 2.7 when the aquifer is anisotropic with **K** (in m/day) given by $\begin{bmatrix} 30 & 8 \\ 8 & 10 \end{bmatrix}$, and the aquifer thickness is $B = 50$ m.
Ans.: 16.8 m/md; $-8°$ from x axis.

2.10. For K of Problem 2.9, determine the principal values and principal directions. Hint: use $\tan 2\theta = 2K_{xy}/(K_{xx} - K_{yy})$

$$\dfrac{K_{x'x'}}{K_{y'y'}} = \dfrac{K_{xx} + K_{yy}}{2} \pm \left[\left(\dfrac{K_{xx} + K_{yy}}{2}\right)^2 - K_{xy}^2\right]^{1/2}$$

Ans.: $K_{x'x'} = 27.55$ m/d, $K_{y'y'} = 12.45$, $\theta = 19.33°$.

2.11. Let $K_x = 36$ m/d and $K_y = 16$ m/d be the principal values of **K** in an anisotropic aquifer, in the x and y directions, respectively, in two-dimensional flow. The hydraulic gradient is 0.004 in a direction making an angle 30° with the +x axis.
Required: (a) Determine **q**. (b) What is the angle between **J** and **q**.
Ans.: (a) $(q_x, q_y) = (0.1247, 0.032)$ m/d, (b) 15.6°.

2.12. A horizontal confined aquifer is made up of three layers: 20 m of $K = 20$ m/d, 30 m of $K = 40$ m/d and 10 m of $K = 50$ m/d. What is the aquifer's transmissivity.
  *Ans.*: 2100 m²/d.


2.13. A leaky-confined aquifer of thickness 20 m (and impervious bottom at $-28$ m) is overlain by a leaky phreatic aquifer, with a water table at $+3$ m. The aquitard separating them has a thickness of 1 m and hydraulic conductivity of 0.1 m/d. The leaky confined aquifer is at a constant head $+15$. The hydraulic conductivity of the upper leaky phreatic aquifer is $K = 20$ m/d.
  *Required*: Determine the leakage rate. *Hint*: assume that the flow in the leaky phreatic aquifer is vertical only, caused by the leakage.
  *Ans.*: 1.2 m/d.


2.14. Two observation wells in the direction of flow in a phreatic aquifer show water levels at $+30$ and $+20$ m above a horizontal impervious bottom at elevation $-10$ m. The distance between the wells is 1200 m.
  *Required*: (a) Determine the rate of the flow in the aquifer for $K = 18$ m/d. (b) Repeat (a) if $K = 10$ m/d for $x = 0$ to $x = 400$ m from the well at $+20$ m, while $K = 30$ m/d for the remaining distance.
  *Ans.*: (a) 5.25 m³/md, (b) 5.25 m³/md.


2.15. Given a cross-section of a homogeneous aquifer between two rivers. The distance between the two rivers is 3000 m. Water levels in the right hand river and in the left hand one are $+25$ and $+35$ m, respectively, above a horizontal impervious bottom at an elevation of $-20$ m. In each river, the depth of the water is approximately 5 m. The hydraulic conductivity is $K = 20$ m/d.
  *Required*: (a) The flow to each river. (b) The elevation of the water table midway between the rivers. (c) The location and elevation of the peak of the phreatic surface. (d) What has to be the rate of water withdrawal from a horizontal gallery located at $x = 1000$ m from the left river, in order to keep it aerated. The elevation of the (aired) gallery is $+15$ m. *Hint*: Use (2.3.9).
  *Ans.*: (a) 3.34 m³/md, (b) 50.25 m, (c) $x = 0$, (d) 22 m³/md.


2.16. In the phreatic aquifer of the Problem 2.15, a horizontal thin impervious clay layer at an elevation of $+5$ starts 600 m from the left river and ends at a distance of 1600 m from that river.
  *Required*: The same as in Problem 2.15.
  *Ans.*: (a) The impervious layer has no effect. (b) As in Problem 2.15.


2.17. A well pumps from a phreatic aquifer. Initially, the water table is at $+30$ m. The horizontal impervious bottom is at $+10$. The aquifer's hydraulic conductivity is $K = 10$ m/d.
  *Required*: Determine the rate of pumping, when the water level in the well is maintained at $+28$ m and an observation well at 200 m from the pumping well shows no drop in the water table elevation. The radius of the well is 0.5 m. (*Hint*: Write (2.3.9) in radial coordinates and integrate.)
  *Ans.*: 25.4 m³/h.


2.18. The same as Problem 2.17, but upon completion of drilling, various techniques were used to increase the hydraulic conductivity in the vicinity of the well, causing the hydraulic conductivity up to 8 m from the well to rise to 30 m/d.
  *Ans.*: 36.6 m/d.

2.19. The same well as in Problem 2.17 is used to artificially recharge the aquifer. As a result, a thin layer (skin) of partly clogged porous medium developes around the well and the rate of recharge under a build-up of 2 m reduces to 60% of the rate of pumping that produces a drawdown of 2 m at the well.
  *Required*: Determine the hydraulic resistance, $c$, of the semipervious skin around the well.
  *Ans.*: $c = 17.7$ days.

## Chapter 3

3.1. In order to prevent seawater intrusion into a confined coastal aquifer, the piezometric surface in the aquifer has to be raised by 8 m over an area of 2 km by 20 km along the coast. The aquifer's storativity is $S = 5 \times 10^{-5}$.
  *Required*: Determine the volume of water that should be injected for this purpose through an array of wells parallel to the coast.
  *Ans.*: $16\,000$ m$^3$.

3.2. The water table of a phreatic aquifer drops 4.5 m over an area of 4 km$^2$. The aquifer's porosity is 0.3 and its specific retention is 0.1.
  *Required*: Determine (a) The aquifer's specific yield. (b) The change in volume of water in the aquifer.
  *Ans.*: (a) 0.2, (b) $9.6 \times 10^6$ m$^3$.

3.3. Water flows upward through a column of sand of length 2 m. The sand ($n = 0.3$, $\gamma_s = 2.6$ gr/cm$^3$) is overlain by a layer of water of thickness 0.5 m. A piezometer inserted into the midpoint of the column indicates a water level 1 m above the sand's top.
  *Required*: (a) Determine the total stress, the effective stress and the pressure head distributions along the column. (b) The values at the midpoint of the column.
  *Ans.*: (b) $\sigma = 2.62$ t/m$^2$, $\sigma' = 0.62$ t/m$^2$, $p_w/\gamma_w = 2.0$ m.

3.4. A confined sand layer of thickness 5 m is overlain by 6 m of clay (specific weight 1.8 t/m$^3$). The piezometric head in the sand is $+5$ m above the sand's upper surface. The clay layer is excavated to a depth of $h$ below ground surface.
  *Required*: Determine the maximum value of $h$ that will not produce 'boiling' (i.e., determine $h$ for $\sigma' = 0$).
  *Ans.*: 3.22 m.

3.5. Rewrite the mass conservation equation (3.3.1) without sources, assuming (a) that the solid matrix is nondeformable, (b) that the fluid is homogeneous and incompressible and that the flow is under isothermal conditions, (c) that both (a) and (b) are valid, (d) that the flow is steady, (e) that $|\mathbf{q} \cdot \nabla \rho| \ll |n \partial \rho/\partial t|$.
  *Ans.*:

(a) $-\nabla \cdot \rho \mathbf{q} = n\, \partial \rho/\partial t$,
(b) $-\nabla \cdot \mathbf{q} = \partial n/\partial t$,
(c) $\nabla \cdot \mathbf{q} = 0$,
(d) $\nabla \rho \mathbf{q} = 0$,
(e) $-\rho \nabla \cdot q = \dfrac{\partial}{\partial t}(\rho n)$.

3.6. Start from the mass balance equation (3.3.22), and assuming that the fluid is homogeneous and $S_0$ is constant, rewrite it in the $xy$ coordinate system for the cases (a) the porous medium is inhomogeneous and anisotropic, (b) same as (a), but $x$, $y$ are principal directions.
Ans.:

(a) $\dfrac{\partial}{\partial x}\left(K_{xx}\dfrac{\partial \phi^*}{\partial x} + K_{xy}\dfrac{\partial \phi^*}{\partial y}\right) + \dfrac{\partial}{\partial y}\left(K_{yx}\dfrac{\partial \phi^*}{\partial x} + K_{yy}\dfrac{\partial \phi^*}{\partial y}\right) = S_0 \dfrac{\partial \phi^*}{\partial t}$,

(b) $\dfrac{\partial}{\partial x}\left(K_{xx}\dfrac{\partial \phi^*}{\partial x}\right) + \dfrac{\partial}{\partial y}\left(K_{yy}\dfrac{\partial \phi^*}{\partial y}\right) = S_0 \dfrac{\partial \phi^*}{\partial t}$.

3.7. Repeat Problem 3.6 but write the equation in cylindrical coordinates $r$, $\theta$, $z$, for (a) the porous medium is inhomogeneous and isotropic, (b) the porous medium is homogeneous and anisotropic, with $K = K_x = K_y \neq K_z$.
Ans.:

(a) $\dfrac{1}{r}\dfrac{\partial}{\partial r}\left(rK\dfrac{\partial \phi^*}{\partial r}\right) + \dfrac{1}{r^2}\dfrac{\partial}{\partial \theta}\left(K \cdot \dfrac{\partial \phi^*}{\partial \theta}\right) + \dfrac{\partial}{\partial z}\left(K\dfrac{\partial \phi^*}{\partial z}\right) = S_0 \dfrac{\partial \phi^*}{\partial t}$,

(b) $K\left[\dfrac{1}{r}\dfrac{\partial}{\partial r}\left(r\dfrac{\partial \phi^*}{\partial r}\right) + \dfrac{1}{r^2}\dfrac{\partial}{\partial \theta}\dfrac{\partial \phi^*}{\partial \theta}\right] + K_z\dfrac{\partial^2 \phi^*}{\partial z^2} = S_0 \dfrac{\partial \phi^*}{\partial t}$.

3.8. Rewrite the Laplace equation (3.3.28) in (a) cylindrical coordinates, $r$, $\theta$, $z$, (b) in spherical coordinates, $r$, $\theta$, $\phi$.
Ans.:

(a) $\dfrac{1}{r}\dfrac{\partial}{\partial r}\left(r\dfrac{\partial \phi^*}{\partial r}\right) + \dfrac{1}{r^2}\dfrac{\partial^2 \phi^*}{\partial \theta^2} + \dfrac{\partial^2 \phi^*}{\partial z^2} = 0$,

(b) $\dfrac{1}{r^2}\dfrac{\partial}{\partial r}\left(r^2\dfrac{\partial \phi^*}{\partial r}\right) + \dfrac{1}{r^2 \sin \theta}\dfrac{\partial}{\partial \theta}\left(\sin \theta \dfrac{\partial \phi^*}{\partial \theta}\right) + \dfrac{1}{r^2 \sin^2 \theta}\dfrac{\partial^2 \phi^*}{\partial \rho^2}$.

3.9. A segment of the boundary of an aquifer domain in the $xy$-plane has the shape of an ellipse $x^2/a^2 + y^2/b^2 = 1$.
*Required*: For an equipotential boundary, determine the direction of the specific discharge vector $\mathbf{q}$, (a) if the aquifer is isotropic, and (b) if the aquifer is anisotropic, with $T_x/T_y = 5$ ($x$ and $y$ principal directions).
Ans.: (a) $\tan \theta = q_y/q_x = y\,a^2/xb^2$, (b) $\tan \theta = y\,a^2/5xb^2$.

3.10. The boundary of Problem 3.9 is impervious.
*Required*: Write the boundary conditions.
Ans.:

(a) $\dfrac{x}{a^2}\dfrac{\partial \phi}{\partial x} + \dfrac{y}{b^2}\dfrac{\partial \phi}{\partial y} = 0$,

(b) $\dfrac{5x}{a^2}\dfrac{\partial \phi}{\partial x} + \dfrac{y}{b^2}\dfrac{\partial \phi}{\partial y} = 0$.

3.11. Determine the angle of refraction of a streamline upon passage through the boundary between two isotropic porous media, when the angle of incidence is 30° (a) when $K_1 = 10\,K_2$, (b) $K_2 = 10\,K_1$.
Ans.: (a) 3.3°, (b) 80°.

3.12. Repeat Problem 3.11 for an interface that is (a) parallel to the y-axis (x and y are principal directions), and (b) parallel to the x axis, when $K_{x_1} = 6 K_{y_1}$, $K_{x_2} = 10\, K_{y_2}$, $K_{x_1} = 5\, K_{x_2}$.
Ans.: (a) $\tan 30°/\tan \theta = 50/6$, (b) $\tan 30°/\tan \theta = 5$.

3.13. Write the complete mathematical model that describes unsteady flow in the cross-section shown in Figure 3.4a, (a) when the cross-section is isotropic, (b) when the cross-section is anisotropic with $x$, $y$ principal directions.
Ans.: (a) The variable is $\phi = \phi(x, z, t)$. The balance equation is

$$\partial(K\, \partial \varphi/\partial x)/\partial x + \partial(K\, \partial \varphi z)/\partial z = S_0\, \partial \phi/\partial t\ (\approx 0).$$

Boundary conditions: on AB: $\phi = H_1(t)$, on BC and DE: $\phi = z$, on EF: $\phi = H_2(t)$ on FG: $-K\, \partial \phi/\partial x = (\phi - H_2(t))/c$, on CD:

$$K\{(\partial \phi/\partial x)^2 + (\partial \phi/\partial z)^2 - \partial \phi/\partial z\} - N(\partial \phi/\partial z - 1) = S_y\, \partial \phi/\partial t,$$

(b) The balance equation is: $\partial(K_x\, \partial \phi/\partial x)/\partial x + \partial(K_z\, \partial \phi/\partial z)/\partial z = S_0\, \partial \phi/\partial t\ (\approx 0)$. Boundary conditions: on AB: $\phi = H_1(t)$, on BC and DE: $\phi = z$, on EF: $\phi = H_2(t)$, on FG: $-K_x\, \partial h/\partial x = (\phi - H_2(t))/c$, on CD:

$$K_x \left(\frac{\partial \phi}{\partial x}\right)^2 + K_z \left( \left(\frac{\partial \varphi}{\partial z}\right)^2 - \frac{\partial \phi}{\partial z} \right) - N(\partial \varphi/\partial z - 1) = S_y\, \partial \phi/\partial t.$$

Initial conditions for (a) and (b): $\phi(x, z, 0) = \phi_0(x, z)$.

3.14. A very long trench of width $w$, is used for artificially recharging an underlying homogeneous isotropic aquifer at a rate $I = I(t)$. The aquifer's bottom is at an elevation $\eta = \eta(x)$ above some datum level. Initially, the water table in the aquifer is at elevation $h = h(x, 0) = h_0(x)$ above the same datum level. Assume that at a large distance, $L$, water levels are not affected by the recharge and are therefore known values of space and time.
Required: Give the complete model for flow in the saturated zone. Hint: assume $h \ll L$ and, hence, essentially horizontal flow.
Ans.: (a) The variable is $h = h(x, t)$ (height of the water level). The balance equation is:

$$\partial(K(h - \eta)\, \partial h/\partial x)/\partial x + I(x, t) = S_y\, \partial h/\partial t.$$

$I(x, t) = 0$ except between some $x = x_0$, $x = x_0 + w$. Boundary conditions: $x = -L$, $h = h_0(-L)$; $x = L$; $h = h_0(L)$.

3.15. Repeat Problem 3.14 for a circular pond (of radius $R$) and write the model in radial coordinates.
Ans.: (a) The variable is $h(r)$. The balance equation is

$$\frac{1}{r}\frac{\partial}{\partial r}\left\{ K(h - \eta)\frac{\partial h}{\partial r} \right\} + I(r, t);\ I = 0\ \text{for}\ r > R.$$

Boundary conditions: $r = 0$, $\partial h/\partial r = 0$; $r = L$, $h = h_0$. Initial conditions: $h(r, 0) = h_0(r)$. Hint: Discontinuity at $r = R$.

3.16. A well of radius $r$ is pumping water at a constant rate, $Q_w$, from an infinite phreatic aquifer with a horizontal bottom. The aquifer is anisotropic, with $K_x = K_y \neq K_z$.
Required: Write the complete exact (i.e., three-dimensional) model for the piezometric head distribution in the saturated zone, assuming that the flow domain is bounded by the water table.

3.17. A pond of extended area is used to artificially recharge an underlying phreatic aquifer. Water in the pond is at a height $H$ above pond's bottom. The phreatic surface is very deep. Assume that behind the downward advancing wetting front we have complete saturation, and neglect elastic storativity.

*Required*: (a) Write the complete mathematical model that is required in order to determine the depth, $\zeta = \zeta(t)$, of the advancing front below the pond's bottom. (b) For $H = 2$ m, $n = 0.4$ and $K = 3$ m/d, determine the value of $\zeta$ at after 3 days. (c) Repeat (a) and (b) if $K = 3$ m/d up to 1 m below the pond and then $K = 1.5$ m/d. (d) Repeat (a) and (b) if due to clogging, a very thin layer of semipervious material of resistance $c = 100$ days has developed on the pond's bottom.

Ans.:

(a) $\dfrac{\partial}{\partial z}\left(K\,\dfrac{\partial \phi(z,t)}{\partial z}\right) = 0;\qquad \phi(0,t) = H;\ \phi(-\zeta,t) = -\zeta;$

$$\dfrac{d\zeta}{dt} = V = -\dfrac{K}{n_e}\dfrac{\partial \phi}{\partial z},$$

(b) 27.9 m, (c) 15.5 m, (d) 0.155 m.

# Chapter 4

4.1. A system of aquifers consists of the following sequence of strata: impervious bottom at 100 m below ground surface, an aquifer of thickness 50 m, above the bottom, a silty sand semipervious layer of 3 m, and a sandy phreatic aquifer with the water table in the range 30–40 m below ground surface. The hydraulic conductivities of these layers are 50 m/d, 0.1 m/d, and 20 m/d. Water is pumped from the lower aquifer through a number of wells. The phreatic aquifer is recharged by rain and artificially through a recharge pond.

*Required*: Write the complete model that is needed in order to determine the future water levels in the phreatic aquifer.

Ans.: Problem variables: for phreatic aquifer: $h = h(x,y,t)$. For confined aquifer: $\varphi = \varphi(x,y,t)$, both from the same datum level. $Q_w$ — Pumping well, $Q_p$ — Infiltration through recharge pond, $N$ — Rain.

PDE — Confined Aq.

$$\dfrac{\partial}{\partial x}\left(T\,\dfrac{\partial \varphi}{\partial x}\right) + \dfrac{\partial}{\partial y}\left(T\,\dfrac{\partial \varphi}{\partial y}\right) - \dfrac{\varphi - h}{B_2/K_2} - \sum_{(i)} Q_{wi}(\mathbf{x}_i, t)\,\delta(\mathbf{x} - \mathbf{x}_i)$$

$$= S_1\,\dfrac{\partial \varphi}{\partial t}.$$

PDE — Phreatic

$$K\left\{\dfrac{\partial}{\partial x}\left((h - B_1 - B_2)\,\dfrac{\partial h}{\partial x}\right) + (h - B_1 - B_2)\,\dfrac{\partial}{\partial y}\left(\dfrac{\partial h}{\partial y}\right)\right\} + N + \dfrac{\varphi - h}{B_2/K_2} + Q_p(x, y, t)$$

$$= S_y\,\dfrac{\partial h}{\partial t}.$$

Add initial and boundary conditions.

4.2. Given the aquifer shown in Figure P.4.2; City A gets water ($16 \times 10^6$ m$^3$/yr) from wells B and from the river. Recently, the water in the river has been polluted and the city has to replace the river water by additional well water, either at location B or elsewhere.

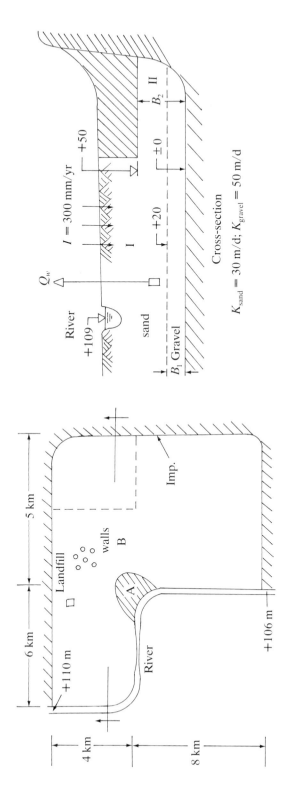

Fig. P.4.2.

*Required*: Construct the mathematical model that will help to determine the location of the additional wells and then the permissible rate of pumping, such that there will be no induced recharge of (polluted) river water into the aquifer.

*Ans.*: Variables: $h = h(x, y, t)$ of phreatic aquifer (I); $\varphi = \varphi(x, y, t)$ for the confined part of aquifer (II). Assume $\varphi > B_2$.

PDE — zone I: $\dfrac{\partial}{\partial x}\left\{[K_g B_1 + K_s(h - B_1)]\dfrac{\partial h}{\partial x}\right\} + \dfrac{\partial}{\partial y}\left\{[K_g B_1 + K_s(h - B_1)]\dfrac{\partial h}{\partial y}\right\} + I -$

$\displaystyle\sum_{(i)} Q_{w_i}(x_i, y_i, t)\,\delta(x - x_i, y - y_i) = S_y \dfrac{\partial h}{\partial t},$

where $B_1 = 20$ m. For confined zone II:

$$\dfrac{\partial}{\partial x}\left(T\dfrac{\partial \varphi}{\partial x}\right) + \dfrac{\partial}{\partial y}\left(T\dfrac{\partial \varphi}{\partial y}\right) = S\dfrac{\partial \varphi}{\partial t},$$

where $B_2 = 50$ m, $T = B_1 K_g + (B_2 - B_1)K_s$. First kind BC along river. Second kind BC along impervious periphery.

4.3. Given an aquifer with the cross-section shown in Figure P.4.3.

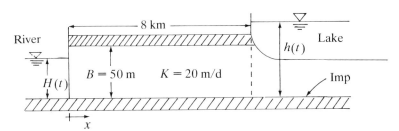

Fig. P.4.3.

*Required*: (a) Write the model that will provide information on the rate of flow from the lake to the river, as the water level fluctuates in the lake. (b) For $h = 70$ m, determine the rate of flow to the river and the elevation of the water table or piezometric head at $x = 1000$ m.

*Ans.*: The answer depends on whether $H(t) < 50$ (phreatic flow), or $H(t) > 50$ (partly confined, partly phreatic). (a) For $H(t) > 50$ m. For phreatic part, $0 < x < L_p$: $\partial(Kh\,\partial h/\partial x) = S_y\,\partial h/\partial t$. For the confined part, $L_p < x < L$; $\partial(T\,\partial\varphi/\partial x) = 0$ (neglecting elastic storage). BC: $x = 0$ $h = H(t)$; $x = L_p$ $h(t) = \varphi(t) = B$; $x = L$, $\varphi = h(t)$. (b) For $H(t) = 40$ m, $Q = 3.62$ m³/m/d; $x = 1000$ m, $\varphi = 44.3$ m.

4.4. Data of Problem 2.15. Natural replenishment occurs over the aquifer strip at a rate $I = I(x, t)$.

*Required*: (a) Write the equation of balance in terms of $h = h(x, t)$. (b) Determine the shape $h = h(x)$ for steady flow with constant $I$, (c) Write the equation of balance as in (a), but with a bottom of variable elevation, $\eta = \eta(x)$.

Ans.:

(a) $K \dfrac{\partial}{\partial x}\left(h \dfrac{\partial h}{\partial x}\right) + I(x, t) = S_y \dfrac{\partial h}{\partial t}$;

$x = 0, h = h_0 = 55$ m, $x = L = 3000$ m, $h = 45$ m.

(b) $h^2(x) = h_0^2 + \dfrac{I}{K} x(L - x) + \dfrac{x}{L}(h_L^2 - h_0^2)$,

(c) $K \dfrac{\partial}{\partial x}\left\{(h - \eta) \dfrac{\partial h}{\partial x}\right\} + I(x, t) = S_y \dfrac{\partial h}{\partial t}$.

4.5. Repeat Problem 4.4 but $K_1 = 20$ m/d from $x = 0$ to $x = 1000$, measured from the left river and $K_2 = 10$ m/d for the remaining distance.
Ans.:

$0 \leqslant x \leqslant 1000$ m, $K_1 \dfrac{\partial}{\partial x}\left(h_1 \dfrac{\partial h_1}{\partial x}\right) + I(x, t) = S_y \dfrac{\partial h_1}{\partial t}$;

$x = 0, h_1 = h_0 = 55$ m; $x = 1000$ m, $h_1 = h_2$, $K_1 \partial h_1/\partial x = K_2 \partial h_2/\partial x$,

$1000 < x \leqslant 3000$,

$K_2 \dfrac{\partial}{\partial x}\left(h_2 \dfrac{\partial h_2}{\partial x}\right) + I(x, t) = S_y \dfrac{\partial h_2}{\partial t}$;    $x = 3000$ m, $h_2 = 45$ m.

4.6. Repeat Problem 4.4a and b, but add an aerated gallery at $+15$ m midway between the two rivers. $I = 0.5$ m/year.
Ans.:

(a) $0 \leqslant x < 1500$ m; $K \partial(h \partial h/\partial x)/\partial x + I(x, t) = S_y \partial h/\partial t$,
$x = 0, h = h_0, x = 1500$ m, $h = 35$ m;
$1500$ m $\leqslant x \leqslant 3000$ m; same equation; $x = 1500$ m, $h = 35$ m, $x = 3000$ m, $h = h_L$,

(b) $x \leqslant 1500$ m, $h^2(x) = (30\,250 - 9.9x - 0.00137x^2)/10$;
$x > 1500$ m, $h^2(x) = (1163 + 8.42x - 0.000685x^2)/10$.

4.7. A 5 m deep trench of width 200 m is used to recharge a perched horizon located 50 m below ground surface. The perched horizon is formed on an extended thin silty sand layer of resistance $c = 20$ d. The semipervious layer is aerated from below.
*Required*: (a) Construct the mathematical model that will help to determine the steady elevations of the water table of the perched horizon. (b) What is the maximum infiltration rate that can be recharged if the water table in the trench is 2 m below ground surface.
Ans.: (a) For $h_{max} < 45$ m, p.d.e. is $K \partial[h \partial h/\partial x] + I - h/c = 0$; $I \neq 0$ for $0 \leqslant x < 200$. BC: $x = 0$, $\partial h/\partial x = 0$; $x = x_{max}$, $h = 0$. For $h_{max} \geqslant 45$, p.d.e. $K \partial[h \partial h/\partial x] - h/c = 0$; $x = 200$, $h = 45$ m; $x = x_{max}$, $h = 0$.

4.8. Figure P.4.8 shows a cross-section through a seasonal water storage reservoir. A system of parallel drains is required in order to prevent flooding of the area adjacent to the reservoir when the latter is full.

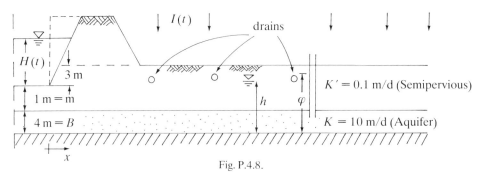

Fig. P.4.8.

*Required*: Construct a model that will help design the drainage network. (*Hint*: The problem can be simplified by assuming horizontal flow in the aquifer and vertical flow only in the overlying semi-pervious material.)

*Ans.*: Variables: $\varphi(x, t)$ in aquifer, $h(x, t)$ in semipervious layer. The p.d.e.'s without drains.

$$\partial(T\, \partial\varphi/\partial x)/\partial x - q_v = S\, \partial\varphi/\partial t \approx 0;\ x = 0,$$
$$x = 0,\ \varphi = m + B + H(t),\ x \to \infty,\ \varphi = \varphi_0.\ q_v = (\varphi - h)/[(h - B)/K']$$
$$I + q_v = S_y\, \partial h/\partial t,\ x = 0,\ h = m + B + H(t);\ x \to \infty,\ h = \varphi_0,$$

With drains: fixed drain elevations serve as boundary conditions.

4.9. Figure P.4.9 shows a map and a cross-section of an aquifer. Two well fields operate in

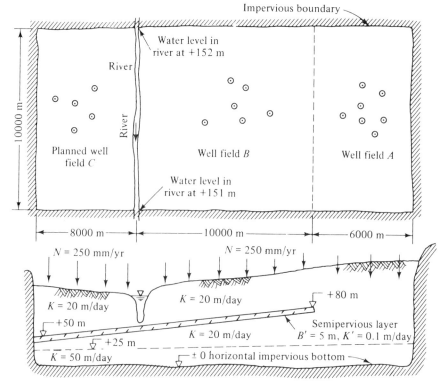

Fig. P.4.9.

PROBLEMS 397

the aquifer. Well field A pumping $10 \times 10^6$ m$^3$/yr and well field B pumping $16 \times 10^6$ m$^3$/yr. All screens of well field B are below the semipervious layer. A new well field, C, is planned for an annual pumping of $15 \times 10^6$ m$^3$/yr and if posisble, up to $25 \times 10^6$ m$^3$/yr.

*Required*: How would you determine the feasibility of the pumping rates proposed for well field C. Determine the methodology of solving the problem. The river itself cannot serve as a source for water, as it is heavily contaminated. Add any further information required to reach a conclusion.

## Chapter 5

5.1. Determine the height above the phreatic surface from which a sample was taken, if a laboratory test of the sample gave the result: $S_w = 0.35$. Use the retention curve:

| $S_w$ % | 100 | 80 | 60 | 40 | 30 | 20 |
|---|---|---|---|---|---|---|
| $h_c$ cm | 0 | 70 | 80 | 90 | 100 | 115 |

$S_{w0} = 16\%$
   Ans.: 95 cm.

5.2. Write the equation of unsaturated flow of water in the *xz*-vertical plane: (a) in terms of pressure, (b) in terms of $S_w$, for a homogeneous isotropic nondeformable soil, under isothermal conditions. (c) What information is required to obtain a solution for a specific soil.

5.3. The downward specific discharge in the unsaturated zone is a constant $10^{-3}$ cm/s, and the unsaturated hydraulic conductivity is given by $K = K_0 \exp(-a\psi)$, where $K_0 = 10^{-2}$ cm/s and $a = 10^{-2}$ cm$^{-1}$.
   *Required*: Determine the distribution of the tension, along the vertical, if the water table is at a depth of 15 m.

5.4. Use the retention curve of Problem 5.1 together with the effective permeability curve of Figure 5.13 to determine the moisture diffusivity curve $D(S_w)$.

5.5. For a column of length $L$, write the complete statement of the model that will enable to predict the soil moisture along the column under the conditions: (a) A fixed water table at the lower boundary and a constant rate of infiltration at the top, and (b) when there is no infiltration, but evaporation may occur.

5.6. The column of Problem 5.5 is made of two layers: an upper layer $(L_1, K_1)$ and a lower one $(L_2, K_2)$. The depth of ponded water above the column is $H$.
   *Required*: (a) The flow rate, assuming full saturation everywhere. (b) For what values of $H$ will the flow become unsaturated. (*Hint*: air entry requires negative pressure of a value equal at least to the air entry value, $p_{crit}$.)
   Ans.: (b) For $p_{crit} = 0$, $H > L_1(K_2/K_1 - 1)$.

5.7. An extended horizontal impervious layer is located at a depth of 8 m below ground surface. The water table fluctuates between 3 and 0.5 m below ground surface. To prevent the water table from rising above permissible levels as a result of excessive rains, an array of parallel drains is planned.
   *Required*: Construct a model that will enable the planner to determine optimal (equal) spacing and depth of the drainage system for a quick removal of rain water (a) taking into account saturated flow only, (b) taking into account flow in both the saturated and the unsaturated zone. In both cases, assume no soil deformation and a homogeneous isotropic soil. List all your assumptions.

5.8. Write the complete model for flow of both air and water in the unsaturated zone, assuming that both fluids are compressible, but the soil is non-deformable.

## Chapter 6

6.1. For uniform flow at a velocity of $V = 2$ m/d that makes an angle of 30° with the $x$-axis in plane flow, and an isotropic aquifer with $a_L = 20$ m, $a_T = 1$ m, determine the components $D_{xx}$, $D_{xy}$, $D_{yy}$, etc. of the coefficient of mechanical dispersion.
 Ans.: $D_{xx} = 30.5$ m²/day, $D_{xy} = D_{yx} = 16.45$ m²/day, $D_{yy} = 11.5$ m²/day.

6.2. Given the following data for two-dimensional flow in a horizontal aquifer

| Observation well | A | B | C |
|---|---|---|---|
| $x$-coordinate | 0 | 0 | 400 |
| $y$-coordinate | 200 | 0 | 0 |
| piezometric head (m) | +21 | +20 | +22 |
| Cl-concentration (p.p.m) | 50 | 150 | 300 |

$K = 40$ m/d, $a_L = 30$ m, $a_T = 3$ m, $D_d^* = 10^{-4}$ cm²/sec, $n = 0.25$.

*Required*: Determine (a) the advective chloride flux at point B, (b) the dispersive flux, (c) the diffusive flux, and (d) the total flux of the chlorides in the aquifer (magnitude and direction).
 Ans.: (a) 42.3 gr/m²d; $\theta = -135°$, (b) (+0.16, +0.90) gr/m²d, (c) $(-0.81 \times 10^{-4}$, $1.08 \times 10^{-4})$ gr/m²d.

6.3. Write the equation of hydrodynamic dispersion for two-dimensional flow in the $xy$-plane, in the absence of sources and sinks, but for a radioactive pollutant that obeys a linear adsorbtion isotherm and undergoes decay.

6.4. Write the complete mathematical model for an ideal tracer that moves in a one dimensional horizontal porous medium column. On the left-hand side, the column is connected to a well mixed reservoir in which the water is maintained at a constant level and a constant concentration $c_0$. On the right-hand side, the column is open to the atmosphere. Initially, the tracer concentration in the column is $c = 0$. The objective of the model is to predict the water concentration at a point in the middle of the column.

6.5. Construct the one-dimensional vertical model that is required in order to determine the movement of a pollutant from a lagoon to an underlying phreatic aquifer. List your assumptions and required data.

6.6. Write the complete mathematical model for a vertical column made up of two porous medium layers: fine sand and gravel. The flow through the column takes place under conditions of full saturation. Water is ponded above the column at a constant depth, $H$. At the lower end, the water drains into the atmosphere. Initially, the water in the column has a concentration $c$ of some inert tracer. At $t = 0$, flow begins, with the water in the pond at zero tracer concentration. The objective of the model is to determine the speed of flushing the tracer from the column. (*Hint*: Divide domain into zones of different porous medium properties and write a complete model for each zone. Conditions on the boundary between zones are equality of concentrations and of the tracer's total flux.)

6.7. Figure P.6.7 shows a map and a cross-section of a liquid waste disposal lagoon of

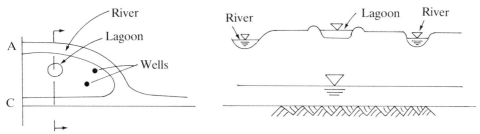

Fig. P.6.7.

radius 100 m located above a phreatic aquifer in which flow takes place. The elevations of the aquifer's bottom range from +5 to +10 m above some datum level. Water table elevations vary in the range of +25 and +30 m. The lagoon's bottom is at +35 m. Water, containing toxic material infiltrates through the lagoon's bottom, moves through the unsaturated zone, and reaches the aquifer. A plume has developed that endangers the wells used for water supply. The toxic material is absorbed to the clay fraction in the soil.

*Required*: Construct a conceptual model (= a list of simplifying assumptions) and a complete mathematical model that can be used to predict the movement of the plume in the saturated zone. The model will be used for determining the location of warning wells and pumping wells that will be used to intercept the plume and prevent it from reaching the pumping well. List all the information required in order to obtain numerical results. Assume that there is enough vertical mixing to justify a two-dimensional approach in the aquifer.

6.8. Complete the model constructed in Problem 4.2 so that it can be used to predict the movement of leachate from the landfill to Well Field B and/or to the river.

6.9. The approximate solution for $c(x, t)$ that satisfies the one-dimensional dispersion equation in the semi-infinite column, $x > 0$, is:

$$\frac{\partial c}{\partial t} = \frac{D_h}{R_d} \frac{\partial^2 c}{\partial x^2} - \frac{q}{nR_d} \frac{\partial c}{\partial x},$$

with initial conditions $c(x, 0) = 0$ and boundary conditions $c(0, t) = c_0$, $c(\infty, t) = 0$, is

$$\frac{c}{c_0} = \frac{1}{2} \operatorname{erfc}\left(\frac{R_d x - (q/n)t}{2(R_d D_h t)^{1/2}}\right).$$

where $R_d$ is the retardation factor and $n$ is porosity.

*Required*: For $c_0 = 100$ ppm, $R_d = 20$, $q = 1$ m/d, $n = 0.25$, $a_L = 10$ m, determine the concentration distribution after 100 days and compare with the distribution in the absence of absorption (recalling that $R_d = 1 + (1 - n)\rho_s K_d/n$). Neglect molecular diffusion. Compare the results with those for movement in the absence of dispersion.

Ans.: $c$ in %.

| $x =$ | 5 | 10 | 20 | 40 | 100 | 200 | 400 | 500 |
|---|---|---|---|---|---|---|---|---|
| $R_d = 20$ | 77.33 | 69.14 | 50 | 15.87 | 0.00317 | 0 | 0.0 | 0.0 |
| $R_d = 1$ | 99.99 | 99.99 | 99.99 | 99.99 | 99.96 | 98.73 | 50 | 13.17 |
| $R = 1, D_h = 0$ | 100 | 100 | 100 | 100 | 100 | 100 | 100 | 0 |

## Chapter 7

**7.1.** Two observation wells, above and below an interface in a confined aquifer, show water levels of +3 m and +1.8 m above sea level.
 *Required*: At what depth is the interface below sea level ($\gamma_s = 1.03$ t/m$^3$).
 *Ans.*: 38.20 m below sea level.

**7.2.** At some distance from the coast, the discharge of fresh water to the sea is 1000 m$^3$/year/m. The steady interface is at a depth of 80 m and the phreatic surface is at +2 m (both above sea level). Determine the slopes of the phreatic surface and of the interface. Assume approximately horizontal flow with $K = 20$ m/d.
 *Ans.*: Interface Slope: 0.0668.

**7.3.** A very long phreatic aquifer strip ($K = 20$ m/d) of width $L = 3$ km is underlain by a horizontal impervious bottom that serves as a datum level. The strip is bounded on both sides by sea water ($\gamma_s = 1.03$ t/m$^3$) reservoirs (down to the impervious bottom) with water levels at $H_s = 50$ m. The aquifer is recharged uniformly at a rate $I$. Assume that the flow is everywhere always horizontal.
 *Required*: (a) Find the steady shapes of the phreatic surface and of the interface. (b) Under what conditions will the fresh water lens reach the aquifer's bottom. (c) Write the general unsteady model which can be used for the determining the time dependent shapes of the phreatic surface and of the interface, when $I = I(x, t)$.
 *Ans.*: $\zeta$ denotes depth of interface.

(a) $\zeta^2 = (Lx - x^2) \dfrac{\delta^2 I}{(1 + \delta)K_f}$  for an interface with $\zeta_{max} \leqslant H_s$.

$h_f = \zeta/\delta$ for the phreatic aquifer's shape,

(b) $I \geqslant 250$ mm/yr,

(c) $\dfrac{\partial}{\partial x}\left[K_f(\zeta + \phi_f)\dfrac{\partial \phi_f}{\partial x}\right] + I = S_y \dfrac{\partial \phi_f}{\partial t} + n_e \dfrac{\partial \zeta}{\partial t};$

$\dfrac{\partial}{\partial x}\left[K_s(H_s - \zeta)\dfrac{\partial}{\partial x}\left(\dfrac{\delta}{1+\delta}\phi_f + \dfrac{\zeta}{1+\delta}\right)\right] = -n_e \dfrac{\partial \zeta}{\partial t}.$

$x = 0$, $\phi_f = 0$; $\zeta = 0$; $x = L/2$ $\partial\phi_f/\partial x = 0$, $\partial\zeta/\partial x = 0$. Results of (a) can be used as.

**7.4.** Repeat Problem 7.2 for a circular island of radius $R = 1500$ m surrounded by sea water ($\gamma_s = 1.03$ t/m$^3$).
 *Ans.*:

(a) $\zeta^2 = \dfrac{I\delta^2}{2(\delta + 1)k_f}(R^2 - r^2)$ for the interface;

$\zeta = \delta h_f$ for the phreatic surface,

(b) $I \geqslant 500$ mm/yr.

(c) $\dfrac{1}{r}\dfrac{\partial}{\partial r}\left[K_f r(\zeta + \phi_f)\dfrac{\partial \phi_f}{\partial r}\right] + I = S_y \dfrac{\partial \phi_f}{\partial t} + n_e \dfrac{\partial \zeta}{\partial t},$  $r = 0 \dfrac{\partial \phi_f}{\partial r} = \dfrac{\partial \zeta}{\partial r} = 0,$

$\dfrac{1}{r}\dfrac{\partial}{\partial r}\left[K_s(50 - \zeta)r\dfrac{\partial}{\partial r}\left(\dfrac{\delta}{1+\delta}\phi_f + \dfrac{\zeta}{1+\delta}\right)\right] = -n_e \dfrac{\partial \zeta}{\partial t},$

$r = R$, $\zeta = \phi_f = 0$,

$r = 0$, $\partial \phi_f/\partial r = \partial \zeta/\partial r = 0$.

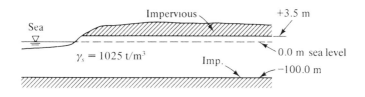

Fig. P.7.5.

7.5. The figure shows a cross-section of a coastal aquifer, perpendicular to the coast. The aquifer ($K = 20$ m/d) is overlain by an impervious clay layer. The piezometric head at a point 10 km from the coast is $+12$ m above sea level.

*Required*: (a) Determine the amount of freshwater flowing to the sea. (b) The length of seawater intrusion.

*Ans.*: (a) $Q = 2.22$ m³/md, (b) $L = 1154$ m.

7.6. The same data as in Problem 7.5, without the impervious clay, but with a uniform rate of natural replenishment ($I = 500$ mm/yr) and a phreatic surface of $+12$ m at 10 km, where aquifer ends.

*Required*: The length of seawater intrusion.

*Ans.*: $L_1 = 188$ m.

## Chapter 9

9.1. Run Program BV9-1 for the problem illustrated in Figure 9.3. Verify that the final solution, shown in that figure, is indeed obtained by the program. If possible, let the output be printed by a printer, using a slightly modified program (with LPRINT replacing PRINT in lines 460—480).

9.2. Run Program BV9-1 for the problem illustrated in Figure 9.4. Verify that the results obtained by the program are as given in Section 9.1. Again, if possible, let the output be printed by a printer, using a slightly modified program (with LPRINT replacing PRINT in lines 460—480).

9.3. Compare the results obtained in Problem 9.2 with the analytical solution for steady flow towards a well in a confined aquifer of radius $R$, $\phi = (Q/2\pi T) \ln(r/R)$, (Bear, 1979, p. 305; Verruijt, 1982, p. 17). Note that Figure 9.4 actually represents only one-quarter of the aquifer.

9.4. Investigate the stability of Program BV9-2 by running the program for the problem illustrated in Figure 9.5, entering a time step larger than the value suggested by the program.

9.5. Verify that Program BV9-3 is unconditionally stable, by running the program, entering a value for the first time step larger than the time step suggested by the program.

9.6. Verify the statement in Subsection 9.2.2 that Program BV9-3 can be used to determine the steady state solution of a problem in a single step, by taking the storativity $S = 0$, and using a time step of arbitrary, nonzero, length. Therefore, run Program BV9-3 for the problem defined in Subsection 9.2.1, and verify that the steady-state solution ($\phi = 0.073$ m) can indeed be obtained in one step. Also verify that Program BV9-2 does not have this feature.

9.7. Modify the boundary conditions used in Programs BV9-2, or BV9-3, to those of the case of a rectangular aquifer, with two parallel impermeable boundaries, for instance the upper and lower boundaries. Then use the modified program to run the problem solved by the finite element method in Program BV10-5 (see Section 10.5). Compare the results with the analytical results shown in Figure 10.11.

9.8. Write a modified version of Program BV9-3, using the interpolation parameter $\varepsilon = \frac{1}{2}$ (the Crank–Nicholson algorithm).

## Chapter 10

10.1. Run Programs BV10-1 or BV10-2 for the case of a nonhomogeneous transmissivity. Take $T = 1$ in the left half and $T = 9$ in the right half, and verify that the program correctly calculates the head in nodes 3 and 4.
 Ans.: $\phi = 1$.

10.2. Extend each of the Programs BV10-1–BV10-3 to an aquifer of anisotropic permeability, assuming that the $x$- and $y$-directions are the principal directions of the permeability tensor. Note that the terms representing flow in $x$- and $y$-direction appear separately in all equations, so that each of them can be given its own multiplication factor.

Fig. P.10.3.

10.3. Use Programs BV10-1 or BV10-2 to determine the distribution of the piezometric head, for the case of radial flow towards a well, in a completely confined aquifer. Because of symmetry only one-quarter of the total flow field needs to be modeled. Compare the results with the analytic solution $\phi = (Q/2\pi T)\ln(r/R)$, (Bear, 1979, p. 305; Verruijt, 1982, p. 17).

10.4. Use Program BV10-3 to determine the distribution of the piezometric head, for the case of radial flow towards a well, in a leaky confined aquifer. Use various values for the ratio $\lambda/R$, where $\lambda$ is the leakage factor ($\lambda = \sqrt{Tc}$), and $R$ is the external radius of the aquifer. Compare the results with the analytic solution for a well in an aquifer of infinite extent, $\phi = -(Q/2\pi T)K_0(r/\lambda)$, where $K_0(x)$ is the modified Bessel function of the second kind and order zero (Bear, 1979, p. 315; Verruijt, 1982, p. 22).

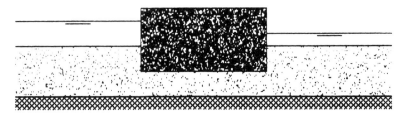

Fig. P.10.5.

# PROBLEMS

**10.5.** Consider the flow of groundwater through the soil underneath a hydraulic structure in a river or canal (Figure P.10.5). Determine the distribution of the piezometric head in this case, using Program BV11-2 or BV11-3.

**10.6.** Write a program (in BASIC) that will generate a datafile, to be stored on disk, containing the input data for a modified version of Program BV11-2. This modified version of Program BV11-2 should first read the data form the datafile, and then execute the finite element calculations.

**10.7.** Run Program BV10-4 for the (theoretical) case of a dam with vertical faces, such that the length is 1.62 m, the water level on the left side is 3.22, and the water level on the right side is 0.84 m. Compare the results with those obtained by an analytical solution (Muskat, 1937, p. 314).

**10.8.** Generalize Program BV10-4 to the case of a nonhomogeneous dam, such that the permeability in each of the sections in which the dam is divided may be different. This enables to study the free surface in a dam with a core of smaller permeability. Run the modified program for the case of a core, having a permeability one-tenth of the rest of the dam.

**10.9.** In Program BV10-4 the total discharge through the dam is not calculated. Note that the value in the right-hand side of the system of linear equations (10.2.2) expresses the discharge entering the aquifer at a certain node. Use this property to modify the program, such that it will also calculate the total discharge through the dam, by summing the discharges at the nodes on the outflow boundary. Verify that the same result may be obtained by summing the discharges on the inflow boundary.

**10.10.** Write a modified version of Program BV10-4, that can be used for the nonsteady flow through a dam, caused by a fluctuating water level on the left side boundary. (*Hint*: For this case it is most convenient to consider the phreatic surface as a boundary with prescribed head ($\phi = y$). The discharge to be supplied at each node to maintain the phreatic surface in this location can be calculated as the right-hand side ($Q$) of the system of linear equations (10.2.2). Because there is no such supply of water, the water level will drop by an amount $Q \, \Delta t / S \, \Delta x$, where $\Delta t$ is the time interval considered, $S$ is the storativity, and $\Delta x$ is the horizontal distance between the centers of the two neighboring boundary segments.)

**10.11.** Use Program BV10-5 to determine the distribution of the piezometric head, for the case of nonsteady flow towards a well, in a completely confined aquifer. Compare the results with the analytic solution (Theis, 1935) for a well in an aquifer of infinite extent, $\phi = -(Q/4\pi T)E_1(r^2 S/4Tt)$, where $E_1(x)$ is the exponential integral, (Bear, 1979, p. 320; Verruijt, 1982, p. 92).

**10.12.** In all programs the transmissivity data of the aquifers are considered to be given *a priori*. In reality for a phreatic aquifer the transmissivity is defined as the permeability multiplied by the depth of the water layer, which is equal to the head in the Dupuit approximation. This means that in such aquifers $T = k\phi$, so that the transmissivity actually depends upon the head $\phi$. This can be taken into account by an iterative procedure, in which the transmissivity is gradually adjusted. Introduce this possibility in the non-steady Program BV10-5, and investigate the influence on the final results, for instance for the case illustrated in Figure 10.11.

**10.13.** If there is no storage ($S = 0$) the nonsteady flow equation reduces to the equation for steady flow. Investigate if the procedure given for nonsteady flow in Section 10.5 includes steady flow as a special case. Verify that Program BV10-5 indeed gives the correct steady flow solution if $S = 0$. Explain why this is the case only for $\varepsilon = 1$.

## Chapter 11

11.1. Run Program BV11-1-2 with the datasets DATA11.1 and DATA11.2, in order to verify the generation of Figures 11.1 and 11.2.

11.2. Modify Program BV11-1-2, so that it uses the Euler method for the displacement in each step, rather than the Runge—Kutta method. Then run the program with the datasets DATA11.1 and DATA11.2, and compare the results with those obtained in Problem BV11-1.

11.3. Eliminate the statements in Program BV11-1-2 that check the position of a point on a streamline with reference to the wells, i.e., eliminate the use of the arrays $TA$ and $TB$, and eliminate the check on the radius $R$. Run the simplified program, and observe the behaviour of the streamlines near a well. Explain the singular behaviour of certain streamlines.

11.4. Use the method of images to create a flow system, for Program BV11-1-2, with a horizontal line of symmetry, acting as an impermeable boundary. (*Hint*: This can be accomplished by introducing imaginary wells of the same discharge, on the image side of the boundary.)

11.5. Use the method of images to create a flow system, for Program BV11-1-2, with a horizontal line of symmetry, acting as a boundary of constant head. (*Hint*: This can be accomplished by introducing imaginary wells of opposite discharge, on the image side of the boundary.)

11.6. Modify the dataset DATA11.2, such that markers will be set on the streamlines emanating from the well, and no markers on the streamlines coming from the left. (This can be accomplished by modifying the last values in the last 29 lines, from 0 to 1, and conversely). Then run the program to show the propagation of a front of injected water. Compare the results with the analytical results (Muskat, 1937, p. 474; Bear, 1979, p. 281).

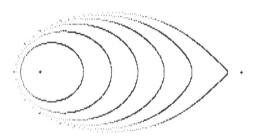

Fig. P.11.7.

11.7. In order to improve the graphical representation of a moving front, omit the drawing of the streamlines (in line 460) from Program BV11-1-2 increase the number of streamlines emanating from the well, and reduced the interval between markers. For the problem of a well and a recharge well in uniform flow a figure as shown in Figure P.11.7 should be drawn on the screen by such a modification of the program.

11.8. Run Program BV11-1-2, or the modified version mentioned in the previous problem, for the problem of a single well in uniform flow, showing the progress of the front of injected water. Compare the results with the analytical solution (Bear, 1979, p. 282).

11.9. Run Program BV11-1-2 with the dataset DATA11.3, in order to verify the generation of Figure 11.3.

11.10. Run Program BV11-1-3 with the dataset DATA11.3, and verify that Figure 11.4 is produced on the screen.

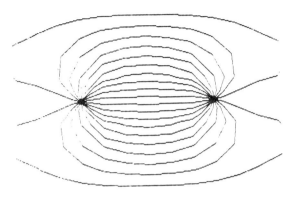

Fig. P.11.11.

11.11. Generate a dataset that can be used to solve the problem of a well and a recharge well, in a region bounded by two parallel impermeable boundaries, by the finite element method, in terms of the stream function. (*Hint*: Such a dataset can be generated, in principle, by Program BV11-2-1, or, preferably, by a text editor, e.g., a word processing program. For a network with 161 nodes and 128 elements, the result may be as illustrated in Figure P.11.11).

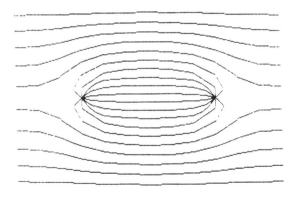

Fig. P.11.12.

11.12. After solving Problem 11.11, modify the dataset, such that it applies to a similar problem in a region with uniform flow, parallel to the impermeable boundaries, see Figure P.11.12.

11.13. Generate a dataset for the problem of a well in a rectangular region, of 1200 m × 800 m. The region is bounded on the left and below by boundaries of constant head, while the upper and right boundaries are impermeable. The well is operating at a distance of 800 m from the left boundary, and 500 m from the lower boundary. After generating the dataset, run Program BV11-2-3 to determine the values of the stream function, and then run Program BV11-2-4 to show the streamlines on the screen. The result should be similar to Figure P.11.13.

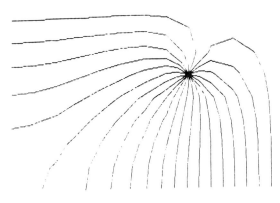

Fig. P.11.13.

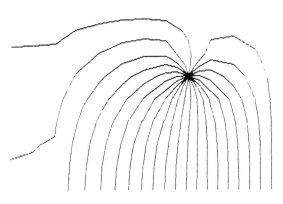

Fig. P.11.14.

11.14. After solving Problem 11.13, modify the dataset used for that problem, in order to investigate the effect of a zone of smaller transmissivity along the left side boundary. For a zone of 200 m width, having a transmissivity 10 times smaller than the remaining part of the aquifer, the result should be similar to that shown in Figure P.11.14.

## Chapter 12

12.1. Run Program BV12-1 with a very small dispersivity, in order to verify that the program will correctly represent pure convection, without dispersion.

12.2. Investigate the stability of Program BV12-2, by using time steps larger than the time step suggested by the program.

12.3. Run Program BV12-2 with a modified dispersion coefficient, using (12.2.8), in order to suppress numerical dispersion.

PROBLEMS

12.4. Run Program BV12-3 with a modified dispersion coefficient, using (12.2.9), in order to suppress numerical dispersion.

12.5. Run Program BV12-4, using the dataset DATA12.5, and verify that Figure 12.5 is indeed produced on the screen.

12.6. Modify Program BV12-4 so that it will print the concentration in a certain given point, as a function of time, as output, in the form of a table on the screen, or a table on the printer, or in the form of a graph on the screen.

12.7. Run Program BV12-4 with a large dataset. If possible, increase the dimensions of the arrays in the DIMENSION statements.

12.8. Run Program BV12-5, using the dataset DATA12.7, in order to verify that Figure 12.7 is produced on the screen.

12.9. Modify the dataset DATA12.7, such that the discharge of the well is 10% larger, and verify that then all the pollution is captured by the well.

12.10. Verify that Program BV12-5 correctly describes pure convection if the dispersivity is zero.

## Chapter 13

13.1. Run Program BV13-1, using the default values of the input data, suggested by the program.

13.2. Run Program BV13-1, using the geometrical input data as suggested by the default data in the program, and a greater discharge of the well, or more time steps. Observe that the program breaks down when upconing becomes too large.

13.3. Modify Program BV13-1, such that the upper and lower boundaries become boundaries of constant head. For this problem, with the lower and right boundaries at infinity, an exact solution for the steady state has been found (Strack, 1972; Verruijt, 1979, p. 73). Verify that the modified program approaches that steady state solution as time increases indefinitely. (This requires a large number of time steps, and a long computation time).

13.4. Write a modified version of Program BV13-1, using input by reading data from a dataset on disk. Generalizations suggested are: (1) Several wells operating in the fresh water, (2) Various possibilities for the boundary conditions along the left, right and upper boundaries, (3) Input of an initial network by specification of location of mesh lines in a rectangular mesh.

13.5. Run Program BV13-2, and verify that the shape of the interface generated after 17 time steps is in agreement with Figure 13.2. Also, verify the behaviour of the interface after a single time step.

13.6. Modify the data in Program BV13-2, such that the program will solve the problem of a single well in the center of a square region. Let the initial position of the interface be horizontal. Observe the gradual upconing of the interface.

13.7. Verify the results shown in Figure 13.9 for an initially vertical interface, by modifying the data in Program BV13-2.

# Index of Subjects

Abrupt front approximation 186
Accuracy 239, 322
Adhesion 30
Adhesion tension 125
Adsorption 170
  isotherm 170, 173
Advection 159, 164, 186, 285, 316
  of pollution particles 188
Advection-dispersion equation 169
Advective flux 68
Aeration, zone of 4
AEV 18
Air entry pressure 130
Analytical elements 223
Analytical solutions 216
Anisotropic porous medium 35
  high Re flow 43
Anisotropy 43, 243
Aquiclude, definition of 1
Aquifer(s)
  artesian 7
  classification of 5
  compaction of 59
  confined 5, 97
  definition of 1
  leaky confined 7
  leaky-phreatic 7, 99, 101
  mining of 8
  natural replenishment of 8
  perched 7
  phreatic 7, 45
  roles of 7
  storage in 9
  storativity of 14
  transmissivity of 14, 43, 44
  yield of 8, 9
Aquifer approach 23
Aquifer type flow 21
Aquitard, definition of 3
Arbitrary elementary area 18
Arbitrary elementary volume 17, 18
Artesian aquifer 7
Artesian well 7
Artificial recharge 8, 9, 10
Average
  intrinsic phase 20

phase 20
Average velocity 29

Backward finite difference 219, 227, 322
Balance, mass 60
Balance equation
  anisotropic confined aquifer 98
  confined aquifer 98
  for a pollutant 167
  inhomogeneous confined aquifer 98
  leaky-phreatic aquifer 100
  nonlinear 101, 146
  unsaturated zone 145
BASCOM compiler 26
Base flow 9
Biot's theory 78, 113, 114
Boundary
  between two porous media 75
  equipotential 70
  impervious 71
  phreatic·surface 72
  semipervious 72
  with finite volume reservoir 74
Boundary condition(s) 13, 66
  aquifer flow 102
  between two unsaturated media 137
  Cauchy 72, 150
  Dirichlet 150, 170, 180
  equipotent 102
  first kind 70, 150
  general 14, 68
  impervious 103, 150
  mixed 72
  Neumann 71
  no-jump 68
  phreatic surface 72, 183
  plannar incremental total stress 112
  prescribed flux 70, 103, 150
  prescribed piezometric head 102
  prescribed pressure 70
  prescribed water content 150
  second kind 71, 150
  semipervious 103
  spring 102
  stress 109
  third kind 72, 150

Boundary condition(s) - *contd.*
　unsaturated zone 150
　well mixed zone 181
Boundary elements 222
Boundary surface 66
　continuum approach 67
　equation of 68
Boussinesq equation 100
Breakthrough curve 157
Bubbling pressure 131, 144
Burdine's equation 145

Capillary
　pressure 126, 128, 140
　pressure head 130
　　critical 131
Capillary diffusivity 140
Capillary forces 87
Capillary fringe 4, 46, 134
Capillary rise 47, 134
Cauchy–Riemann conditions 119
Central finite difference 219, 227, 319
Characteristics 189, 286
Coastal aquifer 196
Coefficient of
　bulk compressibility of porous medium 55
　fluid compressibility 57
　fluid mass storage 59
　retardation 176
　soil compressibility 56, 58
Coefficients 18
Compressibility
　bulk, of porous medium 56
　fluid, coefficient of 57
　soil, coefficient of 58
　solid 56
Compressible fluid 35
Computer code 16
Computer programming 223
Conceptual model 12, 13
Conceptual flow model
　content of 76
Confined aquifer 5
Conjugate gradient method 265, 378
Consolidation 78
Constitutive equations 13
Contact angle 124, 133
Continuity equation
　leaky confined aquifer 88, 90
Continuum 17
Continuum approach 13, 14, 17, 27, 67
Contours 283

Control volume 60
Creep 84

Dam, steady flow through a 272
Darcy (unit) 32
Darcy's formula
　(*see also* Darcy's law)
Darcy's law 14, 27–42
　empirical form 27
　general form of 39, 40
　range of validity 33
　theoretical derivation 39
　three-dimensional flow 34
Dead end pore 30
Delayed yield 87, 94
Desorption curve 131
Diffusion 158, 164
Diffusive flux 68, 159
Dirac delta function 62, 90
Dirichlet condition 70, 150, 180, 228
　(*see* boundary conditions)
Discontinuity
　in aquifer 105
Dispersion 316
Dispersive flux 68, 159
Dispersivity 161
　longitudinal 162
　transversal 162
Displacement, 81
　horizontal 78, 105
　vertical 78
Divergence of flux 61
Double index summation 38
Double layer 34
Downstream 322
Drag 40, 42
Dupuit's assumption 23, 45
　error in 49
　in regional studies 52
　regions not valid 52
Dupuit–Forchheimer discharge formula 51

Effective hydraulic conductivity 140, 142
Effective permeability 139, 141, 142
Effective porosity 30, 73, 87, 135
Effective saturation 144
Effective stress 53, 55, 56, 58, 78, 105
Einstein's summation convention 38, 40
Elastic body 81
Electroosmotic counterflow 34
Entrapped air 133
Equilibrium equation 81, 84
　integrated form 109

# SUBJECT INDEX

Equilibrium water saturation 125
Error function 317
Euler method 287
Eulerian approach 60, 115
Eulerian formulation 62
Evaporation 2
Evapotranspiration 2
Explicit method 234, 319
Extensive quantity 13

Fair and Hatch formula 31
Fick's law 164
Field capacity 5, 135
Film 132
Finite differences 26, 217, 225, 319
Finite elements 26, 221, 247, 326, 355
Fishnet 282
Flow model
   complete mathematical statement 76
Flow models
   confined aquifer 85
   land subsidence, regional 105
   leaky aquifer 85
   phreatic aquifer 85
Flowing well 7
Flownet 120
Forecasting problem 23, 53
Forward finite difference 219, 227, 322
Front 117
Funicular water 125, 126

Gauss elimination 371
Gauss–Seidel method 229, 258, 367
Gaussian distribution 337
Ghyben–Herzberg approximation 202
Grain diameter, effective 30
Groundwater 1, 4
   management of 7, 10
   modeling of 11
   one-time reserve 9

Hook's law 81
Horizontal flow approximation 21–23, 45, 85, 201
Hubbert's potential 35
Hydraulic approach 23, 85, 201
Hydraulic conductivity 30
   anisotropic porous medium 38
   effective 140
   equivalent normal to layers 37
   equivalent parallel to layers 37
   units of 31

Hydraulic gradient 28, 34
Hydraulic radius 19
Hydrodynamic dispersion 9, 25, 155
   coefficient of 165
Hydrologic cycle 1, 2, 27
Hydrological constraints 11
Hygroscopic coefficient 5
Hygroscopic water 5
Hysteresis 125, 132
   effective hydraulic conductivity 141
   ink bottle effect 132
   moisture diffusivity 141
   permeability 144
   raindrop effect 133

Identification problem 15, 23
Imbibition 132
   entrapped air 133
Immiscible fluids 124
Immobile water 30, 177–179
Implicit method 237
Indentification problem 23
Inertial effects 14, 28, 40, 41, 42
Infiltration 2, 268
Inhomogeneous medium 35
Initial conditions 13, 66
Integrated finite differences 244
Integrodifferential equation 96
Interface 117, 344
Interfacial tension 124, 126
Intergranular stress 53, 54, 56
Intrinsic phase average 20
Inverse problem 15
Ion exchange 170
Irreducible moisture content 73
Irreducible water content 132, 135
Irreducible water saturation 132, 139
Irreducible wetting fluid saturation 142
Isochoric flow, macroscopic 117
Isoparametric interpolation 313
Isotropic porous medium 30, 35
Isotherm 171
   equilibrium 171
   Fruendlich 171
   Langmuir 172
   nonequilibrium 171

Jacob's theory 78, 113

Kozeny–Carman equation 31
   high Re 42

Lagrangian approach 115
Lame's coefficient 81
Laminar flow 33
Land subsidence 8, 10, 78, 82
    regional model 105
    two-step approach 84
Laplace equation 65, 98
Laplace formula 127
Layered aquifer 281
LDU-decomposition 374
Leakage 268
    coefficient of 90
Leakage factor 90
Leaky confined aquifer 7, 88
Leaky formation 3
    (see also aquitard)
Leaky phreatic aquifer 7
Leibnitz rule 44, 91
Linear equations 217, 364

Macrodispersion 190, 325
Macroscopic level 14, 17, 66
Macroscopic value 18
Macroscopic variables 13
Management of groundwater
    constraints 10, 11
    decision variables 10
    objective function 10
    state variables 10
Management problem 23
Mass balance equation 60, 62
Mass conservation equation 53
Material surface 70
Mathematical model 12
    contents of 13
    complete statement of 76
Mechanical dispersion 158
Memory function 97
Microscopic level 14, 17
Microsoft BASIC 26
Mining of groundwater 9
Miscible displacement 156
Model 66
    conceptual 12, 13
    definition of 12
    mathematical 12, 13
    numerical 12, 16
    physical 12
    regional, land subsidence 105
    selection of 12
    statement, for aquifer 104
    stochastic 17

Modeling 11
    flow in unsaturated zone 123
    soil displacement 78
Moisture content 123
Moisture diffusivity 140
Moisture retention 130
Molecular level 17
Momentum balance equations 14, 40
    averaging of 41
Motion equation 14
    (see also Darcy's law)
    air 138
    large Re 41
    nonlinear 42
    nonlinearity 141
    unsaturated flow 138
    water 138
Natural replenishment 2, 8
Navier–Stokes equation 17, 42
Neumann condition 181, 228
    (see boundary conditions)
New balance equation 53
Newtonian fluid 42, 43
No-jump condition 68
Non-Darcy behavior 34
Nonhomogeneity 243
Nonlinear motion equation 42
Nonrenewable resource 8, 9
Nonwetting fluid 125
Normal distribution 337
Numerical dispersion 26, 322, 323, 343
Numerical methods 16, 216
Numerical model 12

Objective function 10
Operational yield 9
Optimal yield 9
Osmotic effect 138
Over-relaxation 229, 258

Parameter identification 14
Particle tracing 311, 342
Path of fluid particle 115
Pathline 115
Pendular rings 125, 126, 131, 132
Perched aquifer 7
Permeability 14, 30
    conversion of units 32
    effective 139
    formulae for 31
    relative 142, 143
    units of 31, 32

# SUBJECT INDEX

Phase average 20
Phreatic aquifer 7, 45
Phreatic surface 4, 7, 47, 183
　equation of 47
Piezometric head 5, 28
Piezometric surface 5
Plannar incremental total stress 112
Pointer matrix 262, 369
Pollution 153
　sources of 154
Pore space 3
Porosity 20, 29
　effective 30
Porous medium 17
　continuum model of 18
Porous plate 128
Potential flow 119
Principal directions 39

Quality 153

Radioactive decay 173
Random 19
Random walk 26
　model 336, 340
REA 21
Regional interface problem 355
Relative density difference 353
Relative permeability 142, 143
Relaxation factor 258
Renewable resource 8
Representative elementary volume 13, 20
　size of 18, 19
Representative elementary area 21
Residual nonwetting fluid saturation 139, 142
Resistance of aquitard 90
Retardation factor 176
Retention curve 131
Return flow 2
REV 17, 18, 19
Reynolds number 33
Runge–Kutta 287

Saturation 123
　equilibrium, water 125
　funicular, water 126
　insular, air 126
　zone of 4
Scanning curves 133
Seawater intrusion 196, 334, 351
　regional 208

Seepage face 50, 74
Shape functions 249
Shear stress 55
Soil deformation 54
Soil water zone 4, 5
Solid matrix 3, 17
Sorption curve 131
Sources, point 90
Specific discharge 28, 34
　relative to solids 40
Specific discharge potential 119
Specific retention 87, 136
Specific storativity 63, 78, 79, 85, 113
Specific surface 31
Specific yield 73, 87, 135, 136
　time dependency 137
Stability 239, 322
Stable solution 77
Stagnant water 177
Steady flow 225, 257, 268
Stochastic models 17
Storage, mass 56
　in semipermeable layer 93
Storativity 14
　confined aquifer 78, 85
　of phreatic aquifer 86
Strain 54
Streakline 117
Stream function 114, 118, 300
　confined equifer 120
Streaming potential 34
Streamlines 103, 114, 116, 291, 308
Stress, boundary condition 109
Subsurface water 1
Suction 127
Surface runoff 2
Surface tension 124

Tensiometer 46, 127, 129
Tension 127
Tensor
　hydraulic conductivity 38
　permeability 38
　principal directions 39
Terzaghi's theory 78
Terzaghi–Jacob approach 114
Thermodynamic equilibrium 69
Thermoosmotic effect 138
Threshold gradient 34
Threshold pressure 130
Tortuosity 30, 165
Total stress 55

Transmissivity 14, 43, 44
   horizontally stratified aquifer 45

Uncertainty 16
Undrained test 63
Unsaturated zone 4, 14
   (*see also* aeration, zone of)
   modeling flow in 123
Unsteady flow 233, 276
Upstream 322

Vadose water zone 4
Velocity
   average 29, 30, 34
   intrinsic phase average 29
   of solids 40
Verruijt's theory 78
Void space 3, 17

Volumetric fraction 21

Water balance 8
Water capacity 141, 147
Water divide 51, 103
Water table 4, 46
Wave front method 372
Weighted residuals 247, 250
Well posed boundary value problem 77
Wells 288, 296, 303
   artesian 7
Wettability 124, 125
Wetting fluid 125

Yield of aquifer
   operational 9
   optimal 9
Young's equation 124